BIOETHICS

Walt Wyman
Fall 1982
Whitman

Revised

BIOETHICS

*basic writings on the
key ethical questions that surround
the major, modern biological
possibilities and problems*

edited by
THOMAS A. SHANNON

Cover Design: Tim McKeen

Library of Congress
Catalog Card Number: 76-18054

ISBN: 0-8091-1970-6

Published by Paulist Press
545 Island Road, Ramsey, N.J. 07446

Printed and bound in the   .
United States of America.

Acknowledgments

Abortion: Some Ethical Issues by Daniel Callahan: From *Abortion, Society, and the Law,* edited
by David F. Walbert and J. Douglas Butler, 1973. Reprinted by permission of The Press
of Case Western Reserve University. *Abortion: A Changing Morality and Policy?* by Richard
A. McCormick, S.J.: Reprinted with permission from *Hospital Progress,* February 1979. ©
1979 by The Catholic Health Association. *Ethical Problems of Abortion* by Sissela Bok: Reprinted from *Hastings Center Studies,* Vol. 2, No. 1, January 1974, by permission of the Institute of Society, Ethics and the Life Sciences, Hastings-on-Hudson, N.Y. 10706.
Abortion: On Fetal Indications by Susan Nicholson. Reprinted with permission of the *Journal of Religious Ethics.* Department of Religious Studies, Knoxville, Tn. 37916. *Abortion and
the Roman Catholic Church.* JRE Studies in Religious Ethics, II. *Critical Issues in Newborn Intensive Care: A Conference Report and Policy Proposal* by A. R. Jonsen, et al. Reprinted with
the permission of *Pediatrics* 55: 756-768, June 1975 © American Academy of Pediatrics
1975. *Abortion, Euthanasia, and Care of Defective Newborns* by John Fletcher, Ph.D.: Printed
with permission from *The New England Journal of Medicine,* Vol. 292, No. 2, January 9,
1975, 75-78. *Mongolism, Parental Desires, and the Right to Life* by James M. Gustafson: From
Perspectives in Biology and Medicine, Summer 1973, 529–557 Reprinted by permission of
The University of Chicago Press. *To Save or Let Die* by Richard A. McCormick: Reprinted
with permission of *America,* July 13, 1974. All rights reserved. Copyright © 1974 by
America Press, 106 W. 56th St., N.Y., N.Y. 10019. *Brain Death: 1) A Status Report of Medical
and Ethical Considerations; 2) A Status Report of Legal Considerations* by Frank J. Veith, M.D. et
al. Reprinted with the permission of the *Journal of the American Medical Association,* vol.
238 (10 October 1977) pages 1651-1655 and vol. 238 (17 October 1977) pages 1744-1748.
© 1977, American Medical Association. *The Right To Die* by Hans Jonas. Reprinted with
the permission of the *Hastings Center Report,* August 1978, pages 31-36. © Institute of Society, Ethics and the Life Sciences, Hastings-on-Hudson, N.Y. 10706. *The Freedom To Die*
by Daniel Maguire: From *Commonweal,* August 11, 1972, 423-427. Reprinted by permission of Commonweal Publishing Co., Inc. *The Withdrawal of Treatment: The Cost and Benefits of Guidelines* by Thomas A. Shannon. Reprinted with permission from *The Bioethics
Digest, Summaries of Literature on Biomedical Ethics,* vol. 3, no. 4 (April, 1978). © 1978 by Information Planning Associates, Inc., Gaithersburg, Maryland. *Philosophical Reflections on
Experimenting with Human Subjects* by Hans Jonas, Ph.D.: Reprinted by permission of *Daedalus,* Journal of the American Academy of Arts and Sciences, Boston, Massachusetts,
Spring, 1969, *Ethical Aspects of Experimentation with Human Subjects,* 219-247. *The Ethical Design of Human Experiments* by David D. Rutstein, M.D.: Reprinted by permission of *Daeda-*

Contents

PART THREE
DEATH AND DYING

PART FOUR
RESEARCH AND HUMAN EXPERIMENTATION

PART FIVE
ETHICAL DILEMMAS IN OBTAINING INFORMED CONSENT

PART SIX
GENETIC ENGINEERING AND GENETIC POLICY

Leon R. Kass, M.D.

PART SEVEN
THE ALLOCATION OF SCARCE RESOURCES

PART EIGHT
BEHAVIOR MODIFICATION

Preface

The study, evaluation, and resolution of problems in the field of bioethics are complicated by the basic fact of the interdisciplinary nature of such problems. In few other fields does such a wide range of disciplines converge on particular problems. This disciplinary problem is further complicated by continuing advances in medical-scientific technology. As a result, the questions and problems of bioethics come much easier and faster than resolutions—or even clear statements—of the problems. But the very urgency and significance of the problems demand an almost immediate resolution.

The rise of major institutes, the introduction of bioethical courses in colleges, universities, and medical schools, and a veritable flood of publications indicate the seriousness with which these issues are perceived. They also show a wide variety of interpretations and solutions to the problems. Consensus—a difficult task in any case—is more difficult in bioethics because of both the social and personal dimensions of the issues. In addition, there is the diversity of ethical, philosophical and religious frameworks that are brought to bear in an evaluation of the issues. But it also is causing certain issues to come to the fore and be perceived as thematically important for all problems in bioethics. Examples of these issues are the constitutive elements of personhood, the rights of a person, the rights of society, personal integrity, consent and distributive justice. As these are discussed from various perspectives and in differing contexts, the issues are becoming more sharply defined and the topography of the problems is becoming a little clearer.

This reader in bioethics is presented as a contribution to the ongoing discussion of issues such as these. Each article has a specific ethical argument or position, but the various arguments come from an interdisciplinary perspective which helps to broaden their impact. The articles are also very basic and important contributions to the discussions of a particular problem area in bioethics. In addition, each article frames its argument in such a way as to have a long-term interest and significance. The result is a set of articles that are both critical and informative and which also make an on-going contribution to the analysis of problems in bioethics. The topics selected are the standard problem areas of bioethics; they are also very significant human and ethical problems. The way in which these problems are resolved may very well set the tone for our future.

THE SECOND EDITION

I am pleased that this reader has been thought of highly enough to have it go into three printings and to encourage the editors at Paulist Press to ask me to prepare a revised edition. I hope that this second edition will continue to be as helpful as the first edition has been.

The field of bioethics has grown tremendously since the 1976 edition, and there has been a corresponding increase in the number of articles published, to say nothing of readers and monographs. This has made the task of selecting articles that much more difficult. I am keeping with my original format of organizing the articles around topics, not only because this is a convenient way to do it, but also because many courses are organized this way. It also means that the book can be used in a variety of ways because no one section is logically dependent upon another. I have retained 17 of the 29 articles in the first edition. I think that these articles either state a position well, provide an excellent survey of various positions, or are simply, in my opinion, a classical contribution to the field.

The second edition has a total of 33 articles. Of these 16 new articles in the second edition, four are in a new chapter entitled "Ethical Dilemmas in Obtaining Informed Consent." The other 12 articles are distributed throughout the other chapters and attempt to take account of either new arguments for a position or new discussions and developments in the field. Thus, there are new articles on abortion, recombinant DNA, *in vitro* fertilization, the National Commission for the Protection of Human Subjects, Health Policy, as well as new articles on other topics in the first edition.

This new edition certainly does not claim to be complete in its coverage of all topics. Clearly, many articles that could have been included have been omitted. I think that the articles chosen provide a variety of perspectives on a particular topic as well as introduce the reader to the differing resolutions to a problem.

I would like to acknowledge in a special way the continuing support of my colleagues at Worcester Polytechnic Institute as well as the encouragement given me by my new colleagues in the Department of Psychiatry at the University of Massachusetts Medical Center. Dr. Stanley Walzer, Head of the Department, and his professional and secretarial staff have been most helpful in providing the resources necessary to produce this new edition. I am grateful to them for their support.

Thomas A. Shannon, Ph.D.
Worcester Polytechnic Institute
University of Massachusetts
Medical Center

introduction | *The Tradition of a Tradition: An Evaluation of Roman Catholic Medical Ethics*

THOMAS A. SHANNON

A. INTRODUCTION

One tradition in Roman Catholic moral theology, especially in America, has been the development of a set of ethical teachings related specifically to medicine. While much of this grew out of a genuine sense of pastoral concern for both the patient and physician, there was also a theoretical probing of the issues themselves. All Catholic ethicists will be grateful for the pioneering work of Gerald Kelly in his book *Medico-Moral Problems*. While not all today would agree with his methodology or his solution of cases, nonetheless all have benefited from his summaries of the tradition and the careful statement of issues and problems. Kelly's work has provided a respectable and responsible starting point for many developments in medical ethics.

The thrust of Kelly's work has provided much inspiration for other Roman Catholic ethicists. Kelly worked with John Ford to address other ethical issues related to medicine. Their joint efforts provided another strong addition to the growing literature on medical ethics. Fortunately, this effort has been carried on by many individuals in the tradition of its major founders. Richard McCormick has been a sensitive and reflective commentator on the tradition and most careful in his reinterpretation and reapplication of it. His studies, mainly in *Theological Studies*, have been an excellent model and a significant contribution to a discussion of problems in medical ethics. This type of work is also continued by individuals such as Charles Curran and Daniel Maguire. Through their articles and books, these men have opened new doors and shed bright light on many of the nuances in the tradition. Daniel Callahan has both written profoundly

3

about issues in bioethics and was instrumental in establishing an institute devoted to systematic and long-term research in bioethics.

There are numerous other Catholics who are making significant contributions to this new and extremely important field. But the purpose of this introduction is not to establish who was the first to tackle bioethical issues or who has the best track record. Its purpose, rather, is to help recall the place and significance of such a tradition within Roman Catholicism so that the wisdom of the past can be meaningfully used in understanding and evaluating the problems encountered in our contemporary situation.

B. PRACTICAL PROBLEMS AND ETHICAL ISSUES IN ROMAN CATHOLIC BIOETHICS

1. The Problems

By its very nature, bioethics is problematic. The urgency of problems such as the crisis atmosphere of the intensive care unit, the pathos of the intensive care nursery, and the specter of practically incomprehensible developments and application of new research techniques guarantee that bioethics should never be guilty of armchair philosophy. Bioethics is at the center of the action and has an important role to play. But in doing this, there are several practical problems that must be recognized and dealt with.

The first of these is the very quantity and quality of advances in medicine from both a scientific and technological perspective. These advances have come so fast that one barely has the time to understand a particular advance—let alone evaluate it—before a new and even more significant development occurs. Many times ethics has been slow to respond to such advances. Often this was true because of ignorance of the advances or because the advances came into use as a matter of course without much publicity. The problem of how to understand and evaluate a rapidly expanding field of research is common to all of bioethics. It may be a particular problem for Roman Catholics, especially given the tradition of the Church's being slow and cautious in making moral evaluations. While this slowness is not necessarily a defect—for it does permit time for a variety of opinions to come forth and be evaluated—nonetheless, problems in some way must be evaluated before their consequences overwhelm us. Hence a greater sensitivity and responsiveness to current developments is called for.

A second problem is that of the evaluation of circumstances. Although Roman Catholicism is often accused of not looking at the circumstances and of applying moral principles abstractly and absolutely, it should be recalled that circumstances were traditionally one of the three major criteria for evaluating a moral act. The problem was that many Roman Catholic moral theologians did not fully recognize the significance of the circumstances in the theoretical development of moral theology. Most of the emphasis on circumstances was dealt with in pastoral care—which often showed a marked difference from the clearly stated theory. There is, however, ambivalence regarding the significance of circumstances in Roman Catholic moral theology and this needs to be squarely faced, for in medical ethics circumstances are extremely important and may change the entire moral nature of the case. Although it may not be totally clear what it is in the circumstances that make an act right or wrong as contextualism would have it, nonetheless the circumstances are of critical importance. This is an area which is really underdeveloped in Roman Catholic moral theology in general and in bioethics in particular. Such work is necessary because of the genuine moral nuances present in the specific circumstances that makes an act right or wrong as contextualism would have rendering of principles; it does imply, however, a greater awareness of the full range of moral data within a given situation. This can only help lead to a more profound ethical analysis.

The advances within medicine and science and the complexity of circumstances lead to a third problem: the bringing of the moral tradition up to the current state of the art. One strong element in Roman Catholic medical ethics has been the development, refinement, and the continued critique of principles such as the double effect principle and the principle regulating the use of extraordinary means of treatment. The basic problem is that these principles—of great ethical significance —were developed in an age and circumstances other than our own. The problem is how are these principles to be understood in the light of our current technological advances and the refinements of modern health care. This demands a real, critical rethinking of the tradition from within, as well as attempts to apply the re-evaluated tradition to new problems. This has been the thrust of many developments within contemporary Roman Catholic medical ethics. As such, this is a positive response to a rich tradition, as well as a recognition that a tradition does not remain viable by holding to past glories.

The problem of the reapplication of the tradition leads to a fourth and final problem: the use of language in both medicine, technology, and medical ethics. Although there are many issues and problems which are only semantic in nature, there are also many that may well not be. One clear case is an evaluation of what differences there may be between killing and letting die. Again, the emphasis is one of moral nuance and an attempt to evaluate fully and critically what we are doing. A related issue is the quality of language used in science, medicine, and bioethics. One should always strive for precision and accuracy in one's language, especially when describing a complex situation. This obviously means an increase in the use of scientific terminology, which is all to the good if it does not unconsciously seduce us into believing that a scientific evaluation or description which attempts to be value-free would lead us to imply that the issue itself is value-free. To describe a situation scientifically does not eliminate the possibility of other equally significant descriptions of the same situation. The major danger is to think that by renaming the position and the players we have changed the nature of the game. The task of bioethics here is to remind us constantly that value-free descriptions do not imply value-free issues.

2. Ethical Issues

In addition to the more practical or methodological problems described above, there are also some significant ethical issues that need examination and re-evaluation. Many of these issues are inherent in the nature of Roman Catholic ethical theory, but they are of particular concern to an examination of bioethical theory because of their significance.

The first of these issues is the setting forth of the relationship between natural law and revelation. There is the continual claim that the Magisterium of the Roman Catholic Church is the guardian and interpreter of the natural law; the claim for the authority for this function is based on the New Testament and the teachings of Jesus. Unfortunately, this claim has never been specified other than on the general ground of the Church's having authority to teach. There is no clear teaching in the New Testament on the nature and structure of the Church, let alone on how the Church might relate to a particular theory of natural law.

This issue is more than a restatement of the traditional discussion of reason and revelation. The Church is claiming a privileged position in being able to have a definitive insight on the structure or application of a particular theory of natural law—which is a philosophical problem. That this is the tradition and practice of the Roman Catholic Church is certainly clear; that this tradition and practice has been validated by the New Testament or the words of Jesus is quite unclear. The important issue is not an argument that the Magisterium should not use the wisdom of the Church in attempting to understand and resolve a problem. What is critical is that the basis of such understanding and evaluation must be made clear. If the Magisterium uses a philosophical argument, that argument is only as good (or bad) as the type of philosophy and the strength of the argument that is developed. The unsubstantiated claim of a privileged position adds nothing to such an argument. In fact, such a claim may seriously weaken an argument, especially if the argument is perhaps already poorly developed. Revelation was not intended as a fail-safe mechanism for poor reasoning or weak philosophical analysis. To make it serve such an unworthy purpose cheapens revelation and erodes the integrity of the Church.

A closely related issue is that of Church authority. Even though this authority has already been seriously weakened in the past few decades, obedience is a meaningful virtue—though indeed neither the only nor the highest virtue. Catholicism has been fortunate in moving from rule by imperial fiat to a more collegial understanding of authority. Yet the significance and meaning of authority is only as good (or as bad) as the source of the authority itself.

What this means is that authority and the commands it gives must be continually tested and evaluated. Several issues emerge here. Authority must be clearly separated from social control. The fact that a particular practice may insure good order neither increases nor decreases this practice's ethical significance. If a practice is ordered, its ethical meaning and significance must be given priority. Secondly we must remember that there have been no infallible declarations on issues of morality. Certain acts or practices have been absolutely prohibited—but there has been no infallible statement on their inherent immorality. This means that all declarations of the Magisterium are subject to re-evaluation. This does not imply, of course, that such declarations are meaningless, useless, or irrational. It simply means that they are not the last word on the subject. Finally, there is the problem of testing and relating the data of revelation and the elaboration of a philosophical argument, whether this be from a natural law perspective or some other one. The real problem here is to relate the values found in each. While this does not imply that revelation and philosophy are by definition at crossed swords, neither does it imply that they are automatically or a priori in harmony. Philosophy and revelation must stand in a dialectic relationship so that the values of each may be tested and so that significant issues can be fully and critically evaluated. This dialectic insures the formation of a mature conscience, sensitive to the reality of the living God and to the meaning of the human situation.

Authority also relates to a third major issue: a movement from principle to practice. There is a clear difference between the theoretical or abstract statement of a principle and its specific application to a case. In one situation, one dimension may be emphasized: in another, some other elements may be more important. The important element to remember is that how one comes down in a particular case cannot be totally predicted in advance. One may have a fairly good idea of where one's principles and values will lead one; but to suggest that having principles eliminates the need for reflection on the situation or on how a principle might apply to the case, is to say that all is predetermined. In this case, morality becomes the Procrustian bed in which differences and nuances are dispensed with.

There is a very torturous road leading from principle to practice. To say that authority knows the way is no better than saying that the situation will tell you what to do. Both eliminate the hard, painful work of thinking, evaluating, weighing, testing, and resolving a multiplicity of data, feelings, and values. The important element to remember is that the step from principle to practice must be taken—otherwise ethics becomes a sham. But the how, when, where, and why of the taking of that step cannot be determined a priori. The principle and practice need thorough examination and testing. To deny this process is to deny the dynamic process of life itself and to attempt to hide from the reality of life.

One final issue needs to be discussed. Although it may not be as important as the others, it can be just as troublesome. This is the confusing of ethics with strategy or the collapsing of the two into one reality. In saying this, it is important to remember that strategy is not without an ethical dimension nor is ethics without a strategic dimension. But they are not simply convertible. Oftentimes the strategic elements can gain more attention or importance than the ethical issues themselves. This can become very problematic because then the focus of the discussion of the ethical problem can change dramatically. Hopefully there will always be a coherence of ethics and strategy. When there is not, it may be a case of good ethics with a bad strategy or poor ethics with a good strategy—or any other combination thereof. What is important is that we should remember what we are criticizing, for what reasons we are criticizing it, and whether it is an ethical or a strategic issue. For often there can be great tolerance of poor strategy rather than for poor ethics.

C. Conclusion: METHODOLOGICAL CONSIDERATIONS

Given the above, what can one say about methodology, especially if one is a Roman Catholic?

One important consideration is that all major points of view must be stated clearly and fairly and then critically evaluated. One reality is becoming clear: since bioethics operates at the intersection of science, sociology, law, and ethics—to name just a few—a conflict of rights or

values is a common occurence. The very interdisciplinary nature of reality itself insures that a variety of dilemmas will manifest themselves.
If any ethical methodology is to have integrity the least it can and,
indeed, must do is insure that these dilemmas are exposed and precisely stated. Such an exposition is a critical first step in addressing the
bioethical problem.

A second step is to state clearly and sharply the specific values or
ethical principles involved in the situation. For only if one knows
what is at stake or what is of significance, can one begin a clear value
analysis. Included in this is an ordering of the values involved and an
attempt to have a coherent and harmonious relationship established.

Thirdly, one needs to examine one's membership in various associations to discover what these organizations may or may not have to
say about the particular problem. For a Catholic this would be important for the Roman tradition does speak to a variety of issues and
there is wisdom in the experience of the Church. Turning to a variety
of other traditions and associations will help specific issues to come
forward and this will suggest an ordering of values. This is also important because we are often caught up in the passion of the moment
and forget to gain needed perspective. Of course, a tradition can also
inhibit one from seeing new and significant values and it is at this
point that one must be willing to embrace a particular tradition a little
less tightly. This is not to reject a tradition nor to render it useless; it
is to attempt to integrate the old and the new. If real problems are to
be faced and resolved as best we can and if ethics and even religion are
to be understood and felt to be an important part of our life, then we
cannot go to tradition as one might go to a rack of ready-made clothes
and put on the one that fits the best. Rather we might have to go to a
tailor to have a correct fit insured. In both cases the clothes may be of
the same quality—but tailored clothes are fitted to meet the needs of
the situation. And this is what a responsible ethic will do. By reflecting on the tradition and the situation, as well as the full range of the
interdisciplinary elements, a mature ethical solution can emerge which
is perceived to be appropriate to the case. In doing this one need not
be afraid of making analogies or finding similarities in a variety of
cases. Rather what one must be aware of is the prefabricated solution
that sees all similar cases as identical.

The result of all of this will be the decision of what to or not to
do. Obviously many unmentioned elements enter into this: one's
sense of the fitting, one's temperament, culture, family situation. One
could go on and on. The important thing to remember in bioethics is
that while there are many elements over which we have no control,
there are many over which we do. For these elements we are responsi-

ble and stand accountable before God and one another. If we shirk the task of being responsible for that for which we can be responsible, we have failed at the ethical task and we are at best only opportunists. The assumption of one's proper responsibility does not and cannot insure that mistakes—and possibly horrible mistakes—will not be made. But it will insure that we have been faithful and have pursued integrity. Such values can only serve to heighten the level at which bioethics is pursued and the quality of the decisions that ensue.

This book represents one part of the methodology suggested in this chapter. By presenting a variety of points of view on a number of different topics, the book hopes that different values can emerge and can be examined critically in relationship to one another. In this way one is forced to rethink a particular problem from a variety of points of view to discover and examine a fuller range of interests, values, and dilemmas. From this type of exposure it is then possible to go into a theoretical analysis of the issues themselves and to draw a specific conclusion. However, if one does not first go through the process of the examination of the range of values themselves and the types of problems and perspectives that are proper to a given situation, then the ethical analysis itself is lacking and the solution is at best only partial and at worst totally inadequate. Hopefully the variety of issues and viewpoints presented in this book will serve the purpose of raising the question of the seriousness of these issues and also finding out the wide range of ethical prospectives brought by people who are seriously attempting to think these issues through. It is important to examine these differences for through such examination one is forced to reconcile and resolve as best as one can the inherent dilemmas in bioethical problems. Through this process one hopes to come to a mature resolution of a specific issue. It is in this spirit and hope that these essays are presented.

August, 1976
WORCESTER POLYTECHNIC INSTITUTE
Worcester, Mass.

PART ONE

ABORTION

Daniel Callahan
Richard A. McCormick
Sissela Bok
Susan Nicholson

1 Abortion:
Some Ethical Issues

Daniel Callahan

In this article, Callahan gathers together a selection of often used pro-abortion arguments. Although Callahan disagrees with most of these arguments, he states them fairly and presents careful counter-arguments. As a result of this approach, one is quickly drawn into the mainstream of the abortion debate and is helped in articulating a position. Thus, although Callahan would assume that many will disagree with many of his counter-arguments, one must still respond to them. Through this process, one's own position can become much clearer, more well-developed, and more sensitive to the wide range of ethical issues.

Dr. Daniel Callahan, Ph.D., is the Director of the Institute of Society, Ethics, and the Life Sciences.

A bortion is a peculiarly passionate topic, largely because many people invest their positions with a symbolic weight that transcends immediate social and legal issues. The most obvious examples of this tendency can be found in some segments of the women's liberation movement, on the one hand, and in some factions of those opposed to abortion, on the other. For each the way society solves the abortion problem will be taken to show just what its deepest values are. And those values have implications that extend far beyond abortion.

I. INTRODUCTION

The Women's Liberation movement sees abortion as the most significant liberation of all, from the body and from male domination. The most effective solution to unwanted pregnancy, it removes the final block to full control of reproduction. Unless reproduction can be fully controlled, women will remain in bondage not only to their sexuality but, even more, to those legions of male chauvinists who use female sexuality to their own domineering ends.

By contrast, many of those opposed to abortion see the issue as indicating the kind of respect society will show the most defenseless beings in our midst. If the life of a defenseless fetus is not respected, then there is good reason to believe that the most fundamental of all human rights — the right to life — will have been subverted at its core. The test of the humane society is not the respect it pays to the strongest and most articulate, but that which it accords to the weakest and least articulate.

Of course, these arguments and the symbolic weight they carry simply bypass one another. The opposition seems so fundamental, and the starting premises so different, that any meaningful debate — the kind that leads to give-and-take, concession and adaptation — is ruled out from the start. Moreover, the very charges each side hurls at the other are of a psychologically intolerable nature. No vigorous proponents of abortion are likely to admit, either privately or publicly, that they sanction "murder"; nor are opponents of abortion likely to admit that they sanction the suppression of women. I am using the word "admit" here in a serious sense which implies that one is willing to ponder seriously the possibility that the worst things said about oneself are true. Given that possibility, there remain only two choices: change one's views and confess the errors of one's old ways or violently and aggressively deny the charges.

There is, of course, a third possibility: concede that there may be a grain of truth in what one's opponent says and then undertake the development of a position that tries to meet and integrate the objection in some new position. But this may be the most distasteful solution of all for most people, since it entails a long, drawn-out wrestling with oneself. Abortion is a painful issue, and for just that reason people seem impelled to proceed in all haste to the comfort of "Here I stand," which ends the self-wrestling.

My comments here are not drawn from any hard evidence. They are meant only as reflections on years of trying to discuss abortion in a reasonably calm, rational way, both in public and private. My own professional training is in philosophy, a discipline which (to the despair of many nonphilosophers) places a heavy premium on precision of argument, careful distinctions, developed justifications of ethical positions, methodological elegance, and a cool, temperate mode of discourse. These traits have, to be sure, led to more than one accusation that philosophers are prone to fiddle while the city is burning; and there is probably some truth in this. Nonetheless, I think these traits are still somewhat useful, especially when discussing a topic like abortion, which many take as an invitation to express their unbridled feelings and convictions. Worse still, the politics of abortion seems to pay handsome dividends for such a stance. It can, and has, pushed many an abortion reform bill through reluctant legislatures, just as it has, in different hands, killed many a bill. When argument fails, what better tactic is there than to bring out the fetus-in-a-bottle ("See what you are doing to innocent life!") or the raped mongoloid mother of ten with the drunken husband ("See what you are doing to her!")?

In discussing the quality and weight of ethical arguments here, I hope to achieve the most minimal kind of goal — simply to make plausible the radical notion that there remain some unresolved questions, some hazy areas, and some further points to be thought about. Before proceeding, let me state something for the record about my "position." (Experience has taught me the painful lesson that abortion politicians of either persuasion usually care not at all about one's arguments, but only about one's final "position.") My position is that abortion should be legally available on request up to the twelfth week of pregnancy; that abortion is morally justifiable under a variety of circumstances, but should always be undertaken reluctantly and with a strong sense of tragedy; and that the humane society would be one in which women were neither coerced to go through with pregnancies they do not want nor coerced by social, economic, or psychological circumstances into abortion. I cannot accept the position of those who would deny all respect to the fetus. Nor can I accept the position of those who hold that the right to life of the fetus is sufficient in all cases to override the right of women to choose an abortion. On the contrary, I accord the right of women to control their procreation a high status, as a crucial ingredient of the sanctity or dignity of life.

Callahan's Position

I will not try to defend or fully explain this position here. My intent is, resolutely, to talk about what seem to me good and bad arguments. But I want to note, no less resolutely, that it is perfectly possible for those with bad arguments to come nevertheless to good conclusions. This happens all the time, even if it is a process which does some violence to logic. (Or, as is more likely the case, people begin with good intuitions and then defend them with bad reasons.) Let me begin by laying out what seem to me bad or at least incomplete arguments. For convenience, I shall set them out as propositions, most of which should be readily recognizable to anyone even faintly acquainted with the abortion literature.

II.   Nine Inadequate Arguments

A.   *Abortion is a religious or philosophical issue, best left to the private conscience rather than to public legislation.*

This argument rarely makes much sense. If it means that for some churches and some religious believers their positions are the direct result of religious teachings, this hardly entails the conclusion that the issue is thus intrinsically religious. One might as well say that the Vietnamese war is a religious issue, not subject to legislation because there are some churches which declare that war immoral on religious grounds. Religious groups have taken religious stands on many social issues, including war, race, poverty, population, and ecology, without exempting those problems from public legislation or turning them into "theology."

Nor is it enough to argue "sociologically" that religion plays a large role in what people feel and think about abortion, and that, somehow, this shows the religious issue to be paramount. If that were the case (and the sociological facts are, in any event, more complicated), then everyone, regardless of position, is implicated; for virtually everyone can be identified (if only culturally) with one or another religious heritage. Why is it, however, that the person who comes out of a religious heritage (e.g., Roman Catholicism) that condemns abortion is said to be acting on "religious" grounds, while a person from a heritage which does not condemn abortion is not (particularly when the latter tradition has *theological* reasons for not condemning abortion, as with some branches of Judaism and Protestantism)?

The claim that abortion is not a religious but a "philosophical" issue is surely true. But, then, every serious social question is philosophical. What is justice? What is freedom? Those questions arise all the time, and they are philosophical (and legal) in nature. The answers to them shape legislation in a very decisive way. It is inconsistent to argue that the right of the fetus is exclusively a philosophical problem, to be left to individual conscience, while the right of women is a matter to be protected or implemented legislatively. If it is legitimate to legislate on the latter (which it is), then it should be equally legitimate to legislate on the former.

B. *To remove restrictive abortion laws from the books passes no judgment on the substantive ethical issues; it merely allows individuals to make up their own minds.*

That an absence of legislation allows freedom of individual choice is undoubtedly true. But it would be highly surprising if a social decision to remove restrictive laws did not reflect a significant shift in public moral thinking about the issue at hand. Civil libertarians, for instance, would be outraged if it were proposed to repeal all legislation designed to protect the civil rights of blacks on the ground that this would maximize individual freedom of choice: they would accurately discern any such trend as both moral and constitutional regression.

In the instance of abortion, a public decision to leave the question up to individuals reflects at least three premises of a highly philosophical sort: (1) that private abortion decisions have few if any social implications or consequences; (2) that there are no normative standards whatever for determining the rights of fetuses, except the standard that individuals are free to use or create any standard they see fit; and (3) that changes in law have no effect one way or another on individual moral judgments. My point here is not to judge these premises (though obviously much could be said about them), but only to point out that each involves a philosophical judgment and has philosophical implications. A decision to remove abortion laws from the books is no more ethically neutral than a decision to put such laws on the books or keep them there.

C. *Any liberalization of abortion laws, or a repeal of such laws, will lead in the long run to a disrespect for all human life.*

This is a fundamental premise of those opposed to abortion There is no evidence to support such a judgment, however, and evidence rather than speculation is what is required.

In the first place, it is exceedingly difficult to correlate abortion attitudes throughout the world with any trend toward disrespect for nonfetal life. On the contrary, insofar as liberal abortion laws are designed to promote free choice for women, there is a prima facie case that their intent is to enhance respect for the lives of women.

Secondly, there is no evidence to support a "domino theory" of the kind which predicts a quick move from liberalization of abortion laws to the killing of the defective, the elderly, and the undesirable. This has certainly not happened in Japan or the eastern European countries, which have had liberal abortion laws for a number of years.

Finally, since most of those who support liberal abortion laws either do not believe that fetal life is human life or do not believe that it is life which has reached a stage requiring social protection, it is unfair to accuse them of harboring attitudes which inevitably lead to atrocities against all forms of human life. This kind of judgment reflects more the moral logic of the group leveling such charges than the moral principles of those at whom the charge is leveled.

D. *Scientific evidence, particularly modern genetics, has shown beyond a shadow of a doubt that human life begins at conception, or at least at the time of implantation.*

Scientific evidence does not, as such, tell us when human life begins. The concept "human" is essentially philosophical, requiring both a philosophical and an ethical judgment. Even if it could be shown that human life begins at conception, that finding would not entail the further moral judgment that life at that stage ethically merits full protection. When human life begins and when human life, once begun, merits or requires full respect are two different questions.

J. B.

It is, I think, reasonable to contend that human life begins at conception. But this is as much a philosophical and ethical position as it is scientific. At the same time, however, it is capricious to ignore all scientific evidence. As any elementary textbook in genetics shows, the fusion of sperm and egg marks a decisive first step in the life of any individual. It is only in abortion arguments that one hears vague protestations that life is just one great continuum, with no decisive, significant changes from one stage or condition to another. But if it is bad science to talk that way, it is equally bad science to say that science dictates in some normative manner when human life begins.

E.   *The fetus is nothing more than "tissue" or a "blob of protoplasm" or a "blueprint."*

Definitions of this kind can only be called self-serving. This is not the way a fetus is defined in any dictionary or any embryological text. All life is tissue and protoplasm; that fact alone tells us nothing whatever. Would it be acceptable for a student in a college biology course to define "fetus" with a one-word term "tissue" or "protoplasm"?

It is no less unscientific to call an embryo or a fetus a mere "blueprint." Blueprints of buildings are not ordinarily mixed into the mortar; they remain in the hands of the architects. Moreover, once a building has been constructed, the blueprint can be thrown away, and the building will continue to stand. The genetic blueprint operates in an entirely different way: it exerts a directly causal action in morphological development; as an intrinsic part of the physiological structure, it can at no point be thrown away or taken out.

F.   *All abortions are selfish, ego-centered actions.*

This reflects a strain of thought which runs very deep among those violently opposed to abortion. But the argument manages to ignore the decisions of those who choose abortion out of a sense of responsibility to their living children. It also manages to beg the question of whether individuals have some rights to determine what is in their own welfare, and to choose in favor of themselves some of the time.

Most broadly, this contention typifies the widespread tendency on all sides of the abortion debate to indulge in amateur psychologizing and *ad hominem* argumentation. Those opposed to abortion are adept at reducing all proabortion arguments to their psychological ingredients: homicidal impulses, selfishness, the baneful effects of a decadent culture, genocidal aspirations, a hatred of children, and the like. Those favorable to abortion are no mean masters of the art themselves: since it is well known that all opposed to abortion are dogma-ridden, male chauvinists (or females brainwashed by male chauvinists), insensitive to the quality of life, sadists and/or fascists. In short, don't listen to anyone's arguments; it is more profitable to hunt out hidden pathologies. And don't credit anyone with a mistake in reasoning, too small and human a flaw for propaganda purposes, when it is far more emotively effective to convict them of general crimes against humanity.

G. *Abortions are "therapeutic," and abortion decisions are "medical" decisions.*

Abortion is not notably therapeutic for the fetus — an observation I presume will elicit little disagreement. Even in the instance of a fetus with a grave defect, abortion is not therapeutic. It may be merciful and it may be wise, but, unless I am mistaken, the medical profession does not classify procedures with a 100 percent mortality rate as therapeutic.

Perhaps, then, abortion is therapeutic for the woman who receives it. That it is beneficial to her in some ways seems undeniable; she is relieved of an unwanted social, economic, or psychological burden. But is it proper to employ language which has a very concrete meaning in medicine — the correction or amelioration of a physical or psychological defect — in a case where there is usually no physical pathology at all? Except in the now-rare instances of a direct threat to a woman's life, an abortion cures no known disease and relieves no medically classifiable illness.

Thomas Szasz has been an especially eloquent spokesman for two positions. The first is that abortion should be available on request in the name of individual freedom. The second is that essentially non-medical decisions should not be dressed in the mantle of "medical" language simply because they require medical technology for their execution. "To be sure," he has written, "the procedure is surgical; but this makes abortion no more a medical problem than the use of the electric chair makes capital punishment a problem of electrical engineering."[1]

Szasz's point seems undeniable, yet it is still common to hear abortion spoken of as a medical problem, which should be worked out between the woman and her physician. Even if that is the proper way to handle abortion, that does not make it a medical solution. The reason for this obfuscation is not far to be sought, and Szasz has stressed it as a constant theme in a number of his writings: the predilection in our society to translate value judgments into medical terms, giving them the aura of settled "scientific" judgments and the socially impregnable status of medical legitimation.

H.   *In a just society there would be no abortion problem, since there the social and economic pressures that drive women to abortion would not exist.*

This proposition is usually part of a broader political argument which sees abortion as no more than a symptom of unjust, repressive societies. To concentrate on abortion as a response to poverty, poor housing, puritanical attitudes toward illegitimacy, and racism is a cheap and evasive solution. It achieves no more than reinforcement of unjust political and social structures and institutions.

Up to a point there is some merit in this kind of argument, and that is why I believe that (as in the Scandinavian and eastern European countries) abortion should be handled in a context of full maternal and child-care welfare programs. A woman who wishes to have a child but is not socially and economically free to do so is not a free woman. Her freedom is only superficially enhanced by allowing her, in that kind of repressive context, to choose abortion as a way out. She is not even being given half the loaf of freedom, which requires the existence of a full range of viable options.

At the same time, however, there are some serious limitations to the notion that abortion is nothing but a symptom of an unjust society. It utterly ignores the fact, common enough in affluent countries, that large numbers of women choose abortion because they have decided they want no children at all, or at least no more children than they already have. They are acting not out of social or economic coercion but out of a positive desire to shape and live a life of their own choosing, not dominated by unexpected pregnancies and unwanted children. In addition, it neglects the reality of contraceptive failure, which can and does occur independently of economic and social conditions (though it may of course be influenced by them). Short of the perfect contraceptive perfectly used, some portion of women, against their intentions, will become pregnant.

I. *Abortion is exclusively a women's issue,*
    to be decided by women.

The underpinning of this argument seems to consist of three assumptions. <u>First</u>, that there is no role for male judgment, intervention,
or interference because it is women who get pregnant and who have to
live with the pregnancies. <u>Second</u>, that abortion laws are repressive
because they have been established by male legislators. And, <u>third</u>,
that the fetus is a part of a woman's body and is thus exclusively subject to her judgments and desires.

While I am fully prepared to agree that approval of a male,
whether husband and/or father, should not be a legal condition for a
woman to receive an abortion, this should not be construed to mean
that nothing is owed, in justice, to the male. Even ignoring the well-
known fact that women do not get pregnant by themselves, a few
other considerations remain. At the least, there is an injustice in giving
males no rights prior to birth but then imposing upon them a full
range of obligations after birth. If the obligations toward a child are
mutual after birth, why should there not be a corresponding parity of
rights prior to birth? I have not seen a satisfactory answer to that
question. Moreover, if — to accept the feminist premise — women
have been forced to carry through unwanted pregnancies because of
male domination, the sexist shoe is put on the other foot if all the
rights involved in having a child are ceded exclusively to women. One
injustice is corrected at the expense of creating another, and sexism is
still triumphant.

That legislatures are dominated by males is an obvious fact. That
the history of abortion legislation would have been different had there
been legislative equality or even a female majority, however, has not
been demonstrated. Indeed, it has been a consistent survey finding
that women are less willing than men to approve permissive abortion
laws. There have of course been attempts to explain these rather awk-
ward findings, since they are inconsistent on their face with claims
that resistance to abortion is a male phenomenon. These efforts usually take the form of speculations designed to show that the resistant
women were culturally brainwashed into adopting repressive male attitudes. This is a plausible theory, but one for which there unfortunately is no evidence whatever. And, apart from these speculations,
there is no evidence that the thinking of women on abortion must, of
biological and experiential necessity, be utterly different from that of
men. But that is exactly the premise necessary if the contention is to
be sustained that an exclusively female domination of abortion legisla-

tion would produce a different result from either a male domination or legislative equality.

In this context it should be mentioned that child-bearing and child-rearing have consequences for everyone in a society, both men and women. To imply that women alone should have all the rights, even though the consequences involve the lives of both sexes, is an unfair conclusion. Or are we to overthrow, as the price of abortion reform, the long-honored principle that all of those who will have to bear the consequences of decisions have a right to be consulted? I find that price too high. Yet, since I agree that abortion decisions should not legally require the consent of husband and/or father — for I see no way to include such a requirement in the law without opening the way for a further abuse of women — I am left with the (perhaps pious) hope that there will be some recognition that problems of justice toward the male are real (however new!) and that an ethical resolution will be found.

Finally, a quick word about the contention that a fetus is "part" of a woman's body. That a fetus is *in* a woman's body is an evident biological datum. That it is thereby a *part* of her body, in much the same way that her heart, arm, liver, or leg is part of her body, is biologically false. The separate genetic constitution of the fetus, its rate of growth and development, and its separate organ system clearly distinguish the body of the fetus from the body of the mother.

Some clarity would be brought to the language of the abortion debate if this distinction were admitted. It could still be argued that, because the fetus is in the mother's body, she should have full rights in determining its fate. But that argument is different from likening the fetus to any other part of a woman's body and then transposing the rules concerning the exercise of rights over one's body. Here is a clear instance where, in order to find a constitutional precedent for women's right to control procreation, violence is often done to some elementary facts of biology.

III.   SOME VALID CONCLUSIONS

I have tried to show, using nine different propositions, that some bad, or at least incompletely developed, arguments are too much in currency to be allowed to go by default. Would the abortion debate be significantly altered if these arguments were no longer used by the contending sides? This is a moot question, as there seems to be little likelihood that they will cease being employed. They are too powerful to be set aside, for both good and bad reasons — the good being that they are able to elicit responses which build upon some pervasive feelings about abortion, feelings which are decisive in shaping thought and behavior, even if they are poorly articulated.

The great strength of the general movement for abortion on request (apart from the questionable validity of particular arguments brought to bear in support of it) is that it perceives and seeks to correct two elementary realities, one social and the other biological. The social reality is that women have not had the freedom to make their own choice in a matter critical to their development as persons. Society, through the medium of male domination, has forced its choice upon them. The biological reality is that it is women who become pregnant and bear children; nature gave them no choice. Unless they are given a means to control the biological facts — and abortion is one very effective means — they will be dominated by them.

Without such control, which must be total if it is to have any decisive meaning, women are fated to accept Freud's principle that "anatomy is destiny." That kind of rigid biological determinism is increasingly unacceptable, not only to women in the case of procreation, but to most human beings confronted with the involuntary rigidities of nature. The deepest philosophical issue beneath the abortion question is the extent to which, in the name of freely chosen ends, biological realities can be manipulated, controlled, and set aside. This is a very old problem, and the trend toward abortion on request reflects the most recent tendency in modern thought — namely, the attempt to subordinate biology to reason, to bring it under control, to master it. It remains to be seen whether procreation can be so easily mastered. That question may take centuries to resolve.

The great strength of the movement against abortion is that it seeks to protect one defenseless category of human or potentially human life; furthermore, it strives to resist the introduction into society of forms of value judgments that would discriminate among the worth of individual lives. In almost any other civil rights context, the cogency of this line of reasoning would be quickly respected. Indeed, it has been at the heart of efforts to correct racial injustices, to improve health care, to eradicate poverty, and to provide better care for the aged. The history of mankind has shown too many instances of systematic efforts to exclude certain races or classes of persons from the human community to allow us to view with equanimity the declaration that fetuses are "not human." Historically, the proposition that all human beings are equal, however "inchoate" they may be, is not conservative but radical. It is constantly threatened in theory and subverted in practice.

Although the contending sides in the abortion debate commonly ignore, or systematically deride, the essentially positive impulses lying behind their opponents' positions, the conflict is nonetheless best seen as the pitting of essentially valuable impulses against one another. The possibility of a society which did allow women the right and the freedom to control their own lives is a lofty goal. No less lofty is that of a society which, with no exceptions, treated all forms of human life as equally valuable. In the best of all possible worlds, it might be possible to reconcile these goals. In the real world, however, the first goal requires the right of abortion, and the second goal excludes that right. This, I believe, is a genuine and deep dilemma. That so few are willing to recognize the dilemma, or even to admit that any choice must be less than perfect, is the most disturbing element in the whole debate.

The bad reason why the arguments I have analyzed will endure is that they readily lend themselves to legal use. Nothing in our society has so muddied the ethical issues as its tendency to turn ethical problems into legal matters. The great prize, sought by all sides, is a favorable court decision. Toward that end, the best tactic is to find a way of bringing one's own ethical case under one or more constitutional protections or exclusions. If one can succeed in convincing the courts that abortion is a religious issue (which it is not), then there is a good chance that they will favor private choice and rule against legislation. The same tactic is evident in efforts to show that abortion decisions come under the constitutional protections afforded to "privacy" (which begs the question of the rights of the fetus), or, on the other side, to show that abortion violates the equal protection and due process requirements of the Constitution (which also begs the question of the rights of the fetus). Since the possibility of a legal victory is an irresistible goal, there seems to be no limit to the bad arguments which will be brought to bear to gain it.

NOTE
1. Szasz, *The Ethics of Abortion*, Humanist, Sept.-Oct. 1966, at 148.

2 Abortion: A Changing Morality and Policy?

RICHARD A. MCCORMICK

This article concerns itself with an examination of the legal and ethical grounds for devising a public policy on abortion. To do this McCormick examines the relationship between law, morality, and public policy; examines a variety of bases for establishing the value of fetal life; and presents his perception of the classical Christian moral position on abortion. In the light of this background, McCormick then sets out several procedures, principles, and applications of policy with respect to abortion. He then examines a variety of other issues that will be affected by public policy on abortion.

Rev. Richard A. McCormick S.J., Ph.D., is Rose F. Kennedy Professor of Christian Ethics at the Kennedy Institute of Ethics, Georgetown University

Over a million legal abortions are performed annually in the United States. If this is what many people think it is (unjustified killing of human beings in most cases), then it constitutes the nation's major moral tragedy. In contrast, over many years 50,000 Americans were killed in Vietnam. The present abortion situation is a major stain on the national conscience; and it is particularly upsetting during a time of cultural schizophrenia, that is, a time when we rejoice in the production—at great cost and energy—of a test-tube baby.

One of the things that is particularly tiresome to a theologian is that the national debate on this utterly serious subject has collapsed into slogans. People feel deeply about this subject; and when they do, they are going to marshal every shred of evidence—questionable or not—and bend it to their purposes. As a theologian, I am committed to the thesis that emotions are excellent companions but not always

25

good guides. It is essential, therefore, to submit one's loyalties and value judgments to constant scrutiny and questioning and to those theological criteria that make abortion also (though not only) a theological question, a task not without its risks. For instance, in a July issue of *America* I offered to both sides of this national discussion 10 rules for debate.[1] My own basic value commitments were, in my and almost everyone else's judgment, transparent—with one exception. In an August issue of the *Steubenville Register,* I was mildly shocked to read a column on the article entitled "How to Get Soft on Abortion."[2] I say that I was *mildly* shocked because it is well known that plenty of people in this debate are ready to excommunicate if one does not formulate matters exactly as they do.

In this article I want to discuss public policy on abortion. One cannot do that, of course, without discussing morality, for while public policy and morality are not identical necessarily, they are intimately related. That is, if one regards fetal life as disposable tissue, clearly abortion ought not be on the penal code at all, except to protect women against bungling and incompetent tissue scrapers. If, however, fetal life is to be regarded as human life making claims on society's protection, then the possibility is that it ought to be protected by law.

Morality, then, concerns itself with rightness or wrongness of human conduct. Law, or public policy, on the other hand, is concerned with the common good, the welfare of the community. Clearly, morality and public policy are both related and distinct. They are related because law, or public policy, has an inherently moral character due to its rootage in existential human ends (goods). The welfare of the community cannot be unrelated to what is judged to be promotive or destructive to the individual. Morality and public policy are distinct, because only when individual acts have ascertainable public consequences on the maintenance and stability of society (welfare of the community) are they the proper concern of society. Thus all civilized societies outlaw homicide.

What immoral or wrongful actions affect the welfare of the community in a way that demands legislation is subject to a further criterion: feasibility. Feasibility is that quality whereby a proposed course of action is not merely possible but practicable, adaptable, depending on the circumstances, cultural ways, attitudes, traditions of a people. The criterion of feasibility, therefore, raises questions such as, Will the policy be obeyed? Is it enforceable? Is it prudent to undertake this or that ban in view of possibly harmful effects in other sectors of social life? Can control be achieved short of coercive measures? As John R. Connery, SJ, professor of moral theology, Loyola University of Chicago, words it: "One cannot conclude . . . that because the state does not penalize some action it is not morally wrong. All one can infer is that it was not judged to be harmful to the community, *or if it was judged harmful, the harm was less than that which would result from prohibitive legislation"* [emphasis added].[3]

DIFFICULTY OF DIFFERING VALUES SET ON FETAL LIFE

What makes public policy on abortion such an intractable and divisive issue is that the good (and evil) itself whose legal possibility is under discussion is an object of doubt and controversy. That is, the evaluation of fetal life differs in our society. Some persons see it as living, but disposable (worthless?) tissue. Others see the fetus as human life making claims, but claims that are overridden by a rather broad class of maternal or familial concerns. Still others see nascent life as something to be protected in all but a very few exceptional instances. In other words, some see permissive laws as injustice (and therefore clearly something touching the welfare of the community). Others see restrictive laws as injustice to the woman. To complicate matters, all of these positions are enveloped in some highly questionable rhetoric and question begging. Given this disagreement, it is very difficult to arrive at acceptable or feasible policy, for ultimately public policy must find a basis in the deepest moral perceptions of the majority; or if not, at least in principles the majority is reluctant to modify.

The key issue—both morally and eventually legally—is the evaluation of nascent life, first in itself, then when set against competing claims. By this I do not mean, Is the fetus a person? That is a legitimate question and of no inconsiderable legal importance. But the definition of person is often elaborated with a purpose of mind. One defines personhood and then grants or does not grant personhood according to what one wants to do or thinks it acceptable to do with nonpersons. That this can be a dog-chasing-tail definition is quite clear. As Paul Ramsey of Princeton (NJ) University is fond of questioning, Does one really need a PhD from Harvard to be a person, or is a functioning cerebral cortex quite sufficient? Nor do I mean, When does human life begin? Put 100 biologists and geneticists in a room on that question, and you will get a single answer, At fertilization. What results from fertilization is living, not dead; is human, not a mouse.

So the question is one of evaluation of nascent life, of the moral claims that the nascent human being (what Pope Paul, in a brilliant finesse, referred to as *personne en devenir,* "a person in the process of becoming") makes on us. This is a straightforward moral question, and we must turn to it if reflections on public policy are to be coherent. Daniel Callahan of the Hastings Center has caught this exactly when he says: "The essence of the moral problem in abortion is the proper way in which to balance the rights of the unborn . . . against the right of a woman not to have a child she does not want."[4]

The present public policy of this country in the *Wade* and *Bolton* decisions is that the nascent being has no rights. The underlying moral evaluation is quite clear: Whatever the fetus is, it is not to be evaluated as a higher priority than any competing concern of the woman. To be perfectly blunt, this means that the fetus is a nothing, a zero. Justice Harry Blackmun held that the fetus is only "potential life." Judge Clement Haynsworth carried one step further what was implicit in Blackmun: "The Supreme Court decided that the fetus in the womb is neither alive nor a person within the meaning of the Fourteenth Amendment" (*Floyd v. Anders*, F. Supp. DSC 1977). Callahan is once again absolutely correct when he says: "Under permissive laws, any talk whatever of the 'sanctity of life' of the unborn becomes a legal fiction. By giving women the full and total right to determine whether such a sanctity exists, the fetus is, in fact, given no legal or socially established standing whatever."[5] Under such policy any talk of fetal sanctity is double-talk. Callahan continues: "The law forces a nasty either-or choice, devoid of a saving ethical ambiguity."

It is not surprising, then, that the ethical history of abortion is the history of an evaluation. What has that evaluation been? Over the centuries—as Connery has so painstakingly pointed out in his historical study of abortion—some theologians, because of the biological knowledge available to them, spoke of three types of abortion: prevention of conception, abortion of an inanimate fetus, abortion of an animated fetus.[6] Still others felt that all interventions into the life-giving process were homicides. Others argued that evacuation of a nonanimated fetus was permissible to save the mother's life, for the inanimate fetus was not yet a human being. Some contended that abortion even of an inanimate fetus was immoral either as imperfect homicide or misuse of *semen conceptum*.

As time went on, many of these qualifications vanished and obscurities diminished. Beneath these debates and developments, one finds an evaluation of fetal life that yielded it to very few competing interests—what John Noonan of the University of California at Berkeley calls "an almost absolute value."

In recent years nearly every national Catholic episcopal conference has repeated some such evaluation. By noting this I do not mean that others than Catholics have not made similar statements, nor do I mean that the matter is a "Catholic question." Clearly it is not. For example, the Belgian bishops, in their 1973 pastoral, cite approvingly *Abortus Provocatus*, a study issued by the Center of Demographic and Family Studies of the Ministry of Health: "There is no objective criterion for establishing, in the gradual process of development, a limit between 'non-human' life and 'human' life. In this process each stage is the necessary condition for the following and no moment is 'more important,' 'more decisive,' or 'more essential' than another."[7]

In a truly remarkable document the Catholic bishops of Germany joined with the Evangelical bishops in a statement rejecting *Fristenregelung*, which was before the *Bundestag*. They stated: "The right to life must not be diminished, either by a judgment on the value or lack thereof of an individual life, nor by a decision on when life begins or ends. All decisions that touch human life can only be oriented to the service of life."[8] On this basis they stated that the task of the lawmaker is to identify those conflict situations where interruption of pregnancy will not be punished ("Straflos lassen").

In the past few years, this evaluation has been challenged or at least probed from a number of theological sources. Let me cite just three as examples.

CHALLENGES TO THIS EVALUATION

Bernard Haring. Haring sees probability in several theories about hominization. The data of embryology seem to support those who hold that the most decisive moment is fertilization. Yet Haring sees probability in individualization at a later date, specifically with the development of the cerebral cortex, for this constitutes the biological substratum for personal life. His argument is that since a considerable number of embryos are anencephalic (lacking in essential parts of the typically human brain) and incapable of personal activity before formation of the cerebral cortex, "there exists merely a biological center of life bereft yet of the substratum of a personal principle." Haring sees this as a "theory" only, and the early fetus must enjoy the favor of the doubts that cling to such a theory.[9]

Etudes Dossier. An interdisciplinary group headed by Jesuit Bruno Ribes distinguishes "human life" from "humanized life." The distinction is analyzed somewhat as follows. Being relational beings, we receive our singularity and proper being in relation to others. In this perspective, the existence of the fetus is an injunction to the parents. The parents' recognition of fetal life gathers this injunction into a call: The parents call the child to be born. It is this recognition and call that "humanizes." Prior to that happening, the fetus is "only human." Refusal to humanize, of course, is intolerable. It dissociates the biological from the human, the generating function from the human. They argue, however, that there can be instances where genuine humanization is impossible. In these instances abortion is socially tolerable.[10]

The reactions to this study were swift and predictable. The Archbishop of Rouen referred to it as "inadmissible casuistry," which was "subtle and coarse." The Cardinal Archbishop of Paris attempted to suppress the second half of its two-part publication in *Etudes.*[11]

I believe the study errs in two crucial ways. First, it confuses a distinction with a division. A distinction results from an analysis of a unique process. Distinctions are valid in analyzing intrauterine life. But a division is not valid for this purpose, for in reality there is no stage marking passage from one mode of existence to another. Second, the study uses the word "humanize" in two different senses—an error that corrupts any argument. At one point the word refers to a recognition and call by the parents. At another it refers to the quality of life after birth ("impossibility of humanizing").

Bernard Quelquejeu. In a long study Quelquejeu argues that a change in method is called for in facing the contemporary situation. He suggests that we must consult the perplexed conscience and discover there a new principle. When he does study the perplexed conscience, he discovers his new principle: the fact that we must not prescind from the right to exercise one's sexuality in judging this matter. Concretely, he argues as follows: If the preceding will not to conceive was reasonable and responsible, abortion is justifiable. Otherwise a biological fact prevails over a reasonable will.[12]

I believe it must be said that this confuses two factors: (1) the legitimacy of the will to procreate or not (a moral question) and (2) the relationship of the parents' will to the constitution of fetal humanity (an ontological question). Specifically, the ontological status of a being cannot and does not depend on an intention exterior to its being. In Quelquejeu's analysis, even a born child is only a "biological fact" as long as there was the will not to procreate. I cannot dissolve the impression that in this analysis children are really like property. They exist for the parents. They are really the chattel of the parents' private value judgments.

In my judgment, then, recent theological and philosophical studies have not made serious inroads on the substance of the classical position. I say "substance" deliberately because one must always distinguish the substance from its formulation at a given time. For this reason the eminent theologian Karl Rahner correctly notes that the concrete ethical judgments of the Church's magisterium are inherently "provisional." For instance, when one questions the crucial moral relevance of the direct/indirect distinction in this area—as one can and, I think, ought to do—one is questioning the formulation of Pius XI and Pius XII and even Paul VI and the Sacred Congregation for the Doctrine of the Faith. One is not necessarily tampering with the substance of that teaching. The distinction between substance and formulation is not only present in Vatican II explicitly. It is clear in St. Thomas, specifically in the I–II, q. 100, a. 8, where he discusses the dispensability of the precepts of the decalogue. There he discusses the possible difference between the original or intended senses and the formulated letter, thus laying the basis for *mutatio materiae* (as it was later known) due to circumstances.

THE CLASSICAL CHRISTIAN MORAL POSITION

What is the substance of the classical moral position? Distinguishing the abiding substance from its enveloping and changeable formulation is a tricky and difficult business—even, it would seem, an arrogant one. At a certain point one has to assume the posture of a person standing over history and filtering out the limitations of others when one is at least knee-deep in historical limitations oneself.

With that caution, I would like to attempt to frame in three statements the substance of the classical Christian moral position in our time.

1. Human life as a basic gift and good, the foundation for the enjoyment of all other goods, may be taken only when doing so is the only life-saving and life-serving alternative, or only when doing so is, all things considered (not just numbers), the lesser evil. I have said here "human life," not the "human person." For the word "person," as I suggested, only muddies the moral discussion. "Person," as Albert Outler of Emory University, Atlanta, notes, is a code word for a self-transcending, transempirical reality. Self-transcendence is not, contrary to the notions or wishes of so many, a part of the organism. It is the organism as oriented to its self-transcending matrix. I have said "life-saving and life-serving" for two reasons. First, not every life-saving action is life-serving. (For example, some actions could, while saving numerical lives, actually and simultaneously undermine other basic goods in a way that would be a disservice to life itself by attacking an associated value.) Second, it seems to me that the exceptions historically tolerated (for example, ectopic pregnancies) fit this general category.

2. By "human life" I mean human life from fertilization or at least from the time at or after which it is settled whether there will be one or two distinct human beings. (The phrase is Ramsey's.) There are phenomena in the preimplantation period that generate evaluative doubts about the claims the fetus at this stage makes, at least in some cases. I refer to twinning, the number of spontaneous abortions, the possibility of recombination of two fertilized ova into one (chimeras), the time of the appearance of the primary organizer. These phenomena create problems—doubts only. After all, one could and should

counter, the only thing that stands between an 8-cell embryo in a petri dish and Louise Brown is a uterine home for 266 days. I find it remarkable—*sc.,* worth remarking—that Louise Brown is referred to as the "test tube *baby*." The answer to the question, Where did Louise Brown begin? is clearly, In the petri dish if "baby" means anything. We do not say "test tube tissue." In doubt one generally favors life—but I think not always.

3. For an act to be life-saving and life-serving, to be the lesser evil (all things considered), there must be at stake human life or its moral equivalent, a good or value comparable to life itself. This is not what the traditional formulations say, but it is where the corpus of teachings on life taking lead (for example, just war, capital punishment). For instance, if human beings may go to war and take human life to defend their freedom (political autonomy) against an enemy who would strip them of it, something is being said about human freedom compared with life. I realize that life and liberty cannot be compared, as apples and oranges cannot. But in daily life we somehow manage to parse this incommensurability in many areas. We choose to smoke or drink or eat creamy butter for enjoyment's sake at risk of a shorter life in which to enjoy anything. We elect a heart bypass operation to relieve oppressive and continuous chest pains, although we know the operation is still being investigated and itself may kill us. We choose a contemplative or a more active career for ourselves.

In all these instances it is not inaccurate to say that we somehow manage to weigh incommensurables, to overcome indeterminacy. As philosopher Donald Evans once put it: "Such compromises are a part of everyday morality, but they raise serious problems for ethical theory, for their logic and rationale are obscure." So we may grant with philosopher W. D. Ross that we are faced with great difficulties when we try to commeasure good things of very different types. But I would suggest the difficulties are not insuperable. I make this very slight opening in fear and trembling because it is made in a world and at a time where it can be terribly misused and misunderstood. If that is a serious danger, I withdraw it—and here and now. But I have known cases where failure to terminate a pregnancy has resulted in complete and permanent loss of freedom for the mother (insanity). We die for our freedom, do we not? "Give me liberty or give me death" resonates with all of us, though there are still some who would rather be "red than dead"—but, I suspect, because they think they will really have their liberty after all.

I have presented at least, and perhaps at best, a defensible account of the substance of the Catholic community's evaluation of nascent life over the centuries. It is an evaluation I share. I want to make three points about the notion of evaluation. First, evaluation is a complex concept, involving many dimensions of human insight and judgment. It cannot be reduced simply to rational arguments or religious dogma. Persons with profound religious faith often make judgments that coincide with those of a genuine humanism. Having said that, I would add, however, that the best way to state why I share the traditional evaluation is that I can think of no persuasive arguments that limit the sanctity of human life to extrauterine life. In other words, arguments that justify abortion seem to me equally to justify infanticide—and more.

Second, by saying I share the evaluation, I do not mean to suggest that all problems are solved. They are not. For instance, the moral relevance of the distinction between direct and indirect abortion (really direct/indirect anything) remains a theoretical problem of the first magnitude. I am inclined to agree with John Noonan, Bruno Schuller, Denis O'Callaghan, and, most recently, Susan Teft Nicholson that the distinction is not of crucial moral significance. For instance, O'Callaghan wrote: "If it was honest with itself, it [scholastic tradition] would have admitted that it made exceptions where these depended on chance occurrence of circumstances rather than on free human choice. In other words, an exception was admitted when it would not open the door to more and more exceptions, precisely because the occurrence of the exception was determined by factors of chance outside of human control."[13] He gives intervention into ectopic pregnancy as an example. The casuistic tradition, he believes, accepted what is in principle an abortion because it posed no threat to the general position, though this tradition felt obliged to rationalize this by use of the double effect. Tubal pregnancy, as a relatively rare occurrence and one independent of human choice, does not lay the way open to abuse.

The Belgian hierarchy, perhaps unwittingly, seems to agree. Of those very rare and desperate conflict instances (where both mother and child will die if abortion is not performed) they note: "The moral principle which ought to govern the intervention can be formulated as follows: Since two lives are at stake, one will, while doing everything possible to save both, attempt to save one rather than to allow two to perish."[14]

Similarly, in her recent study *Abortion and the Roman Catholic Church* Susan Nicholson has suggested that abortion might be reconceptualized. If it is viewed dominantly as a killing intervention, we might come out one way. If it is conceptualized above all as withdrawal of maternal assistance, different questions arise—and possibly different answers. Thus the question comes up, Is a woman bound (heroically) to provide assistance (nourishment, etc.) when the pregnancy is the result of rape?[15] I shall not dialogue with Nicholson here except to say that she raises legitimate questions.

The third point to be made about this evaluation is its historico-theological rootage. Albert Outler puts it well: "One of Christianity's oldest traditions is the sacredness of human life as an implication of the Christian convictions about God and the good life. If all persons are equally the creatures of the one God, then none of these creatures is authorized to play God toward any other. And if all persons are cherished by God, regardless of merit, we ought also to cherish each other in the same spirit. This was the ground on which the early Christians rejected the prevalent Graeco-Roman codes of sexuality in which abortion and infanticide were commonplace. Christian moralists found them profoundly irreligious and proposed instead an ethic of compassion (adopted from their Jewish matrix) that proscribed abortion and encouraged 'adoption.' "[16] Thus the value of human life leading to the traditional evaluation was seen in God's special and costing love for each individual—for fetal life, infant life, senescent life, disabled life, captive life, enslaved life, yes, and most of all, unwanted life. These evaluations can be and have been shared by others than Christians, of course. But Christians have particular warrants for resisting any cultural callousing of them.

No Agreement on Which To Base Public Policy

I now want to turn to public policy. What makes this so enormously difficult is that public policy is precisely that—public. It cannot exist successfully if there is not some ground of agreement to support it and the principles on which it stands. And at some point that agreement must be on evaluation of fetal life. There is no such agreement. Indeed there is virulent disagreement, as the recent and unfortunate manifesto "Call to Concern" (issued by a group of over 200 largely Protestant and Jewish scholars) indicated once again.[17] Furthermore, there is already a public policy in place—the *Wade* and *Bolton* decisions—but one I regard as bad law and bad morality.

In the face of such difficulties all one can do is say what one thinks the law ought to be and why. Let me do this in three steps: procedures, principles, applications.

Procedures. We should accept that in a pluralistic society, legal positions tracing back to almost any moral position (whether it be that of Vatican Council II or that of the Supreme Court) are going to be experienced as an imposition of one view on another group. Thus many view the *Wade* and *Bolton* decisions as an imposition, the introduction of legalized killing into their world. Such a matter should be decided in the Congress, not by 9 men in a use of "raw judicial power." Congress is the place where all of us, through our representatives, have a chance to share in the democratic process. This process is often halting, messy, and frustrating, but have we not learned that it is the most adequate way to live with our differences? Furthermore, it gives a reasonable chance from year to year to tighten and purify the policies that have emerged from compromise. Certainly it is a way more adequate than that of a decision framed and finalized by a Court that imposes its own poorly researched and shabbily reasoned moral values as the basis for the law of the land. There are reasons to believe that if this matter were returned to the electorate through its representatives, the nation would have a remarkably different policy. At least there is the chance that the gap between the good and the feasible would be narrowed.

Principles. I believe three general guidelines are called for as a basis for specific policy if one accepts the traditional Christian (and my) evaluation of fetal life and competing interests. First, there should be a strong presumption in law for the legal protection of fetal life. Second, exceptions should be as clearly and precisely delineated as possible, and such exceptions should contain their own substantive and procedural controls. This statement is but an entailment of the strong presumption favoring fetal life and coincides with the sound policy advocated by the Catholic-Evangelical episcopates of the Federal Republic of Germany. Without both substantive and procedural controls, there is simply no strong presumption in law for protection. Thus to leave the abortion decision simply to the woman ("pro-choice") is legal forfeiture of any presumptive claims by the fetus. In bioethics two questions are often distinguished: What is the right decision? Who decides? (At some point a loose answer to the second question destroys in principle the possibility of a disciplined answer to the first.) Third, the policy should be under constant review. The values at stake are fundamental to the continuance of civilized society. That being the case, not only legal protection of nascent life is required, but social protection—getting at the causes of abortion. Abortion exists because of a cluster of factors that make up the quality of a society. It will disappear only when that quality is changed. Optimistically I contend that the quality can be altered and that therefore the legal provisions must be under constant review lest they themselves become a factor corrosive of this quality.

Applications. When I try to fit the Christian evaluation of fetal life into the contemporary American scene and to develop a feasible protective law, I believe it is realistic (feasible) to say that many people would agree that abortion is legally acceptable if the alternative is tragedy, unacceptable if the alternative is mere inconvenience. Furthermore, I believe that Americans are capable of distinguishing the two in policy—certainly not in a way that will satisfy everyone, but in a way that at least very many can live with. In other words, I do not believe the distinction can or ought to be left entirely to individuals. Such a policy would prohibit abortion unless the life of the mother is at stake; there is a serious threat to her physical health and to the length of her life; the pregnancy is due to rape or incest; fetal deformity is of such magnitude that life-supporting efforts would not be considered obligatory after birth. (I hesitate to list this last exception because it is accordionlike and easily abused. Furthermore, it is problematic because the very situation of life supports for disabled neonates is itself problematical.)

This list is for all practical purposes and with a few changes the American Law Institute's proposals in the Model Penal Code. I do not list these indications because I judge them to be exceptions that are morally right. I do so only because at the present time many people believe that continuing the pregnancy in such circumstances is heroic and should not be mandated by law. I list them also because among the evils associated with any law, these seem to represent the lesser evil. I am confident that such a policy will completely satisfy no one—including me. Certainly it will not satisfy the pro-life advocates, and certainly it will not satisfy the advocates of pro-choice. It ought to be more acceptable to those morally opposed to abortion than to others. But again this acceptance is rooted solidly in one's assessment of fetal life.

Several more issues in regard to this policy require attention. First, is it a realistically possible policy now? Probably not, unless some of the cultural changes I will cite occur. Second, does not this policy force some women to bear pregnancies that they do not want or that they find terribly onerous? Yes, and I think it should do so. If it does not do this, if it leaves this decision simply in the hands of the individual, the law has forfeited all sanctity of fetal life to personal value judgments. The law does this in no other area where human life is at stake. Third, would a great number of people suffer under such a policy? Yes they would. But a great number of people are suffering under the present policy. It is a question of measuring suffering and choosing the lesser. In such a measuring one returns to evaluation of fetal life.

In connection with suffering the picture is incomplete without taking seriously—personally and at the level of social policy—genuine alternatives to abortion. There are such and there ought to be more. We have some experience of this at Georgetown; and we know that when genuine, supportive alternatives are provided, the felt need and the actual experiences of abortion are remarkably lower. Any policy on abortion ought to live in an atmosphere where alternatives exist. The policy I have proposed can only be evaluated if such an atmosphere is assumed to be present. I know it is not present in some places—perhaps not in many. But that does not attack the policy; it says only that more work must be done. And one reason such alternatives do not always exist is that some do not see them as alternatives.

I want to raise a final point to anticipate an objection one frequently hears: How can you, a man—and a celibate priest to boot—really know what it is to carry a pregnancy with your back against the wall? It is women, and the individual woman, who ought to make this decision, according to this objection. I want to grant the element of truth in this way of thinking. The more one knows experientially of a situation, the more it is one's own experience, the more sensitive one ought to be to the situation's many-faceted circumstances, though I believe this is frequently not the way things turn out. Self-involved agents are frequently self-interested agents with a one-dimensional view of things obvious to most reasonable and reflective people. Still, it remains true that if I, as a moral theologian, am going to reflect on the moral problem of terribly disabled neonates, I ought to learn all I can from neonatal intensive care units, from nurses, from experienced physicians, from parents.

After that element of truth has been granted, I must respond that if the objection is pushed to the limit—as it often is by extremists—it does away with the possibility of generalization in ethics, that is, with ethics or moral theology itself. Do I really have to be a veteran of Vietnam to judge the disproportion of that war? Do I have to be a politician to recognize obstruction of justice in public office and judge it wrong? Do I have to be a male to know that torture by castration is to be condemned? Hardly. Those who assert that one must be a woman or a married person to judge abortion a moral tragedy are saying, by implication, that only the involved agent is capable of assessing the moral character of his or her action. That is destructive of the entire ethical enterprise and, at the level of policy, leads to the total dissociation of the moral and legal order.

Try an experiment in thinking about this question. If you were a fetus during the great depression, would you want your future to be in the hands of an individual undergoing economic hardship? I dare say you would not want that. Indeed, most of us would appreciate retrospectively the existence of controls (and their social supports) that allowed us to come to be. It is easy for adults undergoing hardship to forget this.

A constitutional amendment is needed to realize such a policy or any policy that legally protects fetuses. I wish to discuss that here only indirectly, to the extent that in order to get a constitutional amendment of any kind, more of a consensus on fetal evaluation must emerge than seems presently to exist, particularly on fetal evaluation with respect to competing claims of the woman (and family). In order to get closer to consensus, the climate of opinion that affects this evaluation must change. I want to list some of the factors that can easily (and corrosively) affect evaluation of nascent life. Attempts to alter policy that ignore and bypass these factors might easily prove shortsighted. Or, more positively, those who concern themselves (in thought and action) with these climate factors are doing something very important to affect policy.

Human sexuality. What Americans as a culture think about sexuality and how they live it will have a strong influence on their evaluation of fetal life and abortion. One need be neither a eunuch nor a Cassandra nor even a phallocrat to think that something has gone wrong in this area and that we must readjust our communal head on the matter. The latest right seems to be the right to sexual pleasure without incumbrance or consequence. Thus we have sex without marriage, sex outside marriage, and a whole variety of couplings and noncouplings that increasingly separate sex and *eros* from covenanted *philia*—the one relationship that offers the best chance to retain quality in sexual expression. In other words the culture has trivialized sexuality. Symbols of this trivialization are all about us, from *Charlie's Angels* and *Soap* to the teenage illegitimacy rate. Not every loosening is a true liberation. Sexual trivialization means sexual irresponsibility, and that must deflate our esteem for the offspring of sexual activity. Unless a genuine cultural change occurs in regard to sexuality, I have little hope for a shift in fetal evaluation.

The concept of privacy. The assumption in American society that abortion is a private affair powerfully undergirds attitudes on the subject. The Supreme Court struck down all prohibitive state laws by appeal to the penumbral right of privacy. Basically the reasoning is that abortion is a private matter because the right to have a child is a private matter. But as John Noonan points out, it is not merely unpersuasive but inaccurate to say destruction of the unborn is private to the mother because it affects only her body. "It affects the child, the father, the grandparents, the physician, the nurse and the hospital. It is not merely unrealistic but false to hold that a decision ending nascent life does not affect basic social attitudes about life, fidelity and responsibility and does not touch the fabric of society."[18]

Abortion is not just a matter of the bedroom, as proponents sometimes argue. The right of privacy is increasingly confused with excessive individualism and automony—the "freedom-to-be-let-alone-to-do-my-own-thing" syndrome. Many of the arguments surrounding the abortion debate reveal this lonely individualism. For instance, the *New York Times* welcomed the 1973 abortion decisions with this editorial statement: "Nothing in the Court's approach ought to give affront to persons who oppose all abortion for reasons of religion or individual conviction. They can stand as firmly as ever for those principles, provided they do not seek to impede the freedom of those with an opposite view."[19] This simply says: "Nobody requires you to kill. Just let others do it." That assumes that I am unrelated to, unaffected by, others' actions. It supposes that you and I are islands—individual and isolated from the society and atomized within it. From such beginnings comes the nonsense about abortions as an exercise of "privacy." The Supreme Court, in its 1976 abortion decision *Planned Parenthood v. Danforth,* further intensified this individualism by declaring Missouri's requirement of spousal consent unconstitutional. This decision just atomizes individuals within the family. Unless the nation can somehow overcome this radical individualism, fetuses are at risk.

   The interventionist mentality. As a highly technological and pragmatic people, Americans are possessed by the interventionist mentality. What Daniel Callahan calls the "power-plasticity" model has shaped our imaginations and feelings.[20] The best solution to the problems of technology is more technology. We obliterate cities to liberate them. We segregate senior citizens in leisure worlds, and retarded "defective" children in institutions. What is efficient becomes the morally good and right. Technology creates its own morality. In this perspective we eliminate the maladapted condition rather than compassionately adjust the environment to it. We are moving toward what Paul Ramsey calls "administered death" much in the way that 20 years of liberated chat about sexuality has brought us "calisthenic sexuality." The society has the "fly now, pay later" attitude. The ultimate in this interventionist mentality is the statement of Joseph Fletcher: "Man is a maker and a selector and a designer, and the more rationally contrived and deliberate anything is, the more *human* it is." He continues: "Laboratory reproduction is radically human compared to conception by ordinary heterosexual intercourse. It is willed, chosen, purposed and controlled; and surely these are among the traits that distinguish *Homo sapiens* from others in the animal genus. . . . Coital reproduction is, therefore, less human than laboratory reproduction."[21] Welcome 1984! That something is wrong in this statement I have no doubt. Its surest symptom is that the fun has gone out of things. The "interventionist mentality" leads us to identify making a problem go away with a human solution. Unless we can somehow overcome this, fetuses are at risk.

Utilitarian attitudes. The prevalent popular morality, or structure of moral thinking, is utilitarian. By this I mean an attitude that measures rightness and wrongness in terms exclusively of "getting results," as Joseph Fletcher puts it. If the results seem urgent enough, then paying the price is morally justified regardless of who or what gets stepped on in the process. Not too thinly disguised in these attitudes is a highly functional concept of persons. This attitude is evident frequently in medical procedures involving experimentation, in military strategy, and in government decision making. It emerged in the reaction of many people in _in vitro_ fertilization with embryo transfer. Results are only one determinant of the moral quality of human conduct; and until this is realized, fetuses—and all of us are but grown-up fetuses—are in trouble.

The influence of the media. I do not wish to join the Agnews of this world in blackening a whole industry. Rather, I want to suggest that via TV we are frequently and mediately exposed to the suffering and deprivation of others. By "mediately" I mean a remove from the actual happening. Thus we daily witnessed the body counts from Vietnam in a cozy chair and comfortable room with standard swank Beefeaters at our sides to soothe any quivering ganglia. In the same way now we experience media presentations of suffering and death in Lebanon, Biafra, Guatemala, El Salvador, Guyana, and a host of other places. (And here it should be added that the very prevalence of legal abortion is part of this picture. We are getting used to it. It is well known that wherever abortion laws have been made very permissive, thousands of women come forward feeling the need for abortion who did not feel it in a different climate. That they do this suggests the subtly coercive character of liberalized laws.) The "mediate" experience of suffering inevitably blunts our moral imaginations and sensitivities, and such blunting can chip away at our grasp on the uniqueness and equality of each of God's children, particularly the poorest of the poor and the weakest of the weak.

Attitudes toward childbearing and the family. A whole series of influences (divorce and instability of the nuclear family, population pressures, financial pressures, the emergence of a "comfort ethic" in a consumer society) has drastically modified the attitude toward childbearing. Many young couples now view children as a nuisance, an incursion on double careers and so on. Sterilization has become the contraceptive method of choice. In some of the nation's hospitals obstetrics units are no longer active. I simply want to note this, not to analyze it.

Many will cheer this development as the American contribution to solving the "population problem." But before they do—thus adding to the antinatalist atmosphere—they would be well advised to take a long look at some statistics. For instance, the nation is now spending $112 billion (25 percent of the federal budget) on social security for persons over 65. In 1977 the ratio of workers to dependents was 6 to 1. With present fertility rates, in the year 2025 the nation will be spending $635 billion (40 percent of the federal budget) for those over 65. The ratio of workers to dependents will be 3 to 1. Therefore, a heavier burder will fall on fewer—the burden being 12 times what it is today. A narrow, unsophisticated attitude toward the "population problem" forms part of our evaluative preconditioning where fetal life is concerned. When some people see a pregnant woman in the supermarket, they groan about another polluter or another college education. Far too few think of social security checks. Until they do, fetuses will remain at risk.

The concepts of health and disease and the role of the medical profession. The concept of disease—and therefore health—has gone through several stages. First, disease referred to an inflammatory or degenerative condition that, if left untreated, would lead to severe illness and eventually death. Next, people who deviated from a statistical norm were seen as diseased. Thus hypoglycemia, hypercholesterolemia, hypertension, etc.—hypo or hyper anything. People who deviated from a norm in such ways were not diseased in the earlier sense but were more than others likely to experience an untoward event, what my colleague Andre Hellegers calls "hyperuntowardeventitis." The next step in enlarging the disease concept was to include one's ability to function in society. Someone having difficulty in this regard was diseased, to be treated. We are a nation of uppers and downers. The fine Jewish nose was a disease in Nazi Germany. Old age is a disease in contemporary American society—so that millions of dollars are spent each year on enlarging breasts, shrinking buttocks, removing wrinkles. Grecian Formula 16 is a kind of medicine. At some point, of course, one has to wonder who is sick here, the individual or society.

The *final* development in the notion of disease comes from the World Health Organization, which defines health as including an overall sense of social well-being. At its logically most absurd, one can imagine a physician prescribing a change from a Dodge Aspen to a Mercedes for one who felt status-threatened.

What is going on is the increasing desomatization of the notions of health and disease. That means that physicians are increasingly "treating" the desires of people in a move toward the discomfortless society. Columnist George Will mentions the woman who had a mastectomy because her left breast bothered her golf swing.[22] That this notion of medical service and of health and disease provides powerful support for an abortion ethic seems clear. For the fetus is a *soma*. These are but a few of the cultural obstacles threatening the proper evaluation of nascent life. I am sure there are many more. In combination they suggest that our liberation from abortion as a "crime" has led many to reject any evaluation of abortion as a moral evil, a moral tragedy. Until we as a nation come seriously to grips with these cultural factors, I am afraid there is very little hope that we will move toward an abortion policy that is truly in our overall best interests. We will continue to protect the redwood, the sperm whale, the snaildarter more than we do nascent life—which is just plain sad.

NOTES

1. "Abortion: Rules for Debate," *America*, July 15–22, 1978, pp. 26–30.

2. *Steubenville Register*, Aug. 11, 1978, p. 4.

3. John R. Connery, SJ, "Difference Between Law and Morality," *Catholic Standard*, April 6, 1978, p. 10.

4. Daniel Callahan, "Abortion: Thinking and Experiencing," *Christianity and Crisis*, Jan. 8, 1973, pp. 295–298.

5. Cf. note 4.

6. John R. Connery, SJ, *Abortion: The Development of the Roman Catholic Perspective*, Loyola, Chicago, 1977.

7. "Déclaration des évêques belges sur l'avortement," *Documentation Catholique*, vol. 70, 1973, pp. 432–438 at p. 434.

8. " 'Fristenregelung' entschieden abgelehnt," *Ruhrwort*, Dec. 8, 1973, p. 6.

9. *Medical Ethics*, Fides, Notre Dame, 1973, pp. 81ff.

10. "Pour une réforme de la législation francaise relative à l'avortement," *Etudes*, January 1973, pp. 55–84.

11. Cf. *Theological Studies*, vol. 35, 1974, p. 333.

12. Bernard Quelquejeu, OP, "La volonté de procréer," *Lumière et vie*, vol. 21, 1972, pp. 57–71.

13. Denis O'Callaghan, "Moral Principle and Exception," *Furrow*, vol. 22, 1971, pp. 686–696.

14. Cf. note 7.

15. Susan Teft Nicholson, *Abortion and the Roman Catholic Church*, Religious Ethics, Inc., Knoxville, 1978.

16. Albert C. Outler, "The Beginnings of Personhood: Theological Considerations," *Perkins Journal,* vol. 27, 1973, pp. 28–34.

17. "A Call to Concern," *Christianity and Crisis,* vol. 37, 1977, pp. 222–224.

18. John T. Noonan, Jr., "Abortion in the American Context" (manuscript).

19. *New York Times,* Jan. 23, 1973.

20. Daniel Callahan, "Living with the New Biology," *Center Magazine,* vol. 5, 1972, pp. 4–12.

21. Joseph Fletcher, "Ethical Aspects of Genetic Controls," *New England Journal of Medicine,* vol. 285, 1971, pp. 776–783 at 781.

22. *Washington Post,* June 25, 1978.

3 Ethical Problems of Abortion

SISSELA BOK

This is a very sensitive and thorough analysis of some critical issues in the ethics of abortion. Bok presents an excellent evaluation of issues such as the relation of mother and fetus, the purposes of the attempt to distinguish human from non-human, and the reasons for protecting human life. The article concludes with an examination of the delicate problem of ethical line drawing. Bok forces the reader to become critically involved in the issues and to re-evaluate one's own position in the light of a thoughtfully developed position.

Dr. Sissela Bok, Ph.D., teaches Medical Ethics at the Radcliffe Institute, Cambridge, MA.

The recent Supreme Court decisions[1] have declared abortions to be lawful in the United States during the first trimester of pregnancy. After the first trimester, the state can restrict them by regulations protecting the pregnant woman's health; and after 'viability' the state may regulate or forbid abortions except where the medical judgment is made that an abortion is necessary to safeguard the life or the health of the pregnant woman. But it would be wrong to conclude from these decisions that no *moral* distinctions between abortions can now be made—that what is lawful is always justifiable. These decisions leave the moral issues of abortion open, and it is more important than ever to examine them.

While abortion is frequently rejected for religious reasons,[2] arguments against it are also made on other grounds. The most forceful one holds that if we grant that a fetus possesses humanity, we must accord it human rights, including the right to live. Another argument invokes the danger to *other* unborn humans, should abortion spread and perhaps even become obligatory in certain cases, and the danger to newborns, the retarded, and the senile should society begin to take the lives of those considered expendable. A third argument stresses the danger that physicians and nurses and those associated with the act of abortion might lose their traditional protective attitude toward

45

life if they become inured to taking human lives at the request of mothers.

Arguments favoring permitting Abortion

Among the arguments made in favor of permitting abortion, one upholds the right of the mother to determine her own fertility, and her right to the use of her own body. Another stresses, in cases of genetic defects of a severe variety, a sympathetic understanding of the suffering which might accompany living, should the fetus not be aborted. And a third reflects a number of social concerns, ranging from the problem of overpopulation *per se* to the desire to reduce unwantedness, child abuse, maternal deaths through illegal abortions, poverty and illness.

In discussing the ethical dilemmas of abortion, I shall begin with the basic conflict—that between a pregnant woman and the unborn life she harbors.

I. Mother and Fetus

Up to very recently, parents had only limited access to birth prevention. Contraception was outlawed or treated with silence. Sterilization was most often unavailable and abortion was left to those desperate enough to seek criminal abortions. Women may well be forgiven now, therefore, if they mistrust the barrage of arguments concerning abortion, and may well suspect that these are rear-guard actions in an effort to tie them still longer to the bearing of unwanted children.

Some bad arguments ("the woman's right to control")

Some advocates for abortion hold that women should have the right to do what they want with their own bodies, and that removing the fetus is comparable to cutting one's hair or removing a disfiguring growth. This view simply ignores the fact that abortion involves more than just one life. The same criticism holds for the vaguer notions which defend abortion on the grounds that a woman should have the right to control her fate, or the right to have an abortion as she has the right to marry. But no one has the clear-cut right to control her fate where others share it, and marriage requires consent by two persons, whereas the consent of the fetus is precisely what cannot be obtained. How, then, can we weigh the rights and the interests of mother and fetus, where they conflict?

The central question is whether the life of the fetus should receive the same protection as other lives—often discussed in terms of whether killing the fetus is to be thought of as killing a human being. But before asking that question, I would like to ask whether abortion can always be thought of as *killing* in the first place. For abortion can be looked upon, also, as the withdrawal of bodily life support on the part of the mother.

A. Cessation of Bodily Life Support

Would anyone, before or after birth, child or adult, have the right to continue to be dependent upon the bodily processes of another against that person's will? It can happen that a person will require a sacrifice on the part of another in order not to die; does he therefore have the *right* to this sacrifice?

Judith Thomson has argued most cogently that the mother who finds herself pregnant, as a result of rape or in spite of every precaution, does not have the obligation to continue the pregnancy:

> I am arguing only that having a right to life does not guarantee having either a right to be given the use of or a right to be allowed the continued use of another person's body—even if one needs it for life itself.[3]

Abortion, according to such a view, can be thought of as the cessation of continued support. It is true that the embryo cannot survive alone, and that it dies. But this is not unjust killing, any more than when Siamese twins are separated surgically and one of them dies as a result. Judith Thomson argues that at least in those cases where the mother is involuntarily pregnant, she can cease her support of the life of the fetus without infringing its right to live. Here, viability—the capability of living independently from the body of the mother—becomes important. Before that point, the unborn life will end when the mother ceases her support. No one else can take over the protection of the unborn life. After the point where viability begins, much depends on what is done by others, and on how much assistance is provided.

It may be, however, that in considering the ethical implications of the right to cease bodily support of the fetus we must distinguish between causing death indirectly through ceasing such support and actively killing the fetus outright. The techniques used in abortion differ significantly in this respect.[4] A method which prevents implantation of the fertilized egg or which brings about menstruation is much more clearly cessation of life support than one which sucks or scrapes out the embryo. Least like cessation of support is abortion by saline solution, which kills and begins to decompose the fetus, thus setting in motion its expulsion by the mother's body. This method is the one most commonly used in the second trimester of pregnancy. The alternative method possible at that time is a hysterotomy, or "small Ceasarean," where the fetus is removed intact, and where death very clearly does result from the interrution of bodily support.

If we learn how to provide life support for the fetus outside the natural mother's body, it may happen that parents who wish to adopt a baby may come into a new kind of conflict with those who wish to have an abortion. They may argue that *all* the aborting mother has a right to is to cease supporting a fetus with her own body. They may insist, if the pregnancy is already in the second trimester, that she has no right to choose a technique which also kills the baby. It would be wrong for the natural parents to insist at that point that the severance must be performed in such a way that others cannot take over the care and support for the fetus. But a conflict could arise if the mother were asked to postpone the abortion in order to improve the chances of survival and well-being of the fetus to be adopted by others.

Are there times where, quite apart from the technique used to abort, a woman has a *special* responsibility to continue bodily support of a fetus? Surely the many pregnancies which are entered upon voluntarily are of such a nature. One might even say that, if anyone ever did have special obligations to continue life support of another, it would be the woman who had *voluntarily undertaken* to become pregnant. For she has then brought about the situation where the fetus has come to require her support, and there is no one else who can take over her responsibility until after the baby is viable.

To use the analogy of a drowning person, one can think of three scenarios influencing the responsibility of a bystander to leap to the rescue. First, someone may be drowning and the bystander arrives at the scene, hesitating between rescue and permitting the person to drown. Secondly, someone may be drowning as a result of the honestly mistaken assurance by the bystander that swimming would be safe. Thirdly, the bystander may have pushed the drowning person out of a boat. In each case the duties of the bystander are different, but surely they are at their most stringent when he has intentionally caused the drowning person to find himself in the water.

These three scenarios bear some resemblance, from the point of view of the mother's responsibility to the fetus, to: first, finding out that she is pregnant against her wishes; second, mistakenly trusting that she was protected against pregnancy; and third, intentionally becoming pregnant.

Every pregnancy which has been intentionally begun creates special responsibilities for the mother.[5] But there is one situation in which these dilemmas are presented in a particularly difficult form. It is where two parents deliberately enter upon a pregnancy, only to find that the baby they are expecting has a genetic disease or has suffered from damage in fetal life, so that it will be permanently malformed or retarded. Here, the parents have consciously brought about the life which now requires support from the body of the mother. Can they now turn about and say that this particular fetus is such that they do not wish to continue their support? This is especially difficult when the fetus is already developed up to the 18th or 20th week. Can they acknowledge that they meant to begin a human life, but not *this* human life? Or, to take a more callous example, suppose, as sometimes happens, that the parents learn that the baby is of a sex they do not wish?[6]

In such cases the justification which derives from wishing to cease life support for a life which had not been intended is absent, since this life *had* been intended. At the same time, an assumption of responsibility which comes with consciously beginning a pregnancy is much weaker than the corresponding assumption between two adults, or the social assumption of responsibility for a child upon birth for reasons which will be discussed in the next section.

To sum up at this point, ceasing bodily life support *of a fetus or of anyone else* cannot be looked at as a breach of duty except where such a duty has been assumed in the first place. Such a duty is closer to existing when the pregnancy has been voluntarily begun. And it does not exist at all in cases of rape. Certain *methods* of abortion, furthermore, are more difficult to think of as cessation of support than others. Finally, pregnancy is perhaps unique in that cessation of support means death for the fetus up to a certain point of its development, so that nearness *to* this point in pregnancy argues against abortion.

I would like now to turn to the larger question of whether the life of the fetus *should* receive the same protection as other lives—whether killing the fetus, by whatever means, and for whatever reason, is to be thought of as killing a human being.

A long tradition of religious and philosophical and legal thought has attempted to answer this question by determining if there is human life before birth, and, if so, when it *becomes* human. If human life is present from conception on, according to this tradition, it must be protected as such from that moment. And if the embryo *becomes* human at some point during a pregnancy, then that is the point at which the protection should set in.

B. Humanity

The point in a pregnancy at which a human individual can be said to exist is differently assigned. John Noonan generalizes the predominant Catholic view as follows:

If one steps outside the specific categories used by the theologians, the answer they gave can be analyzed as a refusal to discriminate among human beings on the basis of their varying potentialities. Once conceived, the being was recognized as a man because he had man's potential. The criterion for humanity, thus, was simple and all-embracing: If you are conceived by human parents, you are human.[7]

Once conceived, he holds, human life has about an 80% chance to reach the moment of birth and develop further. *Conception*, therefore, represents a point of discontinuity, after which the probabilities for human development are immensely higher than for the sperm or the egg before conception.

Others have held that the moment when *implantation* occurs, 6-7 days after conception, is more significant from the point of view of humanity and individuality than conception itself. This permits them to allow the intrauterine device and the 'morning after pill' as not taking human life, merely interfering with implantation.

Another view is advanced by Jérôme Lejeune, who suggests that unity and uniqueness, "the two headings defining an individual" are not definitely established until between two and four weeks after conception.[8] Up to that time it is possible that two eggs may have collaborated to build together one embryo, known as a "chimera," whereas after that time such a combination is no longer possible. Similarly, up to that time, a fertilized egg from which twins may result may not yet have split in two.

Still another approach to the establishing of humanity is to say that *looking* human is the important factor. A photo of the first cell having divided in half clearly does not depict what most people mean when the use the expression "human being." Even the four-week-old embryo does not look human, whereas the six-week-old one is beginning to. Recent techniques of depicting the embryo and the fetus have remarkably increased our awareness of the "human-ness" at this early stage; this new *seeing* of life before birth may come to increase the psychological recoil from aborting those who already look human—thus adding a powerful psychological factor to the medical and personal factors already influencing the trend to earlier and earlier abortions.

Others reason that the time at which electrical impulses are first detectable from the brain, around the eighth week, marks the line after which human life is present. If brain activity is advocated as the criterion for human life among the dying, they argue, then why not use it also at the very beginning?[9] Such a use of the criterion for human life has been interpreted by some to indicate that abortion would not be killing before electrical impulses are detectable, only afterwards. Such an analogy would seem to possess a symmetry of sorts, but it is only superficially plausible. For the lack of brain response at the end of life has to be shown to be *irreversible* in order to support a conclusion that life is absent. The lack of response from the embryo's brain, on the other hand, is temporary and precisely not irreversible.

Another dividing line, once more having to do with our perception of the fetus, is that achieved when the mother can feel the fetus moving. *Quickening* has traditionally represented an important distinction, and in some legal traditions such as the common law, abortion has been permitted before quickening, but is a misdemeanor, "a great misprision," afterwards, rather than homicide. It is certain that the first felt movements represent an awe-inspiring change for the mother, and perhaps, in some primitive sense, a 'coming to life' of the being she carries.

Yet another distinction occurs when the fetus is considered *viable*. According to this view, once the fetus is capable of living independently of its mother, it must be regarded as a human being and protected as such. The United States Supreme Court decisions on abortion established viability as the "compelling" point for the state's "important and legitimate interest in potential life," while eschewing the question of when 'life' or 'human life' begins.[10]

A set of later distinctions cluster around the process of birth itself. This is the moment when life begins, according to some religious traditions, and the point at which 'persons' are fully recognized in the law, according to the Supreme Court.[11] The first breaths taken by newborn babies have been invested with immense symbolic meaning since the earliest gropings toward understanding what it means to be alive and human. And the rituals of acceptance of babies and children have often served to define humanity to the point where the baby could be killed if it were not named or declared acceptable by the elders of the community or by the head of the household, either at birth or in infancy. Others have mentioned as factors in our concept of humanity the ability to experience, to remember the past and envisage the future, to communicate, even to laugh at oneself.

In the positions here examined, and in the abortion debate generally, a number of concepts are at times used as if they were interchangeable. 'Humanity,' 'human life,' 'life,' are such concepts, as are 'man,' 'person,' 'human being,' or 'human individual.' In particular, those who hold that humanity begins at conception or at implantation often have the tendency to say that at that time a human being or a person or a man exists as well, whereas others find it impossible to equate them.

Each of these terms can, in addition, be used in different senses which overlap but are not interchangeable. For instance, humanity and human life, in one sense, are possessed by every cell in our bodies.

Many cells have the full genetic makeup required for asexual reproduction—so-called cloning—of a human being. Yet clearly this is not the sense of those words intended when the protection of humanity or of human life is advocated. Such protection would press the reverence for life to the mad extreme of ruling out haircuts and considering mosquito bites murder.

It may be argued, however, that for most cells which have the potential of cloning to form a human being, extraordinarily complex measures would be required which are not as yet sufficiently perfected beyond the animal stage. Is there, then, a difference, from the point of view of human potential, between these cells and egg cells or sperm cells? And is there still another difference in potential between the egg cell before and after conception? While there is a statistical difference in the *likelihood* of their developing into a human being, it does not seem possible to draw a clear line where humanity definitely begins.

The different views as to when humanity begins are not dependent upon factual information. Rather, these views are representative of different worldviews, often of a religious nature, involving deeply held commitments with moral consequences. There is no disagreement as to what we now know about life and its development before and after conception; differences arise only about the names and moral consequences we attach to the changes in this development and the distinctions we consider important. Just as there is no point at which Achilles can be pinpointed as catching up with the tortoise, though everyone knows he does, so too everyone is aware of the distance traveled, in terms of humanity, from before conception to birth, though there is no one point at which humanity can be agreed upon as setting in. Our efforts to pinpoint and to define reflect the urgency with which we reach for abstract labels and absolute certainty in facts and in nature; and the resulting confusion and puzzlement are close to what Wittgenstein described, in *Philosophical Investigations*, as the "bewitchment of our intelligence by means of language."

Even if some see the fertilized egg as possessing humanity and as being "a man" in the words used by Noonan, however, it would be quite unthinkable to act upon all the consequences of such a view. It would be necessary to undertake a monumental struggle against all spontaneous abortions—known as miscarriages—often of severly malformed embryos expelled by the mother's body. This struggle would appear increasingly misguided as we learn more about how to preserve early prenatal life. Those who could not be saved would have to be buried in the same way as dead infants. Those who engaged in abortion would have to be prosecuted for murder. Extraordinary practical complexities would arise with respect to the detection of early abortion, and to the question of whether the use of abortifacients in the first few days after conception should also count as murder. In view of these inconsistencies, it seems likely that this view of humanity, like so many others, has been adopted for limited purposes having to do with the prohibition of induced abortion, rather than from a real belief in the full human rights of the first few cells after conception.

II. PURPOSES FOR SEEKING TO DISTINGUISH HUMAN AND NON-HUMAN

A related reason why there are so many views and definitions of humanity is that they have been sought for such different *purposes*. I indicated already that many of the views about humanity developed in the abortion dispute seem to have been worked out for one such purpose—that of defending a preconceived position on abortion, with little concern for the other consequences flowing from that particular view. But there have been so many other efforts to define humanity and to arrive at the essence of what it means to be human—to distinguish men from angels and demons, plants and animals, witches and robots. The most powerful one has been the urge to know about the human species and to trace the biological or divine origins and the essential characteristics of mankind. It is magnificently expressed beginning with the very earliest creation myths; in fact, this consciousness of oneself and wonder at one's condition has often been thought one of the essential distinctions between men and animals.

A separate purpose, both giving strength to and flowing from these efforts to describe and to understand humanity, has been that of seeking to define what a *good* human being is—to delineate human aspirations. What ought fully human beings to be like, and how should they differ from and grow beyond their immature, less perfect, sick or criminal fellow men? Who can teach such growth—St. Francis or Nietzsche, Buddha or Erasmus? And what kind of families and societies give support and provide models for growth?

Finally, definitions of humanity have been sought in order to try to set limits to the protection of life. At what level of developing humanity can and ought lives to receive protection? And who, among those many labelled less than human at different times in history—slaves, enemies in war, women, children, the retarded—should be denied such protection?

Of these three purposes for defining 'humanity,' the first is classificatory and descriptive in the first hand (though it gives rise to normative considerations). It has roots in religious and metaphysical thought, and has branched out into biological and archeological and anthropological research. But the latter two, so often confused with the first, are primarily _normative_ or prescriptive. They seek to set norms or guidelines for who is fully human, and who is at least minimally human—so human as to be entitled to the protection of life. For the sake of these normative purposes, definitions of 'humanity' established elsewhere have been sought in order to determine action—and all too often the action has been devastating for those excluded.

It is crucial to ask at this point why the descriptive and the normative definitions have been thought to coincide; why it has been taken for granted that the line between human and non-human or not-yet-human is identical with that distinguishing those who may be killed from those who are to be protected.

One or both of two fundamental assumptions are made by those who base the protection of life upon the possession of 'humanity.' The first is that all human beings are not only different from, but _superior_ to all other living matter. This is the assumption which changes the definition of humanity into an evaluative one. It lies at the root of Western religious and social thought, from the Bible and the Aristotelian concept of the 'ladder of life,' all the way to Teilhard de Chardin's view of mankind as close to the intended summit and consummation of the development of living beings.

The second assumption holds that the superiority of human beings somehow justifies their using what is non-human as they see fit, dominating it, even killing it when they wish to. St. Augustine, in _The City of God_,[12] expresses both of these anthropocentric assumptions when he holds that the injunction "Thou shalt not kill" does not apply to killing animals and plants, since, having no faculty of reason,

therefore by the altogether righteous ordinance of the Creator both their life and death are a matter subordinate to our needs.

Neither of these assumptions is self-evident. And the results of acting upon them, upon the bidding to subdue the earth, to subordinate its many forms of life to human needs, are no longer seen by all to be beneficial.[13] The very enterprise of _basing_ normative conclusions on such assumptions and distinctions can no longer be taken for granted.

Despite these difficulties, many still try to employ definitions of 'humanity' to do just that. And herein lies by far the most important reason for abandoning such efforts: the monumental misuse of the concept of 'humanity' in so many practices of discrimination and atrocity throughout history. Slavery, witchhunts and wars have all been justified by their perpetrators on the grounds that they held their victims to be less than fully human. The insane and the criminal have for long periods been deprived of the most basic necessities for similar reasons, and excluded from society. A theologian, Dr. Joseph Fletcher, has even suggested recently that someone who has an I.Q. below 40 is "questionably a person" and that those below the 20-mark are not persons at all.[14] He adds that:

> This has bearing, obviously, on decision making in gynecology, obstetrics, and pediatrics, as well as in general surgery and medicine.

Here a criterion for 'personhood' is taken as a guideline for action which could have sinister and far-reaching effects. Even when entered upon with the best of intentions, and in the most guarded manner, the enterprise of basing the protection of human life upon such criteria and definitions is dangerous. To question someone's humanity or personhood is a first step to mistreatment and killing.

We must abandon, therefore, this quest for a definition of humanity capable of showing us who has a right to live. To do so must not, however, mean any abandon of concern with the human condition—with the quest for knowledge about human origins and characteristics and with aspirations for human goodness. It is only the use of the concept of 'humanity' as a criterion of *exclusion* which I deplore.

In recent decades, philosophers have devoted much thought to the nature of ethical principles, to the kind of statement they make, and to their internal grammar. Much has been written about the requirement that these principles be universal—that they hold for all mankind, all moral persons, all rational beings. As a rough distinction, such a simple characterization of the *extent* to which ethical principles should hold is undoubtedly natural and relatively unproblematic. It would rule out, for example, the denial of basic rights to some persons while according them to others, whereas it would not prohibit the employment of plant fiber in clothing or lumber in furniture. But I submit that in the many borderline cases where humanity is questioned by some—the so-called 'vegetables,' the severely retarded, or the embryo—even the seemingly universal yardsticks of 'humanity' or rationality are dangerous.

But if we rule out the appeal to a standard of 'humanity' in deciding about the protection of life in such difficult cases, may we not have lost the only criterion of objective decisions? Or could there be other criteria less dangerous and vague than that connected with 'humanity'?

In order to seek such criteria, it is crucial to arrive at an understanding of the harm that comes from the taking of life. Why do we hold life to be sacred? Why does it require protection beyond that given to anything else? The question seems unnecessary at first—surely most people share what has been called "the elemental sensation of vitality and the elemental fear of its extinction," and what Hume termed "our horrors at annihilation."[15] Many think of this elemental sensation as incapable of further analysis. They view any attempt to say *why* we hold life sacred as an instrumentalist rocking of the boat which may endanger this fundamental and unquestioned respect for life. Yet I believe that such a failure to ask what the respect for life ought to protect lies at the root of the confusion about abortion and many other difficult decisions concerning life and death. I shall try, therefore, to list the most important reasons which underlie the elemental sense of the sacredness of life. Having done so, these reasons can be considered as they apply or do not apply to the embryo and the fetus.

III.   Reasons for Protecting Life

1. Killing is viewed as the greatest of all dangers *for the victim.*
  • The knowledge that there is a threat to life causes intense anguish and apprehension.
  • The actual taking of life can cause great suffering.
  • The continued experience of life, once begun, is considered so valuable, so unique, so absorbing, that no one who has this experience should be unjustly deprived of it. And depriving someone of this experience means that all else of value to him will be lost.

2. Killing is brutalizing and criminalizing *for the killer.* It is a threat to others, and destructive to the person engaged therein.

3. Killing often causes *the family of the victim and others* to experience grief and loss. They may have been tied to the dead person by affection or economic dependence; they may have given of themselves in the relationship, so that its severance causes deep suffering.

4. *All of society,* as a result, has a stake in the protection of life. Permitting killing to take place sets patterns for victims, killers, and survivors, that are threatening and ultimately harmful to all.

These are neutral principles governing the protection of life. They are shared by most human beings reflecting upon the possibility of dying at the hands of others. It is clear that these principles, if applied in the absence of the confusing terminology of 'humanity,' would rule out the kinds of killing perpetrated by conquerors, witch-hunters, slave-holders, and Nazis. Their victims feared death and suffered; they grieved for their dead; and the societies permitting such killing were brutalized and degraded.

Turning now to abortions once more, how do these principles apply to the taking of the lives of embryos and fetuses?

A. *Reasons to Protect Life in the Prenatal Period*

Consider the very earliest cell formations soon after conception. Clearly, most of these *reasons* for protecting human life are absent here.

This group of cells cannot suffer in death, nor can it fear death. Its experiencing of life has not yet begun; it is not yet conscious of the loss of anything it has come to value in life and is not tied by bonds of affection to other human beings. If the abortion is desired by both parents, it will cause no grief such as that which accompanies the death of a child. Almost no human care and emotion and resources have been invested in it. Nor is a very early abortion brutalizing for the person voluntarily performing it, or a threat to other members of the human community.[16] The only factor common to these few cells and, say, a soldier killed in war or a murdered robbery victim is that of the *potential* denied, the interruption of life, the deprivation of the possibility to grow and to experience, to have the joys and sorrows of existence.

For how much should this one factor count? It should count *at least* so much as to eliminate the occasionally voiced notion that pregnancy and its interruption involve only the mother in the privacy of her reproductive life, that to have an abortion is somehow analogous with cutting one's finger nails.

At the same time, I cannot agree that it should count enough so that one can simply equate killing an embryo with murder, even apart from legal considerations or the problems of enforcement. For it *is* important that most of the reasons why we protect lives are absent here. It does matter that the group of cells cannot feel the anguish or pain connected with death, that it is not conscious of the interruption of its life, and that other humans do not mourn it or feel insecure in their own lives if it dies.

But, it could be argued, one can conceive of other deaths with those factors absent, which nevertheless would be murder. Take the killing of a hermit in his sleep, by someone who instantly commits suicide. Here there is no anxiety or fear of the killing on the part of the victim, no pain in dying, no mourning by family or friends (to whom the hermit has, in leaving them for good, already in a sense 'died'), no awareness by others that a wrong has been done; and the possible brutalization of the murderer has been made harmless to others through his suicide. Speculate further that the bodies are never found. Yet we would still call the act one of murder. The reason we would do so is inherent in the act itself, and depends on the fact that his life was taken, and that he was denied the chance to continue to experience it.

How does this privation of potential differ from abortion in the first few days of pregnancy? I find that I cannot use words like 'deprived,' 'deny,' 'take away,' and 'harm' when it comes to the group of cells, whereas I have no difficulty in using them for the hermit. Do these words require, if not a person conscious of his loss, at least someone who at a prior time has developed enough to be or have been conscious thereof? Because there is no semblance of human form, no conscious life or capability to live independently, no knowledge of death, no sense of pain, one cannot use such words meaningfully to describe early abortion.

In addition, whereas it is possible to frame a rule permitting abortion which causes no anxiety on the part of others covered by the rule —other embryos or fetuses—it is not possible to frame such a rule permitting the killing of hermits without threatening other *hermits*. All hermits would have to fear for their lives if there were a rule saying that hermits can be killed if they are alone and asleep and if the agent commits suicide.

The reasons, then, for the protection of lives are minimal in very early abortions. At the same time, some of them are clearly present with respect to *infanticide*, most important among them the brutalization of those participating in the act and the resultant danger for all who are felt to be undesirable by their families or by others. This is not to say that acts of infanticide have not taken place in our society; indeed, as late as the nineteenth century, newborns were frequently killed, either directly or by giving them into the care of institutions such as foundling hospitals, where the death rate could be as high as 90 percent in the first year of life.[17] A few primitive societies, at the edge of extinction, without other means to limit families, still practice infanticide. But I believe that the *public acceptance* of infanticide in all other societies is unthinkable, given the advent of modern methods of contraception and early abortion, and of institutions to which parents can give their children, assured of their survival and of the high likelihood that they will be adopted and cared for by a family.

B. *Dividing Lines*

If, therefore, very early abortion does not violate these principles of protection for life, but infanticide does, we are confronted with a new kind of continuum in the place of that between less human and more human: that of the growth in strength, during the prenatal period, of these principles, these reasons for protecting life. In this second continuum, it would be as difficult as in the first to draw a line based upon objective factors. Since most abortions can be performed earlier or later during pregnancy, it would be preferable to encourage early abortions rather than late ones, and to draw a line before the second half of the pregnancy, permitting later abortions only on a clear showing of need. For this purpose, the two concepts of *quickening* and *viability*—so unsatisfactory in determining when humanity begins —can provide such limits.

Before quickening, the reasons to protect life are, as has been shown, negligible, perhaps absent altogether. During this period, therefore, abortion could be permitted upon request. Alternatively, the end of the first trimester could be employed as such a limit, as is the case in a number of countries.

Between quickening and viability, when the operation is a more difficult one medically and more traumatic for parents and medical personnel, it would not seem unreasonable to hold that special reasons justifying the abortion should be required in order to counterbalance this resistance; reasons not known earlier, such as the severe malformation of the fetus. After viability, finally, all abortions save the rare ones required to save the life of the mother[18] should be prohibited, because the reasons to *protect* life may now be thought to be partially present; even though the viable fetus cannot fear death or suffer consciously therefrom, the effects on those participating in the event, and thus on society indirectly, could be serious. This is especially so because of the need, mentioned above, for a protection against infanticide. In the unlikely event, however, that the mother should first come to wish to be separated from the fetus at such a late stage, the procedure ought to be delayed until it can be one of premature birth, not one of harming the fetus in an abortive process.

Medically, however, the definition of 'viability' is difficult. It varies from one fetus to another. At one stage in pregnancy, a certain number of babies, if born, will be viable. At a later stage, the percentage will be greater. Viability also depends greatly on the state of our knowledge concerning the support of life after birth, and on the nature of the support itself. Support can be given much earlier in a modern hospital than in a rural village, or in a clinic geared to doing abortions only. It may some day even be the case that almost any human life will be considered viable before birth, once artificial wombs are perfected.

As technological progress pushes back the time when the fetus can be helped to survive independently of the mother, a question will arise as to whether the cutoff point marked by viability ought also be pushed back. Should abortion then be prohibited much earlier than is now the case, because the medical meaning of 'viability' will have changed, or should we continue to rely on the conventional meaning of the word for the distinction between lawful and unlawful abortion?

In order to answer this question it is necessary to look once more at the reasons for which 'viability' was thought to be a good dividing-line in the first place. Is viability important because the baby can survive outside of the mother? Or because this chance of survival comes at a time in fetal development when the *reasons* to protect life have grown strong enough to prohibit abortion? At present, the two coincide, but in the future, they may come to diverge increasingly.

If the time comes when an embryo *could* be kept alive without its mother and thus be 'viable' in one sense of the word, the reasons for protecting life from the point of view of victims, agents, relatives and society would still be absent; it seems right, therefore, to tie the obligatory protection of life to the present conventional definition of 'viability' and to set a socially agreed upon time in pregnancy after which abortion should be prohibited.

To sum up, the justifications a mother has for not wishing to give birth can operate up to a certain point in pregnancy; after that point, the reasons society has for protecting life become sufficiently weighty so as to prohibit late abortions and infanticide.

IV. MORAL DISTINCTIONS

But moral distinctions ought nevertheless to be made by the mother considering an abortion even during the period when she may lawfully obtain one. In addition to those having to do with the *method* of abortion and the degree to which the pregnancy was voluntary or involuntary (as discussed previously), the *time* in pregnancy, the weightiness of the *reasons* for wanting the abortion, the desires of the *father*, and the possibility of alternatives such as adoption, must all be considered.

1. The *time* in pregnancy at which the abortion takes place is a very important factor. Very early in pregnancy, the reasons for protecting life are clearly absent. Few will have to face the questions which come with aborting a 4 or 5-month-old fetus when early abortions are generally available. But *in* such late abortions, it is especially important to consider what the reasons are for desiring the abortion.

2. Among all of the reasons why a pregnancy is unwanted, it is possible to perceive a gradation from reasons all would recognize as very compelling, such as a threat to the mother's life, to reasons most would think of as frivolous, such as a determination that only a fetus

of a desired sex should be allowed to be born. This gradation among the reasons for wishing not to have a baby will be part of any judgment concerning the *morality* of acts to prevent births. It is also possible to divide the innumerable reasons for not wanting a pregnancy into two main categories. The first one, sometimes called 'selective' unwantedness, refers to those pregnancies which are desired, often planned, by the parents, but during the course of which evidence comes to light concerning a risk, or even a certainty of abnormality in the fetus. If, for example, the mother has Rubella, or German Measles, in the first trimester, there is a probability of fetal abnormality. And it is now possible to learn, through prenatal diagnosis, whether the fetus suffers from a chromosomal abnormality, the most common of which causes mongolism, or from one of a number of genetic diseases which can cause malformation or mental retardation.[19] In all of these cases the parents, while they might ordinarily welcome a pregnancy, may come to the conclusion that they do not wish to give birth in this particular case.

But the determination of such defects through amniocentesis can only be made when the amniotic fluid is present to a sufficient degree, and the final results of the tests may not be available until the fifth month of pregnancy. Only *late* abortions are possible after amniocentesis, and this makes the decision for parents and doctors a much more difficult one.

The reasons for not wanting a malformed baby differ with the capacities of the parents and the severity of the abnormality. Some parents, and families, cope admirably and with great love with children who would prove burdensome and even destructive to other families. A severely disabled fetus, likely to suffer greatly once born and perhaps to die in childhood, could be 'unwanted' out of concern for its own welfare, as well as for the welfare of the family. A great deal depends on the help available from the community, in terms of financial assistance, special schooling, medical resources, and general support. Other factors which can be important are the pride of the family, or even parental prejudice, e.g., parents' wish for an abortion after learning that the fetus is of one sex rather than another.

Perhaps most difficult from a moral point of view are the situations where the parents know beforehand that they are carriers of genetic defects, and where they enter upon a pregnancy determined in advance to abort any fetus which is found to exhibit the defect. I say this with the greatest humility, knowing the strength of the urge to have one's own babies. But I see no difference between starting another human life with such plans, and creating 'test-tube' fetuses only to throw away those deemed undesirable. In cases such as these, other ways of bringing children into the lives of parents must be worked out. At times artificial insemination may provide an answer,[20] at other times adoption, or working with children in the many capacities where help is needed, may be preferable.

But there are many cases where these distinctions cannot be so clearly made. It may be difficult to know whether there was an intention to have a baby, or to risk becoming pregnant. It might be argued that someone who engaged in sexual activity, even using contraceptives, ought to be willing to take the responsibility for a human life which results. Whereas to abort under such circumstances, or even after a voluntarily begun pregnancy, is not murder, it ought not to be taken lightly. For the same reason, it is insensitive to omit contraceptive measures and to rely on the availability of abortion in the case of pregnancy. (Though the availability of methods making abortions possible in the very earliest days of pregnancy and the hazy line between such abortions and contraception may make such a distinction less pointed.)

Another set of criteria which will be difficult to work out when considering reasons for abortion is that which should govern abortions for the sake of the welfare of the fetus. For while almost all would agree about the extreme cases I have mentioned, there will be disagreement as to what to do in those cases where the affliction is not totally debilitating, or where there is merely a *risk* of disease, not a certainty. What if the risk is small? A recent newspaper article stated that there is one chance out of a hundred that a baby will be born retarded if the mother has had the flu in the first trimester of her pregnancy. Whether or not this particular concern turns out to be correct, it is going to be increasingly possible to specify odds of this kind, sometimes with a very low probability of danger. It has been suggested that parents will come to want to take very few chances of defects, so long as the choice is open to them of having abortions.

Even if it is possible, however, to work out criteria concerning the welfare of the baby, there are times when the cost at which this welfare is to be purchased must be weighed against the welfare of other human beings. If for example, a fetus is diagnosed as having a disease which can be controlled after birth so as not to cause suffering, but only at staggering costs to the family or the community—say of millions of dollars each day—abortion would clearly be called for in spite of the theoretical possibility of carrying the baby to term and treating it. The other possibility would be not aborting, and permitting the baby to suffer in the absence of such expensive relief, and then once more, the magnitude of the suffering might have argued in favor of abortion.

All these cases, where certain births are unwanted because of the characteristics of the fetus, differ crucially from those in the second category where no children at all are wanted at the time of the pregnancy. In this larger group are the more familiar cases where there is danger to the mother's physical health or her emotional stability, or where there is not enough food, clothing, or shelter to cope with yet another child. Here, too, are cases where there has been rape, or incest, or where a very young girl is pregnant. There are also the frequent cases where the mother feels she is beyond the age best suited for child-caring, or does not want to accept the great changes in life—the restriction, the financial pinch, and the feeling of being tied down—which often accompany the birth of a child. These changes affect mothers most powerfully in our society of nuclear families where the burdens of child-rearing often fall on them alone. In all of these cases, contraception could have avoided the pregnancy, and an early abortion is possible as a last resort. Adoption is an alternative resort which should always be considered. It must be remembered, however, that with prevailing attitudes it would be exceedingly difficult for a married woman with existing children to give a baby up for adoption.

The distinction between the two *kinds* of reasons for not wanting a pregnancy is crucial. For while the first group of conceptions—unwanted because of the characteristics of the fetus—often require abortions if births are to be prevented (and often late abortions, since prenatal diagnosis takes time and can rarely begin until the second trimester of pregnancy), the second group can usually be prevented through contraception, sterilization, abstinence, or protection of the mother from sexual assault. Abortion is necessary here only as a last resort, where other methods have failed, and an *early* abortion is possible, presenting fewer medical, ethical, and emotional problems than a later one.

3. At times, there are conflicts between mothers and fathers of the unborn. According to one study,[21] about one-half of the pregnancies unwanted by one or both parents were unwanted by *only one* parent. Very often such disagreements are settled amicably, usually in favor of having the baby. But who should make the decision when the mother wishes to have an abortion, and the father wants to restrain her?[22] In a recent Canadian case,[23] a judge prohibited an abortion desired by a mother. The father had brought suit on his own behalf and on that of the 'infant plaintiff.'

It is difficult to see how such a disagreement can be anything but disruptive for the relationship between the two parents, as well as very harmful for the child after birth. Whoever 'wins' in such a conflict will have won a Pyrrhic victory indeed.

Early in pregnancy, the mother has at her disposal methods of abortion which need not involve the father's knowledge of her condition. The same is true if he does not learn of the pregnancy as it progresses. But barring such eventualities, who would decide in the event of a conflict?

In such a conflict, while it is important to ascertain the father's views when possible, there ought not to be a *requirement* that both parents consent to an abortion, as is now often the case.[24] The mother has the burden of pregnancy, and most often of caring for the baby she bears. The decision to interrupt her pregnancy should therefore be hers. But into her decision should go the awareness of the heavy price she will have to pay in the relationship with the father, if she aborts their unborn child against his wishes. And the father's reasons for wishing to continue the pregnancy should be given due weight, so as to counterbalance in her judgment all but the most pressing reasons she has for wishing to have the abortion.

The father's wishes should be given great weight, especially if he wants not only to preserve the life of his unborn child, but also to share responsibility and care after birth. At a future time, when it may be possible to remove a fetus relatively early in pregnancy and protect it artificially until 'birth,' fathers, just as adoptive parents, ought to have the right to declare their intentions to take responsibility for the baby. Mothers at that time, while severing their connections with the fetus, should not be able to demand its death.

Furthermore, if we look back on the reasons for protecting life, one of them concerns the grief felt by family members when someone is killed. If, therefore, a father feels such grief, and if he supports his contention by promising to assume the burdens of child-rearing after birth, this ought to be an important consideration, persuasive to the mother or to her physician or to both. Our society has been moving in the direction of recognizing that men as well as women can provide care and nurturance for children. To permit a father to prevent the abortion of his child on the condition that he bring it up later would seem to be a move in the same direction. If he is unwilling to make such a commitment, however, his grief at the impending death of the fetus is less entitled to respect.

4. The alternatives to abortion differ depending upon whether birth prevention is considered before or after conception. The alternatives open before conception—abstinence, different methods of contraception, and sterilization—do not raise the particular moral problems connected with taking the life of the fetus, or of rejecting the baby after birth.

Once conception *has* occurred, the alternatives to abortion are to accept responsibility for the baby after birth, or to relinquish it to the state or to adoptive parents. It is extremely important to consider these alternatives in the case of each unwanted pregnancy, and only to have recourse to abortion after discarding them. Many pregnant women, whether they are seeking abortions or not, are ambivalent, struggling within themselves in order to reconcile the tenderness normally evoked by the thought of a baby, with fears connected with their pregnancy. The fears may have to do with the future of the baby, or with the future of the family unit into which the baby will come. Sometimes there is no such family unit, and sometimes the relationship with the baby's father is such as to threaten the future of the baby. The decisive point comes when the choice is made to prevent a pregnancy or a birth. And this choice in turn is strongly influenced by social attitudes towards means of birth prevention, and by their availability.

The fact of having children has always been considered 'natural,' and someone not wishing a child, or any children, has been expected to produce reasons in support of such an attitude. It may be that we are now coming closer to a time when choosing not to have a child will be seen to reflect, not necessarily a hostile and niggardly attitude, a 'denial of life,' but a respect for the living, and a correct estimate of what kind of life a baby can be given. In that case, reasons will come to be expected *before* giving birth to a new baby, and thoughts for the welfare of the child to be will come to be seen as an important aspect of child-bearing.

In order for such choices to be possible at all, however, *information* is necessary. All those who are physically able to become parents must have wise and full advice regarding family life, sexual life, and birth prevention. From a moral point of view, contraception is greatly preferable to abortion. The *knowledge* about contraceptive alternatives to childbirth, or to abortion, is therefore crucial to all potential parents. Withholding information in order to preserve 'innocence' among the young is a self-defeating and unjustifiable exercise of paternalistic power, contributing to the birth of unwanted children and to shattered lives.

I have argued that it may be moral to have an abortion under certain circumstances, but that the range of morally justifiable abortions is more restricted than that of those abortions declared lawful by the Supreme Court. But some argue that such views of morality and legality, if widely followed, could lead to great dangers for society.

V.   PROBLEMS OF LINE-DRAWING

A. *Can We Allow Abortion Without Risking Infanticide?*

Foes of abortion argue that a society which permits abortion may not be able to hold the line against infanticide.[25] Once we admit *reasons* for abortions such as fetal malformation or simply not wanting another child, they say, what is to prevent people from acting upon these very same reasons after birth?[26] A baby just before birth, they argue, is identical to one just after birth. What, then, will provide the discontinuity?

I have argued, on the contrary, that another set of *reasons*—the reasons for protecting human life—gain in strength during pregnancy and are such as to prohibit abortions after a certain point and therefore also to prohibit infanticide. While it is true that no theoretical line can be drawn which distinguishes between a baby just before birth and one just after birth, there is no difficulty in distinguishing an aborted embryo from a newborn baby. A time must therefore be set in pregnancy well before birth for the cutting-off point. The discontinuity will then exist between abortion and infanticide. The argument that the reason *for* aborting may still exist at childbirth does not take into account the reasons *against* killing, and the threat which would be felt by all if infanticide as a parental option were thought to be possible.[27]

How can one *know* whether such a discontinuity can be observed in practice? The only way to know is to consider those societies which have already permitted abortion for considerable lengths of time. These countries do not in fact experience tendencies toward infanticide. The infant mortality statistics in Sweden and Denmark are extremely low, and the protection and care given to all living children, including those born with special problems, is exemplary.

Moreover, Nazi Germany, which is frequently cited as a warning of what is to come once abortion becomes lawful, had *very strict laws prohibiting abortion*. In 1943, Hitler's regime made the existing penalties for women having abortions, and those performing them, even more severe by removing the limit on imprisonment and by including the possibility of "hard labor" for "especially serious cases."[28]

The fear of slipping from abortion towards infanticide, therefore, while understandable, does not seem to be grounded in fact.

B. *Is There a Risk of Compulsory Abortion?*

A second type of line-drawing problem is the following: if a beginning is made by permitting amniocentesis and abortion in cases where the mother learns she is expecting a grossly malformed baby, might there not come to be a *requirement* for others to undergo amniocentesis, and to induce abortion if the fetus is found to be defective? And once abortion is no longer reprehensible, might a community not require abortions where expectant mothers are heavily addicted, and where it is not only likely that they will harm or neglect their children after birth, but where they are demonstrably severely harming them even before birth? Might it not be increasingly easy for parents to force their daughters to have abortions should they become pregnant while they are too young, or unmarried?[29] Or even for husbands to require abortions where they judge their wives to be unstable or perhaps ill? And finally, if abortion is permitted for indigent mothers, in part out of sympathy for mother and child, and in part out of a computation of the likely costs to the community of enforcing the births of unwanted children, might there not in the long run be a requirement for abortion where mothers on welfare are concerned, or any others who are judged unable to provide, materially or emotionally or intellectually, for the needs of their children?

One can readily concede that it is important to be vigilant against any such developments. Any inroads upon a pregnant woman's physical integrity, against her will, are very serious and we need strong protection for the control she should be able to exercise over her own body. Great risks of abuse would obviously accompany any provision for obligatory abortion. But to forbid voluntary abortion because of the danger of involuntary abortion would be like forbidding voluntary adoptions on the grounds that they might lead to involuntary adoption policies. The battle against coercion must be fought at all times, with respect to many social options, but this is no reason to prohibit the options themselves.

Conclusion

There are many reasons which may lead a mother not to wish to give birth, but to have an abortion instead. They range from the most compelling to the most trivial. In early pregnancy, society's reasons for protecting life—the suffering and harm to the victim, to the agent, to the family and friends, and to society as a whole—do not apply, either to the zygote or to the embryo. Abortion, for whatever reasons, should then be available upon request. Preventing birth before conception or just after conception, however, presents fewer ethical conflicts than later abortions.

As pregnancy progresses, the social reasons for preventing killing are more and more applicable to the fetus. At the stage where a fetus is viable—capable of independent life outside the mother's body—these reasons begin to be as substantial as at birth and thereafter. In addition, viability represents the time when cessation of bodily support by the mother need not result in fetal death; as a result a viable fetus is capable of protection by others. For these reasons, I believe that after the established time of possible viability, methods separating fetus and mother so as to kill the fetus should be prohibited. But an earlier time

—perhaps 18 rather than 24 weeks—is preferable for all but exceptional cases (such as those occurring after prenatal diagnosis of severe malformation).

Even though abortion may be *lawful* up to this time, however, it is not necessarily an act which an individual may consider right or justifiable. This discrepancy results, I believe, from the fact that the *social* reasons for protecting life may also be looked upon in each case as *individual* reasons. Society may not find that abortion harms either victim or agent, family or social practices. But the individual parent or physician may see risks to himself as a person from such acts and look at them as breaches of personal responsibility toward the unborn. They may then regard abortion as personally distasteful, even though it is lawful.

Some physicians, for example, do feel that they cannot participate in abortions without personal danger of brutalization and without sharing responsibility for killing. This may be especially true when they are called in, as in large hospitals, to perform one abortion after another without any chance to consult with the women involved and to hear their case histories. There should never be a requirement that a physician or nurse must participate in an abortion. Even if women have a right to abortion, they have not therefore the right to force others to perform such acts.

In the same way, a mother or a father may feel personal grief over the death of a fetus, and responsibility for killing it, quite apart from the legality of the act. This grief and this responsibility, which would be present as a matter of course where parents wish for the birth of their baby, may also accompany an unwanted pregnancy. The following factors should then be weighed by the mother before she can be confident that abortion is the right way out of her dilemma, and one she will not come to regret or view with guilt:

- whether or not the pregnancy was voluntarily undertaken.
- the importance and validity of the *reasons* for wanting the abortion.
- the technique to be used in the abortion; the extent to which it can be regarded as 'cessation of bodily life support,' rather than as outright killing.
- the time of pregnancy.
- whether or not the father agrees to the abortion.
- whether or not all other alternatives have been considered, such as adoption.
- her religious views.

And the father, if he weighs these factors differently, may feel the grief and responsibility differently too, and wish to take over the care of the baby after birth.

Abortion is a last resort, and must remain so. It is much more problematic than contraception, yet it is sometimes the only way out of a great dilemma. Neither individual parents nor society should look at abortion as a policy to be encouraged at the expense of contraception, sterilization, and adoption. At the same time, there are a number of circumstances in which it can justifiably be undertaken, for which public and private facilities must be provided in such a way as to make no distinction between rich and poor.

NOTES

1. Roe v. Wade, *United States Law Week 41*, 1973, pp. 4213-33. Doe v. Bolton, *Ibid.*, pp. 4233-40.

2. See *The Morality of Abortion*, ed. by John T. Noonan, Jr. (Cambridge: Harvard University Press, 1970), and G. H. Williams, "Religious Residues and Presuppositions in the American Debate on Abortion," *Theological Studies* 31 (1970), 10-75.

3. Judith Thomson, "A Defense of Abortion," *Philosophy and Public Policy* 1 (1971), 47-66.

4. See Selig Neubardt and Harold Schulman, *Techniques of Abortion* (Boston: Little, Brown and Company, 1972).

5. But lines are hard to draw here. There are many intermediate cases between the pregnancy intentionally begun and, for instance, that resulting from carelessness with contraceptives.

6. See Morton A. Stenchever, "An Abuse of Prenatal Diagnosis," *Journal of the American Medical Association* 221 (July 24, 1972), 408.

7. Noonan, *Morality of Abortion*, p. 51. For a thorough discussion of this and other views concerning the beginnings of human life, see Daniel Callahan, *Abortion: Law, Choice and Morality* (New York: Macmillan Company, 1970).

8. Jérôme Lejeune, "On the Nature of Man," (Lecture at the American Society of Human Genetics at San Francisco, October 2-4, 1969).

9. Paul Ramsey, "Feticide/Infanticide upon Request," *Religion in Life* 39 (July, 1970), 170-86. Arthur J. Dyck, "Perplexities for the Would-Be Liberal in Abortion," *Journal of Reproductive Medicine* 8 (June, 1972), 351-54.

10. Roe v. Wade, *United States Law Week 41*, pp. 4227, 4229.

11. *Ibid.*, p. 4227. For further discussion see L. Tribe, "Foreword: Toward a Model of Roles in the Due Process of Life and Law," *Harvard Law Review* 87 (1973), 1-54.

12. Augustine, *The City of God Against the Pagans*, Book I. Ch. XX (Cambridge: Harvard University Press, 1957).

13. C. D. Stone, "Should Trees Have Standing? Toward Legal Rights for Natural Objects," *Southern California Law Review* 45, 450-501, provides an interesting analysis of the extension of rights to those not previously considered persons, such as children, and a discussion of possible future extensions to natural objects.

14. Joseph Fletcher, "Indicators of Humanhood: A Tentative Profile of Man," *The Hastings Center Report* 2 (November, 1972), 1-4.

15. Edward Shils, "The Sanctity of Life," in *Life or Death: Ethics and Options,* ed. by D. H. Labby (Seattle: University of Washington Press, 1968), p. 12. David Hume, "Of the Immortality of the Soul," *Essays: Moral, Political, and Literary* (London: Longmans, Green, and Co., 1882), II, p. 405.

16. This question will be taken up in detail in Part V. It is because all of the reasons for protecting life are *present* when someone considered to be a slave is murdered that the spate of recent sensationalistic comparisons of abortion and slavery do not make sense, even though it is true that in both cases there are denials of the humanity of the victims. Once again, a confusion in the use of the word humanity' is at fault.

17. William L. Langer, "Checks on Population Growth: 1750-1850," *Scientific American* 226 (February, 1972).

18. Every effort must be made by physicians and others to construe the Supreme Court's statement "If the State is interested in protecting fetal life after viability, it may go so far as to proscribe abortion during that period except when it is necessary to preserve the life or health of the mother" to concern, in effect, only the life or threat to life of the mother. See Alan Stone, "Abortion and the Supreme Court: What Now?" *Modern Medicine,* April 30, 1973, pp. 33-37, for a discussion of this question and what it means for physicians.

19. Theodore Friedmann, "Prenatal Diagnosis of Genetic Disease," *Scientific American* 225 (November, 1971), 34-42 and A. Milunsky, *et al.,* "Prenatal Genetic Diagnosis," *New England Journal of Medicine* 283 (December 17, 1970), 1370-81; (December 24, 1970), 1441-47; (December 31, 1970), 1498-1504.

20. Especially when genetic evaluation of *donors* becomes a common practice. See Walter Wadlington, "Artificial Insemination: The Dangers of a Poorly Kept Secret," *Northwestern University Law Review* 64 [6] (1970), 777-807.

21. See Edward Pohlman, "Unwanted Conceptions: Research on Undesirable Consequences," *Eugenics Quarterly* 14 [2] (June, 1967), 144.

22. I discuss the reverse situation, where the father wishes to force the mother to have an abortion, in the next section.

23. See *New York Times,* Saturday, January 28, 1972.

24. "When abortion is recommended by a physician, the indications should be stated in the patient's records, and informed consent obtained from the patient and her husband, or herself if she is unmarried, or from her nearest relative or guardian if she is under the age of consent." From *Policy on Abortion,* issued in August, 1970 by the Executive Board of the American College of Obstetricians and Gynecologists. See Tribe, *Harvard Law Review,* 38-41.

25. See for example, Noonan, *Morality of Abortion,* p. 258.

26. Some have used the same argument for the opposite conclusion. Since, or if, we allow abortion, they say, we *should* allow infanticide under certain conditions. See Michael Tooley, "Abortion and Infanticide," *Philosophy and Public Affairs* 2 (Fall, 1972), 37-65, and John M. Freeman and Robert E. Cooke, "Is There a Right to Die—Quickly?", *Journal of Pediatrics* 80 (Spring, 1972), 940-5. Once again, such a conclusion fails to take into account the powerful social reasons against infanticide.

27. It is important to be clear here about the differences between active killing of infants and the fact that the battle for life, in those rare cases where an infant is born with a severe malformation, such as the absence of a brain, is not undertaken or not carried as far as it would otherwise be. There *are* difficult borderline cases, but nothing suggests that killing actively in early pregnancy opens the door to the active killing of infants.

28. See *Reichsgesetzblatt*, 1926, Teil I, Nr. 28, 25 May 1926, ¶ 218, and 1943, Teil I, 9 March 1943, Art. I, "Angriffe auf Ehe, Familie, und Mutterschaft."

29. See the Maryland case *in Re Smith* reported in *41 U.S. Law Week* 2202, 1972, where a 16-year-old girl was jailed at the request of a Circuit Court judge in order to undergo the abortion she refused, but which her mother insisted upon. At the last moment, a higher court freed the girl.

4 Abortion: On Fetal Indications

SUSAN NICHOLSON

The problem of aborting a fetus because of fetal indications presents a specific problem in morality; for in this situation the intent of the abortion is to insure the death of the fetus, which may not be part of an abortion for other reasons which could result only in the separation of the fetus from the mother. In this provocative article, Nicholson argues that fetal euthanasia can be compared to the withdrawal of support from congenitally handicapped newborns. On the basis of this analogy, Nicholson argues that it is consistent with Roman Catholic morality to approve fetal euthanasia. In addition to the development of this argument, Nicholson also provides an interesting commentary on the Roman Catholic moral tradition of the obligation to preserve life and the distinction between the use of ordinary and extraordinary means of treatment.

Susan Nicholson is a member of the Department of Philosophy at Simmons College

Preceding sections have argued that it is morally permissible for a woman to terminate bodily life-support of a fetus resulting from rape, or where continuation of support is incompatible with her own life. Abortion in such circumstances is morally justifiable even though it predictably results in the death of what has been assumed to be a human being. In such abortions, fetal death is not the course aimed at. Hence, the preceding arguments would not justify fetal killing in circumstances in which development of an artificial uterus made possible termination of bodily life-support *without* killing the fetus.

In this section we consider a justification offered for abortion which, if accepted, *would* permit fetal killing even in a technologically advanced age of artificial uteruses. I refer to what is sometimes called the "fetal indication" for abortion, namely, that the fetus is seriously malformed. A woman who seeks an abortion on fetal indications typically terminates fetal life-support because she wants the fetus dead. If an artificial uterus were available, she would not want the fetus transferred to it. For a variety of reasons she feels that it would be better not just that the bodily connection between herself and the fetus be terminated, but that the fetus *die.*

The purpose of this section is to attempt to answer with regard to abortion on fetal indications questions raised previously with regard to rape and therapeutic abortion—namely, whether or not Roman Catholic doctrine on abortion is consistent with other aspects of Catholic moral theology, and whether or not it properly commands support from secular morality.

As in preceding sections, the structure of my argument is analogical and I assume, for the sake of argument, the human status of the fetus. I argue that the Roman Catholic condemnation of abortion on fetal indications is incompatible with the principle of Roman Catholic medical ethics permitting omission or termination of extraordinary measures for prolonging life. The logical implications of this principle, while disavowed by Roman Catholic moralists, are surprisingly liberal. In fact, they are so liberal that it is quite possible that Roman Catholic moral theology here goes beyond what is generally approved in our secular society.

This last point is a speculative one. Since the principles involved in determining the morality of abortion on fetal indications are much more controversial than those relevant to therapeutic abortion and abortion of pregnancies resulting from rape, it is not possible to specify the relationship between the Roman Catholic position and secular morality with much assurance.

The technique of *amniocentesis,* developed within the last decade, has made possible identification of a fetus as malformed. Available from approximately 16 weeks of pregnancy onwards, amniocentesis consists of injecting a needle through the woman's abdomen into the amniotic sac and withdrawing some of the fluid which surrounds the fetus. This amniotic fluid contains fetal cells, and a chromosomal analysis is capable at present of detecting the presence of all the major chromosomal abnormalities and more than 60 rare biochemical diseases (Powledge, 1976:7). It is anticipated that further development of amniocentesis and other new methods for monitoring fetal development will add to this list. Amniocentesis replaces the statistical knowledge of fetal anomalies available through genetic analyses with the kind of information hitherto possible only at birth.

95. Several reasons might be given for aborting a fetus whose malformities have been detected by amniocentesis. Broadly speaking, the reasons may be social, familial, or fetal. Social reasons include a

eugenic concern for the "health" of the gene pool, as well as a concern for the social costs involved in the care, support, and education of a severely handicapped person during her/his lifetime. A familiar reason involves concern for the stresses on the particular person or persons to whom the disabled person is related. A "fetal" reason would be concern for the emotional and physical suffering which will be experienced by a person with the chronic, incurable condition detected *in utero*.

I wish to consider the strongest possible case which can be made for abortion on fetal indications. Hence I will not attempt to defend the abortion of a malformed fetus which is performed for social or familial reasons. This simplification will reduce considerably the complexity of the moral issues involved.

In particular, abortion for social or familial reasons involves whether or not an individual's presumed interest in living may justifiably be overridden in the interests of society or the family. If the fetus is aborted because it is believed that it is in an individual's *own* best interests not to live a disabled life, this issue does not arise. The question I wish to address, then, is whether or not it could ever be morally justified to abort a malformed fetus for the fetus' own sake.

96. It is evident that we are confronted here with the question of the morality of *euthanasia,* and of *fetal euthanasia* in particular. Euthanasia will be taken in this study to refer to acts or omissions which result in death, which are aimed at death, and which are motivated primarily by compassion for the person who dies.[1]

Roman Catholic doctrine unequivocally condemns fetal euthanasia. The following passage, taken from an essay on abortion by Congressman Fr. Robert F. Drinan (1973:130) is typical of Roman Catholic opinion:

But can one logically and realistically claim that a defective nonviable fetus may be destroyed without also conceding the validity of the principle that, at least in some extreme cases, the taking of a life by society may be justified by the convenience or greater overall happiness of the society which takes the life of an innocent but unwanted and troublesome person?

I submit that it is illogical and intellectually dishonest for anyone to advocate as morally permissible the destruction of a defective, nonviable fetus but to deny that this concession is not a fundamental compromise with what is surely one of the moral-legal absolutes of Anglo-American law—the principle that the life of an innocent human being may not be taken away simply because, in the judgment of society, nonlife for the particular individual would be better than life.

It is intellectually dishonest to maintain that a defective, non-viable fetus may be destroyed unless one is also prepared to admit that society has the right to decide that for certain individuals, who have contracted physical and/or mental disabilities, nonexistence is better than existence.

Drinan's remarks against fetal euthanasia are telling in that they conjure up visions of physicians who carry out social policy by administering fatal air injections over the protests of disabled persons who want to go on living. This picture is misleading, however, in several respects.

97. In the first place fetal euthanasia, as defined above, is not carried out *on behalf of* societal interests, but on behalf of the person who would otherwise live with the defects detected *in utero.* Nor is the judgment that nonlife for that person would be better than life a judgment made *by* society. Rather, the judgment is generally made by the person's parents.

98. In the second place, fetal euthanasia never brings about the fetus' death contrary to fetal wishes. Even those who believe that the fetus is a human being must acknowledge that the fetus lacks the capacity to desire life or death.

It is true that fetal euthanasia, since it occurs without the fetus' consent, is nonvoluntary euthanasia. Voluntary euthanasia is defined as euthanasia which occurs with the patient's consent; nonvoluntary euthanasia as that which occurs without the consent. However, some distinctions within this category of nonvoluntary euthanasia are called for. Nonvoluntary euthanasia consists of situations in which (1) the patient is considered competent to give consent and withholds it, wishing to live; (2) the patient is not considered competent to give consent, although capable of wanting to live or to die; (3) the patient is presently incapable of wishing either to live or to die. Within this third category we may make still another distinction. Some persons presently incapable of desiring either to live or to die have nonetheless lived long enough to enable close others to offer a judgment, based on their acquaintance with the patient's personality and values, of what the patient would want. With other presumed persons, such as fetuses, this is of course impossible.

It is important to make these distinctions because objections made to nonvoluntary euthanasia may not be applicable to all varieties of it. In particular, some objections to euthanasia which occurs *contrary* to the patient's wishes are not applicable to euthanasia which occurs *in the absence of such wishes.*

For instance, euthanasia which occurs contrary to a patient's desire to go on living involves overriding the patient's judgment of what will advance her/his own interests, or at the very least involves the greatest frustration on a person's desires imaginable. These objectionable features do not characterize fetal euthanasia. It is possible to maintain that a patient's judgment or desires concerning her/his life or death should not be overridden but that when the patient is incapable of either judgment or desire, someone else should make a judgment on the patient's behalf.[2]

99. In the third place, fetal euthanasia differs from Drinan's depiction in so far as it involves the withdrawal of life-support rather than the initiation of lethal measures. Consequently, abortion on fetal indications should be compared with other situations in which life-support is withdrawn in order to bring about the merciful death of the person receiving such support, where that person is incapable of making any judgments or having any wishes in the matter.

100. In accordance with this discussion, I suggest that fetal euthanasia may appropriately be compared to withdrawal of support from congenitally handicapped newborns. Recent medical literature contains discussions of instances in which treatment was withdrawn from infants born with severe handicaps, because it was believed to be in the infant's own best interests that it die.[3] The similarities with fetal euthanasia are obvious. Infants, like fetuses, are incapable of desires concerning the perpetuation or nonperpetuation of their lives, although if they live they may develop opinions as to what should or should not have been done on their behalf. Furthermore, in the case of both infants and fetuses, it is generally the parents in consultation with medical personnel who decide on behalf of their child that support should be withdrawn.

My strategy here is to set out the Roman Catholic doctrine on the steps which must be taken to preserve life. I shall argue that this doctrine entails the permissibility of a merciful withholding of treatment from infants born with severely deforming and incapacitating conditions, and hence that Roman Catholic moralists should, to be consistent, approve fetal euthanasia.

Roman Catholic doctrine on the measures which must be taken to preserve life is not developed with the rigor and thoroughness that mark Catholic doctrine on therapeutic abortion in particular, and the abortion doctrine in general. I was able to find only one book-length treatment of the subject which compares favorably with the treatment given abortion by Callahan (1970), Grisez (1970a), Granfield (1971), or Huser (1942).[4] However, several basic themes emerge from papal addresses to medical societies, Catholic texts in medical ethics,[5] and articles in Catholic theological journals. Where there is disagreement on significant details, this will be noted.

101. The two basic principles of the Catholic doctrine on the

prolongation of life are: (1) Persons are obliged to use *ordinary measures* to preserve their life and health. (2) Persons are not obliged to use *extraordinary measures* except where the preservation of their life or health is required to attain some greater good.

The duties of a physician and relatives of the patient are to provide all ordinary measures of sustaining life, in addition to whatever extraordinary measures are reasonably requested by the patient.[6] In the case of a young child the rule is that the parents are obligated to see to it that the child is provided with all ordinary means of preserving life. Parents may, if they wish, provide extraordinary measures for a child, but they are not obligated to do so. (See, e.g., Healy, 1956:80–90.)

These principles are affirmed by papal statement. According to Pope Pius XII (1957b:395–6):

> Natural reason and Christian morals say that man (and whoever is entrusted with the task of taking care of his fellowman) has the right and the duty in case of serious illness to take the necessary treatment for the preservation of life and health. . . .
>
> But normally one is held to use only ordinary means—according to circumstances of persons, places, times, and culture— that is to say, means that do not involve any grave burden for oneself or another. A more strict obligation would be too burdensome for most men and would render the attainment of the higher, more important good too difficult. Life, health, all temporal activities are in fact subordinated to spiritual ends. On the other hand, one is not forbidden to take more than strictly necessary steps to preserve life and health, as long as he does not fail in some more serious duty.

102. The most commonly used definitions of ordinary and extraordinary measures are those formulated by Kelly, and cited earlier in this study. According to Kelly (1951:550) ordinary means are

> all medicines, treatments, and operations, which offer a reasonable hope of benefit and which can be obtained and used without excessive expense, pain, or other inconvenience.

Extraordinary measures, on the other hand, are those

> which cannot be obtained or used without excessive expense, pain, or other inconvenience, or which, if used, would not offer a reasonable hope of benefit.

It may seem strange that a patient should be *required* to use ordinary measures of sustaining life. A patient would normally want to use such measures to prolong life, and where s/he didn't, why shouldn't it be permissible to dispense with them? It is the patient's own death, not someone else's, which would result from omitting such measures.

To answer this question we must refer to the Catholic view that human beings are not the owners but only the stewards of their bodily lives.[7] According to this view, God alone has complete dominion over a person's bodily life. As a good steward of someone else's possession, one may have to submit to the prolongation of a life one would just as soon relinquish. When the hardships involved in the treatments prolonging life become great enough, however, one is mercifully released from the duty of conserving that which belongs ultimately to God. Of course, one may be sufficiently attached to one's own bodily life to be willing to undergo great hardships to preserve it. In that event Roman Catholic doctrine permits one to do so, provided the effort does not detract from the attainment of higher values.

103. Prior to the development of anaesthetics and antibiotics, surgery was frequently very painful as well as dangerous, and classified by Catholic moralists as extraordinary. However, an operation or therapy which today is considered neither very dangerous nor very painful may be considered extraordinary relative to the physical condition of a particular patient.

Healy (1956:64–67) describes an appendectomy, for example, as ordinary if performed on an otherwise healthy patient, but extraordinary if performed on a patient suffering from cancer who has at most three months to live. Similarly, Kelly (1958:130) describes intravenous injections of glucose and digitalis as extraordinary treatment of a 90-year-old comatose and apparently terminally ill patient.

It is evident from these examples that the designation of a treatment as extraordinary involves a *weighing* of the benefits of the treatment against the hardships involved in the treatment. The minimum expense, inconvenience, and pain associated with intravenous feeding will not vary significantly from one patient to another, but the benefit in length of life made possible by intravenous feeding will vary considerably. This distinction between extraordinary and ordinary measures no doubt differs from the distinction made by non-Catholic physicians, who could be expected to describe intravenous therapy as ordinary treatment even where it was expected to prolong the life of a particular patient by only a very short time.[8]

Even where a treatment can be expected to prolong a patient's life for many years, it may be regarded as extraordinary if it deprives the patient of normal functioning. This is indicated by the remarks made by several Roman Catholic moralists concerning amputations or other mutilating surgery.

Sullivan (1949:65), for example, classifies amputation of both arms and both legs as an extraordinary measure. Another moralist, McFadden (1967:253), while acknowledging that modern medicine has greatly reduced the severe physical and psychological hardships of living with an amputated limb, allows that there might be circumstances in which amputation is not obligatory, as where other bodily afflictions make it impossible to develop a facile use of an artificial limb or continue as one's sole means of support. McFadden (1967:255) says that mutilating surgery is also extraordinary in cases where it is possible to "foresee with clarity the truly severe and permanent handicaps which will be the outcome of what might otherwise be called 'successful' surgery." Healy (1956:68–70) makes a similar point concerning a 40-year-old badly deformed and crippled married man who cannot survive unless his leg is amputated. According to Healy, this man is not obligated to undergo surgery. Moreover, because of the added hardship and sacrifice which would be required of his wife should he lose a leg. Healy says the man might rightly judge that he is obliged to forego the operation.

104. It is clear from these examples that the licitness of foregoing measures necessary for life does not apply solely to terminal patients. A patient may be dying, of course, in the sense that s/he will soon be dead unless preventive measures are taken, but not be dying in the stronger sense that s/he will soon be dead despite preventive measures. The patient in Healy's illustration is not dying in this stronger sense. Although he will indeed die unless his leg is amputated, if the amputation is performed he could live presumably another twenty to thirty years.

105. There are two additional aspects of the Roman Catholic doctrine which are important for our purposes. The first is that in application of the two basic principles cited above, Roman Catholic moralists make no moral distinction between failing to initiate extraordinary measures and terminating extraordinary measures. Nor do they distinguish between terminations of extraordinary measures accomplished by omissions, and those requiring the performance of acts. Kelly (1950; 1951; 1958), for example, does not differentiate between what is from the physician's point of view failure to perform a particular operation and discontinuation of intravenous feeding. Pope Pius XII (1957b) in the address quoted above states his approval of the discontinuation of mechanical respiration in certain situations. Shut-

ting off a respirator presumably requires an act. In terms of the typology described in Figure 1, Roman Catholic moralists thus approve letting die$_1$, letting die$_{2a}$, and killing$_{2b}$, provided that in each case the measures omitted or terminated are extraordinary. I will speak of the *withholding* of extraordinary measures to cover both the physician's failure to initiate treatment, and termination of treatment by either act or omission.

106. The second aspect of the doctrine which should be emphasized here is this: It is licit to withhold extraordinary measures from a patient without the patient's consent, provided the patient is unable to give consent. Comatose patients, for example, cannot consent to the withholding of treatment, but Catholic medical ethics nonetheless permits extraordinary measures to be withheld from them. The decision to withhold is to be made by a close relative of the patient's, or where no relatives are available, by the patient's physician. In the case of a very young child, the parents are to decide whether or not extraordinary measures will be withheld. Healy (1956:81) states that in making such decisions, the parents are obligated to do "whatever they prudently think the child would reasonably request if he were actually able to pass judgment on the matter."

107. To make this last point more concrete, consider the following example. Imagine you are told you have developed a disease which will soon claim your life unless you submit to a series of operations and other therapy which will remove or destroy arms, legs, and nerve tissue, leaving you with almost a normal life-expectancy but completely helpless, blind, deaf, subject to uncontrollable spasms, without control over your bodily functions and vulnerable to periods of intractable pain. You recognize that you would not be remiss in your duty to God to prolong your bodily life if you decided to forego

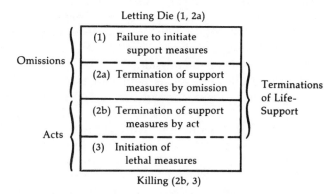

Figure 1. Killing/Letting Die Typology

such treatment, and for your own part, you would find such an existence intolerable. You therefore decide to forego treatment, even though you know that your death will be an inevitable result.

Since this is the decision you make for yourself, you may in good conscience make a similar decision on behalf of your year-old daughter when informed that she has been stricken with the same disease, and requires similar treatment to survive. You decide it is best to spare her such hardship and suffering, even though you are aware that sparing her will inevitably result in her death. You judge for her, as for yourself, that non-life is better than life accompanied by such travail.

108. Now suppose you are confronted with your newly born infant son who is *delivered* with just those same multiple deformities. Due to exposure to a certain drug during pregnancy, the infant is incurably blind and deaf, has no limbs, and will be subject throughout his life to uncontrollable spasms and periods of intractable pain. The doctor informs you that your son's life expectancy is nevertheless only slightly less than normal. Born prematurely, the infant is presently in an incubator.

You realize that you have found yourself in this situation twice before. Again you make the decision that non-life is better than life. Accordingly you request that your infant be removed from the incubator, to die.

109. Roman Catholic moralists would say that you have acted licitly in the first two instances but not in the third. How is this possible? They cannot maintain that you do not have the right to decide for your newborn that non-life is better than life, because we have seen that Catholic doctrine grants you the right to decide precisely that with regard to your year-old daughter. Nor can they object that in the present case your infant son is *killed* by the *act* of removing him from the incubator, because as we have seen, Roman Catholic doctrine makes no moral discrimination between the termination of supportive measures by act, and the termination of or failure to initiate support which occurs through omission.

What they can do, however, is point out that in the first two cases death is only a *foreseen* consequence of withholding therapy whereas in the last case death is *aimed* at. That is, in the first two cases you do *not* arrange that treatment be withheld *in order that death occur.* Rather, you request that treatment be withheld in order to avoid the use of extraordinary measures. In the third case, however, you *do* arrange that treatment be withheld *in order that death occur.* Unable to avert an intolerable existence except through death, you request that your infant be removed from the incubator *in order that he die.*

110. It is thus the *means/foresight* distinction which accounts for the moral discriminations Roman Catholic moralists would make among these three cases. In the third case, the bad effect (death), is a *means* to the good effect, whereas in the first two cases death is merely a *foreseen* consequence of dispensing with extraordinary measures of life-support.[9]

It will be recalled that the means/foresight distinction, while of no use in understanding the Roman Catholic doctrine of therapeutic abortion, is the key to many applications of the Double Effect Principle. In previous applications, the Double Effect Principle distinguished *acts* in which the bad effect is a means from *acts* in which the bad effect is merely foreseen. Here it is evident that the principle is being used to make similar distinctions among *omissions* as well.

111. It will be further recalled that except in cases of therapeutic abortion, Roman Catholic moralists regard a killing as direct if death is aimed at either as means or end, and as indirect if death is merely foreseen. Hence, except in cases of therapeutic abortion, the direct/indirect distinction is identical to what may be called the *aim/foresight* distinction. It will be observed that the means/foresight distinction is subsumed under the aim/foresight distinction.

112. It is instructive to refer to *Figure 2.* Striking a vertical line down the middle of the chart, we may distinguish acts or omissions in which death is aimed at either as means or end (left side of chart) from acts or omissions in which death is merely foreseen (right side of chart). The chart is thus divided vertically by the aim/foresight distinction, and horizontally by the killing/letting die distinction. This is done in Figure 2.

Notice that the killing/letting die distinction crosscuts the direct/indirect distinction. Acts and omissions which *aim* at death are *direct* killing and *direct* letting die, respectively. Acts and omissions which merely *foresee* death are *indirect* killing and *indirect* letting die, respectively.

Now, since euthanasia was defined in this study as acts or omissions which result in death, are aimed at death, and are motivated primarily by compassion for the person who dies, it appears on the left side of the chart. Euthanasia may involve the failure to initiate life-support (letting die$_1$), the termination of life-support by omission (letting die), or by act (killing$_{2b}$), or the initiation of lethal measures (killing$_3$). Hence it appears in all four blocks on the left side of the chart. A distinction is sometimes made between *active* or *positive* euthanasia, and *passive* or *negative* euthanasia. The distinction between active and passive euthanasia appears to parallel the distinction between killing and letting die.

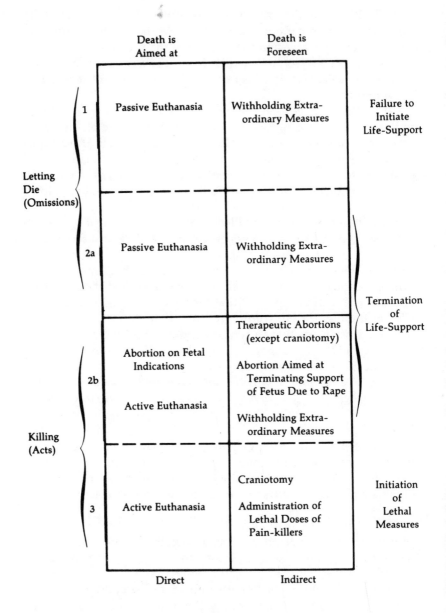

	Death is Aimed at	Death is Foreseen	
1	Passive Euthanasia	Withholding Extra-ordinary Measures	Failure to Initiate Life-Support
2a	Passive Euthanasia	Withholding Extra-ordinary Measures	Termination of Life-Support
2b	Abortion on Fetal Indications Active Euthanasia	Therapeutic Abortions (except craniotomy) Abortion Aimed at Terminating Support of Fetus Due to Rape Withholding Extra-ordinary Measures	
3	Active Euthanasia	Craniotomy Administration of Lethal Doses of Pain-killers	Initiation of Lethal Measures

Letting Die (Omissions)

Killing (Acts)

Direct   Indirect

Figure 2. Typology of the Cross-Cutting of the Killing/Letting Die Distinction by the Aim/Foresight Distinction

84

Notice that each time euthanasia appears in Figure 2, there is a comparable type of indirect letting die or indirect killing which is approved by Catholic moral theology. This chapter presents the Catholic doctrine giving approval to the merciful withholding of extraordinary measures of prolonging life. It will be recalled that Catholic moral theology approves the compassionate initiation of increased dosages of palliatives which are predictably lethal, provided death is not aimed at but is only foreseen. This is comparable to an instance of active euthanasia such as mercifully giving a suffering patient an air injection so that s/he may die.

We may now return to the hypothetical examples of your year-old daughter to whose mutilating surgery you refuse consent, and your congenitally deformed infant son whom you order removed from the incubator. Both examples involve the merciful withholding of measures necessary for life, undertaken without the child's consent. However, the former, in which death is merely foreseen, belongs on the right side of Figure 2. The latter, in which death is a means, is a type of euthanasia, and belongs on the left side of Figure 2. It is thus by insisting on the moral relevance of the means/foresight distinction that orthodox Catholic moralists are able to approve the compassionate withholding of support measures, as well as the merciful initiation of lethal doses of palliatives, while condemning the practice of euthanasia in all its forms.[10]

Having said this, I should point out that there are traditional Catholic moralists who *do* give their approval to the withholding of extraordinary measures in order that death might occur. Sullivan is one, and Kelly, somewhat tentatively, another. Sullivan (1949:72) presents the following illustration.

A cancer patient is in extreme pain and his system has gradually established what physicians call "toleration" of any drug, so that even increased doses give only brief respites from the ever-recurring pain. The attending physician knows that the disease is incurable and that the person is slowly dying, but because of a good heart, it is possible that this agony will continue for several weeks. The physician then remembers that there is one thing he can do to end the suffering. He can cut off intravenous feeding and the patient will surely die. He does this and before the next day the patient is dead.

It is evident that in this case the physician cuts off intravenous feeding in order that the patient die. In thus approving the physician's action in this case, Sullivan clearly condones a cessation of treatment directed at ending a patient's suffering by death. Kelly (1950:219) states that on purely speculative grounds he is in agreement with Sullivan's analysis, but that on practical grounds he hesitates to recommend it lest "the abrupt ceasing to nourish a conscious patient might appear to be a sort of 'Catholic euthanasia' to many who cannot appreciate the fine distinction between omitting an ordinary means and omitting a *useless* ordinary means."

It would appear that Kelly's fears are well-founded. Sullivan has indeed approved an instance of euthanasia. The uselessness of the treatment does not alter the fact that the treatment is terminated in order that the patient die.

It is not entirely clear what Kelly has in mind when he distinguishes here between an ordinary means and a useless ordinary means.[11] Most likely he means by useless, a treatment which cannot forestall death for more than a very short period of time. In that case, "Catholic euthanasia" would be available only to those dying in the strong sense of dying discussed in section 104.

113. I submit, however, that even without Kelly's concession Roman Catholic doctrine is committed to the licitness of euthanasia, and to far more than euthanasia of the terminally ill alone. Consider again the examples of your daughter who requires mutilating surgery to live, and your congenitally deformed son in the incubator. What you want in *both* cases is to avoid subjecting your child to an existence so severely burdened that not even God requires that this bodily existence be preserved. If preserving a severely deformed bodily existence were owed to God then you could not in the first case licitly withhold mutilating treatment from your daughter while foreseeing that she would die untreated. In the second case, again your ultimate end is avoidance of a grossly disabled bodily existence. Only here it is not the therapy provided which will create such disabilities for your son. On the contrary, the disabilities are already irremediably present. Consequently, the only way to avoid such an existence is to arrange for your son's death.

Now, I do not see how the avoidance of a handicapped existence can be so desirable an end that it justifies withholding support measures with death as a foreseen result, yet fail to be a desirable enough end to justify withholding support measures in order that death occur. It would of course be immoral to use death as a means in circumstances where the end in view could be accomplished some other way. But this is not the case here. Moreover, it is similarly immoral to adopt a means in which death is a certain foreseen side-effect unless the end in view cannot be accomplished in any other way.

Note that I am not urging the general teleological principle that it is morally justifiable to adopt death as a means where doing so will increase the total good. I contend only that if two acts of omissions are similar in all relevant respects except that in one death is merely a foreseen consequence whereas in the other death is a means, then if the first is morally permissible so is the second.

It should be emphasized that this does not mean that in circumstances in which one is permitted to act or fail to act while foreseeing a person's death, one may also aim at the death of *that same person.* Consider again the parent who withholds mutilating surgery from a child while foreseeing that the child will die untreated. That this is permissible does not imply that the parent may aim at that same child's death as a means. (Similar points have been made with regard to abortion of pregnancies resulting from rape, and therapeutic abortion. A woman who aborts a fetus resulting from rape foresees but may not aim at fetal death. Should her fetus unexpectedly survive the abortion, she may not kill or let the fetus die.) It may be thought that this demonstrates the moral significance of the means/foresight distinction.

We may account for these moral judgments, however, without invoking the means/foresight distinction. The ultimate end of the parent who withholds mutilating surgery while foreseeing the child's death is avoidance of an existence intolerable to the child. The child's death is not, of course, a means to that end. Hence the parent who aims at that same child's death must have some *other* end in view. Perhaps the parent seeks the child's death in order to be rid of unwanted parental duties.

In some situations, then, the means/foresight distinction is associated with a difference which *is* morally relevant, for instance a difference in ultimate ends. This does not, of course, demonstrate that the means/foresight distinction itself possesses moral significance.

The following reflections suggest another reason why some people attribute moral significance to the means/foresight distinction. If a parent withholds mutilating treatment from a child, and the physician is mistaken in believing that the child will die untreated, then nothing is lost. The child lives, and lives without debilitating handicaps. If, on the other hand, a parent withholds treatment from a malformed child in order that the child die, and the physician is mistaken about the irremediability of the child's physical problems, then all is lost. The child could have lived without debilitating handicaps, but now is dead. Some persons may conclude that the fallibility of medical predictions provides a basis for distinguishing acts or omissions in which death is a foreseen result, from acts or omissions in which death is a means.

It would be a mistake, however, to draw this conclusion. The fallibility of medical predictions works both ways. Suppose the doctor is correct in predicting that the child will die without mutilating surgery, but wrong in estimating the severity of the disabilities which would result from the surgery. Or suppose new techniques are on the horizon for enabling persons with such disabilities to lead satisfying lives. Then if a parent withholds mutilating surgery while merely foreseeing death as a result, all is lost. The child could have lived without intolerable handicaps, and now is dead. I conclude that the fallibility of medical predictions does not provide a basis for imputing moral significance to the means/foresight distinction.

114. It is instructive at this point to consider defenses of the means/foresight distinction offered by two Roman Catholic moralists. Consider the following passage from Grisez (1970b:76).

> In this nonutilitarian moral outlook, whether or not another person's death is admitted within the scope of our intention is extremely important. A difference of intention can relate identical behavior in quite different ways to our moral attitude, and to the self being created through our moral attitude. If one intends to kill another, he accepts the identity of killer as an aspect of his moral self. If he is to be a killer through his own self-determination, he must regard himself in any situation as the lord of life and of death. The good of life must be treated as a measurable value, not as an immeasurable dignity. Others' natural attitudes toward their own lives must be regarded as an irrational fact, not as a starting point for reasonable community.

It may be argued, however, that Grisez's criticisms of direct killing are equally applicable to indirect killing.

In the first place, Grisez claims that a person who aims at death accepts the identity of killer and regards himself as the lord of life and death. Is this any less true of a doctor who deliberately administers increased dosages of a palliative to a patient while merely foreseeing the patient's death as an inevitable result? In administering dosages which s/he knows to be fatal, the physician accepts the identity of (compassionate) killer and, at least in this situation, regards her/himself as lady/lord of life.

In the second place, Grisez maintains that a person who aims at death regards life as a measurable value. But surely the permissibility of measuring life against other values is inherent in the fourth condition of the Double Effect Principle, which requires that the good intended be commensurate with the evil foreseen.

That is, the physician who administers increased dosages of a palliative while merely foreseeing that death will result has, in conformity with the Double Effect Principle, judged that the good effect of freedom from pain is commensurate with the bad effect of this particular patient's death. Similarly, a person who decides to forego extraordinary measures of life-support while foreseeing her/his own death as an inevitable result has judged that the value of avoiding burdensome life-support measures outweighs the value of continuation of her/his own life. But to weigh life and death against other values is surely to regard life as a measurable value.

Finally, it should be obvious that the attitudes of others towards their own lives may be taken into account while aiming at death, and disregarded where death is merely a foreseen result.

In a comprehensive and insightful study of the Double Effect Principle, Richard McCormick (1973) tentatively proposes the following modified understanding of the moral relevance of the means/foresight distinction. Both acts in which the bad effect is merely a foreseen consequence and acts in which the bad effect is a means may be justified by their good consequences. This is, of course, a departure from traditional Roman Catholic moral theology which forbids use of a bad means to an end, no matter how good the end. Nonetheless, McCormick does not regard the means/foresight distinction as morally superfluous. He contends that while the immediate consequences of direct and indirect killing are the same, their long-range consequences may be very different. That is, the over-all bad consequences of direct killing may be worse than those of indirect killing. Hence, the teleological assessment of the two may differ.

Why should the long-range consequences of direct killing differ from those of indirect killing? McCormick answers that the person who kills directly is more closely associated with death, more willing that death occur, than one who kills indirectly. This difference in psychological awareness of the bad effect may, in the long run, lead to significantly different consequences.

McCormick provides little elaboration or defense of his view that there is a psychological difference between aiming at death and merely foreseeing death as a result of one's actions. However, his view may derive plausibility from the following reflections.

Consider again the parent who arranges that mutilating surgery be withheld from a child while foreseeing the child's death as a result. Were the child to live without the surgery, the parent presumably would be pleased. This may be contrasted with the example of a parent who arranges that support measures be withheld from a malformed infant in order that the infant die. Were the infant to live without the support measures, the parent presumably would be displeased. Thus it may appear that the parent who merely foresees a child's death as a result of withholding support measures is less willing that death occur than the parent who, in withholding support aims at a child's death.

Observe, however, that in both instances the parent is willing that death occur *only insofar as death is associated with accomplishment of the end in view.* For a fair comparison, one must imagine the reaction of the parent of the malformed infant were s/he to learn, after support measures had been withheld, of new techniques to ameliorate the infant's disabilities. Were the infant to live without the support measures, presumably this parent too would be pleased. The point is that it would gratify both parents if the good end could be accomplished without bringing about the child's death.

115. I conclude, therefore, that if it is morally permissible for a parent to refuse mutilating surgery for a child while foreseeing that the child will die untreated, then it is morally permissible for a parent to request that a congenitally malformed infant be removed from an incubator in order that the infant die. But the case of the congenitally malformed infant was introduced as a parallel to fetal euthanasia. Hence if a parent may have a malformed infant mercifully removed from an incubator, then a woman may have a malformed fetus mercifully removed from her uterus.

It would appear, then, that the Roman Catholic doctrine of the licitness of withholding extraordinary measures of life-support entails the permissibility of fetal euthanasia. Accordingly, Roman Catholic moralists should approve of abortion on fetal indications where the abortion is performed in the best interests of the fetus. The argument I have presented here for fetal euthanasia may be summarized by a slightly altered version of the passage previously quoted from Drinan:

> It is intellectually dishonest to deny that a defective, nonviable fetus may be destroyed while admitting that parents have the right to decide that for certain very young children who have contracted physical and/or mental disabilities, nonexistence is better than existence.

It must be reiterated that since the moral principles relevant to nonvoluntary euthanasia are much more controversial than those relevant to therapeutic abortion and abortion of pregnancies due to rape, it is not claimed that the Roman Catholic doctrine is without support from a common secular morality concerning the sanctity of life. The claim made here is solely that Roman Catholic moralists cannot, in keeping with their doctrine of the prolongation of life, condemn abortion on fetal indications.

116. Two questions arise naturally from the foregoing analysis. While they cannot be dealt with here, they suggest directions for future discussion and research.

The first concerns the difference between withholding support from malformed fetuses and infants, and initiating lethal measures against them. Once it is admitted that it is permissible to withhold supportive measures from a defective fetus or infant in order that a merciful death may occur, is there any reason why lethal measures could not be initiated for the same purpose? In his study of the well-publicized Johns Hopkins case in which the parents of a Down's syndrome infant born with an intestinal blockage refused to have the blockage corrected by simple surgery, the Protestant moralist Gustafson (1973:547) raises this question.

Once a decision is made not to engage in a life-sustaining and lifesaving procedure, has not the crucial corner been turned? If that is a reasonable and moral thing to do, on what grounds would one argue that it is wrong to hasten death?

It might, in fact, be argued that it is sometimes preferable to hasten death rather than to allow death to come naturally. The Down's syndrome infant in this particular case took 15 days to die after all the supportive measures had been withdrawn. The Roman Catholic theologian Maguire (1974:11) points out that "though the death of this child *may* have been merciful, its dying was not," and suggests that this might be a case in which withholding therapy is harder to justify than initiating lethal measures.

The second question is in what circumstances can it truly be said that nonexistence is better than existence with certain disabilities? As far as I know, no longitudinal studies have been made of persons born with serious malformities. Do these people wish they had not been born? In the absence of empirical research into the conditions under which a person would prefer not to have lived, we can expect that parental perception of the manner of child it would be pleasing to rear will often fill the void. To combat this tendency, it would be necessary to establish empirically based limits on the range of malformities with reference to which a parent may choose death for a child. Until such limits have been established, a practice of infant and fetal euthanasia cannot seriously be recommended.

NOTES

1. This characterization of euthanasia is drawn from an unpublished doctoral dissertation on voluntary euthanasia by Bok (1970).

2. The same point is made by Smith (1974). A similar point is accepted, furthermore, where the use of experimental procedures is at issue. It is not considered proper to use experimental medical procedures on a competent adult without her/his consent. It is considered proper, however, for a parent to consent to the use of such procedures on a child where the parent judges this to be in the child's own best interest. See, e.g., Ramsey (1970:1–58).

3. A recent article in the *New England Journal of Medicine* describes itself as breaking a taboo of silence by disclosing that in the special care nursery of the Yale-New Haven Hospital from January 1970 to June 1972, there were 43 deaths related to decisions by physicians and parents to withhold treatment from newborns whose "prognosis for meaningful life was extremely poor or hopeless." (Duff and Campbell, 1973.) A physician responsible for an English clinic specializing in treatment of babies born with spina bifida argues that the past policy of his clinic to treat all infants is not in the interest of the patients, and urges selective treatment only. (Lorber, 1971, 1973.) For additional medical literature, consult Foltz *et al.* (1972) and Shaw (1973). For a discussion by an ethicist of the moral issues involved in the well-publicized Johns Hopkins case in which the parents of a Down's syndrome infant withheld consent to simple corrective surgery, see Gustafson (1973). Additional discussion of ethical issues occurs in Bard and Fletcher, Joseph (1968); Cooke (1972); Fletcher, John (1974); Freeman, E. (1972); Freeman, J. and Cooke (1972); Reich and Smith, H. (1973); Smith, D. (1974); and Zachery (1968).

4. This is *Death by Choice* (1974), by the theologian Daniel C. Maguire. Maguire argues for the liberalization of the orthodox Catholic doctrine. An exposition of the orthodox doctrine, presently out of print and having little depth, is given by Sullivan (1949).

5. Texts consulted were: Häring (1973); Healy (1956); Kelly (1958); Kenny (1962); Marshall (1964); McFadden (1967); and O'Donnell (1959).

6. See, e.g., Kelly (1950:216). Kelly also mentions that the physician's professional ideal may create obligations which extend beyond the duties and wishes of the patient.

7. For a good exposition of this doctrine, see Häring (1973:66–73).

8. With regard to glucose and digitalis for a 90-year-old comatose patient, Kelly (1958:130) says that moralists would generally say this was ordinary means if it were merely a matter of tiding a patient over a temporary crisis, yet "in the present case the actual benefit they confer on the patient is so slight in comparison with the continued cost and difficulty of hospitalization and care that their use should be called an *extraordinary* means of preserving life."

9. Häring (1973:146), for example, writes that he trusts the non-Catholic reader will not consider it "hairsplitting" to distinguish between cases in which the direct objective is to dispense with extraordinary measures and cases where treatment is omitted or stopped in order to allow the patient to die.

10. Kenny (1962:134), for example, claims that "the proponents of euthanasia have little, if any, conception of morality."

11. In Kelly's 1951 article, he defines ordinary means as means which, among other things, offers "a reasonable hope of benefit." Thus in this later article he appears to have incorporated the notion of usefulness into his concept of an ordinary means.

Abortion/Further Reading

Bracken, Michael B., et al. "Abortion, Adoption, or Motherhood; An Empirical Study of Decision Making During Pregnancy." *American Journal of Obstetrics and Gynecology* 130 (February 1, 1978), 251–62.

Cohen, Marshall, et al., eds. *The Rights and Wrongs of Abortion.* Princeton: Princeton University Press, 1974.

Humber, James M. "Abortion, Fetal Research, and the Law." *Social Theory and Practice* 4 (Spring 1977), 127–47.

Legalized Abortion and the Public Health. Report of a study by a committee of the Institute of Medicine. Washington, D.C.; National Academy of Sciences, May 1975.

Lee, Luke T., and Paxman, John M. "Pregnancy and Abortion in Adolescence: A Comparative Legal Survey and Proposals for Reform." *Columbia Human Rights Law Review* 6 (Fall–Winter 1974–75) 307–55.

Manier, Edward; Liu, William; and Soloman, David. *Abortion: New Directions for Policy Studies.* Notre Dame: University of Notre Dame Press, 1977.

Mohr, James C. *Abortion in America: The Origins and Evolution of National Policy.* New York: Oxford University Press, 1978.

Potts, Malcolm; Diggory, Peter; and Peel, John. *Abortion.* Cambridge: Cambridge University Press, 1977.

Tribe, Laurence H. "Toward a Model of Roles in the Due Process of Life and Law." *Harvard Law Review* 87 (November 1973), 1–53.

Warren, Mary Anne. "Do Potential People Have Moral Rights?" *Canadian Journal of Philosophy* 7 (June 1977), 275–89.

PART TWO

SEVERELY HANDICAPPED CHILDREN

A. R. Jonsen
John Fletcher
James A. Gustafson
Richard A. McCormick

5 Critical Issues in Newborn Intensive Care: A Conference Report and Policy Proposal

A. R. JONSEN AND OTHERS

The legal, ethical, personal, and policy dimensions of dealing with newborns in an intensive care nursery continue to trouble many individuals. This article is a report of a conference which dealt with several dimensions of these dilemmas: the development of neonatal intensive care; legal and policy perspectives; early indicators of development; family concerns; and economic issues. Using the materials developed in these presentations, participants in the conference discussed five cases and attempted to come to some conclusions on how they should be handled. As a result of these discussions several procedural recommendations were made and provide the basis for further reflection upon this very painful and sensitive problem of modern medical care.

Albert Jonsen is Professor of Bioethics at the University of San Francisco

Ethical problems concerning newborn infants have received considerable attention in recent years.[1-3] Two particular types of case have been much discussed: aggressive management of spina bifida[4-8] and the advisability of correction of physical anomalies in infants with genetic defects.[9-13] Another less dramatic, but more common and particularly troubling, problem concerns clinical decisions regarding resuscitation and maintenance of small, severely asphyxiated preterm infants, especially those with respiratory distress who require prolonged assisted ventilation. Just a few years ago most very small infants died of asphyxia in the first moments of life or during the first days of life from hyaline membrane disease. However, developments in newborn intensive care have markedly reduced the fatality rate from both asphyxia and hyaline membrane disease.

Life-saving intervention in an infant's existence inevitably raises certain questions about the desirability of saving certain lives, *e.g.*, those that may be marked by physical and intellectual abnormalities. Should doctors (or parents, or society) discriminate among endangered neonates, attempting to save some and not others? If so, what norms should guide this discrimination?

While these and other such questions cannot be answered solely on the basis of medical information, they cannot be considered apart from clear knowledge of the possibilities and limits of medical efforts on behalf of the infants in question. This is true both on the level of individual decisions and on the level of larger policy decisions.

On May 19, 1974, a small conference was convened in the Sonoma Valley, in California, to consider the ethical problems raised by neonatal intensive care. It brought together 20 persons (Appendix A) from different disciplines: medicine, nursing, law, sociology, psychology, ethics, economics, social work, anthropology, and the news media, some of whom have direct responsibility for newborn intensive care, but all with a professional interest in problems of childhood.

Part I of this article summarizes the materials presented to the Conference participants: five illustrative cases, papers on the major considerations, and four clinical questions. While the discussions of cases, papers, and questions are not reported here, the tables showing participant responses to the four clinical questions are given in Appendix B.

Part II presents a "moral policy for neonatal intensive care." This section, consisting of certain ethical propositions and procedural recommendations, was not developed at the Conference but written subsequently by the authors, who take full responsibility for it. The substance of this moral policy arises from our reflections on those two days of conversation and argument. We sent this to the participants for review. Their response was generally favorable, but there were several significant dissenters. However, we felt that the participants, whose contributions formed the matrix in which the policy was shaped, would accept it as an accurate reflection of the mood and tone of the Conference.

The purpose of the moral policy is not to close the agonizing debate over the moral issues of neonatal intensive care. On the contrary, it is intended to stimulate that debate and to propose a framework within which certain questions can be posed more precisely and pointedly than in the past. We have limited our discussions and the subsequent policy statement to problems of decision faced by those who are involved in the clinical care of newborns.

Neonatal intensive care units exist and decisions are being made constantly. Those responsible for these decisions are beginning to ask certain questions. It is to these questions that we address ourselves, quite consciously excluding the major social, political, and economic decisions which must be made about neonatal intensive care. However, we refer to them, especially in the procedural recommendations, and they always lurk in the background. The data required for their discussion are still scarce, and the gradual accumulation of this information will be profoundly affected by stances taken on the nature of clinical decisions. These must be clarified first.

Part I. Conference Report

Case Studies

Five cases were selected to illustrate ethical problems that might emerge in the course of providing intensive care to newborn infants.

In the first two cases both infants were born after a 28- to 29- week gestation. In each, at least one parent urged that the infant not be resuscitated but be allowed to die on the grounds that prematurely born infants are likely to end up retarded or otherwise markedly handicapped. In both cases, the attending physicians argued in favor of providing intensive care, and won the consent of the parents to this line of action. The infants were both vigorously resuscitated and did well in the neonatal period. Now, one is normal at 4 years of age while the other is severely retarded and probably autistic. These cases provide background for the first clinical question: *Is It Ever Right Not To Resuscitate an Infant at Birth?* Because of the initial parental stance, these cases also raise the problem of identifying the proper locus of decision-making authority and of dealing with the tensions that can accompany the exercise of that authority.

In the third and fourth cases the infants were delivered after 32 weeks' gestation, weighing 1,750 and 1,800 grams. They were severely depressed at birth, with Apgar scores of 1 at 1 minute and an arterial blood pH of 6.90 at 5 minutes of age. Both were vigorously resuscitated and now one is normal at 5 years of age. The other required prolonged ventilatory assistance, had a large intracranial hemorrhage, and developed hydrocephalus, with only 1 mm of cortex visible by pneumoencephalography. Although both these infants appeared the same at birth, by 2 weeks of life one had clear evidence of severe, irreversible brain damage, a rapidly enlarging head, was still receiving assisted ventilation, and was a candidate for a ventriculoperitoneal shunt. These two cases highlight the problem of early detection of severely handicapping problems and stress how difficult it is to judge from an infant's initial condition whether he will be healthy or defective ultimately. These cases provide the basis for discussion of the second clinical question: _Is It Ever Right To Withdraw Life Support From a Clearly Diagnosed Poor-Prognosis Infant?_

Case 5   The fifth case concerned an infant who was the first of twins, born after a 28-week gestation, weighing 1,000 grams. He developed severe respiratory distress and had clear evidence of large intracranial hemorrhages at 2 and 4 days of age. Because of a large shunt, the ductus arteriosus was ligated at 10 days of age. He had retrolental fibroplasia and was blind at 2 months of age. He required continuous assisted ventilation to maintain a carbon dioxide below 60 mm Hg. By 4½ months of age, inspiratory pressure of 40 mm Hg and an inspired oxygen concentration of 60% were required to adequately ventilate his diseased lungs. At 4½ months of age, respiratory support was withdrawn. The infant's parents and physicians agreed that should he have a cardiac or respiratory arrest, he would not be resuscitated. His arterial carbon dioxide tension gradually rose to 150 mm Hg at 5 months of age; he developed seizures, and died at 5½ months of age. This case illustrates the problems associated with a decision not to continue life support and is the basis for the third clinical question: _Is It Ever Right To Intervene Directly To Kill a Dying Infant?_

Major Considerations

In addition to the cases, background material was organized around five themes: (1) development of neonatal intensive care; (2) legal and policy perspectives on neonatal intensive care; (3) early indicators of development; (4) family concerns; and (5) economics of neonatal intensive care. These themes were used later to focus the Conference discussions.

Development of Neonatal Intensive Care. Dr. Clement Smith, former editor of _Pediatrics,_ prepared a summary of the development of neonatal intensive care and its attendant ethical concerns.

He stated that recent progress in newborn intensive care has extended the concern of neonatologists beyond mere prevention of death to the question of the value of that life for which infants may now be saved. The longer these once unpreventable deaths are deferred, and the more the factor of expense aggravates the problem, the less acceptable is any course except continuation of treatment. Ideally, reduction of neonatal death should be accompanied by reduction in the persistence of permanent handicaps. Yet, some infants are saved from early death only for an existence which few persons would consider worth living. Dr. Smith concluded that so many variable circumstances surround these problems that many physicians find it difficult to meet them with any single moral rule—except perhaps the Golden one.

Legal and Policy Perspectives. F. Raymond Marks, J.D., a lawyer with the Childhood and Government Project at Boalt Hall School of Law, University of California at Berkeley, submitted an essay as background for this theme (with assistance from Lisa Salkovitz, also of the Childhood and Government Project).

He maintained that the dilemma facing parents, doctors, and society in neonatal intensive care units is similar to that faced and resolved by the United States Supreme Court in the abortion decision. A defective child, like a fetus, may be unwanted. The maintenance of a defective child, like carrying a fetus to term, may involve not only broad social costs but a threat to its family's viability. The decisions to be made in the intensive care nursery are human and societal, based on medical estimates. Marks' paper identified the actors, their interests, and their decision-making capacities. In the question of the survival of a defective child, parents are the true risk-takers and burden-bearers, and this problem calls for informed and reflective consent. Parents must speak for the child, whether for the claim of the right to live or the right to die, as well as for the family unit in balancing medical estimates with economic and other family considerations. Marks argued for a social policy that would withhold legal personhood from certain carefully defined categories of high-risk infants until a clear diagnosis and prognosis can be made concerning them and until their parents have made an informed decision whether or not they want to keep and nurture these infants.

Early Indicators of Development. Dr. Jane Hunt, Developmental Psychologist with the Institute of Human Development, who has followed children from the University of California-San Francisco Intensive Care Nursery, prepared material on this third theme.

She indicated that neonatal intensive care has improved the chances for survival and has reduced the numbers of survivors with severe brain damage. An accurate early prognosis of development is not generally possible, because some subtle but permanent insults to the brain are not evident until later in life, while other demonstrable insults have transient or reversible effects. However, increasingly accurate prognostic evaluations can be made between 1 and 4 years of age. Accurate prognosis depends on better understanding of the loci and extent of brain damage, the behavioral implications of this damage, the potential of undamaged brain areas, and the environmental factors that determine the realization of that potential. Environmental variables include such disparate elements as nutrition, sensory and motor stimulation, and social interaction. Comprehensive special education programs, instituted early in life and focused on specific disabilities and abilities, can yield good results.

Family Concerns. Marna Cohen, M.S.W., Social Worker with the University of California-San Francisco Department of Pediatrics, presented material on the fourth theme.

There is a wide range of individual, initial reactions to the birth of a premature, sick, or defective baby. The initial parental response is partly determined by the nature of the infant's illness or defect, proceeds in stages from shock to adaptation, and encompasses denial, anger/depression, bargaining, acceptance/grief, and adaptation. The authenticity of parental participation in decisions about how far to extend treatment is related to the stage which parents have reached in this process. Timing is a key ingredient to parental involvement in necessary decisions and bears directly on the ultimate resolution of the grief process. The use of available community support for the defective child and his family is important. Unfortunately, these supports are generally inadequate, although the level and quality of assistance varies from state to state.

Economics of Neonatal Intensive Care. Marcia J. Kramer, M.A., of the National Bureau of Economic Research, Inc., of New York City, and the Department of Economics at Swarthmore College, presented material on the economics of neonatal care.

She emphasized that provision of neonatal intensive care necessarily diverts limited resources from other uses. Because the activities given up by choosing to develop neonatal intensive care may be of greater moral worth than the provision of this care itself, it is morally imperative that cost criteria be established for this, as for any other, service. However, the means of doing so lie outside the province of technical economic analysis. The principle of consumer sovereignty is basic both to the market mechanism and cost-benefit analysis, yet an infant is not morally equivalent to a commodity whose value derives solely from the utility it yields to others.

Ethical problems are by no means confined to the determination of moral cost ceilings; in fact, they permeate every phase of the economic analysis of neonatal intensive care. Computation of the expected net cost associated with a decision to offer treatment, for instance, is heavily conditioned by judgments about the duration of the commitment implied by neonatal intensive care, the anticipated quality of post-neonatal special care, and the alternative to neonatal intensive care. Cost-effective decision-making also requires that relative values be attached to each possible outcome, since rates of survival and long-term disability vary greatly from one condition to the next. Because a positive neonatal intensive care decision presupposes that the incidence of costs is acceptable, both interpersonally and intergenerationally, distributional norms must be made explicit.

Four Clinical Questions

See p. 165,
p. 130

Four questions were raised which focused on the clinical situation. They were asked in the broadest possible terms in order to allow the participants to define the limits of their own responses.

The questions address the issues of initial intervention, withdrawal of life support already initiated, active lethal intervention, and allocation of the limited resources of a neonatal intensive care unit.

_(1)_ Many infants are born in need of vigorous resuscitation. The decision whether or not to treat them must be made immediately. Given a situation where such therapy is readily available, *Is It Ever Right Not To Resuscitate an Infant at Birth?*

_(2)_ Many infants who are candidates for intensive care therapy at birth have underlying anomalies, lesions, or disease states that are not readily apparent when the decision to initiate therapy is made; also, complications sometimes develop in the course of therapy that result in severe neurological damage. Given the situation of an infant who is receiving intensive care but who has a clearly recognized defect, *Is It Ever Right To Withdraw Life Support From a Clearly-Diagnosed, Poor-Prognosis Infant?*

_(3)_ Assuming that a decision is made to let a given infant die, on the basis of a clear diagnosis and poor prognosis, and that the infant is now, and will be for some period of time, self-sustaining, *Is It Ever Right To Intervene Directly To Kill the Dying Infant?* (Insofar as one accepts as valid the distinction between *active* and *passive* euthanasia, this question concerns *active* euthanasia, *i.e.,* taking some directly lethal action against the life of the infant.)

_(4)_ An intensive care nursery may have no more room when an infant is born who needs intensive care and no equivalent facilities are available within a reasonable distance. Suppose that one of the infants who is already receiving intensive care is clearly diagnosed and has a poor prognosis, and that the other infant is judged with reasonable certitude to have a better prognosis, *Is It Ever Right To Displace Poor-Prognosis Infant A in Order To Provide Intensive Care to Better-Prognosis Infant B?*

All participants approved, in principle, not resuscitating some infants and withdrawing life support in certain cases; 17 said "yes" to active euthanasia in some circumstances, 2 answered "no," and 1 was "uncertain." The question of choosing between endangered neonates on the basis of poor prognosis versus good prognosis drew 18 "yes" and 2 "no" responses.

Consensus in these responses is on the surface only; much variability appears in the conditions employed to establish boundaries for the responses (Appendix B). Conditions focus variously on the child's status, consequences to the family of a defective child, costs to society of caring for defectives, and such procedural matters as peer consultation and review, open policy in the delivery room and nursery, and informed consent of parents. Some participants objected that the questions focused unwisely on individualistic problems of conscience in isolation from pertinent social issues, such as priorities for resource allocation.*

PART II. A MORAL POLICY FOR NEONATAL INTENSIVE CARE

When individuals face decisions on matters about which they have moral convictions, they act in accord with those convictions or violate them, or find compromises, excuses, or extenuating circumstances to resolve the dilemmas of conscience. However, when many individuals with diverse moral convictions face a series of decisions about similar cases, there should be a way to accommodate the diversity of private beliefs within some degree of broad agreement about how such cases should be managed. We call this effort making a moral policy.[14] This policy should describe not only substantive moral principles with which the majority can agree, but also the social arrangements which would facilitate discussion and action on the basis of those principles. Thus, a moral policy mingles statements of principle with procedure.

This moral policy presents the elements with which one can make a reasonable ethical argument, the elements of which are certain moral rules, attributions of responsibility and duty, and medical, psychological, social, and economic facts.

A word of warning: this moral policy may seem unreal. This is

---

* Dr. Laura Nader, Professor of Anthropology at the University of California at Berkeley, was critical of the Conference on this score: "Let me touch upon some problems. We cannot dismiss the economics of neonatal intensive care by simply stating 'an infant is not simply a commodity whose value is defined by its utility.' Questions should be raised: who benefits economically from neonatal intensive care—the companies that produce the machinery, the doctors who work at this labor, insurance companies, the hospital, the parents, the families of the neonate, the neonate? Furthermore, what are the preventative possibilities, and why was this not relevant? Can the number of such 'infants' be reduced by monitoring drug, genealogy, and environmental inputs? What of the process affecting a chain of professionals and clients alike, decisions being made in and out of the health field? The scarce resources question was dismissed without adequate discussion—it was 'off the track.' We never pursued the question, how has our society come to be spending so much time and money on neonatal intensive care without similar attention to born healthy, but later not so healthy, deprived children—is this development related to special interests that may be ours although we are unaware of them? If so, what are they?" (personal communication).

N.B.

the inevitable result of considering moral decisions apart from the agony of living through the decisions. It reflects abstraction from the actualities of fear, self-interest, exhaustion, the dominance of some and the truancy of others charged with responsibility and duty. But the air of unreality is, we believe, the necessary cool moment which philosophers say should precede any reasonable judgment. That judgment will have to be made amid the hard realities, but it may be better made in the light of reflection on these propositions.

Ethical Propositions

(1). Every baby born possesses a *moral value* which entitles it to the medical and social care necessary to effect its well-being.

(2). Parents bear the principal *moral responsibility* for the well-being of their newborn infant.

(3). Physicians have the *duty* to take medical measures conducive to the well-being of the baby in proportion to their fiduciary relationships to the parents.

(4). The State has an *interest* in the proper fulfillment of responsibilities and duties regarding the well-being of the infant, as well as an interest in ensuring an equitable apportionment of limited resources among its citizens.

(5). The responsibility of the parents, the duty of the physician, and the interests of the State are conditioned by the medicomoral principle, "do no harm, without expecting compensating benefit for the patient."

(6). Life-preserving intervention should be understood as doing harm to an infant who cannot survive infancy, or will live in intractable pain, or cannot participate even minimally in human experience.

(7). If the court is called upon to resolve disagreement between parents and physicians about medical care, prognosis about quality of life for the infant should weigh heavily in the decision whether or not to order life-saving intervention.

(8). If an infant is judged beyond medical intervention, and if it is judged that its continued brief life will be marked by pain or discomfort, it is permissible to hasten death by means consonant with the moral value of the infant and the duty of the physician.

(9). In cases of limited availability of neonatal intensive care, it is ethical to terminate therapy for an infant with poor prognosis in order to provide care for an infant with a much better prognosis.

Commentary

These propositions identify four moral "fields of force" of which cognizance must be taken in decisions about sustaining neonatal life.

Each field is designated by a term with strong ethical connotations: *value, responsibility, duty,* and *interest.* These terms suggest that the various parties in the neonatal situation have diverse roles and relationships.

#1

Moral value indicates that the infant, although unable to comprehend, decide, communicate, or defend its existence, requires by its very existence to be approached with attitudes of respect, consideration, and care. The infant is designated as a being in its own right and morally, if not physically, autonomous. Its life is not merely a function of others.

#2

The term *responsibility* signifies that those who engender and willingly bring an infant to birth are morally accountable for its well-being. They are closest to the infant and must bear the burdens of its nurture, especially if it is ill or defective. This principle is stated with full recognition that some parents will not or cannot exercise this responsibility. Nonetheless, it states an ideal and a demand which medical professionals should acknowledge in their attitude and in their institutional arrangements.

#3

The term *duty* applies to the professional relationships of a physician who has two clients, the infant and the parents.* This relationship is fiduciary, entered into freely by the physician with the parents who entrust their infant to the medical judgment of the physician for the sake of promoting the infant's well-being. Informed consent usually controls fiduciary relationships. However, an infant, the proper patient in this relationship, is unable to be a consenting partner. Thus, parental decisions normally control the relationship. However, the physician, responding directly to the moral value of the infant-patient, may at times be duty-bound to resist a parental decision.

Physicians may feel that their duty extends not only to a particular infant under care but also to all children. For such a reason, some physicians may be devoted to scientific research aimed at improving the quality and effectiveness of neonatal care for all. While this dedication is necessary and praiseworthy, it may, on occasion, influence decisions about the care of a particular patient. The need for observation and data may push a clinician, even unconsciously, to extend a course of care beyond reasonable limits of benefit to the patient.

The designation of fields of force of parental responsibility and

---

* Nurses have a special professional relationship to the infant, the parents, and physicians in the provision of intensive care. Their role in decisions about continuation of therapy for endangered infants merits full discussion and articulation with an understanding of the roles of physicians and parents. Other, nonmedical, persons, such as social workers and clergy, also play important roles in the decision-making process; these become apparent when one focuses on the decision process itself. For the present discussion we have set aside these other considerations in order to focus directly on the parent-physician relationship.

physician's duty means that the ultimate decisions, morally, lie with the parents. This does not, in fact, mean that parents will make those decisions always. They may absent themselves physically or psychologically. They may even abdicate their moral right to make decisions by failure to acknowledge the well-being of the infant upon which their responsibility is predicated. In such cases, the duty of the physician is expanded to include the heavy burden of rendering final decisions.

Interest designates the concern of the State, in particular, and society at large that actions of individuals respect certain values and fulfill certain responsibilities and duties. The State also has an interest in the fair and efficient distribution of benefits throughout that society as well as the promotion of health and well-being of its citizens. If promotion of the child's well-being unavoidably jeopardizes other equally worthy endeavors, a reconciliation of the competing interests must be sought. These concerns remain in the background unless a perceived threat to the common good requires remedial or preventive intervention.

We conceive of these moral fields of force as attracting and repelling certain kinds of actions. Thus, the value of the infant attracts respect, consideration, and care and repels indifference, violence, and neglect. Responsibility attracts specific forms of care for the infant and repels unconcern. The fields of force converge in decisions about neonatal survival, so that the valued infant is the focus of parental responsibility, physician duty, and State interest. Each of these has its limits; each is subordinate to the moral value of the infant.

Responsibilities, duties, and interests require many specific actions. For example, it is the responsibility of parents to nourish the infant, the duty of the physician to cure the infant's illness, and the interest of the State to punish neglect of the infant. However, we propose that the medicomoral principle "do no harm" is most appropriate to guide decisions regarding neonatal survival. Its appropriateness rests on the following considerations:

First, the principle, stated in the negative, admits of no exceptions. Positive formulations of moral obligation, such as "preserve life," admit of exceptions and must be qualified by listing grounds for exception. The traditional medical principle, "do no harm" is a universal. The problem, then, is not in finding grounds for exception but in defining harm.[15]

Secondly, most medical interventions effect some harm, either transient or permanent; that harm is usually justified by an expected compensatory benefit. If no benefit can be reasonably expected or if the benefit does not compensate for the harm, the intervention is unethical. The assessment of whether the benefit does compensate for the harm lies principally with the patient, who must suffer the harm. In the case of infants, the assessment must be made by those who bear responsibility and duty within a context of broad social understanding.

In the context of certain irremediable life conditions, intensive care therapy appears harmful. These conditions are identified in the sixth proposition as inability to survive infancy, inability to live without severe pain, and inability to participate, at least minimally, in human experience.

The first condition recognizes the possibility that some infants may be born with irreparable lesions incompatible with life. They are already in the dying state and, while care should never be neglected, efforts aimed at prolongation of life are best viewed as harming rather than helping such an infant. The second condition envisions the case of an infant who is in constant severe pain which cannot be alleviated either by immediate treatment or as the result of a long course of treatment. The third condition is perhaps the most controversial. Participation in human experience means the assessed expectation that the infant has some inherent capability to respond affectively and cognitively to human attention and to develop toward initiation of communication with others.[12] It deals with presence or absence of capacity, not degrees of deviation from statistical normality.

Our concept of the abilities which count as signs of these human qualities is quite broad. While we are reluctant to quantify or describe in detail the levels of affective and cognitive activity, we would prefer to err on the side which favors the life of the child. A baby with Down's syndrome would fulfill the criteria, whereas one with a trisomy 18 will not.[9,16]

The eighth principle intimates that the question of the morality of active euthanasia is far from settled. We do not intend to settle it here. The formulation of this principle allows for the opinion that the moral value of the infant represents a sanctity against which no lethal action can be judged ethical. The principle also recognizes the opinion that the primary duty of the physician, healing, is incompatible with any lethal action. This formulation allows for the invocation of the double effect doctrine, acceptable to many, which distinguishes between *intended* and *permitted* effects of an action. It also allows for the distinction, favored by some, between acts of commission and acts of omission. All these issues can be defended or criticized.[17-23]

We suggest that there may be a significant moral difference between an infant whose therapy has been terminated and an adult whose condition is diagnosed as hopeless, or (in a rapidly vanishing example) condemned to death. For the adult, the time intervening between verdict and death may be of great personal value. For the infant, the intervening time has no discernible personal value.

The ninth proposition responds to the problem of the allocation of limited resources as it sometimes occurs in the intensive care nursery. Up to this point, moral considerations have focused on the well-being of the individual infant. Now, a new element is introduced, *i.e.,* the comparison between two individuals. It becomes difficult to apply the rule "do no harm," since either decision will effect some harm without providing a compensatory benefit to the one harmed.

The traditional medicomoral rule of triage (screening of the wounded or the ill to determine priority for treatment) may illuminate this dilemma. Casualties in military and civil disasters are divided into those who will not survive even if treated, those who will survive without treatment, and the priority group of those who need treatment in order to survive. A further triage among the priority group would give preference to those who can be reactivated quickly or who are in crucial positions. Considerations of the common good become relevant in such decisions.

Similarly, in the comparative selection of infants for treatment, the interest of the State can be invoked as an ethical consideration since the State has an interest in the recognition of values, in fulfillment of responsibilities and duties, in the fair and efficient distribution of benefits, and in the promotion of a healthy population. These interests are directed toward a common good which, in situations such as this, may be the predominant consideration. Thus, given the impossibility of treating all infants in need, those should have preference who give the greatest hope of surviving with maximal function.

The use of this principle must be approached with grave caution. First, common good considerations are, in practice, often disguised special interest considerations. Favored treatment of certain persons or classes is judged, by those identified with those persons or classes, to contribute to the common good. Secondly, the hope of survival with maximal function is predicated not only on physical potential of the infant but on the socioeconomic world into which it enters. Thus, estimates of the quality of future care may bias selection. Thirdly, selection of better-prognosis infants can be strongly motivated by the physician's interest in compiling favorable statistics and a more rapid selection decision than the condition of an infant may warrant. Thus, the principle of neonatal triage, while instructive in general, is fraught with the risk of serious bias.*

---

* Dr. Jane Hunt, Ph.D., Psychologist at the University of California at Berkeley, registered vigorous opposition to this proposition. "I maintain (neonatal triage) turns out

Procedural Recommendations

A moral policy requires both ethical propositions and procedural recommendations. Procedural recommendations suggest certain institutional and social arrangements which will facilitate deliberation and action on the basis of the ethical propositions. These recommendations discuss the neonatal problem on a different level than the ethical propositions. There, we were concerned with the *microdecisions* made by those bearing responsibility and duty about particular infants at risk. Here, we move to the *macrodecisions* about social and institutional arrangements. Here, issues of State interest, social costs, and economic considerations are most relevant.

We had excluded direct and extensive consideration of public policy questions concerning allocation of scarce medical resources, priorities between preventive and curative medicine, and between medical care and other forms of care affecting personal health from the Conference deliberations. Nonetheless, they must be mentioned.

(1) Research in neonatology should be coordinated at the national level in the interests both of efficiency and caution.

(2) Neonatal intensive care should be organized on a regional basis so that its quality and access are relatively equal among various communities, so that continuing information on techniques can be shared, and so that adequate epidemiological data can be gathered and compared.

(3) On the basis of clinical experience, professionals in neonatal intensive care should refine those converging clinical criteria which render more specific the general conditions of prolonged life without pain and the potential for human communication. These criteria should be communicated broadly within the pediatric, obstetric, and mental health community.

(4) Resuscitation criteria should be established with full awareness of the economic and medical implications of providing this care. Estimates should be made of the financial cost to society of prolonging life, at a humane level, depending upon the condition at birth.

(5) Delivery room policy, based on certain criteria, should state conditions for which resuscitation is not indicated. This policy should be made known to health professionals and to parents who may be at particular risk (possibly to all parents).

(6) Parents at risk should be counseled about the possibilities.

---

not to be a proper ethical question because there is *no* ethical solution. One rule is as good as another, unlike the military triage. You can never articulate a policy which separates "common good" from "special interest" because the latter can be put forward as cogent arguments for the former. For example, consider the comparison between the wanted and the unwanted infant" (personal communication).

Since they bear primary responsibility for their infant, explanatory and supportive counseling is mandatory before and, in the event of a sick infant, after birth. While recognizing that parents often will be unable or unwilling to make decisions, medical professionals must always accept, in principle, that the responsibility should be borne by the parents and attempt to facilitate but not force them to make the decision.*

(7) The decision to terminate care for the infant requires sufficient time for observation, mature assessment and parental involvement in the decision. Thus, it is more ethical, although perhaps more agonizing, to terminate care after a period of time than to withhold resuscitative measures at the moment of birth. This should be an accepted and publicly acknowledged policy in pediatric and obstetric practice.

(8) Regional neonatal intensive care units should establish an advisory board consisting of health professionals and other involved and interested persons. This board is not to be charged with particular decisions about specific infants; this remains the responsibility of the parents with advice and concurrence of the physician. The Board would discuss the problems of the unit and make a periodic retrospective review of the difficult decisions. They would assess the criteria for diagnosis and prognosis in terms of medical validity and social acceptability. They would provide, by bringing a variety of experience, belief, and attitude, a wider human environment for decision making than might otherwise be available. To implement such a procedure, a neonatal intensive care unit could review its experience prospectively for each case in which a decision had to be made. All data pertinent to the decision could be recorded in a case summary, which would be reviewed monthly, describe the current process of ethical decision making in the unit, and provide a basis for any changes which might be planned.

(9) Since some infants may be abandoned by their parents or, because of their condition, be maintained in an institution, neonatology must concern itself with the adequacy of such institutions. The advocates of intensive care must become the advocates for the devel-

---

*Dr. Clement Smith noted the unanimity with which nonmedical members of the conference insisted that life-and-death decisions must be made by the parents, while doctors of medicine with almost equal unanimity saw that as an avoidance of the physician's own responsibility—which may inflict a lifetime of regret whichever course the parents decided upon. He would prefer, and in this he believes most physicians would agree, that the doctor, through intimate participation and full discussion with the parents, interpret their beliefs or wishes clearly enough to act according to those indications rather than confront parents directly with the act of decision (personal communication).

opment of humane continuing care and for sufficient funding of programs to support families whose children require special attention at home or in institutions. Neonatology cannot be developed in isolation from the continuing specialized care which, unfortunately, will be needed by some of the survivors of life-threatening neonatal disorders.

In the neonatal intensive care situation some people have to act in the best interests of others who cannot act for themselves. Rational assessment of such a situation is an important contribution to the work of safeguarding the rights of these infants. We have espoused a definite position, but do not presume to have spoken the final word, in order to invite reflection and debate. We hope this will promote more sensitive appreciation of the needs and rights of all the participants in the drama of newborn intensive care.

REFERENCES

1. Report of the 65th Ross Conference on Pediatric Research: Ethical Dilemmas in Current Obstetric and Newborn Care. Columbus, Ohio: Ross Laboratories, 1973.

2. Duff, R. S., and Campbell, A. G. M.: Moral and ethical dilemmas in the special care nursery. N. Engl. J. Med., 289:890, 1973.

3. Shaw, A.: Dilemmas of "informed consent" in children. N. Engl. J. Med., 289:885, 1973.

4. Lorber, J.: Criteria for selection of patients for treatment. Read before the Fourth International Conference on Birth Defects, Vienna, 1973.

5. Smith, G. K., and Smith, E. D.: Selection for treatment in spina bifida cystica. Br. Med. J., 4:189, 1973.

6. Freeman, J.: To treat or not to treat: Ethical dilemmas of treating the infant with myelomeningocele. Clin. Neurosurg., 20:134, 1973.

7. Freeman, J. M.: Is there a right way to die—quickly? J. Pediatr., 80:904, 1972.

8. Cooke, R. E.: Whose suffering? J. Pediatr., 80:906, 1972.

9. Gustafson, J. M.: Mongolism, parental desires, and the right to life. Perspect. Biol. Med., 16:529, 1973.

10. Smith, D. M.: On letting some babies die. Hastings Center Stud., 2:37, 1974.

11. Tooley, M.: Abortion and infanticide. Philos. Public Affairs, 2:37, 1973.

12. McCormick, R. A.: To save or let die—the dilemma of modern medicine. JAMA, 229:172, 1974.

13. Englehardt, H. T., Jr.: Euthanasia and children: The injury of continued existence. J. Pediatr., 83:170, 1973.

14. Callahan, D.: Abortion: Law, Choice, and Morality. London: Macmillan, 1970, p. 341.

15. Gert, B.: The Moral Rules. New York: Harper & Row, 1966, pp. 60 and 104.

16. Gustafson, J. M.: A Christian approach to the ethics of abortion. Dublin Rev., Winter, 1967–1968, p. 358.

17. Bennett, J.: Whatever the consequences. Analysis, 26:82, 1966.

18. Foot, P.: The problem of abortion and the doctrine of the double effect. Oxford Rev., 5:5, 1967.

19. McCormick, R. A.: Ambiguity in Moral Choice. Milwaukee: Marquette University Press, 1973.

20. Bok, S.: Euthanasia and the care of the dying. Bioscience, 23:, 1973.

21. Myers, D. W.: The legal aspects of medical euthanasia. Bioscience, 23:467, 1973.

22. Crane, D.: Physicians' attitudes toward the treatment of critically ill patients. Bioscience, 23:471, 1973.

23. Cassell, E. J.: Permission to die. Bioscience, 23:475.

APPENDIX A

Conference Participants

Directors
Jonsen, Albert R., Ph.D. University of California, San Francisco
Phibbs, Roderic H., M.D. University of California, San Francisco
Tooley, William H., M.D. University of California, San Francisco

Coordinator
Garland, Michael J., Ph.D. University of California, San Francisco

Participants
Brewer, Eileen, M.D. University of California, San Francisco
Clausen, John, Ph.D. University of California, Berkeley
Clouser, Danner, Ph.D. Hershey Medical Center, Pennsylvania State University
Cohen, Marianna A., M.S.W. University of California, San Francisco
Creasy, Robert K., M.D. University of California, San Francisco
Davis, Morris, J.D., M.P.H. Editor, *Masks, Journal of Black Health Perspectives*
Hunt, Jane V., Ph.D. University of California, Berkeley
Jaffe, Robert, M.D. University of California, San Francisco
Kramer, Marcia, M.A. Swarthmore College, and National Bureau of Economic Research
Margolis, Alan, M.D. University of California, San Francisco
Marks, F. Raymond, J.D. University of California, Berkeley
Nader, Laura, Ph.D. University of California, Berkeley
Nelson, Nicholas, M.D. Hershey Medical Center, Pennsylvania State University
Perlman, David, Science Editor, *San Francisco Chronicle*
Poirier, Teresa, R.N. University of California, San Francisco
Powell, Gloria, M.D. University of California, Los Angeles
Smith, Clement, M.D. Harvard Medical School, Boston

APPENDIX B

Collated Responses to Four Clinical Questions
    The collated responses of the Conference participants have been assembled in Question I to IV. These constitute a complete and organized listing of all the points of reference used to define particular positions while obliterating the internal consistency of each of the positions. This is done in order to display a spectrum of reference

points applicable to the problems. They are organized into five categories. The first three refer directly to the infant's condition: *general physical status, general human status,* and *medical indications.* The fourth and fifth categories refer to conditions external to the infant: *family situation* and *miscellaneous conditions* (*e.g.,* review, consultation, informed consent, and open policy).

Each condition has been given equal weight and has been phrased as if qualifying a *yes* response. All the participants answered *yes* to questions one and two, and nearly everyone answered *yes* to questions three and four. Where the answers were *no* or *uncertain,* the reasons for these are given in footnotes.

QUESTION 1:
"Is It Ever Right Not to Resuscitate an Infant at Birth?"[a]

Child's Situation
General Physical Status
   1. If baby is dying (or is "dead") and there is no hope of correcting the present lethal situation or foreseeable related complications so that, if resuscitated, the baby would probably die in infancy
   2. If baby is in pain which resuscitation would only prolong

General "Human" Status
   1. If the quality of life is and will be intolerable as judged by most reasonable men (the infant's life will predictably involve greater suffering than happiness and it will probably be without self-awareness or socializing capacities)
   2. If the infant has no chance (or small chance) of normal life
   3. If the infant is clearly below human standards for meaningful life

Medical Indications
   If the infant:
   1. is anencephalic
   2. has severe central nervous system disorders
   3. has gross physical anomalies (*e.g.,* no limbs)
   4. has a (large) myelomeningocele
   5. has Down's syndrome (and other chromosomal abnormalities)
   6. is extremely premature or has an extremely low birthweight (3 cut-off points proposed: 900 gm; 750 gm or 26 weeks' gestation; 500 gm or 22 weeks' gestation)
   7. has (major) hydrocephaly

a Response: Unanimous "Yes."

8. has experienced a catastrophe in the birth canal
9. is porencephalic
10. has multiple absence of sense organs
11. is dead as evidenced by tissue decay
12. had no fetal heartbeat for more than five minutes before birth

Family Situations[b]
If the death of the infant would:
1. minimize the suffering of the parents
2. avoid unbearable financial costs to the family
3. avoid emotional burdens on its siblings
If the parents already have a defective child

Miscellaneous Conditions[b]
1. If, insofar as possible, the parents participate in the decision
2. If there is an open and consistent delivery room policy about nonresuscitation
3. If there is prior informed consent from the parents not to resuscitate under specific conditions
4. If delivery room policy of nonresuscitation is kept flexible in response to the state of the art
5. If, insofar as possible, an advocate for the infant assists in the decision whether or not to resuscitate
6. If costs to the State of the infant's survival are considered
7. If, in cases where gross and obvious structural anomalies are not present, decisions not to resuscitate are reviewed by a board of physicians and others

QUESTION 2:
"IS IT EVER RIGHT TO WITHDRAW LIFE SUPPORT
FROM A CLEARLY DIAGNOSED POOR-PROGNOSIS INFANT?"[c]
Child's Situation

General Physical Status
If the infant:
1. is diagnosable as having gross defects (coupled with item 2 or 3)
2. is slowly dying and continued therapy would only delay death
3. is and will remain unable to sustain itself
4. will suffer continued pain as a result of therapy

General "Human" Status
1. If the quality of life of the infant is and will be intolerable as judged by most reasonable men (the infant's life will predictably in-

b All conditions listed here presuppose that the infant is seriously defective.
c Response: Unanimous "Yes."

volve more suffering than happiness and it will probably be without self-awareness or socializing capacities)
2. If the infant will be totally handicapped and dependent
3. If the infant will be markedly impaired with small chance for a normal existence
4. If the infant is defective and *unwanted* by its parents and unneeded by society

Medical Indications
If the infant:
1. has suffered irreparable damage to crucial organs, especially the brain
2. has myelomeningocele with no cord or bladder function
3. has hypoplastic (dysplastic) kidneys
4. is unable to be weaned from respirator
5. has a genetic defect linked to severe mental retardation requiring institutionalization
6. has hypoplastic lungs
7. has cardiac abnormalities for which no corrective or palliative treatment is possible
8. suffers from short gut syndrome
9. is anencephalic

Family Situations[d]
1. If the survival of the infant would threaten the quality of life of the parents and the family as a whole
2. If the survival of the infant would impose excessive financial costs on the family
3. If the parents desire a more speedy death for the dying infant
4. If the parents do not want to rear a severely handicapped child
5. If the parents are judged unable to nurture a severely handicapped infant

Miscellaneous Conditions
1. If the parents participate in the decision
2. If court arbitration is employed to resolve conflict between the physician and the family regarding the decision
3. If strong and continued support is available to parents who decide to have life support withdrawn
4. If there is an open and consistent nursery policy regarding such decisions
5. If the obligation to give care and comfort is understood to continue until death occurs

d All conditions listed here presuppose that the infant is severely defective.

6. If the death of the infant would serve to overcome the demoralizing effect on the nursery staff of prolonged treatment of a hopeless case

QUESTION 3:
"IS IT EVER RIGHT TO INTERVENE
DIRECTLY TO KILL A SELF-SUSTAINING INFANT?"[e]

Child's Situation

General Physical Status
If the infant:
1. is irretrievably dying a lingering death
2. is dying painfully or is in extreme pain or its life would be pain-ridden
3. has gross physical anomalies

General "Human" Status
1. If the quality of life of the infant is and will be intolerable as judged by most reasonable men (the infant's life will predictably involve more suffering than happiness and it will probably be without self-awareness and socializing capacities)
2. If the infant will be totally (or markedly) handicapped and dependent
3. If the infant is defective and unwanted by parents and unneeded by society

Medical Indications
If the infant:
1. is anencephalic
2. is hydrocephalic with little or no cortex
3. has massive brain damage
4. has a flat EEG
5. has severe central nervous system disorders
6. has uncorrectable cardiac abnormalities, *e.g.,* hypoplastic left heart syndrome
7. has Down's syndrome
8. has chromosomal disorders
9. has short gut syndrome

---

e Response: 12 "Yes"; 2 "No"; 2 "Uncertain." Many, especially the physicians who reponded "Yes," indicated that they would not do it themselves, but would not condemn another for doing it. Others indicated that their "Yes" was intellectual, but that they were emotionally uncomfortable with the action. Those who responded "No" cited two reasons: *Subjectively* impossible for respondent; only passive euthanasia is permissible. The respondent who was "uncertain" felt that, although the act intends mercy, society seems wisely unwilling to approve of this kind of power in the hands of physicians.

Family Situations[f]
1. If the quality of parental life is threatened by the continued survival of the infant
2. If the quality of familial life is threatened by the survival of the infant
3. If the parents desire the death of the infant

Miscellaneous Conditions[f]
1. If the parents consent to the action
2. If the obligation to provide care and comfort is understood to continue to the moment of death
3. If the decision is understood to be the responsibility of the physician, advised by a neutral party, and made known to the parents
4. If, *in loco parentis,* the decision is made by the physician and a court-appointed guardian
5. If where feasible, there is prior review of the decision by a committee
6. If, where feasible, the responsible physician consults with an experienced colleague
7. If consideration is given to overcoming the demoralizing effect on the nursery staff of prolonged or cruel dying
8. If the means is chosen primarily because it offers the least painful death to the infant and secondarily because it offers the least suffering to those around the dying infant
9. If the parents administered the syringe of KC1 prepared by the judge with all the lawyers, priests, economists, psychologists, and journalists within a 50-mile radius as witnesses and no physicians, nurses, or medical or nursing students were allowed to be present

QUESTION 4:
"IS IT EVER RIGHT TO DISPLACE POOR-PROGNOSIS INFANT A IN ORDER
TO PROVIDE INTENSIVE CARE TO BETTER-PROGNOSIS INFANT B?"[g]

Child's Situation

General Physical Status
1. If there is a gross difference in prognosis between the two infants
2. If infant A is *dying* and infant B is truly viable (intact or nearly so)
3. If infant B would die without intensive care but otherwise has a good chance for intact survival

f All conditions listed here presuppose that the infant is severely defective.
g Response: 18 "Yes": 2 "No." The two respondents who answered "No" cited that this is not a moral problem but a practical one with no right-or-wrong and that policies favoring such actions can be too easily abused. All conditions listed here presuppose that the infant is severely defective.

4. If infant A is in pain, has been treated for a reasonable period, and has a certain prognosis for a grossly abnormal life if it survives

General "Human" Status
1. If infant A would certainly have a poor-quality, grossly abnormal life
2. If infant A has at best a 5% chance for meaningful existence

Medical Indications
1. If infant A is less than 1,000 gm, has severe respiratory distress syndrome, and has had a central nervous system hemorrhage, and infant B is 1,800 gm and has severe respiratory distress syndrome[1]
2. If infant A is hopelessly toxoplasmic and infant B has neonatal tetanus[1]
3. If infant A is older, exposed to prior stress, less mature, and diagnosed as having presumed brain damage and infant B is younger, more mature, and has no presumed brain damage[1]

Family Situation
1. If infant A's survival would have a negative impact on its family's social and economic condition

Miscellaneous Conditions
1. If parental consent is obtained where possible
2. If certain *policy matters* are operative: (a) Some form of prior review mechanism is required (*e.g.,* peer, arbitration panel, neutral party, court order); (b) there is demonstrable certitude that displacement is necessary to provide care for infant B and that there is a vast difference in prognosis for the two infants; (c) the policy for allocating resources is clear, open, and cautiously applied; (d) there is a clear impartially applied formula for calculating who stands to gain the most from the therapy
3. If the situation is an emergency, the physician must decide and act on the basis of putting his efforts where they are likely to produce the greatest benefit to the greatest number

Additional Comments
1. This situation is not a *moral* problem, but a tragic condition which must be approached practically (like the classic life-boat situation with too many wanting to board). The moral problem would appear on the larger social scale if an affluent society permits such needless, tragic scarcity of resources to be chronic.
2. Displacement is permissible in this situation but would not be so for adults who must be given security that they will not be "bumped" from life support machines to accommodate "better-prognosis" late-comers.

6 Abortion, Euthanasia, and Care of Defective Newborns

JOHN FLETCHER, PH.D.

Although short, Fletcher's article presents a summary statement of basic ethical issues involved in the question of selective euthanasia of the newly born infant. The core problem is will a willingness to accept arguments for genetically indicated abortion tend to influence our treatment of defective neonates. The point is an important one, and Fletcher develops his view through an evaluation of an approving view, a disapproving view, and his own middle position. The article is particularly good in indicating the social context and potential ethical implications of such decisions which present an alternative to a strict right-to-life argument.

Rev. John Fletcher is the Director of Interfaith Metropolitan Theological Education, Inc. in Washington, D.C.

The medical literature has increasingly dealt with the utility and versatility of prenatal diagnosis for the detection of a variety of inborn errors of metabolism,[1-6] chromosomal abnormalities and variants[7-12] and polygenic conditions (e.g., spina bifida anencephaly).[13-14] Most of these diagnostic procedures entail amniocentesis, but some, such as the recent maternal serum test for indexes of neural-tube abnormalities,[15,16] require confirmation by invasive technics only if the test is positive. In every instance, the information obtained about the fetus affords the parents and attending physician data that they may use to decide whether or not to abort the fetus.

Although the proper use of these diagnostic findings in deciding on elective abortion is controversial,[17,18] my purpose here is to consider a different issue: does the ethical reasoning that is applied to prenatal management bear any relation to decision-making about survival after the birth of the infant? The ethical problem facing the medical profession is simply this: how should physicians and parents now understand their obligation to care for the newborn defective infant in the light of arguments for genetically indicated abortion after amniocentesis? To be consistent, a person might ask whether the arguments that support abortion after prenatal diagnosis of genetic disease also support euthanasia of the same infants who slip through that screen and are born. The debate in ethics is at present polarized between a disapproving view, which tends to equate genetically indicated abortion with infanticide,[19-21] and an approving view, which tends toward equating the morality of abortion and selective euthanasia of the defective newborn.[22,23]

A third position, for which I argue, accentuates parental freedom to participate in life-and-death decisions independently in both the prenatal and postnatal situations, accepts abortion of a seriously defective fetus, but disapproves euthanasia of defective newborns.

Paul Ramsey, an exponent of the first view,[20] rejects arguments for abortion that are based upon a positive prenatal diagnosis of a severe fetal disease and the socioeconomic harm that will be done to the family, because he holds that the same arguments might be used under similar circumstances to justify "infanticide." Infanticide is his term for deliberately bringing about the death of a newborn defective infant. His position is built upon the presupposition that there are no clear-cut moral differences between abortion and infanticide in the same disease. In this perspective, abortion for Lesch-Nyhan syndrome, Tay-Sachs disease, or other lethal diseases would be invalid since these justifications could be used for killing the same infant born without benefit of prenatal diagnosis.

In an earlier essay, Ramsey explained that his method of testing right and wrong action in such cases was fashioned upon the ethical measure of "universalizability." This test asks, "what would be the case if everyone in a morally relevant, like situation did as I am doing" (whether there is any tendency for them to do so or not)?[24] Ramsey is primarily interested in using this method for ethical appraisal of genetically indicated abortions. He is not saying that we will necessarily begin to commit infanticide because we do abortions to prevent genetic disorders. He is saying that if we would not do the infanticides, we should not be doing the abortions. In short, there are means that are in themselves wrong regardless of the good ends that may be desired through them.

Joseph Fletcher's arguments, leading the other side of the debate, not only support abortion after prenatal diagnosis, but also advocate setting aside traditional restraints against euthanasia for defective infants. In an article on ethics and euthanasia,[25] Fletcher reasons that a decision to abort a defective fetus, which is "subhuman life," is logically of the same order as a decision to end a "subhuman life in extremis" in old age. Discussing euthanasia, he abhors moral distinctions between acts of commission and omission ("allowing to die"), whether concerned with the newborn or the terminally ill older person. His thought clearly relates the ethical arguments for genetically indicated abortion to justifications for euthanasia of newborn defective infants:

> If we are morally obliged to put an end to a pregnancy when an amniocentesis reveals a terribly defective fetus, we are equally obliged to put an end to a patient's hopeless misery when a brain scan reveals that a patient with cancer has advanced brain metastases.
>
> Furthermore . . . it is morally evasive and disingenuous to suppose that we can condemn or disapprove positive acts of care and compassion but in spite of that approve negative strategies to achieve exactly the same purpose. This contradiction has equal force whether the euthanasia comes at the fetal point on life's spectrum or at some terminal point post-natally.[25]

The bearing of Fletcher's thought on the problem here, in contrast to Ramsey's, is that if we would do abortions based on prenatal diagnosis, we should be active in ending the suffering of infants born with the same condition. To do less — to allow a newborn with severe disease to die by withholding support — is in this view hypocritical, since we would be active in ending the same life in utero.

He comes to his conclusions in ethics through the reasoning of consequentialism, for which "only the end (a proportionate good) makes sense of what we do (the means)."[22] He does not mean that any end justifies any means, and he so states earlier in the passage just quoted. When a fetus or newborn is seriously incapacitated, however, the human harm prevented and suffering relieved by abortion and euthanasia justifies these actions.

In spite of the polar difference between the two positions, the positions exhibit an interesting similarity. Each is based in part on the assumption that the prenatal and postnatal situations are so similar that ethical behavior in the former determines ethical guidance in the latter. Each ethicist, in his own terms, sees the fetus on the same level of value and dignity as he sees the newborn infant. Ramsey is unequivocal about the status of a fetus as a fellow human being whose irreducible dignity derives from God.[19,24] Fletcher's reflections on "indicators of humanhood" lead him to delay conferral of human status upon the fetus or the infant until qualitative and quantitative measures can be made.[26] In the former view, humanhood is a free gift of God; in the latter view, humanhood is a human choice. In both views, moral guidance is given to be consistent in actions before and after birth.

My task is to show that there are morally relevant differences between abortion and euthanasia, even when one considers them for the same depth of impairment. The task is complex because an understanding of human development and growth as a process militates against lifting up one sequence of development too far out of relation with another. The defective newborn infant is the same being, but at a different stage of development, as a fetal candidate for prenatal diagnosis. Yet the post-natal situation in which parents and physicians face life-and-death decisions for extremely ill infants has features of a kind different from decision making about abortion.

What are the differences? One is that the separate physical existence of the infant, apart from the mother, confronts parents, physicians and legal institutions with independent moral claims for care and support. A newborn infant is clearly a patient. The movement of the fetus prepares the parents emotionally for the acceptance of the infant as a separate individual, but before extra-uterine viability the well-being of the fetus should not be considered independently from the mother's condition. H. T. Englehardt has argued persuasively on the difference between a fetus and an infant, noting that "as soon as the fetus actualizes its potential viability, it can play a full social role and can be understood as a person."[27] A moderate ethical stance will avoid the one extreme of regarding every fetus as already a human being with rights and the other extreme of withholding human status until quality-of-life standards have been passed. The former extreme provides no rational grounds for the legitimate interests of parents, family and society to be expressed and guided in abortion decisions. The latter provides no rational ground for the interest of the newborn infant to be expressed in medical decision making. The decision of the United States Supreme Court on abortion responds to the moral imperative that new human life requires a social protection wider than prenatal and medical care, but it observes that before viability, claims for protection cannot be made compellingly without violating the privacy of the mother.[28] In law, the source of the difference between a wider parental freedom before fetal viability and a larger social responsibility after viability lies in the growing moral claims that the new human life makes upon society.

The second major difference is the fact that after birth the disease in the infant is more available to physicians for palliation or perhaps even cure. Confrontation with disease in an independently existing life requires physicians to respond within their obligations to heal and to relieve suffering. The most noteworthy disease at present that can be successfully treated in utero is erythroblastosis fetalis (Rh disease). The moral claim to relieve suffering in a diseased fetus may be more answerable in the future when genetic therapies are possible. For the present, however, the real situation for parents and physicians is that they must wait until birth to respond to the specificity of a disease with decisions to treat or not to treat.[29]

Thirdly, parental acceptance of the infant as a real person is much more developed at birth than in the earlier stages of pregnancy.

 ③

The medical literature amply describes pregnancy as a crisis bringing about in the parents, especially in the mother, a series of behavioral changes that prepare them for caring for the infant.[30-33] We should expect loyalty to the developing life to grow, change, and moderate the ambivalence about the fetus usually present in the parents. An example of the depth of parental loyalty can be seen when parents of a defective newborn "mourn" the loss of the expected healthy child and reconsider acceptance of the child with a defect.[34] Since acceptance of the new life undergoes a development of its own, it should be readily apparent that increasing parental loyalty to the infant constitutes a major difference when one compares abortion of a fetus to euthanasia of an infant.

The effect of these three differences is to establish the newborn infant, even with a serious defect, as a fellow human being who deserves protection on both a legal and an ethical basis, and thus each of the differences contributes to an argument against euthanasia.

Two Father Considerations

Two additional reasons round out the case for opposition to euthanasia. The first is the potential brutalization of those who participate in it. This point has been persuasively made by Bok in a discussion of infanticide,[35] and it is the source of Lorber's rejection of euthanasia as "an extremely dangerous weapon in the hands of unscrupulous individuals."[36] The second reason is the destructive social consequences of changing the ethical ambience of the birth of infants from one of thorough caring for life to one in which the public accepted a policy of euthanasia and supported its legalization. I see no way of making this social change that would not undermine the optimal moral condition for the beginning of life: the experience of trust. Erikson described the "basic trust" required between mother and infant necessary for the first task of forming healthy personality.[37] His thesis was that the task of mothering is strengthened, among other things, by a world view based on confidence that life is good, even though death and tragedy are part of life.

A society that supports acceptance of defective newborns, where reasonably possible, does more to nurture patterns of acceptance in parents and thus reinforces the child's basic trust in the world's trustworthiness.

A brief comment here is appropriate regarding the fact that physicians, parents, and ethicists are confronted with real cases of terribly damaged newborns for whom death is the desirable outcome when therapy either is not available or will only prolong the ordeal without definite ground for hope.[38] In such cases, if we would reject euthanasia because of a negative ethical assessment of the action and its consequences, what should be done? Allowing the infant to die by withholding support while relieving pain is a decision, in my view, that can be ethically justified for reasons of mercy to the infant and relief of meaningless suffering of the parents and medical team. If death is understood as a good outcome in such cases, however difficult the emotional acceptance of the infant's death, parents and physicians do not "do harm" to the infant by the decision, assuming that every reasonable therapeutic step has been taken or evaluated negatively. The crucial difference between euthanasia and allowing to die is that the self-restraint imposed by the latter choice is more consistent with ethical and legal norms that physicians and parents do no harm to the infant.

I have argued here that a decision for abortion after prenatal diagnosis does not necessarily commit parents to one course of action in the care of an infant born with the same degree of illness. The structure of my argument is based upon three differences between the fetus and the infant that are sufficiently grounded in human experience to be verified by observation. A defective newborn is a separate individual, whose disease is available for treatment, and whose parents are prepared by the process of a typical pregnancy to accept the infant. When these human experiences interact with the beliefs and values of the religious and humanistic communities that provide our culture with visions of the ultimately desirable, this interaction produces strong ethical and theologic backing for caring for the infant. If we choose to be shaped by Judeo-Christian visions of the "createdness" of life within which every creature bears the image of God, we ought to care for the defective newborn as if our relation with the Creator depended on the outcome. If we choose to be shaped by visions of the inherent dignity of each member of the human family, no matter what his or her predicament, we ought to care for this defenseless person as if the basis of our own dignity depended on the outcome. We are in a very hazardous situation ethically if we allow abortion for medical reasons in utero and then attempt to usher new life into a thoroughly caring context at birth, unless we are fully responsive to the imperatives to treat disease in the newborn when it appears.

REFERENCES

1. Milunsky A, Littlefield JW, Kanfer JN, et al: Prenatal genetic diagnosis, N Engl J Med 283:1370-1381, 1441-1447, 1498-1504, 1970

2. Gerbie AB, Nadler HL, Gerbie MV: Amniocentesis in genetic counseling: safety and reliability in early pregnancy. Am J Obstet Gynecol 109:765-770, 1971

3. Epstein CJ, Scheider EL, Conte FA, et al: Prenatal detection of genetic disorders. Am J Hum Genet 24:214-226, 1972

4. Valenti C: Antenatal detection of hemoglobinopathies: a preliminary report. Am J Obstet Gynecol 115:851-853, 1973

5. Antenatal Diagnosis. Edited by A Dorfman. Chicago, University of Chicago Press, 1972

6. Antenatal Diagnosis of Genetic Disease. Edited by AEH Emery. Edinburgh, Churchill Livingstone, 1973

7. Stelle MW, Breg WR Jr: Chromosome analysis of human amniotic-fluid cells. Lancet 1:383-385, 1966

8. Jacobson CB, Barter RH: Intrauterine diagnosis and management of genetic defects. Am J Obstet Gynecol 99:796-807, 1967

9. Nadler HL: Antenatal detection of hereditary disorders. Pediatrics 42:912-918, 1968

10. Nadler HL, Gerbie A: Present status of amniocentesis in intrauterine diagnosis of genetic defects. Obstet Cynecol 38:789-799, 1971

11. Milunsky A, Atkins L, Littlefield JW: Polyploidy in prenatal genetic diagnosis. J Pediatr 79:303-305, 1971

12. Idem: Amniocentesis for prenatal genetic studies. Obstet Gynecol 40:104-108, 1972

13. Emery AEH, Eccleston D, Scrimgeour JB, et al: Amniotic fluid composition in malformations of the central nervous system. J Obstet Gynaecol Brit Commonw 79:154-158, 1972

14. Allan LD, Ferguson-Smith MA, Donald I, et al: Amniotic-fluid alphafetoprotein in the antenatal diagnosis of spina bifida. Lancet 2:522-525, 1973

15. Brock DJH, Bolton AE, Monaghan JM: Prenatal diagnosis of anencephaly through maternal serum-alphafetoprotein measurement. Lancet 2:923-924, 1973

16. Brock DJH, Bolton AE, Scrimgeour JB: Prenatal diagnosis of spina bifida and anencephaly through maternal plasma alpha-fetoprotein measurement. Lancet 1:767-769, 1974

17. Gustafson JM: Genetic counseling and the uses of genetic knowledge—an ethical overview, Ethical Issues in Human Genetics. Edited by B Hilton, D Callahan, et al. New York, Plenum Press, 1973, pp 101-112

18. Ramsey P: Screening: an ethicist's view, Ethical Issues in Human Genetics. Edited by B Hilton, D Callahan, et al. New York, Plenum Press, 1973, pp 147-161

19. *Idem:* Feticide/Infanticide upon request. Religion in Life 39:170-186, 1970

20. *Idem:* Abortion. Thomist 37:174-226, 1973

21. Dyck AJ: Perplexities for the would-be liberal in abortion. J Reprod Med 8:351-354, 1972

22. Fletcher J: The Ethics of Genetic Control. Garden City, New York, Doubleday, 1974, pp 121-123, 152-154, 185-187

23. Tooley M: Abortion and infanticide. Phil Pub Affairs 2:37-65, 1972

24. Ramsey P: Reference points in deciding about abortion, The Morality of Abortion. Edited by JT Noonan Jr. Cambridge, Harvard University Press, 1970, pp 60-100

25. Fletcher J: Ethics and euthanasia. Am J Nursing 73:670-675, 1973

26. Fletcher J: Indicators of humanhood: a tentative profile of man. Hastings Center Rep 2(5):1-4, 1972

27. Englehardt HT Jr: Viability, abortion, and the difference between a fetus and an infant. Am J Obstet Gynecol 116:429-434, 1973

28. Roe vs Wade: 410 US 113 (1973)

29. Duff RS, Campbell AGM: Moral and ethical dilemmas in the special-care nursery. N Engl J Med 289:890-894, 1973

30. Kennell JH, Klaus MH: Care of the mother of the high-risk infant. Clin Obstet Gynecol 14:926-954, 1971

31. Bibring GL: Some considerations of the psychological processes in pregnancy. Psychoanal Study Child 14:113-121, 1959

32. Caplan G: Emotional Implication of Pregnancy and Influences on Family Relationships in the Healthy Child. Cambridge, Harvard University Press, 1960

33. Nadelson C: "Normal" and "special" aspects of pregnancy. Obstet Gynecol 41:611-620, 1973

34. Solnit AJ, Stark MH: Mourning and the birth of a defective child. Psychoanal Study Child 16:523-537, 1961

35. Bok S: Ethical problems of abortion. Hastings Center Studies 2:33-52, 1974

36. Lorber J: Selective treatment of myelomeningocele: to treat or not to treat? Pediatrics 53:307-308, 1974

37. Erikson E: The healthy personality. Psychol Issues 1:56-65, 1959

38. McCormick RA: To save or let die: the dilemma of modern medicine. JAMA 229:172-176, 1974

7 Mongolism, Parental Desires, and the Right to Life

JAMES M. GUSTAFSON

This article is a reflection on the well-publicized Johns Hopkins case of a mentally retarded infant with duodenal atresia who was allowed to die because of a parent's decision not to keep the child. Gustafson presents an analysis of the parents' decision and an ethical evaluation of the elements in this decision. Attention is also paid to some of the legal issues involved. In the last half of the article, Gustafson develops two points: whether what one ought to do is determined by what one desires and whether a mentally retarded child has a right to life. Included in this is an ethical evaluation of suffering which adds a new dimension to the general debate.

Dr. James M. Gustafson, Ph.D., is University Professor of Theological Ethics at the University of Chicago.

The Problem

THE FAMILY SETTING

Mother, 34 years old, hospital nurse.
Father, 35 years old, lawyer.
Two normal children in the family.

In late fall of 1963, Mr. and Mrs. ——— gave birth to a premature baby boy. Soon after birth, the child was diagnosed as a "mongoloid" (Down's syndrome) with the added complication of an intestinal blockage (duodenal atresia). The latter could be corrected with an operation of quite nominal

129

risk. Without the operation, the child could not be fed and would die.

At the time of birth Mrs. —— overheard the doctor express his belief that the child was a mongol. She immediately indicated she did not want the child. The next day, in consultation with a physician, she maintained this position, refusing to give permission for the corrective operation on the intestinal block. Her husband supported her in this position, saying that his wife knew more about these things (i.e., mongoloid children) than he. The reason the mother gave for her position—"It would be unfair to the other children of the household to raise them with a mongoloid."

The physician explained to the parents that the degree of mental retardation cannot be predicted at birth—running from very low mentality to borderline subnormal. As he said: "Mongolism, it should be stressed, is one of the milder forms of mental retardation. That is, mongols' IQs are generally in the 50-80 range, and sometimes a little higher. That is, they're almost always trainable. They can hold simple jobs. And they're famous for being happy children. They're perennially happy and usually a great joy." Without other complications, they can anticipate a long life.

Given the parents' decision, the hospital staff did not seek a court order to override the decision (see "Legal Setting" below). The child was put in a side room and, over an 11-day period, allowed to starve to death.

Following this episode, the parents undertook genetic counseling (chromosome studies) with regard to future possible pregnancies.

The Legal Setting

Since the possibility of a court order reversing the parents' decision naturally arose, the physician's opinion in this matter—and his decision not to seek such an order—is central. As he said: "In the situation in which the child has a known, serious mental abnormality, and would be a burden both to the parents financially and emotionally and perhaps to society, I think it's unlikely that the court would sustain an order to operate on the child against the parents' wishes." He went on to say: "I think one of the great difficulties, and I hope [this] will be part of the discussion relative to this child, is what happens in a family where a court order is used as the means of correcting a congenital abnormality. Does that child ever really become an accepted member of the family? And what are all of the feelings, particularly guilt and coercion feelings that the parents must have following that type of extraordinary force that's brought to bear upon them for making them accept a child that they did not wish to have?"

Both doctors and nursing staff were firmly convinced that it was "clearly illegal" to hasten the child's death by the use of medication.

One of the doctors raised the further issue of consent, saying: "Who has the right to decide for a child anyway? . . . The whole way we handle life and death is the reflection of the long-standing belief in this country that children don't have any rights, that they're not citizens, that their parents can decide to kill them or to let them live, as they choose."

The Hospital Setting

When posed the question of whether the case would have been taken to court had the child had a normal IQ, with the parents refusing permission for the intestinal operation, the near unanimous opinion of the doctors: "Yes, we would have tried to override their decision." Asked why, the doctors replied: "When a retarded child presents us with the same problem, a different value system comes in; and not only does the staff acquiesce in the parent's decision to let the child die, but it's probable that the courts would also. That is, there is a different standard. . . . There is this tendency to value life on the basis of intelligence. . . . [It's] a part of the American ethic."

The treatment of the child during the period of its dying was also interesting. One doctor commented on "putting the child in a side room." When asked about medication to hasten the death, he replied: "No one would ever do that. No one would ever think about it, because they feel uncomfortable about it. . . . A lot of the way we handle these things has to do with our own anxieties about death and our own desires to be separated from the decisions that we're making."

The nursing staff who had to tend to the child showed some resentment at this. One nurse said she had great difficulty just in entering the room and watching the child degenerate—she could "hardly bear to touch him." Another nurse, however, said: "I didn't mind coming to work. Because like I would rock him. And I think that kind of helped me some—to be able to sit there and hold him. And he was just a tiny little thing. He was really a very small baby. And he was cute. He had a cute little face to him, and it was easy to love him, you know?" And when the baby died, how did she feel?—"I was glad that it was over. It was an end for him."

The Resolution

This complex of human experiences and decisions evokes profound human sensibilities and serious intellectual examination. One sees in and beyond it dimensions that could be explored by practitioners of various academic disciplines. Many of the standard questions about the ethics of medical care are pertinent, as are questions that have been long discussed by philosophers and theologians. One would have to write a full-length book to plow up, cultivate, and bring to fruition the implications of this experience.

I am convinced that, when we respond to a moral dilemma, the way in which we formulate the dilemma, the picture we draw of its salient features, is largely determinative of the choices we have. If the war in Vietnam is pictured as a struggle between the totalitarian forces of evil seeking to suppress all human values on the one side, and the forces of righteousness on the other, we have one sort of problem with limited choice. If, however, it is viewed as a struggle of oppressed people to throw off the shackles of colonialism and imperialism, we have another sort of problem. If it is pictured as more complex, the range of choices is wider, and the factors to be considered are more numerous. If the population problem is depicted as a race against imminent self-destruction of the human race, an ethics of survival seems to be legitimate and to deserve priority. If, however, the population problem is depicted more complexly, other values also determine policy, and our range of choices is broader.

One of the points under discussion in this medical case is how we should view it. What elements are in the accounts that the participants give to it? What elements were left out? What "values" did they seem to consider, and which did they seem to ignore? Perhaps if one made a different montage of the raw experience, one would have different choices and outcomes.

Whose picture is correct? It would not be difficult for one moral philosopher or theologian to present arguments that might undercut, if not demolish, the defenses made by the participants. Another moralist might make a strong defense of the decisions by assigning different degrees of importance to certain aspects of the case. The first might focus on the violation of individual rights, in this case the rights of the infant. The other might claim that the way of least possible suffering for the fewest persons over the longest range of time was the commendable outcome of the account as we have it. Both would be accounts drawn by external observers, not by active, participating agents. There is a tradition that says that ethical reflection by an ideal external observer can bring morally right answers. I have an observer's perspective, though not that of an "ideal observer." But I believe that it is both charitable and intellectually important to try to view the events as the major participants viewed them. The events remain closer to the confusions of the raw experience that way; the passions, feelings, and emotions have some echo of vitality remaining. The parents were not without feeling, the nurses not without anguish. The experiences could become a case in which x represents the rights of the infant to life, y represents the consequences of continued life as a mongoloid person, and z represents the consequences of his continued life for the family and the state. But such abstraction has a way of oversimplifying experience. One would "weigh" x against y and z. I cannot reproduce the drama even of the materials I have read, the interviews with doctors and nurses, and certainly even those are several long steps from the thoughts and feelings of the parents and the staff at that time. I shall, however, attempt to state the salient features of the dilemma for its participants; features that are each value laden and in part determinative of their decisions. In the process of doing that for the participants, I will indicate what reasons might justify their decisions. Following that I will draw a different picture of the experience, highlighting different values and principles, and show how this would lead to a different decision. Finally, I shall give the reasons why I, an observer, believe they, the participants, did the wrong thing. Their responsible and involved participation, one must remember, is very different from my detached reflection on documents and interviews almost a decade later.

The Mother's Decision

Our information about the mother's decision is secondhand. We cannot be certain that we have an accurate account of her reasons for not authorizing the surgery that could have saved the mongoloid infant's life. It is not my role to speculate whether her given reasons are her "real motives"; that would involve an assessment of her "unconscious." When she heard the child was probably a mongol, she "expressed some negative feeling" about it, and "did not want a retarded child." Because she was a nurse she understood what mongolism indicated. One reason beyond her feelings and wants is given: to raise a mongoloid child in the family would not be "fair" to the other children. That her decision was anguished we know from several sources.

For ethical reflection, three terms I have quoted are important: "negative feeling," "wants" or "desires," and "fair." We need to inquire about the status of each as a justification for her decision.

What moral weight can a negative feeling bear? On two quite different grounds, weight could be given to her feelings in an effort to sympathetically understand her decision. First, at the point of making a decision, there is always an element of the rightness or wrongness of the choice that defies full rational justification. When we see injustice being done, we have strong negative feelings; we do not need a sophisticated moral argument to tell us that the act is unjust. We "feel" that it is wrong. It might be said that the mother's "negative feeling" was evoked by an intuition that it would be wrong to save the infant's life, and that feeling is a reliable guide to conduct.

Second, her negative response to the diagnosis of mongolism suggests that she would not be capable of giving the child the affection and the care that it would require. The logic involved is an extrapolation from that moment to potential consequences for her continued relationship to the child in the future. The argument is familiar; it is common in the literature that supports abortion on request—"no unwanted child ought to be born." Why? Because unwanted children suffer from hostility and lack of affection from their mothers, and this is bad for them.

The second term is "wants" or "desires." The negative feelings are assumed to be an indication of her desires. We might infer that at some point she said, "I do not want a retarded child." The status of "wanting" is different, we might note, if it expresses a wish before the child is born, or if it expresses a desire that leads to the death of the infant after it is born. No normal pregnant woman would wish a retarded child. In this drama, however, it translates into: "I would rather not have the infant kept alive." Or, "I will not accept parental responsibilities for a retarded child." What is the status of a desire or a want as an ethical justification for an action? To discuss that fully would lead to an account of a vast literature. The crucial issue in this case is whether the existence of the infant lays a moral claim that supersedes the mother's desires.

If a solicitor of funds for the relief of refugees in Bengal requested a donation from her and she responded, "I do not want to give money for that cause," some persons would think her to be morally insensitive, but none could argue that the refugees in Bengal had a moral claim on her money which she was obligated to acknowledge. The existence of the infant lays a weightier claim on her than does a request for a donation. We would not say that the child's right to surgery, and thus to life, is wholly relative to, and therefore exclusively dependent upon, the mother's desires or wants.

Another illustration is closer to her situation than the request for a donation. A man asks a woman to marry him. Because she is asked, she is under no obligation to answer affirmatively. He might press claims upon her—they have expressed love for each other; or they have dated for a long time; he has developed his affection for her on the assumption that her responsiveness would lead to marriage. But none of these claims would be sufficient to overrule her desire not to marry him. Why? Two sorts of reasons might be given. One would refer to potential consequences: a marriage in which one partner does not desire the relationship leads to anxiety and suffering. To avoid needless suffering is obviously desirable. So in this case, it might be said that the mother's desire is to avoid needless suffering and anxiety: the undesirable consequences can be avoided by permitting the child to die.

The second sort of reason why a woman has no obligation to marry her suitor refers to her rights as an individual. A request for marriage does not constitute a moral obligation, since there is no prima facie claim by the suitor. The woman has a right to say no. Indeed, if the suitor sought to coerce her into marriage, everyone would assert that she has a right to refuse him. In our case, however, there are some differences. The infant is incapable of expressing a request or demand. Also, the relationship is different: the suitor is not dependent upon his girl friend in the same way that the infant is dependent upon his mother. Dependence functions in two different senses; the necessary conditions for the birth of the child were his conception and *in utero* nourishment—thus, in a sense the parents "caused" the child to come into being. And, apart from instituting adoption procedures, the parents are the only ones who can provide the necessary conditions for sustaining the child's life. The infant is dependent on them in the sense that he must rely upon their performance of certain acts in order to continue to exist. The ethical question to the mother is, Does the infant's physical life lay an unconditioned moral claim on the mother? She answered, implicitly, in the negative.

What backing might the negative answer be given? The most persuasive justification would come from an argument that there are no unconditioned moral claims upon one when those presumed claims go against one's desires and wants. The claims of another are relative to my desires, my wants. Neither the solicitor for Bengal relief nor the suitor has an unconditioned claim to make; in both cases a desire is sufficient grounds for denying such a claim. In our case, it would have to be argued that the two senses of dependence that the infant has on the mother are not sufficient conditions for a claim on her that would morally require the needed surgery. Since there are no unconditioned claims, and since the conditions in this drama are not sufficient to warrant a claim, the mother is justified in denying permission for the surgery.

We note here that in our culture there are two trends in the development of morality that run counter to each other: one is the trend that desires of the ego are the grounds for moral and legal claims. If a mother does not desire the fetus in her uterus, she has a right to an abortion. The other increasingly limits individual desires and wants. An employer might want to hire only white persons of German ancestry, but he has no right to do so.

The word "fair" appeals to quite different warrants. It would not be "fair" to the other children in the family to raise a mongoloid with them. In moral philosophy, fairness is either the same as justice or closely akin to it. Two traditional definitions of justice might show

how fairness could be used in this case. One is "to each his due." The other children would not get what is due them because of the inordinate requirements of time, energy, and financial resources that would be required if the mongoloid child lived. Or, if they received what was due to them, there would not be sufficient time, energy, and other resources to attend to the particular needs of the mongoloid; his condition would require more than is due him. The other traditional definition is "equals shall be treated equally." In principle, all children in the family belong to a class of equals and should be treated equally. Whether the mongoloid belongs to that class of equals is in doubt. If he does, to treat him equally with the others would be unfair to him because of his particular needs. To treat him unequally would be unfair to the others.

Perhaps "fairness" did not imply "justice." Perhaps the mother was thinking about such consequences for the other children as the extra demands that would be made upon their patience, the time they would have to give the care of the child, the emotional problems they might have in coping with a retarded sibling, and the sense of shame they might have. These consequences also could be deemed to be unjust from her point of view. Since they had no accountability for the existence of the mongoloid, it was not fair to them that extra burdens be placed upon them.

To ask what was due the mongoloid infant raises harder issues. For the mother, he was not due surgical procedure that would sustain his life. He was "unequal" to her normal children, but the fact of his inequality does not necessarily imply that he has no right to live. This leads to a matter at the root of the mother's response which has to be dealt with separately.

She (and as we shall see, the doctors also) assumed that a factual distinction (between normal and mongoloid) makes the moral difference. Factual distinctions do make moral differences. A farmer who has no qualms about killing a runt pig would have moral scruples about killing a deformed infant. If the child had not been mongoloid and had an intestinal blockage, there would have been no question about permitting surgery to be done. The value of the infant is judged to be relative to a quality of its life that is predictable on the basis of the factual evidences of mongolism. Value is relative to quality: that is the justification. Given the absence of a certain quality, the value is not sufficient to maintain life; given absence of a quality, there is no right to physical life. (Questions about terminating life among very sick adults are parallel to this instance.)

What are the qualities, or what is *the* quality that is deficient in this infant? It is not the capacity for happiness, an end that Aristotle and others thought to be sufficient in itself. The mother and the doctors knew that mongoloids can be happy. It is not the capacity for pleasure, the end that the hedonistic utilitarians thought all men seek, for mongoloids can find pleasure in life. The clue is given when a physician says that the absence of the capacity for normal intelligence was crucial. He suggested that we live in a society in which intelligence is highly valued. Perhaps it is valued as a quality in itself, or as an end in itself by some, but probably there is a further point, namely that intelligence is necessary for productive contribution to one's own well-being and to the well-being of others. Not only will a mongoloid make a minimal contribution to his own well-being and to that of others, but also others must contribute excessively to his care. The right of an infant, the value of his life, is relative to his intelligence; that is the most crucial factor in enabling or limiting his contribution to his own welfare and that of others. One has to defend such a point in terms of the sorts of contributions that would be praiseworthy and the sorts of costs that would be detrimental. The contribution of a sense of satisfaction to those who might enjoy caring for the mongoloid would not be sufficient. Indeed, a full defense would require a quantification of qualities, all based on predictions at the point of birth, that would count both for and against the child's life in a cost-benefit analysis.

The judgment that value is relative to qualities is not implausible. In our society we have traditionally valued the achiever more than the nonachievers. Some hospitals have sought to judge the qualities of the contributions of patients to society in determining who has access to scarce medical resources. A mongoloid is not valued as highly as a fine musician, an effective politician, a successful businessman, a civil rights leader whose actions have brought greater justice to the society, or a physician. To be sure, in other societies and at other times other qualities have been valued, but we judge by the qualities valued in our society and our time. Persons are rewarded according to their contributions to society. A defense of the mother's decision would have to be made on these grounds, with one further crucial step. That is, when the one necessary condition for productivity is deficient (with a high degree of certitude) at birth, there is no moral obligation to maintain that life. That the same reasoning would have been sufficient to justify overtly taking the infant's life seems not to have been the case. But that point emerges later in our discussion.

The reliance upon feelings, desires, fairness, and judgments of qualities of life makes sense to American middle-class white families, and anguished decisions can very well be settled in these terms. The choice made by the mother was not that of an unfeeling problem-solving machine, nor that of a rationalistic philosopher operating from these assumptions. It was a painful, conscientious decision, made apparently on these bases. On can ask, of course, whether her physicians should not have suggested other ways of perceiving and drawing the contours of the circumstances, other values and ends that she might consider. But that points to a subsequent topic.

The Father's Decision

The decision of the father is only a footnote to that of the mother. He consented to the choice of not operating on the infant, though he did seek precise information about mongolism and its consequences for the child. He was "willing to go along with the mother's wishes," he "understood her feelings, agreed with them," and was not in a position to make "the same intelligent decision that his wife was making."

Again we see that scientific evidence based on professional knowledge is determinative of a moral decision. The physician was forthright in indicating what the consequences would be of the course of action they were taking. The consequences of raising a mongoloid child were presumably judged to be more problematic than the death of the child.

The Decision of the Physicians

A number of points of reference in the contributions of the physicians to the case study enable us to formulate a constellation of values that determined their actions. After I have depicted that constellation, I shall analyze some of the points of reference to see how they can be defended.

The constellation can be stated summarily. The physicians felt no moral or legal obligation to save the life of a mongoloid infant by an ordinary surgical procedure when the parents did not desire that it should live. Thus, the infant was left to die. What would have been a serious but routine procedure was omitted in this instance on two conditions, both of which were judged to be necessary, but neither of which was sufficient in itself: the mongolism and the parents' desires. If the parents had desired the mongoloid infant to be saved, the surgery would have been done. If the infant had not been mongoloid and the parents had refused permission for surgery to remove a bowel obstruction, the physicians would at least have argued against them and probably taken legal measures to override them. Thus, the value-laden points of reference appear to be the desires of the parents, the mongolism of the infant, the law, and choices about ordinary and extraordinary medical procedures.

1. Desires of Parents

One of the two most crucial points was the obligation the physicians felt to acquiesce to the desires of the parents. The choice of the parents not to operate was made on what the physicians judged to be adequate information: it was an act of informed consent on the part of the parents. There is no evidence that the physicians raised questions of a moral sort with the parents that they subsequently raised among themselves. For example, one physician later commented on the absence of rights for children in our society and in our legal system and on the role that the value of intelligence seems to have in judging worthiness of persons. These were matters, however, that the physicians did not feel obligated to raise with the distressed parents. The physicians acted on the principle that they are only to do procedures that the patient (or crucially in this case, the parents of the patient) wanted. There was no overriding right to life on the part of a mongoloid infant that led them to argue against the parents' desires or to seek a court order requiring the surgical procedure. They recognized the moral autonomy of the parents, and thus did not interfere; they accepted as a functioning principle that the parents have the right to decide whether an infant shall live.

Elaboration of the significance of parental autonomy is necessary in order to see the grounds on which it can be defended. First, the physicians apparently recognized that the conscientious parents were the moral supreme court. There are grounds for affirming the recognition of the moral autonomy of the principal persons in complex decisions. In this case, the principals were the parents: the infant did not have the capacities to express any desires or preferences he might have. The physicians said, implicitly, that the medical profession does not have a right to impose certain of its traditional values on persons if these are not conscientiously held by those persons.

There are similarities, but also differences, between this instance and that of a terminal patient. If the terminally ill patient expresses a desire not to have his life prolonged, physicians recognize his autonomy over his own body and thus feel under no obligation to sustain his life. Our case, however, would be more similar to one in which the terminally ill patient's family decided that no further procedures ought to be used to sustain life. No doubt there are many cases in which the patient is unable to express a preference due to his physical conditions, and in the light of persuasive medical and familial reasons the physician agrees not to sustain life. A difference between our case and that, however, has to be noted in order to isolate what seems to be the crucial point. In the case of the mongoloid infant, a decision is made at the beginning of his life and not at the end; the effect is to cut off a life which, given proper care, could be sustained for many years, rather than not sustaining a life which has no such prospects.

Several defenses might be made of their recognition of the parents' presumed rights in this case. The first is that parents have authority over their children until they reach an age of discretion, and in some respects until they reach legal maturity. Children do not have recognized rights over against parents in many respects. The crucial difference here, of course, is the claimed parental right in this case to determine that an infant shall not live. What grounds might there be for this? Those who claim the moral right to an abortion are claiming the right to determine whether a child shall live, and this claim is widely recognized both morally and legally. In this case we have an extension of that right to the point of birth. If there are sufficient grounds to indicate that the newborn child is significantly abnormal, the parents have the same right as they have when a severe genetic abnormality is detected prenatally on the basis of amniocentesis. Indeed, the physicians could argue that if a mother has a right to an abortion, she also has a right to determine whether a newborn infant shall continue to live. One is simply extending the time span and the circumstances under which this autonomy is recognized.

A second sort of defense might be made: that of the limits of professional competence and authority. The physicians could argue that in moral matters they have neither competence nor authority. Perhaps they would wish to distinguish between competence and authority. They have a competence to make a moral decision on the basis of their own moral and other values, but they have no authority to impose this upon their patients. Morals, they might argue, are subjective matters, and if anyone has competence in that area, it is philosophers, clergymen, and others who teach what is right and wrong. If the parents had no internalized values that militated against their decision, it is not in the province of the physicians to tell them what they ought to do. Indeed, in a morally pluralistic society, no one group or person has a right to impose his views on another. In this stronger argument for moral autonomy no physician would have any authority to impose his own moral values on any patient. A social role differentiation is noted: the medical profession has authority only in medical matters—not in moral matters. Indeed, they have an obligation to indicate what the medical alternatives are in order to have a decision made by informed consent, but insofar as moral values or principles are involved in decisions, these are not within their professional sphere.

An outsider might ask what is meant by authority. He might suggest that surely it is not the responsibility (or at least not his primary responsibility) or the role of the physician to make moral decisions, and certainly not to enforce his decisions on others. Would he be violating his role if he did something less determinative than that, namely, in his counseling indicate to them what some of the moral considerations might be in choosing between medical alternatives? In our case the answer seems to be yes. If the principals desire moral counseling, they have the freedom to seek it from whomsoever they will. In his professional role he acknowledges that the recognition of the moral autonomy of the principals also assumes their moral self-sufficiency, that is, their capacities to make sound moral decisions without interference on his part, or the part of any other persons except insofar as the principals themselves seek such counsel. Indeed, in this case a good deal is made of the knowledgeability of the mother particularly, and this assumes that she is morally, as well as medically, knowledgeable. Or, if she is not, it is still not the physician's business to be her moral counselor.

The physicians also assumed in this case that the moral autonomy of the parents took precedence over the positive law. At least they felt no obligation to take recourse to the courts to save the life of this infant. On that issue we will reflect more when we discuss the legal point of reference.

Another sort of defense might be made. In the order of society, decisions should be left to the most intimate and smallest social unit involved. That is the right of such a unit, since the interposition of outside authority would be an infringement of its freedom. Also, since the family has to live with the consequences of the decision, it is the right of the parents to determine which potential consequences they find most desirable. The state, or the medical profession, has no right to interfere with the freedom of choice of the family. Again, in a formal way, the argument is familiar; the state has no right to interfere with the determination of what a woman wishes to do with her body, and thus antiabortion laws are infringements of her freedom. The determination of whether an infant shall be kept alive is simply an extension of the sphere of autonomy properly belonging to the smallest social unit involved.

In all the arguments for moral autonomy, the medical fact that the infant is alive and can be kept alive does not make a crucial difference. The defense of the decision would have to be made in this way: if one grants moral autonomy to mothers to determine whether they will bring a fetus to birth, it is logical to assume that one will grant the same autonomy after birth, at least in instances where the infant is abnormal.

We have noted in our constellation of factors that the desire of the parents was a necessary but not a sufficient condition for the decisions of the physicians. If the infant had not been mongoloid, the physicians would not have so readily acquiesced to the parents' desires. Thus, we need to turn to the second necessary condition.

The second crucial point is that the infant was a mongoloid. The physicians would not have acceded to the parents' request as readily if the child had been normal; the parents would have authorized the surgical procedure if the child had been normal. Not every sort of abnormality would have led to the same decision on the part of the physicians. Their appeal was to the consequences of the abnormality of mongolism: the child would be a burden financially and emotionally to the parents. Since every child, regardless of his capacities for intelligent action, is a financial burden, and at least at times an emotional burden, it is clear that the physicians believed that the quantity or degree of burden in this case would exceed any benefits that might be forthcoming if the child were permitted to live. One can infer that a principle was operative, namely, that mongoloid infants have no inherent right to life; their right to life is conditional upon the willingness of their parents to accept them and care for them.

Previously we developed some of the reasons why a mongoloid infant was judged undesirable. Some of the same appeals to consequences entered into the decisions of the physicians. If we are to seek to develop reasons why the decisions might be judged to be morally correct, we must examine another point, namely, the operating definition of "abnormal" or "defective." There was no dissent to the medical judgment that the infant was mongoloid, though precise judgments about the seriousness of the child's defect were not possible at birth.

Our intention is to find as precisely as possible what principles or values might be invoked to claim that the "defectiveness" was sufficient to warrant not sustaining the life of this infant. As a procedure, we will begin with the most general appeals that might have been made to defend the physician's decision in this case. The most general principle would be that any infant who has any empirically verifiable degree of defect at birth has no right to life. No one would apply such a principle. Less general would be that all infants who are carriers of a genetic defect that would have potentially bad consequences for future generations have no right to life. A hemophiliac carrier would be a case in point. This principle would not be applicable, even if it were invoked with approval, in this case.

Are the physicians prepared to claim that all genetically "abnormal" infants have no claim to life? I find no evidence that they would. Are they prepared to say that where the genetic abnormality affects the capacity for "happiness" the infant has no right to live?

Such an appeal was not made in this case. It appears that "normal" in this case has reference to a capacity for a certain degree of intelligence.

A presumably detectable physical norm now functions as a norm in a moral sense, or as an ideal. The ideal cannot be specified in precise terms, but there is a vague judgment about the outer limits beyond which an infant is judged to be excessively far from the norm or ideal to deserve sustenance. Again, we come to the crucial role of an obvious sign of the lack of capacity for intelligence of a certain measurable sort in judging a defect to be intolerable. A further justification of this is made by an appeal to accepted social values, at least among middle- and upper-class persons in our society. Our society values intelligence; that value becomes the ideal norm from which abnormality or deficiencies are measured. Since the infant is judged not to be able to develop into an intelligent human being (and do all that "normal" intelligence enables a human being to do), his life is of insufficient value to override the desires of the parents not to have a retarded child.

Without specification of the limits to the sorts of cases to which it could be applied, the physicians would probably not wish to defend the notion that the values of a society determine the right to life. To do so would require that there be clear knowledge of who is valued in our society (we also value aggressive people, loving people, physically strong people, etc.), and in turn a procedure by which capacities for such qualities could be determined in infancy so that precise judgments could be made about what lives should be sustained. Some members of our society do not value black people; blackness would obviously be an insufficient basis for letting an infant die. Thus, in defense of their decision the physicians would have to appeal to "values generally held in our society." This creates a different problem of quantification: what percentage of dissent would count to deny a "general" holding of a value? They would also have to designate the limits to changes in socially held values beyond which they would not consent. If the parents belonged to a subculture that valued blue eyes more than it valued intelligence, and if they expressed a desire not to have a child because it had hazel eyes, the problem of the intestinal blockage would not have been a sufficient condition to refrain from the surgical procedure.

In sum, the ideal norm of the human that makes a difference in judging whether an infant has the right to life in this case is "the capacity for normal intelligence." For the good of the infant, for the sake of avoiding difficulties for the parents, and for the good of society, a significant deviation from normal intelligence, coupled with the appropriate parental desire, is sufficient to permit the infant to die.

3. The
Law

a)   A third point of reference was the law. The civil law and the courts figure in the decisions at two points. First, the physicians felt no obligation to seek a court order to save the life of the infant if the parents did not want it. Several possible inferences might be drawn from this. First, one can infer that the infant had no legal right to life; his legal right is conditional upon parental desires. Second, as indicated in the interviews, the physicians believed that the court would not insist upon the surgical procedure to save the infant since it was a mongoloid. Parental desires would override legal rights in such a case. And third (an explicit statement by the physician), if the infant's life had been saved as the result of a court order, there were doubts that it would have been "accepted" by the parents. Here is an implicit appeal to potential consequences: it is not beneficial for a child to be raised by parents who do not "accept" him. The assumption is that they could not change their attitudes.

If the infant had a legal right to life, this case presents an interesting instance of conscientious objection to law. The conscientious objector to military service claims that the power of the state to raise armies for the defense of what it judges to be the national interest is one that he conscientiously refrains from sharing. The common good, or the national interest, is not jeopardized by the granting of a special status to the objector because there are enough persons who do not object to man the military services. In this case, however, the function of the law is to protect the rights of individuals to life, and the physician-objector is claiming that he is under no obligation to seek the support of the legal system to sustain life even when he knows that it could be sustained. The evidence he has in hand (the parental desire and the diagnosis of mongolism) presumably provides sufficient moral grounds for his not complying with the law. From the standpoint of ethics, an appeal could be made to conscientious objection. If, however, the appropriate law does not qualify its claims in such a way as to (a) permit its nonapplicability in this case or (b) provide for exemption on grounds of conscientious objection, the objector is presumably willing to accept the consequences for his conscientious decision. This would be morally appropriate. The physician believed that the court would not insist on saving the infant's life, and thus he foresaw no great jeopardy to himself in following conscience rather than the law.

b)   The second point at which the law figures is in the determination of how the infant should die. The decision not to induce death was made in part in the face of the illegality of overt euthanasia (in part, only, since also the hospital staff would "feel uncomfortable" about hastening the death). Once the end or purpose of action (or inaction) was judged to be morally justified, and judged likely to be free from legal censure, the physicians still felt obliged to achieve that purpose within means that would not be subject to legal sanctions. One can only speculate whether the physicians believed that a court that would not order an infant's life to be saved would in turn censure them for overtly taking the life, or whether the uncomfortable feelings of the hospital staff were more crucial in their decision. Their course of decisions could be interpreted as at one point not involving obligation to take recourse to the courts and at the other scrupulously obeying the law. It should be noted, however, that there is consistency of action on their part; in neither instance did they intervene in what was the "natural" course of developments. The moral justification to fail to intervene in the second moment had to be different from that in the first. In the first it provides the reasons for not saving a life; in the second, for not taking a life. This leads to the last aspect of the decisions of the physicians that I noted, namely, that choices were made between ordinary and extraordinary means of action.

There is no evidence in the interviews that the language of ordinary and extraordinary means of action was part of the vocabulary of the medical staff. It is, however, an honored and useful distinction in Catholic moral theology as it applies to medical care. The principle is that a physician is under no obligation to use extraordinary means to sustain life. The difficulty in the application of the principle is the choice of what falls under ordinary and what under extraordinary means. Under one set of circumstances a procedure may be judged ordinary, and under another extraordinary. The surgery required to remove the bowel obstruction in the infant was on the whole an ordinary procedure; there were no experimental aspects to it, and there were no unusual risks to the infant's life in having it done. If the infant had had no other genetic defects, there would have been no question about using it. The physicians could make a case that when the other defect was mongolism, the procedure would be an extraordinary one. The context of the judgment about ordinary and extraordinary was a wider one than the degree of risk to the life of the patient from surgery. It included his other defect, the desires of the family, the potential costs to family and society, etc. No moralists, to my knowledge, would hold them culpable if the infant were so deformed that he would be labeled (nontechnically) a monstrosity. To heroically maintain the life of a monstrosity as long as one could would be most extraordinary. Thus, we return to whether the fact of mongolism and its consequences is a sufficient justification to judge the livesaving procedure to be extraordinary in this instance. The physicians would argue that it is.

The infant was left to die with a minimum of care. No extraordinary means were used to maintain its life once the decision not to operate had been made. Was it extraordinary not to use even ordinary procedures to maintain the life of the infant once the decision not to operate had been made? The judgment clearly was in the negative. To do so would be to prolong a life that would not be saved in any case. At that point the infant was in a class of terminal patients, and the same justifications used for not prolonging the life of a terminal patient would apply here. Patients have a right to die, and physicians are under no moral obligation to sustain their lives when it is clear that they will not live for long. The crucial difference between a terminal cancer patient and this infant is that in the situation of the former, all procedures which might prolong life for a goodly length of time are likely to have been exhausted. In the case of the infant, the logic of obligations to terminal patients takes its course as a result of a decision not to act at all.

To induce death by some overt action is an extraordinary procedure. To justify overt action would require a justification of euthanasia. This case would be a good one from which to explore euthanasia from a moral point of view. Once a decision is made not to engage in a life-sustaining and lifesaving procedure, has not the crucial corner been turned? If that is a reasonable and moral thing to do, on what grounds would one argue that it is wrong to hasten death? Most obviously it is still illegal to do it, and next most obviously people have sensitive feelings about taking life. Further, it goes against the grain of the fundamental vocation of the medical profession to maintain life. But, of course, the decision not to operate also goes against that grain. If the first decision was justifiable, why was it not justifiable to hasten the death of the infant? We can only assume at this point traditional arguments against euthanasia would have been made.

The Decisions of the Nurses

The nurses, as the interviews indicated, are most important for their expressions of feelings, moral sensibilities, and frustrations. They demonstrate the importance of deeply held moral convictions and of profound compassion in determining human responses to ambiguous circumstances. If they had not known that the infant could have survived, the depth of their frustrations and feelings would have not been so great. Feelings they would have had, but they would have been compassion for an infant bound to die. The actual range of decision for them was clearly circumscribed by the role definitions in the medical professions; it was their duty to carry out the orders of the physicians. Even if they conscientiously believed that the orders they were executing were immoral, they could not radically reverse the course of events; they could not perform the required surgery. It was their lot to be the immediate participants in a sad event but to be powerless to alter its course.

It would be instructive to explore the reasons why the nurses felt frustrated, were deeply affected by their duties in this case. Moral convictions have their impact upon the feelings of persons as well as upon their rational decisions. A profound sense of vocation to relieve suffering and to preserve life no doubt lies behind their responses, as does a conviction about the sanctity of human life. For our purposes, however, we shall leave them with the observation that they are the instruments of the orders of the physicians. They have no right of conscientious objection, at least not in this set of circumstances.

Before turning to another evaluative description of the case, it is important to reiterate what was said in the beginning. The decisions by the principals were conscientious ones. The parents anguished. The physicians were informed by a sense of compassion in their consent to the parents' wishes; they did not wish to be party to potential suffering that was avoidable. Indeed, in the way in which they formulated the dilemma, they did what was reasonable to do. They chose the way of least possible suffering to the fewest persons over a long range of time, with one exception, namely, not taking the infant's life. By describing the dilemma from a somewhat different set of values, or giving different weight to different factors, another course of action would have been reasonable and justified. The issue, it seems to me, is at the level of what is to be valued more highly, for one's very understanding of the problems he must solve are deeply affected by what one values most.

The Dilemma from a Different Moral Point of View

Wallace Stevens wrote in poetic form a subtle account of "Thirteen Ways of Looking at a Blackbird." Perhaps there are 13 ways of looking at this medical case. I shall attempt to look at it from only one more way. By describing the dilemma from a perspective that gives a different weight to some of the considerations that we have already exposed, one has a different picture, and different conclusions are called for. The moral integrity of any of the original participants is not challenged, not because of a radical relativism that says they have their points of view and I have mine, but out of respect for their conscientiousness. For several reasons, however, more consideration ought to have been given to two points. A difference in evaluative judgments would have made a difference of life or death for the infant, depending upon: (1) whether what one ought to do is determined by what one desires to do and (2) whether a mongoloid infant has a claim to life.

To restate the dilemma once again: If the parents had "desired" the mongoloid infant, the surgeons would have performed the operation that would have saved its life. If the infant had had a bowel obstruction that could be taken care of by an ordinary medical procedure, but had not been a mongoloid, the physicians would probably have insisted that the operation be performed.

Thus, one can recast the moral dilemma by giving a different weight to two things: the desires of the parents and the value or rights of a mongoloid infant. If the parents and the physicians believed strongly that there are things one ought to do even when one has no immediate positive feelings about doing them, no immediate strong desire to do them, the picture would have been different. If the parents and physicians believed that mongoloid children have intrinsic value, or have a right to life, or if they believed that mongolism is not sufficiently deviant from what is normatively human to merit death, the picture would have been different.

Thus, we can redraw the picture. To be sure, the parents are ambiguous about their feelings for a mongoloid infant, since it is normal to desire a normal infant rather than an abnormal infant. But (to avoid a discussion of abortion at this point) once an infant is born its independent existence provides independent value in itself, and those who brought it into being and those professionally responsible for its care have an obligation to sustain its life regardless of their negative or ambiguous feelings toward it. This probably would have been acknowledged by all concerned if the infant had not been mongoloid. For example, if the pregnancy had been accidental, and in this sense the child was not desired, and the infant had been normal, no one would have denied its right to exist once it was born, though some would while still *in utero*, and thus would have sought an abortion. If the mother refused to accept accountability for the infant, alternative means of caring for it would have been explored.

To be sure, a mongoloid infant is genetically defective, and raising and caring for it put burdens on the parents, the family, and the state beyond the burdens required to raise a normal infant. But a mongoloid infant is human, and thus has the intrinsic value of humanity and the rights of a human being. Further, given proper care, it can reach a point of significant fulfillment of its limited potentialities; it is capable of loving and responding to love; it is capable of realizing happiness; it can be trained to accept responsibility for itself within its capacities. Thus, the physicians and parents have an obligation to use all ordinary means to preserve its life. Indeed, the humanity of mentally defective children is recognized in our society by the fact that we do not permit their extermination and do have policies which provide, all too inadequately, for their care and nurture.

If our case had been interpreted in the light of moral beliefs that inform the previous two paragraphs, the only reasonable conclusion would be that the surgery ought to have been done.

The grounds for assigning the weights I have to these crucial points can be examined. First, with reference simply to common experience, we all have obligations to others that are not contingent upon our immediate desires. When the registrar of my university indicates that senior grades have to be in by May 21, I have an obligation to read the exams, term papers, and senior essays in time to report the grades, regardless of my negative feelings toward those tasks or my preference to be doing something else. I have an obligation to my students, and to the university through its registrar, which I accepted when I assumed the social role of an instructor. The students have a claim on me; they have a right to expect me to fulfill my obligations to them and to the university. I might be excused from the obligation if I suddenly become too ill to fulfill it; my incapacity to fulfill it would be a temporarily excusing condition. But negative feelings toward that job, or toward any students, or a preference for writing a paper of my own at that time, would not constitute excusing conditions. I must consider, in determining what I do, the relationships that I have with others and the claims they have on me by virtue of those relationships.

In contrast to this case, it might be said that I have a contractual obligation to the university into which I freely entered. The situation of the parents is not the same. They have no legal contractual relationship with the infant, and thus their desires are not bound by obligations. Closer to their circumstances, then, might be other family relationships. I would argue that the fact that we brought our children into being lays a moral obligation on my wife and me to sustain and care for them to the best of our ability. They did not choose to be; and their very being is dependent, both causally and in other ways, upon us. In the relationship of dependence, there is a claim of them over against us. To be sure, it is a claim that also has its rewards and that we desire to fulfill within a relationship of love. But until they have reached an age when they can accept full accountability (or fuller accountability) for themselves, they have claims upon us by virtue of our being their parents, even when meeting those claims is to us financially costly, emotionally distressing, and in other ways not immediately desirable. Their claims are independent of our desires to fulfill them. Particular claims they might make can justifiably be turned down, and others can be negotiated, but the claim against us for their physical sustenance constitutes a moral obligation that we have to meet. That obligation is not conditioned by their IQ scores, whether they have cleft palates or perfectly formed faces, whether they are obedient or irritatingly independent, whether they are irritatingly obedient and passive or laudably self-determining. It is not conditioned by any predictions that might be made about whether they will become the persons we might desire that they become. The infant in our case has the same sort of claim, and thus the parents have a moral obligation to use all ordinary means to save its life.

An objection might be made. Many of my fellow Christians would say that the obligation of the parents was to do that which is loving toward the infant. Not keeping the child alive was the loving thing to do with reference both to its interests and to the interests of the other members of the family. To respond to the objection, one needs first to establish the spongy character of the words "love" or "loving." They can absorb almost anything. Next one asks whether the loving character of an act is determined by feelings or by motives, or whether it is also judged by what is done. It is clear that I would argue for the latter. Indeed, the minimal conditions of a loving relationship include respect for the other, and certainly for the other's presumption of a right to live. I would, however, primarily make the case that the relationship of dependence grounds the claim, whether or not one feels loving toward the other.

The dependence relationship holds for the physicians as well as the parents in this case. The child's life depended utterly upon the capacity of the physicians to sustain it. The fact that an infant cannot articulate his claim is irrelevant. Physicians will struggle to save the life of a person who has attempted to commit suicide even when the patient might be in such a drugged condition that he cannot express his desire—a desire expressed already in his effort to take his life and overridden by the physician's action to save it. The claim of human life for preservation, even when such a person indicates a will not to live, presents a moral obligation to those who have the capacity to save it.

A different line of argument might be taken. If the decisions made were as reliant upon the desires of the parents as they appear to be, which is to say, if desire had a crucial role, what about the desire of the infant? The infant could not give informed consent to the nonintervention. One can hypothesize that every infant desires to live, and that even a defective child is likely to desire life rather than death when it reaches an age at which its desires can be articulated. Even if the right to live is contingent upon a desire, we can infer that the infant's desire would be for life. As a human being, he would have that desire, and thus it would constitute a claim on those on whom he is dependent to fulfill it.

I have tried to make a persuasive case to indicate why the claim of the infant constitutes a moral obligation on the parents and the physicians to keep the child alive. The intrinsic value or rights of a human being are not qualified by any given person's intelligence or capacities for productivity, potential consequences of the sort that burden others. Rather, they are constituted by the very existence of the human being as one who is related to others and dependent upon others for his existence. The presumption is always in favor of sustaining life through ordinary means; the desires of persons that run counter to that presumption are not sufficient conditions for abrogating that right.

The power to determine whether the infant shall live or die is in the hands of others. Does the existence of such power carry with it the moral right to such determination? Long history of moral experience indicates not only that arguments have consistently been made against the judgment that the capacity to do something constitutes a right to do it, or put in more familiar terms, that might makes right. It also indicates that in historical situations where persons have claimed the right to determine who shall live because they have the power to do so, the consequences have hardly been beneficial to mankind. This, one acknowledges, is a "wedge" argument or a "camel's nose under the tent" argument. As such, its limits are clear. Given a culture in which humane values are regnant, it is not likely that the establishment of a principle that some persons under some circumstances claim the right to determine whether others shall live will be transformed into the principle that the right of a person to live is dependent upon his having the qualities approved by those who have the capacity to sustain or take his life. Yet while recognizing the sociological and historical limitations that exist in a humane society, one still must recognize the significance of a precedent. To cite an absurd example, what would happen if we lived in a society in which the existence of hazel eyes was considered a genetic defect by parents and physicians? The absurdity lies in the fact that no intelligent person would consider hazel eyes a genetic defect; the boundaries around the word defect are drawn by evidences better than eye color. But the precedent in principle remains; when one has established that the capacity to determine who shall live carries with it the right to determine who shall live, the line of discussion has shifted from a sharp presumption (of the right of all humans to live) to the softer, spongier determination of the qualities whose value will be determinative.

Often we cannot avoid using qualities and potential consequences in the determination of what might be justifiable exceptions to the presumption of the right to life on the part of any infant—indeed, any

person. No moralist would insist that the physicians have an obligation to sustain the life of matter born from human parents that is
judged to be a "monstrosity." Such divergence from the "normal"
qualities presents no problem, and potential consequences for its continued existence surely enter into the decision. The physicians in our
case believed that in the absence of a desire for the child on the part of
the parents, mongolism was sufficiently removed from an ideal norm
of the human that the infant had no overriding claim on them. We are
in a sponge. Why would I draw the line on a different side of mongolism than the physicians did? While reasons can be given, one must
recognize that there are intuitive elements, grounded in beliefs and
profound feelings, that enter into particular judgments of this sort. I
am not prepared to say that my respect for human life is "deeper,"
"profounder," or "stronger" than theirs. I am prepared to say that the
way in which, and the reasons why, I respect life orient my judgment
toward the other side of mongolism than theirs did.

First, the value that intelligence was given in this instance appears to me to be simplistic. Not all intelligent persons are socially
commendable (choosing socially held values as the point of reference
because one of the physicians did). Also, many persons of limited intelligence do things that are socially commendable, if only minimally
providing the occasion for the expression of profound human affection and sympathy. There are many things we value about human
life; that the assumption that one of them is the *sine qua non*, the necessary and sufficient condition for a life to be valued at all, over-
simplifies human experience. If there is a *sine qua non*, it is physical
life itself, for apart from it, all potentiality of providing benefits for
oneself or for others is impossible. There are occasions on which other
things are judged to be more valuable than physical life itself; we
probably all would admire the person whose life is martyred for the
sake of saving others. But the qualities or capacities we value exist in
bundles, and not each as overriding in itself. The capacity for self-determination is valued, and on certain occasions we judge that it is
worth dying, or taking life, for the sake of removing repressive limits
imposed upon persons in that respect. But many free, self-determining
persons are not very happy; indeed, often their anxiety increases with
the enlargement of the range of things they must and can determine
for themselves. Would we value a person exclusively because he is
happy? Probably not, partly because his happiness has at least a
mildly contagious effect on some other persons, and thus we value
him because he makes others happy as well. To make one quality we
value (short of physical life itself, and here there are exceptions) determinative over all other qualities is to impoverish the richness and vari-

ety of human life. When we must use the sponge of qualities to determine exceptions to the presumption of the right to physical life, we need to face their variety, their complexity, the abrasiveness of one against the other, in the determination of action. In this case the potentialities of a mongoloid for satisfaction in life, for fulfilling his limited capacities, for happiness, for providing the occasions of meaningful (sometimes distressing and sometimes joyful) experience for others are sufficient so that no exception to the right to life should be made. Put differently, the anguish, suffering, embarrassment, expenses of family and state (I support the need for revision of social policy and practice) are not sufficiently negative to warrant that a mongoloid's life not be sustained by ordinary procedures.

Second, and harder to make persuasive, is that my view of human existence leads to a different assessment of the significance of suffering than appears to be operative in this case. The best argument to be made in support of the course of decisions as they occurred is that in the judgment of the principals involved, they were able to avoid more suffering and other costs for more people over a longer range of time than could have been avoided if the infant's life had been saved. To suggest a different evaluation of suffering is not to suggest that suffering is an unmitigated good, or that the acceptance of suffering when it could be avoided is a strategy that ought to be adopted for the good life, individually and collectively. Surely it is prudent and morally justifiable to avoid suffering if possible under most normal circumstances of life. But two questions will help to designate where a difference of opinion between myself and the principals in our drama can be located. One is, At what cost to others is it justifiable to avoid suffering for ourselves? On the basis of my previous exposition, I would argue that in this instance the avoidance of potential suffering at the cost of that life was not warranted. The moral claims of others upon me often involve emotional and financial stress, but that stress is not sufficient to warrant my ignoring the claims. The moral and legal claim of the government to the right to raise armies in defense of the national interest involves inconvenience, suffering, and even death for many; yet the fact that meeting that claim will cause an individual suffering is not sufficient ground to give conscientious objection. Indeed, we normally honor those who assume suffering for the sake of benefits to others.

The second question is, Does the suffering in prospect appear to be bearable for those who have to suffer? We recognize that the term "bearable" is a slippery slope and that fixing an answer to this question involves judgments that are always hypothetical. If, however, each person has a moral right to avoid all bearable inconvenience or suffering that appears to run counter to his immediate or long-range self-interest, there are many things necessary for the good of other individuals and for the common good that would not get done. In our case, there appear to be no evidences that the parents with assistance from other institutions would necessarily find the raising of a mongoloid child to bring suffering that they could not tolerate. Perhaps there is justifying evidence to which I do not have access, such as the possibility that the mother would be subject to severe mental illness if she had to take care of the child. But from the information I received, no convincing case could be made that the demands of raising the child would present intolerable and unbearable suffering to the family. That it would create greater anguish, greater inconvenience, and greater demands than raising a normal child would is clear. But that meeting these demands would cause greater suffering to this family than it does to thousands of others who raise mongoloid children seems not to be the case.

Finally, my view, grounded ultimately in religious convictions as well as moral beliefs, is that to be human is to have a vocation, a calling, and the calling of each of us is "to be for others" at least as much as "to be for ourselves." The weight that one places on "being for others" makes a difference in one's fundamental orientation toward all of his relationships, particularly when they conflict with his immediate self-interest. In the Torah we have that great commandment, rendered in the New English Bible as "you shall love your neighbour as a man like yourself" (Lev. 19:18). It is reiterated in the records we have of the words of Jesus, "Love your neighbor as yourself" (Matt. 22:39, and several other places). Saint Paul makes the point even stronger at one point: "Each of you must regard, not his own interests, but the other man's" (1 Cor. 10:24, NEB). And finally, the minimalist saying accredited both to Rabbi Hillel and to Jesus in different forms, "Do unto others as you would have others do unto you."

The point of the biblical citations is not to take recourse to dogmatic religious authority, as if these sayings come unmediated from the ultimate power and orderer of life. The point is to indicate a central thrust in Judaism and Christianity which has nourished and sustained a fundamental moral outlook, namely, that we are "to be for others" at least as much as we are "to be for ourselves." The fact that this outlook has not been adhered to consistently by those who professed it does not count against it. It remains a vocation, a calling, a moral ideal, if not a moral obligation. The statement of such an outlook does not resolve all the particular problems of medical histories such as this one, but it shapes a bias, gives a weight, toward the well-being of the other against inconvenience or cost to oneself. In this case, I believe that all the rational inferences to be drawn from it, and all the emotive power that this calling evokes, lead to the conclusion that the ordinary surgical procedure should have been done, and the mongoloid infant's life saved.

8 To Save or Let Die

Richard A. McCormick

In the context of a careful review of contemporary literature on the topic, McCormick develops a nuanced argument for a quality of life position within the framework of natural law. McCormick argues that life is a basic
good, but a good to be preserved only as a condition for
other values. From this it follows that life is a relative
good and the duty to preserve it a limited one. The importance of the article is that McCormick takes very
seriously both the clinical and ethical data of terminal situations, as well as developing a quality of life position
within the context of natural law. As such, McCormick's
article provides a challenging counterpoint to Gustafson's presentation, especially since they both utilize a
form of natural law argumentation.

*Rev. Richard A. McCormick, S.J., Ph.D., is the Rose
Kennedy Professor of Christian Ethics at the Kennedy
Center for Bioethics at Georgetown University.*

On February 24, the son of Mr. and Mrs.
Robert H. T. Houle died following court-or-
dered emergency surgery at Maine Medical Center. The child was
born February 9, horribly deformed. His entire left side was malformed; he had no left eye, was practically without a left ear, had a
deformed left hand; some of his vertebrae were not fused. Furthermore, he was afflicted with a tracheal esophageal fistula and could not
be fed by mouth. Air leaked into his stomach instead of going to the
lungs, and fluid from the stomach pushed up into the lungs. As Dr.
André Hellegers recently noted: "It takes little imagination to think
there were further internal deformities" (*Obstetrical and Gynecologi-
cal News*, April, 1974).

As the days passed, the condition of the child deteriorated. Pneu-
monia set in. His reflexes became impaired and because of poor circulation, severe brain damage was suspected. The tracheal esophageal
fistula, the immediate threat to his survival, can be corrected with rela-

tive ease by surgery. But in view of the associated complications and deformities, the parents refused their consent to surgery on "Baby Boy Houle." Several doctors in the Maine Medical Center felt differently and took the case to court. Maine Superior Court Judge David G. Roberts ordered the surgery to be performed. He ruled: "At the moment of live birth there does exist a human being entitled to the fullest protection of the law. The most basic right enjoyed by every human being is the right to life itself."

'Meaningful Life'

Instances like this happen frequently. In a recent issue of the *New England Journal of Medicine* (289 [1973], pp. 890-94), Drs. Raymond S. Duff and A. G. M. Campbell reported on 299 deaths in the special-care nursery of the Yale-New Haven Hospital between 1970 and 1972. Of these, 43 (14 percent) were associated with discontinuance of treatment for children with multiple anomalies, trisomy, cardiopulmonary crippling, meningomyelocele and other central nervous system defects. After careful consideration of each of these 43 infants, parents and physicians in a group decision concluded that the prognosis for "meaningful life" was extremely poor or hopeless, and therefore rejected further treatment. The abstract of the Duff-Campbell report states: "The awesome finality of these decisions, combined with a potential for error in prognosis, made the choice agonizing for families and health professionals. Nevertheless, the issue has to be faced, for not to decide is an arbitrary and potentially devastating decision of default."

In commenting on this study in the Washington *Post* (Oct. 28, 1973), Dr. Lawrence K. Pickett, chief of staff at the Yale-New Haven Hospital, admitted that allowing hopelessly ill patients to die "is accepted medical practice." He continued: "This is nothing new. It's just being talked about now."

It has been talked about, it is safe to say, at least since the publicity associated with the famous "Johns Hopkins Case" some three years ago. (See James M. Gustafson's "Mongolism, Parental Desires and the Right to Life," *Perspectives in Biology and Medicine,* XVI [1973], pp. 529-59.) In this instance, an infant was born with Down's syndrome and duodenal atresia. The blockage is reparable by relatively easy surgery. However, after consultation with spiritual advisers, the parents refused permission for this corrective surgery, and the child died by starvation in the hospital after 15 days. For to feed him by mouth in his condition would have killed him. Nearly everyone who has commented on this case has disagreed with the decision.

It must be obvious that these instances—and they are frequent—

raise the most agonizing and delicate moral problems. The problem is best seen in the ambiguity of the term "hopelessly ill." This used to, and still may, refer to lives that cannot be saved, that are irretrievably in the dying process. It may also refer to lives that can be saved and sustained, but in a wretched, painful or deformed condition. With regard to infants, the problem is, which infants, if any, should be allowed to die? On what grounds or according to what criteria, as determined by whom? Or again, is there a point at which a life that can be saved is not "meaningful life," as the medical community so often phrases the question? If our past experience is any hint of the future, it is safe to say that public discussion of such controversial issues will quickly collapse into slogans such as: "There is no such thing as a life not worth saving"; or "Who is the physician to play God?" We saw, and continued to see, this far too frequently in the abortion debate. We are experiencing it in the euthanasia discussion. For instance, "death with dignity" translates for many into a death that is fast, clean, painless. The trouble with slogans is that they do not aid in the discovery of truth; they co-opt this discovery and promulgate it rhetorically, often only thinly disguising a good number of questionable value judgments in the process. Slogans are not tools for analysis and enlightenment; they are weapons for ideological battle.

Thus far, the ethical discussion of these truly terrifying decisions has been less than fully satisfactory. Perhaps this is to be expected, since the problems have only recently come to public attention. In a companion article to the Duff-Campbell report, Dr. Anthony Shaw of the Pediatric Division of the Department of Surgery, University of Virginia Medical Center, Charlottesville, speaks of solutions "based on the circumstances of each case rather than by means of a dogmatic formula approach." Are these really the only options available to us? Dr. Shaw's statement makes it appear that the ethical alternatives are narrowed to dogmatism (which imposes a formula that prescinds from circumstances) and pure concretism (which denies the possibility or usefulness of any guidelines).

Are Guidelines Possible?

Such either-or extremism is understandable. It is easy for the medical profession, in its fully justified concern with the terrible concreteness of these problems and with the issue of who makes these decisions, to trend away from any substantive guidelines. As *Time* remarked in reporting these instances: "Few, if any, doctors are willing to establish guidelines for determining which babies should receive lifesaving surgery or treatment and which should not" (March 25, 1974). On the other hand, moral theologians, in their fully justified concern to avoid total normlessness and arbitrariness wherein the right is "discovered," or really "created," only in and by brute decision, can easily be insensitive to the moral relevance of the raw experience, of the conflicting tensions and concerns provoked through direct cradleside contact with human events and persons.

But is there no middle course between sheer concretism and dogmatism? I believe there is. Dr. Franz J. Ingelfinger, editor of the *New England Journal of Medicine*, in an editorial on the Duff-Campbell-Shaw articles, concluded, even if somewhat reluctantly: "Society, ethics, institutional attitudes and committees can provide the broad guidelines, but the onus of decision-making ultimately falls on the doctor in whose care the child has been put." Similarly, Frederick Carney of Southern Methodist University, Dallas, and the Kennedy Center for Bioethics stated of these cases: "What is obviously needed is the development of substantive standards to inform parents and physicians who must make such decisions" (Washington *Post*, March 20, 1974).

"Broad guidelines," "substantive standards." There is the middle course, and it is the task of a community broader than the medical community. A guideline is not a slide rule that makes the decision. It is far less than that. But it is far more than the concrete decision of the parents and physician, however seriously and conscientiously this is made. It is more like a light in a room, a light that allows the individual objects to be seen in the fullness of their context. Concretely, if there are certain infants that we agree ought to be saved in spite of illness or deformity, and if there are certain infants that we agree should be allowed to die, then there is a line to be drawn. And if there is a line to be drawn, there ought to be some criteria, even if very general, for doing this. Thus, if nearly every commentator has disagreed with the Hopkins decision, should we not be able to distill from such consensus some general wisdom that will inform and guide future decisions? I think so.

This task is not easy. Indeed, it is so harrowing that the really tempting thing is to run from it. The most sensitive, balanced and penetrating study of the Hopkins case that I have seen is that of the University of Chicago's James Gustafson (the article quoted above). Mr. Gustafson disagreed with the decision of the Hopkins physicians to deny surgery to the mongoloid infant. In summarizing his dissent, he notes: "Why would I draw the line on a different side of mongolism than the physicians did? While reasons can be given, one must recognize that there are intuitive elements, grounded in beliefs and profound feelings, that enter into particular judgments of this sort." He goes on to criticize the assessment made of the child's intelligence as too simplistic, and he proposes a much broader perspective on the meaning of suffering than seemed to have operated in the Hopkins decision. I am in full agreement with Mr. Gustafson's reflections and conclusions. But ultimately, he does not tell us where he would draw the line or why, only where he would *not*, and why.

This is very helpful already, and perhaps it is all that can be done. Dare we take the next step, the combination and analysis of such negative judgments to extract from them the positive criterion or criteria inescapably operative in them? Or more startlingly, dare we *not* if these decisions are already being made? Mr. Gustafson is certainly right in saying that we cannot always establish perfectly rational accounts and norms for our decisions. But I believe we must never cease trying, in fear and trembling, to be sure. Otherwise, we have exempted these decisions in principle from the one critique and control that protects against abuse. Exemption of this sort is the root of all exploitation, whether personal or political. Briefly, if we must face the frightening task of making quality-of-life judgments—and we must—then we must face the difficult task of building criteria for these judgments.

Facing Responsibility

What has brought us to this position of awesome responsibility? Very simply, the sophistication of modern medicine. Contemporary resuscitation and life-sustaining devices have brought a remarkable change in the state of the question. Our duties toward the care and preservation of life have been traditionally stated in terms of the use of ordinary and extraordinary means. For the moment and for purposes of brevity, we may say that, morally speaking, ordinary means are those whose use does not entail grave hardships to the patient. Those that would involve such hardships are extraordinary. Granted the relativity of these terms and the frequent difficulty of their application, still the distinction has had an honored place in medical ethics and medical practice. Indeed, the distinction was recently reiterated by the House of Delegates of the American Medical Association in a policy statement. After disowning intentional killing (mercy killing), the AMA statement continues: "The cessation of the employment of extraordinary means to prolong the life of the body when there is irrefutable evidence that biological death is imminent is the decision of the patient and/or his immediate family. The advice and judgment of the physician should be freely available to the patient and/or his immediate family" (*JAMA*, 227 [1974], p. 728).

This distinction can take us just so far—and thus the change in the state of the question. The contemporary problem is precisely that the question no longer concerns only those for whom "biological death is imminent" in the sense of the AMA statement. Many infants who would have died a decade ago, whose "biological death was imminent," can be saved. Yesterday's failures are today's successes. Contemporary medicine, with its team approaches, staged surgical techniques, monitoring capabilities, ventilatory support systems and other methods, can keep almost anyone alive. This has tended gradually to shift the problem, from the means to reverse the dying process, to the quality of the life sustained and preserved. The questions, "Is this means too hazardous or difficult to use?" and "Does this measure only prolong the patient's dying?"—while still useful and valid—now often become: "Granted that we can easily save the life, what kind of life are we saving?" This is a quality-of-life judgment. And we fear it. And certainly we should. But with increased power goes increased responsibility. Since we have the power, we must face the responsibility.

A Relative Good

In the past, the Judeo-Christian tradition has attempted to walk a balanced middle path between medical vitalism (that preserves life at any cost) and medical pessimism (that kills when life seems frustrating, burdensome, "useless"). Both of these extremes root in an identical idolatry of life—an attitude that, at least by inference, views death as an unmitigated, absolute evil, and life as the absolute good. The middle course that has structured Judeo-Christian attitudes is that life is indeed a basic and precious good, but a good to be preserved precisely as the condition of other values. It is these other values and possibilities that found the duty to preserve physical life and also dictate the limits of this duty. In other words, life is a relative good, and the duty to preserve it a limited one. These limits have always been stated in terms of the *means* required to sustain life. But if the implications of this middle position are unpacked a bit, they will allow us, perhaps, to adapt to the type of quality-of-life judgment we are now called on to make without tumbling into vitalism or a utilitarian pessimism.

A beginning can be made with a statement of Pope Pius XII in an allocution to physicians delivered Nov. 24, 1957. After noting that we are normally obliged to use only ordinary means to preserve life, the Pontiff stated: "A more strict obligation would be too burdensome for most men and would render the attainment of the higher, more important good too difficult. Life, death, all temporal activities are in fact subordinated to spiritual ends." Here it would be helpful to ask two questions. First, what are these spiritual ends, this "higher, more important good"? Second, how is its attainment rendered too difficult by insisting on the use of extraordinary means to preserve life?

(1) "Spiritual" end

The first question must be answered in terms of love of God and neighbor. This sums up briefly the meaning, substance and consummation of life from a Judeo-Christian perspective. What is or can easily be missed is that these two loves are not separable. St. John wrote: "If any man says, 'I love God' and hates his brother, he is a liar. For he who loves not his brother, whom he sees, how can he love God whom he does not see?" (1 Jn. 4:20-21). This means that our love of neighbor is in some very real sense our love of God. The good our love wants to do Him and to which He enables us, can be done only for the neighbor, as Karl Rahner has so forcefully argued. It is in others that God demands to be recognized and loved. If this is true, it means that, in a Judeo-Christian perspective, the meaning, substance and consummation of life are found in human *relationships*, and the qualities of justice, respect, concern, compassion and support that surround them.

(2) Attainment too difficult

Second, how is the attainment of this "higher, more important [than life] good" rendered "too difficult" by life-supports that are gravely burdensome? One who must support his life with disproportionate effort focuses the time, attention, energy and resources of himself and others not precisely on relationships, but on maintaining the condition of relationships. Such concentration easily becomes overconcentration and distorts one's view of, and weakens one's pursuit of, the very relational goods that define our growth and flourishing. The importance of relationships gets lost in the struggle for survival. The very Judeo-Christian meaning of life is seriously jeopardized when undue and unending effort must go into its maintenance.

I believe an analysis similar to this is implied in traditional treatises on preserving life. The illustrations of grave hardship (rendering the means to preserve life extraordinary and nonobligatory) are instructive, even if they are outdated in some of their particulars. Older moralists often referred to the hardship of moving to another climate or country. As the late Gerald Kelly, S.J., noted of this instance: "They [the classical moral theologians] spoke of other inconveniences, too: e.g., of moving to another climate or another country to preserve one's life. For people whose lives were, so to speak, rooted in the land, and whose native town or village was as dear as life itself, and for whom, moreover, travel was always difficult and often dangerous—for such people, moving to another country or climate was a truly great hardship, and more than God would demand as a 'reasonable' means of preserving one's health and life" (*Medico-Moral Problems*, [1957], p. 132).

Similarly, if the financial cost of life-preserving care was crushing, that is, if it would create grave hardships for oneself or one's family, it was considered extraordinary and nonobligatory. Or again, the

Discussion of the Means

grave inconvenience of living with a badly mutilated body was viewed, along with other factors (such as pain in pre-anesthetic days, uncertainty of success), as constituting the means extraordinary. Even now, the contemporary moralist, Marcellino Zalba, S.J., states that no one is obligated to preserve his life when the cost is "a most oppressive convalescence."

The Quality of Life

In all of these instances—instances where the life could be saved— the discussion is couched in terms of the means necessary to preserve life. But often enough it is the kind of, the quality of, the life thus saved (painful, poverty-stricken and deprived, away from home and friends, oppressive) that establishes the means as extraordinary. That type of life would be an excessive hardship for the individual. It would distort and jeopardize his grasp on the overall meaning of life. Why? Because, it can be argued, human relationships—which are the very possibility of growth in love of God and neighbor—would be so threatened, strained or submerged that they would no longer function as the heart and meaning of the individual's life as they should. Something other than the "higher, more important good" would occupy first place. Life, the condition of other values and achievements, would usurp the place of these and become itself the ultimate value. When that happens, the value of human life has been distorted out of context.

In his *Morals in Medicine* (1957), Thomas O'Donnell, S.J., hinted at an analysis similar to this. Noting that life is a relative, not an absolute, good, he asks: Relative to what? His answer moves in two steps. First, he argues that life is the fundamental natural good God has given to men, "the fundamental context in which all other goods which God has given man as means to the end proposed to him, must be exercised" (p. 66). Second, since this is so, the relativity of the good of life consists in the effort required to preserve this fundamental context and "the potentialities of the other goods that still remain to be worked out within that context."

Can these reflections be brought to bear on the grossly malformed infant? I believe so. Obviously there is a difference between having a terribly mutilated body as the result of surgery, and having a terribly mutilated body from birth. There is also a difference between a long, painful, oppressive convalescence resulting from surgery, and a life that is from birth one long, painful, oppressive convalescence. Similarly, there is a difference between being plunged into poverty by medical expenses and being poor without ever incurring such expenses. However, is there not also a similarity? Cannot these conditions, whether caused by medical intervention or not, equally absorb attention and energies to the point where the "higher, more important good" is simply too difficult to attain? It would appear so. Indeed, is this not precisely why abject poverty (and the systems that support it) is such an enormous moral challenge to us? It simply dehumanizes.

Life's potentiality for other values is dependent on two factors, those external to the individual, and the very condition of the individual. The former we can and must change to maximize individual potential. That is what social justice is all about. The latter we sometimes cannot alter. It is neither inhuman nor unchristian to say that there comes a point where an individual's condition itself represents the negation of any truly human—i.e., relational—potential. When that point is reached, is not the best treatment no treatment? I believe that the *implications* of the traditional distinction between ordinary and extraordinary means point in this direction.

In this tradition, life is not a value to be preserved in and for itself. To maintain that would commit us to a form of medical vitalism that makes no human or Judeo-Christian sense. It is a value to be preserved precisely as a condition for other values, and therefore insofar as these other values remain attainable. Since these other values cluster around and are rooted in human relationships, it seems to follow that life is a value to be preserved only insofar as it contains some potentiality for human relationships. When in human judgment this potentiality is totally absent or would be, because of the condition of the individual, totally subordinated to the mere effort for survival, that life can be said to have achieved its potential.

Human Relationships

If these reflections are valid, they point in the direction of a guideline that may help in decisions about sustaining the lives of grossly deformed and deprived infants. That guideline is the potential for human relationships associated with the infant's condition. If that potential is simply nonexistent or would be utterly submerged and undeveloped in the mere struggle to survive, that life has achieved its potential. There are those who will want to continue to say that some terribly deformed infants may be allowed to die *because* no extraordinary means need to be used. Fair enough. But they should realize that the term "extraordinary" has been so relativized to the condition of the patient that it is this condition that is decisive. The means is extraordinary because the infant's condition is extraordinary. And if that is so, we must face this fact head-on—and discover the substantive standard that allows us to say this of some infants, but not of others.

Here several caveats are in order. First, this guideline is not a detailed rule that preempts decisions; for relational capacity is not subject to mathematical analysis but to human judgment. However, it is the task of physicians to provide some more concrete categories or presumptive biological symptoms for this human judgment. For instance, nearly all would very likely agree that the anencephalic infant is without relational potential. On the other hand, the same cannot be said of the mongoloid infant. The task ahead is to attach relational potential to presumptive biological symptoms for the gray area between such extremes. In other words, individual decisions will remain the anguishing onus of parents in consultation with physicians.

Second, because this guideline is precisely that, mistakes will be made. Some infants will be judged in all sincerity to be devoid of any meaningful relational potential when that is actually not quite the case. This risk of error should not lead to abandonment of decisions; for that is to walk away from the human scene. Risk of error means only that we must proceed with great humility, caution and tentativeness. Concretely, it means that, if err we must at times, it is better to err on the side of life—and, therefore, to tilt in that direction.

Third, it must be emphasized that allowing some infants to die does not imply that "some lives are valuable, others not" or that "there is such a thing as a life not worth living." Every human being, regardless of age or condition, is of incalculable worth. The point is not, therefore, whether this or that individual has value. Of course he has, or rather is, a value. The only point is whether this undoubted value has any potential at all, in continuing physical survival, for attaining a share, even if reduced, in the "higher, more important good." This is not a question about the inherent value of the individual. It is a question about whether this worldly existence will offer such a valued individual any hope of sharing those values for which physical life is the fundamental condition. Is not the only alternative an attitude that supports mere physical life as long as possible with every means?

Fourth, this whole matter is further complicated by the fact that this decision is being made for someone else. Should not the decision on whether life is to be supported or not be left to the individual? Obviously, wherever possible. But there is nothing inherently objectionable in the fact that parents with physicians must make this decision at some point for infants. Parents must make many crucial decisions for children. The only concern is that the decision not be shaped out of the utilitarian perspectives so deeply sunk into the consciousness of the contemporary world. In a highly technological culture, an individual is always in danger of being valued for his function, what he can do, rather than for who he is.

It remains, then, only to emphasize that these decisions must be made in terms of the child's good, this alone. But that good, as fundamentally a relational good, has many dimensions. Pius XII, in speaking of the duty to preserve life, noted that this duty "derives from well-ordered charity, from submission to the Creator, from social justice, as well as from devotion towards his family." All of these considerations pertain to that "higher, more important good." If that is the case with the duty to preserve life, then the decision not to preserve life must likewise take all of these into account in determining what is for the child's good.

Any discussion of this problem would be incomplete if it did not repeatedly stress that it is the pride of Judeo-Christian tradition that the weak and defenseless, the powerless and unwanted, those whose grasp on the goods of life is most fragile—that is, those whose potential is real but reduced—are cherished and protected as our neighbor in greatest need. Any application of a general guideline that forgets this is but a racism of the adult world profoundly at odds with the gospel, and eventually corrosive of the humanity of those who ought to be caring and supporting as long as that care and support has human meaning. It has meaning as long as there is hope that the infant will, in relative comfort, be able to experience our caring and love. For when this happens, both we and the child are sharing in that "greater, more important good."

Were not those who disagreed with the Hopkins decision saying, in effect, that for the infant, involved human relationships were still within reach and would not be totally submerged by survival? If that is the case, it is potential for relationships that is at the heart of these agonizing decisions.

Severely Handicapped Children Further Reading

Duff, Raymond S. "On Deciding the Care of Severely Handicapped or Dying Persons: With Particular Reference to Infants." *Pediatrics* 57 (April 1976) 487–92.

Frankel, Lawrence S.; Damme, Catherine J.; and Van Eys, Jan. "Childhood Cancer and the Jehovah's Witness Fate." *Pediatrics* 60 (December 6, 1977), 916–.

Garland, Michael J. "Care of the Newborn: The Decision Not to Treat." *Perinatology/Neonatology* 1 (September–October 1977), 14ff.

Heymann, Philip B., and Holz, Sarah. "The Severely Defective Newborn: The Dilemma and the Decision Process." *Public Policy* 23 (Fall 1975), 381–418.

Jonsen, A. R., and Garland, Michael, eds. *Ethics of Newborn Intensive Care.* San Francisco and Berkeley: University of California School of Medicine and Institute of Governmental Studies, 1976.

Robertson, John A. "Involuntary Euthanasia of Defective Newborns: A Legal Analysis." *Stanford Law Review* 27 (January 1975), 213–67.

Shaw, Anthony; Randolf, Judson G.; and Manard, Barbara. "Ethical Issues in Pediatric Surgery: A Nationwide Survey of Pediatricians and Pediatric Surgeons." *Pediatrics* 60 (October 1977, 588–99).

Swinyard, Chester A., ed. *Decision Making and the Defective Newborn.* Springfield, Illinois.: Charles C. Thomas, publisher, 1977.

Veatch, Robert M. "The Technical Criteria Falacy." *Hastings Center Report* 7 (August 1977), 15–16.

Weber, Leonard J. *Who Shall Live? The Dilemma of Severely Handicapped Children and Its Meaning for Other Moral Questions.* New York: Paulist Press 1976.

PART THREE

DEATH AND DYING

Frank J. Veith
Hans Jonas
Daniel Maguire
Thomas A. Shannon

9 *Brain Death* 1977

FRANK J. VEITH AND OTHERS

Many debates on brain death have occurred over the past decade. Some debates were brought about by the fact of the use of medical technology which would allow circulation and respiration to be artificially maintained even though the person's brain is destroyed; other debates have focused around the utility of a brain death definition so that organs could be harvested for transplantation more readily; also a revised definition of brain death would ease many of the dilemmas associated with turning off life support systems. This article summarizes information on clinical and laboratory criteria for brain death; argues that the concept of brain death is in accord with philosophy and three major western religions; shows the need for legislation that would allow the use of brain death as a criterion of death; and reviews the present status of judicial and statutory law relating to the determination of death. The authors provide a variety of perspectives which help us come to terms with many of the critical issues in accepting or rejecting the use of a particular definition of brain death as a useful criterion for determining death.

Frank J. Veith, M.D., is with the Department of Surgery Montefiore Hospital, New York

Brain Death is a term commonly used to describe a condition in which the brain is completely destroyed and in which cessation of function of all other organs is imminent and inevitable. The concept of brain death is important to consider, since advances in medical technology have resulted in the artificial prolongation of the overall process of dying. In the past, cessation of heartbeat and spontaneous respiration always produced prompt death of the brain, and, similarly, destruction of the brain resulted in prompt cessation of respiration and circulation. In this context, it was reasonable that absence of pulse and respiration became the traditional criteria for pronouncement of death. Recently, however, technological advances have made it possible to sustain brain function in the absence of spontaneous respiratory and cardiac function, so that the death of a person can no longer be equated with the loss of these latter two natural vital functions. Furthermore, it is now possible that a person's brain may be completely destroyed even though his circulation and respiration are being artificially maintained by mechanical devices.

I. A Status Report of Medical and Ethical Considerations

A number of authors have argued persuasively that a person whose brain is totally destroyed is in fact dead,[1-4] and this premise has gained considerable acceptance throughout the world from the public and from professionals in various relevant fields. Accordingly, the pronouncement of death on the basis of irreversible cessation of all brain function has become common. Nevertheless, this use of the concept of brain death has caused considerable controversy among physicians, lawyers, legislators, philosophers, and theologians. This controversy is founded partly on the failure of some to accept the concept that death may be pronounced on brain-related criteria,[5] and partly on the contention that statutory recognition of such pronouncements is neither necessary nor desirable.[6,7] Groups subscribing to either one or both of these positions actively oppose passage of statutory definitions of death and render enactment of such legislation difficult in the 32 states presently without such laws.

The purposes of this communication are to contribute to a resolution of the controversy and thereby to facilitate passage of statutes recognizing brain death by accomplishing several objectives. First, it will summarize information that establishes the ability to determine the state of complete destruction of the brain with certainty on the basis of available clinical and laboratory criteria. Second, it will demonstrate that total destruction of the brain constitutes a determinant of death that is not in conflict with sound secular philosophic considerations, Orthodox Judaic law, traditional Catholic ethics, or the mainstream of Protestant theology. Third, it will document the need for legislative recognition that death may be pronounced on the basis of complete and irreversible destruction of the brain. And fourth, it will review the present status of judicial and statutory law relating to the determination of death in the United States (in a later issue).

Validity of Criteria for Determining Complete Destruction of the Brain

Any ethical or legal considerations concerning pronouncements of death on a neurologic basis must be founded on the certainty that a person who meets the clinical and laboratory criteria has had actual

complete destruction of the brain. In 1968, guidelines were formulated by an Ad Hoc Committee of the Harvard Medical School to permit
the determination of irreversible coma.[8] These Harvard criteria require that neurologic examinations disclose unreceptivity, unresponsiveness, absence of spontaneous movements and breathing, absent
reflexes, fixed dilated pupils, and persistence of all these findings
over a 24-hour period in the absence of intoxicants or hypothermia. A
persistently isoelectric EEG over the same period is also required to
confirm the clinical examination. Since 1968, the validity of these
widely used criteria has been established in several ways.

These validations include the substantial morphologic evidence
that, when the criteria have been fulfilled, there is widespread destruction of the brain. Richardson has found that the brains of 128
patients meeting the Harvard criteria showed extensive destructive
changes (oral communication, March 1976).[9] In a larger series of autopsy studies, however, the exact nature and distribution of these fatal morphologic lesions in the brain were also shown to be dependent
on the etiology and on the interval between fulfillment of the Harvard criteria and pathologic examination.[10] The latter observation is
consistent with the well-known finding in other organs that time
must often elapse before morphologic evidence of cellular destruction
can be detected.

In addition, patients who fulfill the Harvard criteria have been
shown by isotopic techniques to have no significant intracranial
blood flow,[11] and absent intracranial blood flow over a 10- to 15-
minute interval is uniformly associated with subsequent necrosis and
liquefaction of the brain.[12] The latter finding is based on autopsy
studies from several Scandinavian hospitals of more than 120 patients
who had nonvisualization of intracranial arteries after cerebral angiography with contrast injections repeated over a 10- to 15-minute
interval. In related studies from several centers, clinical and EEG
evidence of complete brain destruction was almost always associated with angiographic evidence of cessation of intracranial blood
flow.[13-16]

Another validation of the Harvard criteria derives from cooperative studies of the value of EEG and neurologic examination in the
determination of complete brain destruction.[17,18] In these studies,
members of the American Electroencephalographic Society and of
EEG societies in Europe were questioned. Of the 2,642 cases under
study, there was no instance of recovery in a patient who fulfilled the
Harvard criteria. Furthermore, since 1970, there have been no adequately documented examples in which the Harvard criteria could be
considered invalid.[19] Moreover, many authorities presently consider

these criteria too strict in at least two regards.[19-24] First, it has been shown that spinal reflexes including withdrawal movements may persist after complete destruction of the brain. Second, it is believed that certain determination that the brain is totally destroyed can be made even when the period of clinical and EEG evidence of absent brain function is reduced to less than 24 hours. The latter is consistent with the opinion that methods for measuring intracranial blood flow will allow a sure determination of complete brain destruction to be made with periods of observation less than the 24 hours proposed in the original Harvard criteria.[11-16,22] In this regard, it should be noted that the immature brain is more resistant to all forms of insult. Therefore, altered, less restrictive criteria for determining total brain destruction in patients under 14 years of age may differ from those in adults.

Further support for the use of less restrictive criteria is provided by the recently completed Collaborative Study on Cerebral Survival, which was based on an analysis of 503 unresponsive apneic patients. From this experience it was concluded that, if all appropriate diagnostic and therapeutic procedures had been performed to exclude reversible conditions, brain destruction was always present if certain criteria were observed for at least 30 minutes six hours or more after the cerebral insult had occurred. The specified criteria were unresponsivity, apnea, dilated pupils and absent cephalic reflexes, electrocerebral (EEG) silence, and confirmation of absent cerebral blood flow by angiography, isotopic bolus techniques, or echo encephalography.[23,25,26] The confirmatory test for absent cerebral blood flow was not deemed necessary in cases where the obvious etiologic factor was known to be a nontreatable condition, such as massive brain trauma.

Although some groups have indicated that EEG is not required to determine that brain death has occurred,[21,27] and although many neurologists and neurosurgeons would agree that brain death can safely be pronounced in the absence of electrocerebral silence in the occasional patient, the recommendation that EEG criteria be met before brain death is pronounced is probably best for general usage.[22-25,28] This recommendation appears advisable at present in light of a report that a patient who ultimately recovered had met the clinical criteria of brain death but never had electrocerebral silence,[28] and in view of the current trend toward increasingly frequent medical malpractice suits.

A final validation of the criteria for measuring total destruction of the brain has been an attempt on our part to explore purported anecdotal exceptions. In every instance where recovery of brain function was claimed, the criteria had not been fulfilled. Thus, the validity of the criteria must be considered to have been established with as much certainty as is possible in biology or medicine.

Philosophical and Religious Acceptability

It is one thing to know that we now possess the technical capacity to determine accurately that a human brain has been completely and irreversibly destroyed. It is quite another matter to make the social policy judgment that it is acceptable to use complete and irreversible destruction of the human brain as a basis for treating the person as a whole as if he or she were dead. We are convinced that society now has sufficient philosophical certainty, based on the main stands of secular philosophical thought and the major Western religious traditions, to use destruction of the brain as an indicator that the person has died.

It has been suggested that *one* reason for changing society's concept of death to one oriented to brain function is that it would provide desperately needed organs for transplantation and other useful medical purposes. However, the fact that someone would be useful to others if pronounced dead should not alone be a sufficient reason for considering that person dead and cannot be the sole basis for changing to the use of brain-oriented criteria. Rather, there must be sound reasons independent of that if society is going to alter its definition of death.

The principal reason for deciding that a person is dead should be based on a fundamental understanding of the nature of man. Our present conceptualization of man almost reflexly draws a distinction between a person whose organs are under nervous system influence and the remnant of a person or his corpse in which residual and nonhomeostatic functions may or may not have completely ceased. Without a brain, the body becomes the convenient medium in which the energy-requiring states of organs run down and the organs decay. These residual activities do not confer an iota of humanity or personality. Thus, in the circumstance of brain death, neither a human being nor a person any longer exists.

Although all members of society will not be able to agree precisely on an acceptable formulation of man's nature, fortunately all that is necessary to establish a public policy is agreement on some widely acceptable, general statements about the nature of man. Almost all segments of society will agree that some capacity to think, to perceive, to respond, and to regulate and integrate bodily functions is essential to human nature. Thus, if none of these brain functions are present and will ever return, it is no longer appropriate to consider a person as a whole as being alive.

If there were no offense, no moral or social costs in treating dead persons as if they were alive, then the safer course would be to continue to do so. Quite clearly, however, this is not the case. In addition to reflecting an inadequate understanding of the nature of man, it is

[margin note: To treat the dead as living is morally wrong]

an affront to the individual person or that person's memory to treat a human being who has irreversibly lost all brain function as if he were alive. It confuses the person with his corpse and is morally wrong.

Furthermore, maintenance of a dead person on life support systems for no reason is an irresponsible squandering of our economic and social resources. Such a practice places an unnecessary financial burden on society and an additional emotional burden on the person's family and is thereby also morally wrong. Thus, even without consideration of the use of the body or its organs for transplantation or other altruistic purposes, there are sound moral and social reasons for treating a body that has lost significant thinking, perceiving, responding, regulating, and integrating capacities as dead. Of course, it is a waste of human resources and a further wrong to continue treating a corpse as if it were alive when such treatment may deprive other living persons of needed organs. Thus, from a moral and ethical perspective, persons who have lost all brain function and who are certainly dead should be treated accordingly. Before adopting this conclusion as a public policy, however, it is important to examine how such a position accords with the major religious traditions of our society.

[margin note: Note—the point here on transplant, we would just as expected etc.]

[margin note: Orthodox Judaism]

The Orthodox Jewish response to the premise that death may be pronounced on brain-related criteria is, like much of the moral conscience of Western civilization, based on biblical and talmudic ethical imperatives. According to these, it is axiomatic that human life is of infinite worth. A corollary of this is that a fleeting moment of life is of inestimable worth because a piece of infinity is also infinite. The taking or shortening of a human life is, therefore, ethically wrong, and premature termination of life or euthanasia is no less murder for the good intentions that were the motivation for the immoral act.

The indices of life are many. Which of them can be viewed, in ethical or religious terms, as the definition or sine qua non of the living state rather than a mere confirmation that the patient is still living? It is first important to point out that absent heartbeat or pulse was *not* considered a significant factor in ascertaining death in any early religious sources.[30] Furthermore, the scientific fact that cellular death does not occur at the same time as the death of the human being is well recognized in the earliest biblical sources. The twitching of a lizard's amputated tail or the death throes of a decapitated man were never considered residual life but simple manifestations of cellular life that continued after death of the entire organism had occurred.[31] In the situation of decapitation, death can be defined or determined by the decapitated state itself as recognized in the Talmud and the Code of Laws.[31-33] Complete destruction of the brain, which includes loss of all integrative, regulatory, and other functions of the brain, can be considered physiological decapitation and thus a determinant per se of death of the person.

Loss of the ability to breathe spontaneously is a crucial criterion for determining whether complete destruction of the brain has occurred. Earliest biblical sources recognized the ability to breathe independently as a prime index of life.[30,34] The biblical verse in Genesis records: "And the Lord had fashioned man of dust of the earth and instilled in his nostrils the breath of life and man became a living creature."[34] Spontaneous respiration is thus an indicator of the living state. However, it cannot be considered its definition, since a respirator patient whose sole defect is paralysis of the motor neurons to the muscles of respiration due to neurologic disease is surely fully alive despite his inability to breathe spontaneously. Therefore, to define death in biblical terms, loss of respiration must be combined with other more obvious evidence of the nonliving state. Such evidence would be provided by the clinical and laboratory criteria that allow a physician to determine that complete and irreversible destruction of the brain or physiological decapitation has occurred.

The higher integrative functions of the brain are carried out by portions of the brain other than the brainstem. Irreversible loss of these functions, signifying destruction of corresponding parts of the brain, does not alone constitute a determinant of death in biblical terms. Coincident loss of vegetative functions, represented by loss of spontaneous respiration and indicating destruction of the brainstem, is also a requisite. Thus, destruction of the entire brain or brain death, and only that, is consonant with biblical pronouncements on what constitutes an acceptable definition of death, i.e. a patient who has all the appearances of lifelessness and who is no longer breathing spontaneously. Patients with irreversible total destruction of the brain fulfill this definition even if heart action and circulation are artificially maintained. This definition is also fulfilled in patients who die with or from irreversible cessation of heart action, because this results in a failure to perfuse the brain, which produces total brain destruction. Thus, cessation of heart action is a cause of death rather than a component of its definition. In the light of these considerations, the Harvard criteria or other neurologic criteria for determining death can be viewed as the scientific expression of those observations that, until recently, were the actual way a patient was known to be dead.

The tumult that has greeted the suggestion that brain death be given legal recognition is partly the reaction of an uninformed public who envisions the possibility that a man who can move, feel, and think or can possibly recover these functions could be declared dead.

The realization that brain death is only professional jargon to describe a patient who exhibits a permanent loss of signs of life, such as spontaneous movement and responsivity, and has permanently lost the ability to breathe spontaneously would facilitate society's acceptance of the concept of brain death and would help to gain public support for legislation recognizing that death may be pronounced on the basis of total and irreversible destruction of the brain.

Since the distinction between cellular and organismal death is valid, once death of the person has occurred and can be determined, there is no biblical obligation to maintain treatment or artificial support of the corpse. Thus, according to M. Feinstein, there is no religious imperative to continue to use a respirator to inflate and deflate the lungs and thus maintain the cellular viability of other organs in an otherwise dead patient (written communication, May 5, 1976).

This Orthodox Jewish position is not alone among major Western religious traditions in supporting a concept of death based on irreversible loss of brain function. In the Roman Catholic Church, there is no definitive, authoritative pronouncement, but Catholic theologians interested in moral questions associated with the definition of death issue have generally accepted a concept of death based on brain function. The traditional Roman Catholic understanding of the moment of real death has been based on the time of departure of the soul from the body. Since this separation is not an observable phenomenon, it must be related to physically measurable signs defining apparent death. Because the only certain signs have been the appearance of rigor mortis and the beginning of bodily decomposition, it has been recognized that real death may not coincide with apparent death. Use of such signs as cessation of heartbeat and breathing places the moment of apparent death in greater proximity to the time of true theological death. For practical reasons, theologians have accepted these signs of apparent death as reasonably accurate indicators of irreversible cessation of all vital bodily functions adequate for allowing such processes as embalming and autopsy. When artificial life support systems are used to maintain heart and lung function and when the brain is irreversibly destroyed, there is also no reasonable hope of restoring vital bodily functions to a person. Accordingly, "it would seem that death is more certain under these conditions than it was at the [time of] cessation of spontaneous heart and lung function. If theologians were willing to accept the latter as signs of apparent death, they should be more willing to accept the irreversible cessation of brain function."[35]

A similar position has been reached by the Catholic theologian, Rev. Bernard Haring,[3] who after analysis of the theological arguments concludes, "I feel that the arguments for the equation of the total death of the person with brain death are fully valid." In the same vein, the prominent author on Roman Catholic interpretations of medical ethics, Charles J. McFadden, argues that "once the fact of brain death has been established, *the person is dead*, even though heart beat and respiration are continued by mechanical means."[36] These statements are consistent with the discourse of Pope Pius XII who, in discussing patients who are terminally unconscious, said

> one can refer to the usual concept of separation . . . of the soul from the body; but on the practical level, one needs to be mindful of the connotation of the terms "body" and "separation". . . . As to the pronouncement of death in certain particular cases, the answer cannot be inferred from religious and moral principles, and consequently, it is an aspect lying outside the competence of the Church.[37]

We understand the papal point to be that determination of the criteria for deciding the moment of death requires technical measures that can only be established by those with the appropriate medical expertise.

Among Protestant theologians, there are no consistent positions on questions of medical ethics including the definition of death. However, leading spokesmen of widely diverging traditions accept brain-related criteria for pronouncing death.[2,38-43] The body is an essential element of the person according to Christian theology; but, as many of these authors emphasize, mere cellular and organ system activity alone is not sufficient to treat a human body as if it were alive. Even more conservative thinkers such as Paul Ramsey accept the use of brain-oriented criteria for pronouncement of death. He recognizes proposals for updating the definition of death as, in reality,

> proposals for updating our procedures for determining that death has occurred, for rebutting the belief that machines or treatments are the patient, for withdrawing the notion that artificially sustained signs of life are in themselves signs of life, for telling when we should stop ventilating and circulating the blood of an unburied corpse because there are no longer any vital functions really alive or recoverable in the patient.[2]

Thus, the complete and permanent absence of any brain-related vital bodily function is recognized as death by Jewish, Roman Catholic, and Protestant scholars even if they may disagree among themselves on the precise theoretical foundations of this judgment.

II. A Status Report of Legal Considerations

Part I of this article established the scientific validity of current clinical and laboratory criteria for determining complete destruction of the brain or brain death. It also showed that total destruction of the brain constitutes a determinant of death, which is in accord with secular philosophy and the three major Western religions. In part II, legal issues that arise from use of brain-related criteria to pronounce death are considered.

Need For Statutory Recognition of Brain Death

The fact that physicians can recognize total and irreversible destruction of the brain on the basis of clinical and laboratory criteria is accepted and commonly utilized in many areas of the world. The need to make such pronouncements is based primarily on the requirement of society to respond appropriately to two recent advances in medical technology. The first is the hardware that can artificially maintain lung and heart action in the absence of spontaneous respiration and circulation. Although these devices may be lifesaving in many situations, their use in maintaining respiration and circulation in a human body that is dead by virtue of total destruction of the brain serves no useful purpose. In such instances it is reasonable to terminate these artificial support systems.

The second advance that requires pronouncement of death on brain-related criteria is cadaver organ transplantation. Most suitable donor organs come from patients who die from injury or disease of the brain. Only in such patients may the donor's circulation be artificially maintained after death so that needed organs can be removed with minimal ischemic damage. Since destruction of the brain is the cause of the donor's death, there is no reason not to remove these organs before cessation of the donor's artificially maintained circulation. This requires recognition that destruction of the brain is the basis for death of the donor and pronouncement of death on brain-related criteria.

Since the responsibility for pronouncing death resides with physicians, it has been suggested that no statute giving legal recognition to any particular criterion for determining death is necessary or desirable.[1] However, there is a potential dilemma in the absence of legal recognition of the medically accepted practice of pronouncing death on neurologic criteria. Physicians who pronounce death on this basis may be disputed in a judicial proceeding with the contention that death occurs only when spontaneous respiration and heartbeat cease. This contention could be based on the common law definition of death (to follow, under "Legal Status of Brain Death")[2] which is gen-

erally held applicable to jurisdictions without specific statutes. Without statutory or case law recognition of the use of brain-related
criteria for pronouncing death, it is possible that a valid medical declaration of death could be considered illegal and lead to difficulty in
the prosecution of a murderer or criminal or civil liability on the part
of a physician or hospital. These possibilities have made many neurologists and neurosurgeons reluctant to pronounce death on brain-
related criteria and have given rise to judicial actions in several locales. These cases have been a major factor leading to the passage, in
many states, of statutes recognizing the use of brain-oriented criteria
for pronouncing death.

Case law recognition of the legal validity of pronouncing death
on brain-related criteria, although helpful, is an inadequate solution
to the dilemma that arises from the potential discrepancy between
medically accepted practice and legally accepted practice for two important reasons. First, case law is fluid and subject to appeal and
change by subsequent judicial action. Second, court decisions to recognize the use of brain-related criteria for pronouncing death may relate to certain special circumstances, such as transplant organ
donation. A statute giving general recognition to this concept for all
purposes would avoid future inconsistencies under the law and
would prevent repeated anguish-producing court cases. Such a law
would allow a physician to terminate artificial respiratory support for
a patient who is clearly dead by accepted and validated criteria. It
would obviate the possibility that the physician, other health professionals, next of kin, guardians, and institutions might be held criminally or civilly liable for actions consistent with standards of current
medical practice.

In addition, even if physicians agree that brain-related criteria
should be used in death pronouncements, it is still an open question
what the rest of society would choose to have as its public policy.
Public policy can only be determined by some public act such as legislation. Thus, a statutory definition of death would serve as a vehicle
to translate a generally accepted medical standard into a form that is
accepted by most if not all members of society. As such, it will help to
minimize some of the burdens placed on the family and physicians of
the dead person by facilitating honest relationships and communication between them in many ways.

Until such public policy recognition by legislation has occurred,
the family confronted with the loss of a loved one who is dead by virtue of brain-related criteria is forced to deal with the confusing and
misleading assumption, supported by an out-of-date common law,
that, while the heart beats, there is life and hope of recovery. Of all

the reasons for establishing a statutory definition of death, the simplest and the most important is that it will help the family of the dead patient to appreciate the reality of his death, and to reassure them that the medical determination of death is valid. Such a law will also facilitate relief of the family from financial and emotional pressures and will enable them to confront death with more dignity and understanding.

In a similar way, the physician's onerous task of conveying to the family in such a situation that their loved one is in fact dead would be aided and supported by the passage of an up-to-date statutory definition of death. This would reflect a public policy that recognizes that when the brain is dead, the person as a whole is dead and there is neither life nor hope despite the mechanically supported respiration and heartbeat. The presence of such a statute will remove from the physician the fear of unjust litigation and thereby allow him to practice medicine in a manner consistent with present scientific knowledge and standards. It will allow him to do this openly after honest discussion with the patient's family, and it will permit him to cooperate in efforts to procure cadaver organs in optimal condition for transplantation into other patients.

A further advantage of having a statutory definition of death is that it would help to guarantee that the highest standards of medical science would be used to make this determination. The recent New Jersey Supreme Court decision[3] in the Karen Quinlan case underscores the need to assure the public that pronouncements of death will be based on standardized and thoroughly validated indices. Even though Ms. Quinlan did not fulfill the criteria of brain death, the court held that her parent acting as guardian might authorize cessation of life-sustaining treatment if a physician and a hospital "ethics committee" agreed there was no reasonable possibility of recovery to a cognitive, sapient state. Although this decision and the resulting discontinuation of respiratory support did not alter Ms. Quinlan's course because she was able to breathe spontaneously, there has been substantial confusion between the issues in this case and the debate about the definition of death. Such misunderstandings could result in less than optimal nonstandardized determinations of death. These would be prevented by the existence of a statute that mandates use of the best standards of current medical practice to pronounce death.

A statutory definition of death would also have other advantages from a legal point of view. It would provide a clear and precise definition within which legal rights and relationships after death could be determined. It would facilitate the prosecution of murderers and permit the organs of murder victims to be used as transplants without jeopardizing conviction of the murderer. Such a statute would provide for consistency under the law in various jurisdictions, and it would avoid reliance on jury systems to make medical and legal decisions that might be inconsistent with present scientific knowledge.

Many physicians have suggested that the specific criteria for pronouncing brain death should not be placed in a statutory form. This is a reasonable position, since there is always the possibility that the criteria might change. This would mean that the law would have to be changed prior to utilizing any new and improved criteria. This is obviously a good reason not to legislate *specific* neurologic criteria for pronouncing death. However, it is an inadequate reason for opposing a law that, while leaving the specific criteria flexible, recognizes that death may be pronounced when irreversible cessation of brain function occurs.

Reason for legal not specific criteria

Legal Status of Brain Death

Until recently, the traditional legal definition of death has been consistent with the prevailing medical concept that death is determined by cessation of the vital functions of respiration and heartbeat. This is reflected in the common law definition of death as stated in *Black's Law Dictionary*[2]: "The cessation of life; the ceasing to exist; defined by physicians as a total stoppage of the circulation of the blood, and a cessation of the animal and vital functions consequent thereon, such as respiration, pulsation, etc." With the exception of several notable recent decisions, traditional case law has similarly concentrated on the cardiovascular and respiratory functions as prime determinants of the occurrence of death.

In *Smith vs Smith*,[4] the Supreme Court of Arkansas, in a case that turned on the issue of simultaneous death, adopted *Black's* definition of death verbatim. The court took judicial notice of the fact that "one breathing, though unconscious, is not dead." Similarly, in *Thomas vs Anderson*,[5] a California District Court of Appeals also cited *Black's* definition and stated that ". . . death occurs precisely when life ceases and does not occur until the heart stops beating and respiration ends. Death is not a continuous event and is an event that takes place at a precise time." Other jurisdictions have also relied on this definition.[6,7] In addition, other cases have upheld the premise that death has not occurred until cessation of heartbeat and respiration even in circumstances where the courts have noted complete destruction of the brain.[8-10]

In all the cases cited, the determination of death was dealt with as a question of fact for a jury to decide in connection with the demise of individuals for the purposes of construing and applying "simultaneous death" clauses in testamentary documents. The factual question of the time of death in these cases was judged on the basis of circumstantial evidence relating to the cessation of heartbeat and respiration. This evidence was provided by the testimony of lay persons rather than physicians. These cases, which constitute the leading precedents in this area, predated the landmark report of the Ad Hoc Committee of the Harvard Medical School to Examine the Definition of Brain Death,[11] which is now generally regarded as the first widely recognized index that current medical concepts about the definition of death were changing.

Emerging Case Law

In contrast with these traditional opinions, several recent cases have considered the issue of death in the light of expert testimony by physicians about the irreversible cessation of brain function and have often incorporated such testimony in jury charges. These cases give explicit or implicit legal recognition to a pronouncement of death based on a determination of irreversible cessation of brain function in accordance with the customary standards of medical practice. These legal actions were relevant because the medical determination of the timing and occurrence of death in these cases were based on brain-related criteria, and legal application of traditional criteria of death would have been inappropriate. Thus, judicial action fortunately kept the law apace of scientific developments.

The first such judicial action occurred in the Oregon case, *State vs Brown*,[12] in which the defendant had been convicted on a charge of second-degree murder. On appeal, he contended that the victim's death was caused by termination of life-support systems rather than by the cranial gunshot wound that he had inflicted. The court held that the defendant's contention was without merit on the basis of expert medical testimony that the gunshot wound with resultant brain damage was the cause of death.

One year later, the impact of current medical thinking on case law was clearly evident in a Virginia case, *Tucker vs Lower*.[13] In a wrongful death action, it was alleged that an individual was not dead at the time when his heart and kidneys were removed for purposes of transplantation. The court rejected a motion for a summary judgment in favor of the defendants on the grounds that the Court was bound by the common law definition of death until it was changed by the state legislature. However, after considerable debate, the court instructed the jury that it might properly consider, as a substitute for the traditional criteria for determining the time of death, "the time of complete and irreversible loss of all function of the brain; and, whether or not the aforesaid functions [respiration and circulation] were spontaneous or were being maintained artificially or mechanically." The jury then decided that the transplant surgeons were not guilty of causing a wrongful death. Whether or not this decision was based on the jury's acceptance of the brain death concept is not known. However, this case has been widely publicized as supporting the use of brain-related criteria for pronouncing death. Furthermore, in commenting on this case, one legal scholar points out that "the jury instructions represent an admission by the courts that the old legal definition of death needs modification in the light of advances in medical science. The new definition—'brain death'—which is gaining recognition, reflects the consensus of informed medical opinion."[14]

Similarly, in a widely publicized California case, *People vs Lyons*,[15] a victim had suffered a gunshot wound of the head and had been declared neurologically dead before a transplant team headed by Norman Shumway, MD, had removed his heart. The defendant pleaded not guilty to a charge of murder, contending that the death of the victim had been caused by the removal of his heart rather than by the gunshot wound inflicted by the defendant. On the basis of expert testimony, the jury was instructed as a matter of law that "the victim was legally dead before removal of the organs from his body." The court thereby removed from the jury its traditional task of having to determine the exact time of death. The brain death standard was explicitly accepted.

However, a contrasting ruling was made in the initial phase of another California criminal prosecution, emphasizing the inconsistency that can occur with case law. In this case the defendant, who had been driving on the wrong side of a freeway while intoxicated, had caused an accident that severely injured a 13-year-old girl. She was pronounced dead on brain-related criteria, and her heart was used as a transplant. On the basis of these facts, a municipal court judge at a preliminary hearing did not hold the defendant to answer to a manslaughter charge, apparently determining that the pronouncement of death on neurologic criteria and the subsequent removal of the heart created substantial doubt as to the proximate cause of death. The Court concluded that "the evidence is not certain as to the cause of death of Colenda Ward, certain enough to charge this defendant with manslaughter."[16] On a subsequent appeal by the district attorney, the Superior Court authorized the filing of a manslaughter charge and made reference to "unimpeached medical testimony," which conclusively established "that the [victim's] heart could not beat nor could she breathe without artificial support."[16] The defendant was convicted of both manslaughter and felony drunk driving but received a sentence of less than five months. In commenting on this result, the deputy district attorney observed, "I cannot escape the firm belief that the uncertain state of the case and statutory law on the subject of brain death was a substantial factor in the imposition of such a light sentence" (written communication from Steven T. Tucker, deputy district attorney, Sonoma County, Calif., May 19, 1977).

A rather novel approach to the legal question of when does death occur was taken in 1975 in New York,[17] where a court was requested to set forth, in an action for declaratory judgment, a legal definition of the terms "death" and "time of death" as used in the New York State Anatomical Gifts Act. The court was asked to include in such definition not only the common law criteria of cardiac and respiratory failure, but also the concept of "brain death." Following extensive uncontroverted testimony concerning brain death criteria, the court held that "death" as used in the Anatomical Gifts Act "implies a definition consistent with the generally accepted medical practice of doctors primarily concerned with effectuating the purposes of this statute." Having confined its decision legally recognizing brain-related criteria for pronouncing death to the Anatomical Gifts Act, the court concluded by urging the state legislature "to take affirmative action to provide a Statewide remedy for this problem."

These cases have been helpful in resolving particular controversies. However, they have probably had a greater impact by serving as a vehicle for increasing public awareness of the need for a statutory definition of death. In all but one instance in which litigation has arisen, legislation that recognizes the validity of brain death as a legally accepted standard for determining death has been enacted shortly thereafter. The single exception is New York where proposals are currently pending before the state legislature.

Legislation

At the present time, 18 states have enacted a statutory definition of death: Kansas, Maryland, Virginia, New Mexico, Alaska, California, Georgia, Michigan, Oregon, Illinois, Oklahoma, West Virginia, Tennessee, Louisiana, Iowa, Idaho, Montana, and North Carolina.[18,19] All 18 statutes recognize that death may be pronounced on the basis of irreversible cessation of brain function, and none describes in detail the specific criteria for determining brain death. However, the laws vary in certain major and minor ways. In general, they conform to one of three major types or patterns.

The first of these includes laws providing alternative definitions of death. Typical of this pattern is the first statute enacted in 1970 by Kansas:

> A person will be considered medically and legally dead if, in the opinion of a physician, based on ordinary standards of medical practice, there is the absence of spontaneous respiratory and cardiac function and, because of the disease or condition

which caused, directly or indirectly, these functions to cease, or because of the passage of time since these functions ceased, attempts at resuscitation are considered hopeless; and, in this event, death will have occurred at the time these functions ceased; or

A person will be considered medically and legally dead if, in the opinion of a physician, based on ordinary standards of medical practice, there is the absence of spontaneous brain function; and if based on ordinary standards of medical practice, during reasonable attempts to either maintain or restore spontaneous circulatory or respiratory function in the absence of aforesaid brain function, it appears that further attempts at resuscitation or supportive maintenance will not succeed, death will have occurred at the time when these conditions first coincide. Death is to be pronounced before artificial means of supporting respiratory and circulatory function are terminated and before any vital organ is removed for purposes of transplantation.

These alternative definitions of death are to be utilized for all purposes in this state, including the trials of civil and criminal cases, any laws to the contrary notwithstanding.

An identical statute was enacted by Maryland in 1972. In 1973, Virginia passed a similar law that differed only in that death on brain-related criteria can only be declared by two physicians, one of whom is a specialist in neurology, neurosurgery, or electroencephalography. The Virginia law also mandates that absence of spontaneous respiratory functions accompanies "absence of spontaneous brain functions." The New Mexico statute, passed in 1973, and the Alaska statute, passed in 1974, are similar to the Kansas law. The Oregon statute, enacted in 1975, is the simplest and clearest of the alternative definition type of law: "When a physician licensed to practice medicine acts to determine that a person is dead, he may make sure a determination if irreversible cessation of spontaneous respiration and circulatory function or irreversible cessation of spontaneous brain function exists."

All six of these alternative definition of death statutes suffer the disadvantage of providing two different definitions of death. The choice of which to use in a specific instance is left to the physician. The major flaw with this type of legislation is that it appears to be based on the misconception that there are two separate types of death. This is particularly unfortunate because it seems to relate to the need to establish a special definition of death for organ transplant donors. These laws could lend support to the fear that a prospective transplant organ donor would be considered dead at an earlier point in the dying process than an identical patient who was not a potential donor.

In addition, such laws suffer the legal disadvantage of possibly permitting a physician, either inadvertently or intentionally, to influence the outcome of a will. If, for example, a husband and wife are fatally injured in the same accident, survivorship and consequent inheritance may be determined by the physician's choice of which of the alternative criteria to use in the pronouncements of death.

② The second major type of law was suggested by Capron and Kass to remedy this defect and to provide one definition of death that recognizes that death is a single phenomenon that can be determined by brain related criteria only in situations where artificial support of respiratory and circulatory functions is being maintained[20]:

> A person will be considered dead if in the announced opinion of a physician, based on ordinary standards of medical practice, he has experienced an irreversible cessation of spontaneous respiratory and circulatory functions. In the event that artificial means of support preclude a determination that these functions have ceased, a person will be considered dead if in the announced opinion of a physician based on ordinary practice, he has experienced an irreversible cessation of spontaneous brain functions. Death will have occurred at the time when the relevant functions ceased.

This model statute takes cognizance of the fact that the medical standards for pronouncing death may vary with circumstances. However, unlike the previous laws, it does not leave as an arbitrary decision for the physician the choice of which standard to apply, but defines under what circumstances the new or secondary brain-related criteria may be used. This bill avoids establishing a separate kind of death, brain death, and as pointed out by Capron and Kass provides "two standards gauged by different functions for measuring different manifestations of the same phenomenon. If cardiac and pulmonary functions have ceased, brain functions cannot continue; if there is no brain activity and respiration has to be maintained artificially, the same state [i.e. death] exists."[20] The Capron-Kass Bill, which clearly appears to be satisfactory if not ideal, was adopted by Michigan and West Virginia in 1975 and Louisiana in 1976. The latter law specifies that when organs are to be used as a transplant, an additional physician unassociated with the transplant team must also pronounce death. Iowa in 1976 and Montana in 1977 enacted laws based on the

Capron-Kass model with the additional requirement that brain death pronouncements must be made by two physicians.

The third major type of law follows the suggestion of the American Bar Association, which recognized the need for a standardized statutory definition of death that minimized the risk of confusion from misunderstandings of semantics, medical technology, and legal sophistication and that took into account recent developments in transplantation, supportive therapy, and resuscitation. The suggested law was developed by the Law and Medicine Committee of the American Bar Association in 1974 and approved by the House of Delegates of that organization in 1975:[21] "For all legal purposes, a human body with irreversible cessation of total brain function, according to usual and customary standards of medical practice, shall be considered dead." This Committee states[21] that the advantages of its simple, direct definition are that it (1) permits judicial determination of the ultimate fact of death, (2) permits medical determination of the evidentiary fact of death, (3) avoids religious determination of any facts, (4) avoids prescribing the medical criteria, (5) enhances changing medical criteria, (6) enhances local medical practice tests, (7) covers the three known tests (brain, beat, and breath deaths), (8) covers death as a process (medical preference), (9) covers death as a point in time (legal preference), (10) avoids passive euthanasia, (11) avoids active euthanasia, (12) covers current American and European medical practices, (13) covers both civil law and criminal law, (14) covers current American judicial decisions, and (15) avoids nonphysical sciences.

Some have objected that this simple model statute fails to recognize the still common practice of pronouncing death on the basis of cessation of heartbeat and respiration. However, in practice, death is only pronounced when the functions of circulation and respiration have ceased long enough to cause destruction of the brain and produce other signs of lifelessness. In these instances, cessation of circulation and respiration represents the specific criteria by which irreversible cessation of brain function or death is determined. Thus, this model statute recognizes traditional as well as brain-related criteria for determining death. It is, therefore, also satisfactory and has formed the basis for the California statute enacted in 1974, the Georgia statute enacted in 1975, and the Idaho statute passed in 1977. All three laws require that deaths pronounced on brain-related criteria be confirmed by a second physician. The Illinois statute, enacted in 1975, also resembles the American Bar Association's suggestion in its simplicity. It does not require concurrence of a second physician, although it has the flaw of restricting the use of brain-related criteria to instances involving the Uniform Anatomical Gift Act. It, therefore, has the disadvantage of implicitly establishing alternative types of death with special definition to be used for transplant organ donors. The Oklahoma law, enacted in 1975, also seems to be based on the American Bar Association model but is rendered confusing by the addition of a number of qualifying clauses and phrases mandating that it must also appear "that further attempts at resuscitation and supportive maintenance will not succeed." The Tennessee statute, enacted in 1976, avoids these flaws and complexities and follows exactly the recommendation of the American Bar Association.

Many factors underlie the variability between the statutes enacted in the different states and account for the difficulty in reaching agreement on what constitutes the wording of a single ideal statutory definition of death. Prominent among these factors is the present climate of public mistrust of the medical profession. This has prompted legislators to enact more complicated laws in an attempt to protect patients from erroneous or premature declarations of death.

Hopefully the present article, by summarizing the overwhelming evidence supporting the validity of brain death, will help to allay these concerns and facilitate drafting of simple, effective statutes defining death. Furthermore, by showing that pronouncements of death on brain-related criteria are in accord with secular philosophy and principles of the three major Western religions, it is hoped that the present article will help overcome opposition to legislation from those who previously failed to accept the brain death concept. And lastly, by documenting the compelling reasons to have a statutory definition of death, the present article will hopefully help to influence the American Medical Association and others, who have felt that legislation defining death is unnecessary, to adopt a supportive position, as several of the state medical societies have already done. Such support would greatly facilitate passage of appropriate statutes in the 32 states presently without them. This, in turn, would make the law in regard to brain death consistent with current medical practice throughout the entire United States.[22]

NOTES

Part I
1. High D: Death: Its conceptual elusiveness. *Soundings* 55:438–458, 1972.

2. Ramsey P: *The Patient as Person: Explorations in Medical Ethics.* New Haven, Conn, Yale University Press, 1970, pp. 101–112.

3. Haring B: *Medical Ethics.* Notre Dame, Ind, Fides Publishers Inc, 1973.

4. Geelhoed GW: Life and death: Who decides? *The Pharos,* January 1977, pp 7–12.

5. Jonas H: Against the stream: Comments on the definition and redefinition of death, in Jonas H (ed): *Philosophical Essays: From Ancient Creed to Technological Man.* Englewood Cliffs, NJ, Prentice Hall, 1974, pp 132–140.

6. Definition of death. *JAMA* 227:728, 1974.

7. Tobin CJ: Statement in behalf of the New York State Catholic Conference delivered at the public hearing related to death legislation held by the New York State Assembly Subcommittee on Health Care, Albany, New York, Nov 30, 1976. *Origins* 6:413–415, 1976.

8. A definition of irreversible coma, Report of the Ad Hoc Committee of the Harvard Medical School to Examine the Definition of Brain Death. *JAMA* 205:337–340, 1968.

9. Refinements in criteria for the determination of death: An appraisal, report by the Task Force on Death and Dying of the Institute of Society, Ethics and the Life Sciences. *JAMA* 221:48–53, 1972.

10. Walker AE, Diamond EL, Moseley J: The neuropathological findings in irreversible coma: A critique of the "respirator brain." *J Neuropathol Exp Neurol* 34:295–323, 1975.

11. Korein J, Braunstein P, Kricheff I, et al: Radioisotopic bolus technique as a test to detect circulatory deficit associated with cerebral death: 142 studies on 80 patients demonstrating the bedside use of an innocuous IV procedure as an adjunct in the diagnosis of cerebral death. *Circulation* 51:924–939, 1975.

12. Crafoord C: Cerebral death and the transplantation era. *Dis Chest* 55:141–145, 1969.

13. Heiskanen O: Cerebral circulatory arrest caused by acute increase of intracranial pressure: A clinical and roentgenological study of 25 cases. *Acta Neurol Scand* 40(suppl 7):1–57, 1964.

14. Bergquist E, Bergstrom K: Angiography in cerebral death. *Acta Radiol (Diagn)* 12:283–288, 1972.

15. Lofstedt S, von Reis G: Diminution or obstruction of blood flow in the internal carotid artery. *Opuscula Med* 4:345–360, 1959.

16. Greitz T, Gordon E, Kolmodin G, et al: Aortocranial and carotid angiography in determination of brain death. *Neuroradiology* 5:13–19, 1973.

17. Silverman D, Saunders MG, Schwab RS, et al: Cerebral death and the electroencephalogram, Report of the Ad Hoc Committee of the American Electroencephalographic Society on EEG Criteria for Determination of Cerebral Death. *JAMA* 209:1505–1510, 1969.

18. Silverman D, Masland RL, Saunders MG, et al: Irreversible coma associated with electrocerebral silence. *Neurology* 20:525–533, 1970.

19. Masland RL: When is a person dead? *Resident Staff Physician,* April 5, 1975, pp 49–52.

20. Korein J, Maccario M: On the diagnosis of cerebral death: A prospective study on 55 patients to define irreversible coma. *Clin Electroencephalogr* 2:178–199, 1971.

21. DeMere M, Alexander T, Auerbach A, et al: *Report on Definition of Death, From Law and Medicine Committee. Chicago, American Bar Association,* Feb 25, 1975.

22. This study was supported in part by US Public Health Service grants HL 16476 and HL 17417 and the Manning Foundation.

23. Walker AE: Cerebral death, in Tower DB (ed): *The Nervous System: The Clinical Neurosciences.* New York, Raven Press, 1975, vol 2, pp 75–87.

24. Ueki K, Takeuchi K, Katsurada K: Clinical study of brain death, presentation 286. Read before the Fifth International Congress of Neurologic Surgery, Tokyo, Oct 7–13, 1973.

25. Walker AE: The neurosurgeon's responsibility for organ procurement. *J Neurosurg* 44:1–2, 1976.

26. An appraisal of the criteria of cerebral death: A summary statement, A collaborative study. *JAMA* 237:982–986, 1977.

27. Diagnosis of brain death: Summary of Conference of Royal Colleges and Faculties of the United Kingdom. *Lancet* 2:1069–1970, 1976.

28. Molinari GF: Death: The definition: III. Criteria for death. *Encyclopedia of Bioethics,* to be published.

29. Bolton CF, Brown JD, Cholod E, et al: EEG and "brain life." *Lancet* 1:535, 1976.

30. Bab Talmud Tractate Yoma 85A.

31. Bab Talmud Tractate Chullin 21A and Mishnah Oholoth 1:6.

32. Maimonides: Tumath Meth 1:15.

33. Code of Laws: Who is considered as dead although yet living? Yoreh Deah:370:1.

34. Genesis 2:7.

35. Connery JR: Comment on the proposed act to amend the public health law of the State of New York in relation to the determination of the occurrence of death. Read before the New York State Legislature, 1975–1976 session, March 25, 1975.

36. McFadden CJ: *The Dignity of Life: Moral Values in a Changing Society.* Huntington, Ind, Our Sunday Visitor Inc, 1976, p 202.

37. Pius XII, Acta Apostolicae Sedia 45, November 1957, pp 1027–1033.

38. Nelson J: *Human Medicine: Ethical Perspective on New Medical Issues.* Minneapolis, Augsburg, 1973, pp 125–130.

39. Fletcher J: Our shameful waste of human tissue: An ethical problem for the living and the dead, in Cutler DR (ed): *Updating Life and Death: Essays in Ethics and Medicine.* Boston, Beacon Press, 1969, pp 1–30.

40. Fletcher J: New definitions of death. *Prism* 2:13ff, January 1974.

41. Vaux K: *Biomedical Ethics.* New York, Harper & Row, 1974, pp 102–110.

42. Smith H: *Ethics and the New Medicine.* Nashville, Tenn, Abingdon, 1970.

Part II

1. Definition of death. *JAMA* 227:728, 1974.

2. *Black's Law Dictionary,* ed 4. St Paul, West Publishing Co, 1968, p 488.

3. *In the matter of Karen Quinlan,* 355 A 2d 647, 1976.

4. *Smith vs Smith,* 229 Ark 579, 317 SW 2d 275, 1958.

5. *Thomas vs Anderson,* 211 P 2d 478, 1950.

6. *United Trust Co vs Pyke,* 427 P 2d 67, 1967.

7. *Schmidt vs Pierce,* 344 SW 2d 120, 1961.

8. *Vaegemast vs Hess,* 280 NW 641, 1938.

9. *Gray vs Sawyer,* 247 SW 2d 496, 1952.

10. *In Re Estate of Schmidt,* 67 Cal Reptr 847, 1968.

11. A definition of irreversible coma, Report of the Ad Hoc Committee of the Harvard Medical School to Examine the Definition of Brain Death. *JAMA* 205:337–340, 1968.

12. *State vs Brown,* 8 Oreg App 72, 1971.

13. *Tucker vs Lower,* No. 2381, Richmond Va. L & Eq Ct, May 23, 1972.

14. Kennedy I: The legal definition of death. *Medico-Legal Journal* 14:36–41, 1973.

15. *People vs Lyons,* 15 Crm L Rprt 2240, Cal Super Ct 1974.

16. *People vs Flores,* Cal Super Ct County, 7246-C, 1974, pp 1–2.

17. *New York City Health & Hosp Corp vs Sulsona,* 81 Misc 2d 1002, 1975.

18. State Laws: *Kan Stat Ann* § 77–202, Supp 1974; *Md Code Ann* § 32–364.3:1, Cum Suppl 1975; *New Mex State Ann* § 1–2–2.2, Supp 1973; *Alaska Stat* § 9.65.120, Supp 1974; *Va Code Ann* § 32–364.3:1, Supp 1975; *Cal Health and Safety Code Ann* § 7180–81, West Supp 1975; *Ill Ann Stat* Ch 3, § 552, Smith-Hurd Supp 1975; *Ga Code Ann* § 88–1715.1, 1975; *Mich Stat* PA 158, Laws 1975; *Ore Rev Stat* Ch 565, § 1, Laws 1975; Ch 91, Laws 1975, amending *Okla Stat Ann* tit 63, § 1–301 (g), 1971; *W Va Code Ann* § 16–19–1, Supp 1975; *Tenn Stat,* HB No. 1919 Ch 780, Laws 1976; *La Acts 1976,* No. 233, § 1; *Laws of 66th Iowa General Assembly,* Ch 1245, (1976 Senate File 85); Mont HB No. 371, Ch 228, Laws 1977; *1977 Idaho Session Laws,* No. 1197, Ch 30; *N Carol Laws 1977,* Ch 815, § 90–322.

19. Stuart FP: Progress in legal definition of brain death and consent to remove cadaver organs. *Surgery* 81:68–73, 1977.

20. Capron AM, Kass LR: A statutory definition of the standards for determining human death: An appraisal and a proposal. *U Penn L Rev* 121:87–118, 1972.

21. DeMere M, Alexander T, Auerbach A, et al: *Report on Definition of Death, From Law and Medicine Committee. Chicago, American Bar Association,* Feb 25, 1975.

22. This study was supported in part by US Public Health Service grants HL 16476 and HL 17417 and the Manning Foundation.

10 *The Right To Die*

HANS JONAS, PH.D

The introduction of technology into medicine has provided it with the ability to prolong the lives of individuals. Nonetheless, that ability has raised several problems that Jonas grapples with in this essay: suicide; aiding in suicide; the troublesome distinction between killing and letting die; and the right to die. Recognizing the legal rule that individuals have a right to refuse treatment, Jonas presses on to the more difficult moral dimensions of the problem and discusses issues of responsibility to one's family and society, problems surrounding a duty to force treatment on others, and the recognition of how to deal with people's autonomy in terms of informing them properly about their condition. Jonas pursues these issues by evaluating two different cases: a conscious, suffering patient who has terminal cancer, and an unconscious comatose patient. After drawing his conclusions on these cases, Jonas then concludes by evaluating the task of medicine in the light of these particular problems

Hans Jonas, Ph.D., is a member of the Graduate Faculty of The New School of Social Research

In spite of the almost commonplace ring which the title of this article has acquired in the course of recent public debate, one's first reaction should still be wonder. What an odd combination of words! How strange that we should nowadays speak of a right to *die*, when throughout the ages all talk about rights has been predicated on the most fundamental of all rights—the right to *live*. Indeed, every other right ever argued, claimed, granted, or denied can be viewed as an extension of this primary right, since every particular right concerns the exercise of some faculty of life, the access to some necessity of life, the satisfaction of some aspiration of life.

Life itself exists not by right but by natural fiat: my being alive as a sheer fact whose sole natural endorsement is its innate powers to preserve itself. But among men the fact, once there, needs the sanction of a *right,* because to live means to make demands on the environment and is thus conditional on being accommodated by that environment. Insofar as the environment is human, and the accommodation it accords has a voluntary element in it, such inclusive accommodation underlying all communal life amounts to an implicit granting by the many of the individual's *right* to live and his granting the same to all others. Every further right, equal or not, in natural or positive law, derives from this cardinal one and from the mutual recognition of it by its claimants. Justly, therefore, is "life" named first among the inalienable rights in the Declaration of Independence. And surely, mankind has had an arduous task, and still has, with discovering, defining, debating, obtaining, and protecting the various rights that enlarge upon the right to live.

How exceedingly odd then that we should lately find ourselves engrossed in the question of a right to die! All the more odd, since rights are espoused to secure a blessing, and death is counted as a curse or at best something to which one must be resigned. And most odd when we reflect that with death we are not making a demand on the world, where a question of right might arise, but on the contrary are quitting every possible demand. So how can the very idea of "right" here apply?

But what if my dying or not dying were to some extent in the area of choice—mine or others'? And what if not only a right but also a duty for me to live could be construed? And for others, that is, "society," not only a duty toward my right to live, but also a right to hold me to my duty to live, that is, to prevent me from dying sooner than I must, even if I so will? What, in short, if the whole syndrome of dying becomes amenable to control, and mine is arguably not the sole voice to be heard in exercising that control? Then a "right to die" does become an issue. It has in fact always been a moral and religious issue in the matter of suicide, where the element of choice is most clearly present; and there, it is also a legal issue in many systems of law that mandate or authorize various forms of intervention in this most private of all acts, some even going so far as to make suicide a crime. However, the present concern with a right to die is not about suicide, the deed of an active subject, but about a moribund patient's being passively subjected to the death-delaying ministrations of modern medicine. And although some aspects of the ethics of suicide do intrude into this question too, the presence of a fatal malady as the chief agency of death enables us to distinguish submitting to death from killing oneself, and also permitting to die from causing death.

The novel problem is this: <u>medical technology, even when it can-</u>
<u>not cure or relieve or purchase a further, if short-term, lease on a</u>
<u>worthwhile life, can still put off the terminal event of death beyond</u>
<u>the point where the patient himself may value the life thus pro-</u>
<u>longed, or even is still capable of any valuing at all.</u> This often marks
a therapeutic stage where the line between life and death wholly coincides with that between continuance and discontinuance of the
treatment—in other words, <u>where the treatment does nothing but</u>
<u>keep the organism going, without in any sense being ameliorative, let</u>
<u>alone curative.</u> This case of the hopelessly suffering or comatose pa-
tient is only the extreme in a spectrum of medical knowledge which,
allied to the institutional power of the hospital and backed by the
law, creates situations in which it becomes a question whether the
rights of the (typically powerless and somehow captive) patient are
observed or violated; and among them would be a right to die. In addition, when treatment becomes identical with keeping alive, there
arises for physician and hospital the spectre of killing by discontinuing treatment, for the patient the spectre of suicide with demanding
it, for others that of complicity in one or the other with mercifully facilitating or not resisting it. This aspect of the matter, which alloys its
purely ethical resolution with legal constraints and fears, I will discuss later. <u>As to the rights of the patient, a novel "right to die" does</u>
<u>seem to have emerged with these developments;</u> and because of the
novel, merely sustaining types of treatment, this right is clearly subsumed under the general right to accept or reject treatment. Let us
first discuss this wider and little contested right, which always, if in a
less direct manner, also includes death as a possible and perhaps certain outcome of its choices. In this, as in our whole discourse, we shall
have to distinguish between legal and moral rights (and likewise duties).

The Right to Refuse Treatment

Now legally, in a free society, there is no question that everyone
(except minors and incompetents) is entirely free to seek or not to
seek medical advice or treatment for any illness, and equally free to
withdraw from a treatment at any time other than in the midst of a
critical phase.[1] The only exception is illness that poses a danger to
others, as do contagious diseases and certain mental disorders: there,
treatment and confinement and also preventive measures like inocu-
lation can be made mandatory. Otherwise, without such direct implication of the public interest, my sickness or health is my wholly
private affair, and I freely contract for a physician's services. This, I
believe, is the *legal* position here and generally in nontotalitarian societies.

Morally, the matter is more complex. I may have responsibilities for others whose well-being depends on mine, for example, as provider for a family, mother of small children, crucial performer of a public task, which put moral restraints on my freedom to decide against treatment for myself. They are essentially the same as those that ethically restrict my right to suicide even when the religious strictures no longer count for me. With some kinds of treatment, such as the dialysis machine for kidney failure, rejection in effect amounts to suicide. Yet there is a significant difference from doing violence to oneself: others, including public powers, in fact any bystander, have the right (even considered a duty) to thwart an *active* suicide attempt by timely intervention, not excluding force. This, admittedly, is interference with a subject's most private freedom, but only a momentary one and in the longer perspective an act in behalf of that very freedom. For it will merely restore the status quo of a free agent with the opportunity for second thoughts, in which he can revise what may have been the decision of a moment's despair—or can persist in it. Persistence will in the end succeed anyhow. The time-bound intervention treats the time-bound act like an accident, from which to be saved, even against himself, and can be presumed as the victim's own more enduring, if temporarily eclipsed, wish (sometimes betrayed by the very fact of imperfect secrecy that made the intervention possible). The saved one has the power to prove the presumption wrong. I am not discussing the ethics of suicide itself, only the rights (or duties) of others to interfere with it, and there, what counts in the present discourse is just this: that counterviolence at the moment of suicidal violence will not force the subject to go on living but will merely reopen the issue.

Clearly, it is a different matter to force the hopelessly sick and suffering to keep submitting to a sustaining treatment that buys him a life he deems not worth living. Nobody has a right, let alone a duty, to do this in a protracted denial of self-determination. Temporizing restraint is required to shield the irrevocable from rashness. But beyond such brief external delay, only the inward pull of responsibilities—"I must spare myself for them"—may restrain the subject, through his own will, from doing what all by himself he would choose to do.[2] But the same kind of consideration, let us remember, can also lead to the opposite conclusion: "The treatment is financially ruinous for my family, and for their sake I quit." If a duty—albeit a nonenforceable duty—to live on for others against one's wish is asserted, then at least a right to die for them must also be conceded. But not a duty! The two opposite pulls of responsibility are not of equal moral weight, as is evident when we ask what those with a claim on a person, the objects of his responsibility, can decently plead with him for: surely only his staying alive, never his agreeing to die. Death must be the most uninfluenced of all choices; life may have its advocates, even from self-interest, certainly from love. Yet even life's case must not be pressed too hard in any such pleading. Especially love should acknowledge, against the clamor of self-interest, that no duty-to-live, though it may *overcome* in me the *desire* to die, can really nullify my right to choose death in the circumstances here assumed. Whatever the claims of the world, that right is (outside religion) morally and legally as inalienable as the right to life, although the exercise of either right may by choice—but only by free choice—be subject to other concerns. The matching of the two in a pair means that either right sees to it that the other cannot be turned into an unconditional duty: neither to live nor to die.[3]

Does public law have a place in all this? Yes, in two supportive respects: first, as part of its function to protect the right to life, it must also sanction the right to medical treatment by ensuring equal access to it; and second, given the scarcity of medical resources, it must establish equitable standards of priority for that access. This latter function of public control, as is well known from the dialysis example, can amount to decisions on who is to live and who is to die, and among the priorities governing that agonizing decision can be an individual's responsibilities for dependents, which may give him, other things being equal, an edge of eligibility over the lone individual. Thus the same thing we have found to militate from within against a person's wish or right to refuse treatment, that is, the dependence of others on him, appears now from without as an increased title to that treatment—at the cost of a third party's right to live. But what public authority can give, it can also later take away under the same principle of equity, in favor of a better claim. We shall return to this contingency as an indirect legal resort in aid of the right to die.

The dialysis example is an extreme one. Usually, the right to decline treatment or ignore doctor's orders does not involve the right to die, unless in a most abstract and indirect way, but rather the right to take risks, to gamble a bit with one's health, to trust nature and distrust medical art, to be fatalistic, or simply to come to terms with a disability, even with a shorter life expectancy, in exchange for freedom from a restrictive regimen; or just the right not to be bothered. The dialysis example was chosen because, there, continuous treatment is tantamount to keeping alive, its interruption spelling certain death, and thus opting against it is not "taking a chance" but indeed a straightforward and directly effective decision for death. It is still not quite the type of case where the "right to die" poses itself as the worrisome issue it has lately become. For here the patient is usually un-impaired in his mental powers to decide for himself, and also physically a free agent who can take himself off the machine, and no-body can force him back. Thus his right to die does not draw in the cooperation of others and can be exercised all by himself. The same is true for many other life-sustaining therapies, like insulin use for dia-betics. In such cases, the power both to make the decision and to im-plement it oneself is present, and the right to die is not seriously contested nor effectively hindered from outside, whatever its inward ethics may be. The "worrisome" cases are those of the more or less captive, such as the hospitalized patient with terminal illness, whose helplessness necessarily casts others in the role of accessories to real-izing his option for death, even to the point of substituting for him in making the choice.

I shall discuss two examples: the *conscious,* suffering patient with a disease like terminal cancer; and the irrecoverably *unconscious,* coma-tose patient. The latter example has been capturing the headlines lately because of the legal drama involved, but the former is more rel-evant intrinsically, more common, and also more problem-ridden.

The Conscious Terminally Ill Patient

Consider the following scenario. The doctor says, perhaps after a first or second operation, "We must operate again." The patient says, "No." The doctor says, "Then you will surely die." The patient says, "So be it." Since surgery requires the patient's consent, this would seem to close the matter and raise neither ethical nor legal problems. However, the reality is not quite that simple. In the first place, the patient's refusal must be based on the same enabling condition as his consent: it must be "informed" to be valid. In fact, a person's consent is informed only if he also knows of the adverse prospects on which a "no" might be based. Thus the right to die (if considered to be exer-cised by its competent subject himself, not by a proxy in his behalf) is inseparable from a right to truth and is in effect voided by deception. But such deception is almost part of medical practice, and not only for humane but often also for outright therapeutical reasons.

Imagine the following enlargement of the above dialogue, after the doctor has announced the necessity for another operation. Pa-tient: "What, if successful, will it give me? How long a survival and what sort of survival? As a chronic patient or with a return to normal life? In pain or free from it? How long to the next onset with a repeti-tion of the present emergency?" (Remember, we speak of an incur-able, intrinsically "terminal" condition.) All these questions, to be sure, are about reasonable *chances* according to available medical knowledge—no more, but also no less.

THE RIGHT TO DIE   201

Obviously the patient is entitled to an honest answer. But no less obviously, the doctor is in a quandary when honesty means cruelty. Does the patient really want the unmitigated truth? Can he face it? What will it do to his state of mind in the precious remainder of his doomed life, whether he decides for or against the temporary reprieve? And more vexing still: may the dire truth of my estimate not be self-fulfilling in that it saps the spiritual resources, the famous "will to live," with which the patient might succor my ministration, his "giving up" actually worsening the prognosis? Hope after all is a force by itself, and to stress it more than the opposite is a means not only of persuading to the therapy but also of strengthening its chances. In short, might truth not be actually injurious to the patient and deception not be beneficial in *some* sense, subjective and objective? Thus we find ourselves, in meditating on the right to die, confronted with the much older and familiar question: should the doctor tell? That question indeed arises prior to our imagined situation. Should the doctor have told the patient in the first place that his condition, short of a medical miracle, is incurable? Even "terminal" in the sense that it admits only of brief delay?

Pat answers to these questions would betray insensitivity to their complexity. For myself, I venture this statement of principle: ultimately, the patient's autonomy should be honored, that is, not be prevented by deception from making its best-informed supreme choice—unless he *wants* to be deceived. To find *this* out is part of the true physician's competence, not learned in medical school. He has to size up the person—no mean feat of intuition. Satisfied that the patient really wants the truth—his mere saying so is not proof enough—the physician is morally and contractually bound to give it to him.[4] Comforting deception, if noticeably desired, is fair; so is encouraging deception of direct therapeutical import, which anyway presupposes a situation not calling yet for the supreme choice. Otherwise, and especially if there *is* a choice to be made, the mature subject's right to full disclosure, earnestly claimed and convincingly apparent as his will, ought to have its way *in extremis,* overruling mercy and whatever custodial authority a doctor may have on behalf of the patient's presumed good.

This right to disclosure, by the way, extends beyond the needs of informed decision to a state of affairs where no decision is to be made. What is then involved is not the "right to die," an issue in the practical domain, but the contemplative right to one's death, an issue in the domain not of doing but of being. This needs some explaining. Even in the absence of therapeutical options, and thus with no "right to die" at stake, the terminal patient's right to truth is a right by itself, a sacred right for its own sake, besides its practical bearing on a person's extra-medical arrangements in response to such a truth. Taking a cue from the last sacrament of the Catholic Church, the physician should be ready to honor the essential meaning of death to finite life (against the modern debasing of it to an unmentionable misadventure) and not deny a fellow-mortal his prerogative to come to terms with its drawing near—to his very own terms of appropriation, be they submission, reconciliation, or protest, but anyway in the dignity of knowledge. Unlike the priest who acts *in loco Dei,* the physician in his purely secular role cannot thrust the knowledge on him, but must heed the patient's secret choice as he divines it. Mercy can allow the indignity of ignorance but must not inflict it. In other words, besides the "right to die," there is also the right to "own" one's death in conscious anticipation—really the seal on the right to life as one's own, which must include the right to one's own death. That right is truly inalienable, although human weakness more often than not will prefer to forego it—which again is a right to be respected and nursed along by compassionate deceit. But compassion must not become arrogance. Lying to the stricken in disregard of his credibly evinced will is cheating him of the transcendent possibility of his selfhood to be face to face with his mortality when it is about to meet him. My premise here is that mortality is an integral trait of and not a fortuitous insult to life.

But back to the right to die. Assuming then that the patient has been told and has decided against therapeutical protraction of his moribund state and in favor of letting matters take their course: in enabling him to make the decision and in abiding by it, his right to die has been respected. But then a new problem arises. The option against protracting was, among other things, an option against suffering, and thus it includes the wish to be spared suffering, either by hastening the end or by minimizing pain during the remaining span, the latter in effect sometimes amounting to the former by the heavy dosage of drugs it requires. Heeding such wishes appears to be part of what has already been granted the patient with the "right to die" as such, when his decision was first allowed. Mercy in addition joins with these wishes in proportion as the suffering is acute. But their fulfillment involves the cooperation of others, perhaps even their sole agency, and here the institutionalization of dying by hospital confinement, added to the helplessness of the patient's condition, creates problems of the gravest sort. Usually, discharge into home care is impracticable, and we need not discuss what might or might not be done in the intimate privacy of compassionate love—even this is not free of serious constraints. But the hospital, at any rate, puts the patient squarely in the public domain and under its norms and controls.

Now, as to outright hastening death by a lethal drug, the doctor cannot fairly be asked to make any of his ministrations with this *purpose,* nor the hospital staff to connive by looking the other way if someone else provides the patient the means. The law forbids it, but more so (the law being changeable) it is prohibited by the innermost meaning of the medical vocation, which should never cast the physician in the role of a dispenser of death, even at the subject's request. "Euthanasia" at the doctor's hand is arguable only in the case of a lingering, residual life with the patient's personhood already extinguished. If we rule out euthanasia at the doctor's hand, so as to preserve the integrity of his calling even at the expense of the patient's right to die, then we must add that putting the means in the patient's hands falls little short of administering them. If nothing else, it would be contrary to the premise of the physician's privileged access to such means—a privilege jeopardized by the best-meant abuse.

But there is a difference between "killing" and "permitting to die," and again between permitting to die and aiding suicide. In the case of the suffering *conscious* patient we are speaking of, permitting ought to be freed from the fears of legal or professional reproof if it consists in acceding to his steadfast request (not the request of a despairing moment) to take him off the respirator that keeps him alive with no prospect of recovery or significant improvement. Formally, the right is his, and his alone, by his status of contractual principal in a service relationship, and the jurisdictional problem arises only from the quasi-surrender of rights to institutional trusteeship implicit in hospitalization. Such surrender in matters of medical routine remains conditional on the subject's continuing primary intent and does not extend to his right of revising that intent and making another ultimate choice. As to the *morality* of complying with it by discontinuing the life-support, that is, desisting from its further use, only sophistry can in this case equate desisting with doing, thus permitting to die with killing. And after all, a patient's helplessness that puts him at the mercy of the doctor's compliance should not make his right inferior to that of the mobile patient who simply can get up and leave without hindrance. The captive patient must not be penalized by disenfranchisement for his physical impotence. If he says "enough," he should be obeyed; and social obstructions to that should be removed.[5]

But passing from desisting to doing, what can be said about administering painkillers, which is a positive act on the doctor's part? Here, the oath "not to harm" can come into conflict with the duty to relieve, when harmful doses become necessary to cope with the torture of intractable constant pain. Which duty should prevail? My intuitive answer is: in a terminal condition, the clamor for relief overrules the ban on harming and even on shortening life, and it ought to be heeded if the patient has been told the price of relief. To hasten death in this manner, as a byproduct of the quite different purpose of making the remainder of a doomed life tolerable, is morally right and should be held unimpeachable by law and professional ethics alike, even though it adds another lethal component to the given lethal condition. The latitude, thus carefully circumscribed, does not open the door to mercy killing, and it seems to me that it requires no euthanasia legislation, only a judicial refinement of the malpractice concept that removes such requested terminal alleviation from its scope. Morally and conceptually, there is no confounding this consensual trade-off between tolerableness and length of an expiring stage with "killing." (For nonterminal cases, the issue is debatable, and much depends on the kind of survival at stake.)

The Patient in Irreversible Coma

Finally, consider the patient in irreversible coma, the case of a lingering, artificially sustained residue of life, where not even an imaginary "free agent" is left whose presumed own will a deputy might carry out. Failing such a virtual agent and putative chooser in his own cause, a _right_ to die is, strictly speaking, not involved, since this of all rights must be something that its owner could possibly claim, if not by himself exercise. One could not properly tell whose right is upheld or violated with any decision—that of the former person or that of the present, impersonal remainder. (As only a person can be a subject of rights, it would have to be the former person, whose "posthumous" rights, as it were, could indeed be invoked.) What is in question is rather the right or duty of others to perpetuate the given state, and alternately their right or duty to terminate it by withdrawing its artificial support. Reason, sanity, and humanity, it is safe to assert, overwhelmingly favor the second alternative: let the poor shadow of what once was a person die, as the body is ready to do, and end the degradation of its forced lingering. Yet powerful obstacles stand in the way of this counsel. There is the human reluctance to kill, as the letting die can here be construed, since it involves my ceasing to prevent it. There is the conception of the doctor's duty to be on the side of life under all circumstances. And there is the law, which forbids causing death intentionally and makes culpable even the causing it by neglect or omission. Though none of this touches properly on a right to die, and all of it at best tenuously on a right to live—there being no subject even implicitly claiming the one or the other and being hurt by its denial—yet the case of the comatose has in public argument become entangled with the right to die, and one hears this right invoked in support of one course of action, allowing to die. For this reason, I include the issue here.

There are two ways of getting out of the ethical-legal impasse I have described. One is redefining "death" so that a comatose condition of a certain kind constitutes death—the so-called "brain death" definition, which takes the whole matter (death being already an accomplished fact) out of the realm of decision and makes it a mere matter of ascertaining whether the criteria of the definition are fulfilled. If they are, withdrawal of support would appear to be not just permissible, but a matter of course and even mandatory, as it would be indefensible to waste precious medical resources on a corpse. Or would it? Might not the withdrawal, that is, making the corpse even more of a corpse, cause a waste in another direction? Is not the deceased, if residually kept functioning, a precious medical resource himself? In that case, pronouncing dead and continuing the life support would not seem contradictory; they rather would be parts of one syndrome with concerns besides the patient's. I have described in my essay "Against the Stream" (in *Philosophical Essays,* 1974) my grave misgivings about this kind of a resolution to the problem—removing it by way of a definition that is tailored to the particular situation and its practical quandary, tainted with the suspicion of expediency, and giving rise to apprehensions about the extraneous uses to which it lends itself, for example, to secure the most perfect material for organ transplants. Some of those misgivings have meanwhile come true; and the definition itself has suffered a serious setback as to its helpfulness in meeting the challenge of irreversible coma. For in the Quinlan case, where spontaneous respiration unexpectedly set in after forced respiration was discontinued by permission of the court, the girl was *not* dead by the criteria of the Harvard definition (which, incidentally, was not invoked by the court), yet still in a coma, and the question of further artificial life support (e.g., forced intravenous feeding) was posed again in all its original force. The shift from the moral to the technical ground diminishes our capacity to meet the question.

But there is another way of getting out of the impasse than by definitional semantics about life and death, namely, by squarely facing the issue of the *rightness* of continuing, solely by our artifice, what may perhaps, for all we know, still be called "life," but is only that kind of life, and this totally by grace of our artifice. Here I agree with the papal ruling that "when deep unconsciousness is judged to be permanent, extraordinary means to maintain life are not obligatory. They can be terminated and the patient allowed to die." I go a step further and say: they *ought* to be terminated because the patient *ought* to be allowed to die; stoppage of the sustaining treatment should be mandatory, not just permitted. For something like a "right to die" can, after all, be construed on behalf and in defense of the past dignity of the person that the patient once was, and the memory of which is tainted by the degradation of such a "survival." That memorial, "posthumous" right (extralegal as it is) places a duty on us, the guardians of its integrity by virtue of our mastery over it and thereby the executors of its claim. And if this is too "metaphysical" to convince our pragmatic conscience as to where our duty lies, then a down-to-earth principle of social justice, extraneous to the patient, but more likely to persuade the legislator, can come to the aid of this internal reason for mandating the termination: fair allocation of scarce medical resources (excluding the patient himself from such a classification!).

I have referred before to the life-and-death decisions necessitated by this circumstance, which is most likely to occur with respect to that elaborate apparatus whose life-preserving application must be continual. My former consideration was with the initial *admission* to such facilities when demand for them exceeds supply (the dialysis example). For the sombre choices that have to be made, priorities must be set by criteria as "just" as we can make them. The rough-and-ready battlefield "triage" system that the French adopted in the mass slaughter of World War I was the first instance of such a rule of selection. Under noncatastrophic conditions, the grading of more or less "deserving" cases becomes a complex and controversial matter, which in view of the many imponderables must often, at the upper end of the selection, be resolved arbitrarily. But although it must remain moot which is the "most" deserving case in a spectrum, at its simplifying lower end it is not moot which is the *least* deserving: the one that can least *profit* from the extraordinary treatment in short supply. This granted, the question remains whether such a selectivity extends past admission into the further course of things and can later apply to the patient's *retention* in the treatment, when a "better" candidate comes along. In general, I would deny this and recognize here a preemptive right of prior occupancy. Once underway, it would be an unconscionable cruelty to revoke the life support (still desired by the patient) for whatever extraneous interests: as with an individual once born, the place once allotted to the patient is simply not up for bidding anymore. But the irreversibly comatose is beyond the reach of cruelty as he is beyond the enjoyment of benefit, and "his" profit from the treatment is literally nil, if "his" refers to a subject capable of reaping the profit. In this unique borderline case, therefore, the criterion of "least profit" can indeed apply and ethically rule for discontinuing what has been begun, so as not to withhold from others a life-preservation from which they can profit. To me, as I have made clear, this consideration is secondary to the *internal* merits of the case which I consider sufficient and mandatory ground for termination— indeed *the* genuine ground. But as this internal aspect is, notoriously, not above controversy, distributive social justice—a more pragmatic and hence more securely entrenched principle—may be invoked to the same effect. For me to do so is what Plato called "second sailing" *(deuteros plous):* the next best way.

The Task of Medicine

Not to end with this particular subject, let me recall that my topic is "the right to die," to which the case of the comatose is marginal at best. Not only is it too rare, too extreme, and too sharply edged to serve as a representative paradigm, it is (as noted before) not properly a "rights" issue at all, with which, however, it is conflated in the public imagination. The real locus of that right to agonize about is the much more common and treacherous shadow land of the conscious terminal sufferer. It is he, not the body already lost to all conscious life, whose plight poses the ethically troubling problems. What the two have nonetheless in common is this: that beyond the compass of "rights," they pose the question as to the ultimate *task* of medical art. They force us to ask: is merely keeping a naturally doomed body this side of expiring among the genuine goals or duties of the physician? *N. B.* As to actual goals served by the art we must note that, at one end of the spectrum, the once severe definition of medicine's goals has become much loosened and now includes services (mostly surgical) that are not "medically indicated" at all: contraception, abortion and sterilization on nonmedical grounds, sex change, not to speak of cosmetic surgery for vanity or occupational advantage. Here, "service to life" has been extended beyond the ancient tasks of healing and alleviating to performing functions of a general "body technician" for diverse ends of social or personal choice.

But at the upper end of the spectrum, where our "right to die" issue is located, the august and ancient commitments still determine the doctor's task. It is therefore important to define the underlying "commitment to life" itself and thereby the lengths to which medical art must or must not go in honoring this commitment. Now we have already laid it down as a rule that even a transcendent duty to live on the patient's side does not justify a coercion to life on the doctor's side. But at present, the doctor himself is forced into such a coercion, partly by the ethics of the profession, partly by existing law or judicial usage. Through the institutionalization of sickness and especially

That is, one has a right to refuse treatment

of dying, once the doctor has plugged the patient into the machine at the hospital, he too is plugged in. It is notoriously easier to get a court ruling for imposing treatment (children of Jehovah's Witnesses) than one for breaking off treatment (Quinlan). To defend the right to die, therefore, the real vocation of medicine must be reaffirmed, so as to free both patient and physician from their present bondage. The novel condition of the patient's impotence coupled with the power of life-prolonging technologies prompts such a reaffirmation. I suggest that the trust of medicine is the wholeness of life. Its commitment is to keep the flame of life burning, not its embers glimmering. Least of all is it the infliction of suffering and indignity. How to translate such a statement of principle into legally viable policy is another matter, and however well it is done, we cannot hope it will eliminate twilight zones with anxious choices. But with the principle as such affirmed, there is hope that the doctor can become again a humane servant instead of the tyrannical and himself tyrannized master of the patient.

It is thus ultimately the concept of *life*, not the concept of death, which rules the question of the "right to die." We have come back to the beginning, where we found the right to life standing as the basis of all rights. Fully understood, it also includes the right to death.

NOTES

1. A "critical phase" would be that between two linked operations or during post-operative care, or similar situations where only the complete therapeutical sequence is medically sane. It must then be considered contracted for as an indivisible whole. Physician and hospital would not have performed the first steps without the patient's commitment to the remainder.

2. If a believer, he might even "all by himself" reject the choice as falling within the sin of suicide. I would deny that it does (surrender to the sentence already pronounced is no more suicide than the prisoners on death-row ceasing to seek further reprieve), but we are here considering only the secular ethics of these questions, without preempting their possible theological aspects.

3. Secular and religious ethics here agree. No religion, however strongly it condemns suicide as sin because it holds life to be a duty toward God, makes thereby the preservation of one's own life an unconditional duty—which would indeed lead to horrendous moral consequences.

4. Dr. Eric Cassell gave me this point as being his own policy, with his conviction from experience that the requisite discernment of will is possible.

5. The present state of the law seems to be that such an "enough" by the patient, given his competence, can indeed not be denied but would also, given present "malpractice" policy, force the doctor to resign from the case. As this would deprive the patient of further medical and hospital care (which he still needs for dying tolerably), the threat effectively blocks the option.

11 *The Freedom To Die*

1972

DANIEL MAGUIRE

Maguire presents a discussion of whether we may actively intervene to achieve death by choice or whether we must wait for nature to take its course. Four situations are examined: 1) the irreversibly comatose patient artificially sustained; 2) the conscious patient artificially supported; 3) the conscious patient with a terminal illness supported by natural means; 4) suicide in a nonmedical context. Although much of Maguire's analysis comes from a Catholic perspective, he also uses contemporary Protestant ethical thought to develop his position. This provides a multi-faceted approach to the problem of euthanasia.

Dr. Daniel Maguire, Ph.D., is Professor of Theology at Marquette University.

> Of old when men lay sick and sorely tried,
> The doctors gave them physic and they died:
> But here's a happier age, for now we know
> Both how to make men sick and keep them so!
> Hilaire Belloc

Man is the only animal who knows he is going to die and he has borne this privileged information with uneven grace. On the one hand poets and philosophers have gazed at death and proclaimed the significance of death-consciousness. Hegel saw the awareness of mortality as such a stimulus to human achievement that he could define history as "what man does with death." Schopenhauer called death "the muse of philosophy" and Camus saw man's capacity for suicide as evoking the most fundamental philosophical questions. Poets have called death such things as "gentle night," "untimely frost," or "the great destroyer."

The average person, however, would rather forget it. This is especially true if the average person is an American since in this happiness-oriented land, death (outside of a military context) is seen as something of an un-American activity. It happens, of course, but it is disguised and *sub rosa*, like sex in Victorian England. Most Americans now die in hospitals. And they die without the benefit of the liturgies of dying that attend this natural event in cultures which accept death as a fact of human life. The dying process is marked by deceit where everything except the most important fact of impending death can be addressed. When the unmentionable happens, the deceit goes on as the embalmers embark on their *post mortem* cosmetics to make the dead man look alive. Mourners, chemically fortified against tears that would betray the farce, recite their lines about how well the dead man looks when, in point of fact, he is not well and does not look it.

All of this does not supply the atmosphere in which man's moral right to die with dignity can receive its needed re-evaluation. Nevertheless, the re-evaluation must go on. Technology has already moved ahead of both ethics and law. As Johns Hopkins professor Diana Crane notes, "the nature of dying has changed qualitatively in recent years because of advances in medical knowledge and technology." These qualitative changes have dissipated older definitions of death, given greater power to doctors over life and death, shaken the never too fine art of prognosis, and presented all us mortals with options that old law and ethical theory did not contemplate . . . which options offer frontal challenges to some long-tenured traditions and taboos.

As if this were not complicated enough, the technological revolution has also caused a notable shift in our moral universe. The interplay between technology and morality is, of course, as old as man's first primitive tool. With technological advance comes power and with power, responsibility, and the question of whether this power can be used without intrusion into the realm of the gods. Prometheus, the philanthropic god who stole fire for men, was judged by Zeus to have gone too far. Bellerophon in the Iliad came to a sorry end for trying to ride to the Olympus of the gods and for thinking "thoughts too great for man." And Icarus, exulting in the technology of man-made wings, flew too close to the sun and perished. These myths, like the myths of Babel and Adam's sin, are relevant to the moral dilemma of inventive man. What thoughts and what initiatives are "too great for man?" Where are the sacred borders between the *can do* and the *may do*? Where does presumption enter into the knowledge of good and evil?

Medical ethics has long known the strain of this dilemma. Man has a tendency to consider the physical and biological to be ethically

normative and inviolable. Blood transfusions were resisted by many
on these grounds. Birth control was also, for a very long time, imped-
ed by the physicalist ethic that left moral man at the mercy of his biology. He had no choice but to conform to the rhythms of his physical
nature and to accept its determinations obediently. Only gradually did
technological man discover that he was morally free to intervene creatively and to achieve birth control by choice.

Can We Intervene?

The question now arising is whether, in certain circumstances, we
may intervene creatively to achieve death by choice or whether mortal
man must in all cases await the good pleasure of biochemical and organic factors and allow these to determine the time and the manner of
his demise. In more religious language, can the will of God regarding a
person's death be manifested only through the collapse of sick or
wounded organs or could it also be discovered through the sensitivities and reasonings of moral men? Could there be circumstances
when it would be acutely reasonable (and therefore moral if one uses
"reason" in a Thomistic sense) to terminate life through either positive action or calculated benign neglect rather than to await in awe the
dispositions of organic tissue?

The discussion should proceed in the realization that the simpler
days are past in which the ethics of dying was simple. Most importantly, the definition of death was no problem until recently. When
the heart stopped beating and a person stopped breathing, he was
dead. (Interestingly, the old theology manuals were not too sure of
this since they advised that the sacrament of final anointing could be
administered conditionally up to about two hours after "death.")
Medical technology has changed all of that. It is now commonplace to
restore palpitation to a heart that has fully stopped beating. Even
when the natural capacity for cardiopulmonary activity is lost, these
systems can be kept functioning through the aid of supplementary
machinery. Heartbeat and breathing can be artificially maintained in a
person whose brain is crushed or even deteriorated to the point of liquefaction. Heartbeat is no longer a safe criterion of human life!

This has turned medical people to the concept of "brain death."
Indeed, Dr. Robert Glaser, the Dean of Stanford's School of Medicine
says (somewhat prematurely, I judge): "Insofar as it involves organ
donors and their rights, the technical question of death has been resolved, by popular consensus, in the recognition that the brain and
not the heart is the seat of human life." Brain death, however, is not
an answer without problems. A person with a hopelessly damaged
brain could still have spontaneous heartbeat and respiration. Har-

vard's Dr. Henry Beecher asks the inevitable question here: "Would you bury such a man whose heart was beating?" And lawyers will be quick to tell you that the removal of vital organs from a body that has a dead brain but a live heart would raise the specter of legal liability.

All of this illumines the unsettling fact that death admits of degrees. Some organs and cells die before others. As Paul Ramsey observes, "the 'moment' of death is only a useful fiction." It may however be more fiction than useful if it leaves us hoping for a definition of death that will make it unnecessary for us to judge that a particular "life" should now be completely terminated by a positive act of omission or commission. Science will not give us a litmus test for death that will relieve us of all need for moral judgment and action. That day has passed and that is what makes for the qualitative change in dying and the ethics thereof.

It seems well to pursue this issue by considering four different dying situations: 1) the case of the irreversibly comatose patient whose life is sustained by artificial means; 2) the conscious patient whose life is supported artificially by such means as dialysis (for kidney patients) or iron lung; 3) the conscious patient who is dying of a terminal illness whose life is supported by natural means; 4) the case of self-killing in a non-medical context.

First, then, to the case of the irreversibly comatose, artificially sustained patient. It is a useful beginning here to look at the remarkable address of Pius XII on the prolongation of life, given in 1957. In this talk, the Pope stated that the definition of death is an open question to which "the answer cannot be deduced from any religious and moral principle. . . ." He allows that in hopelessly unconscious patients on respirators "the soul may already have left the body" and he refers to such patients by the ambiguous but interesting term "virtually dead." The Pope points out that only ordinary means need be used to preserve life and he supplies a broad definition of this slippery term saying that ordinary means are "means that do not involve any grave burden for oneself or another." The Pope is dealing with the unconscious patient of whom it can be said that "only automatic artificial respiration is keeping him alive." In these hopeless cases, the Pope concludes that after several days of unsuccessful attempts, there is no moral obligation to keep the respirator going.

The respirator is, of course, an extraordinary means especially when it is used on the irreversibly unconscious. For years now theologians have also argued that even intravenous feeding in this type of case is extraordinary and may be discontinued. (Law lags behind ethics on this.) I would argue further that in such a case where the personality is permanently extinguished, one needs a justifying cause to continue artificial, supportive measures. To maintain bodily life at a vegetative level without cause is irrational, immoral and a violation of the dignity of human life. It is a burden on the family; also, there may be need to allocate these medical resources to curable awaiting patients. It is, moreover, macabre, irreverent and crudely materialistic to preserve by medical pyrotechnics the hopeless presence of what could best be described as a breathing corpse. Of these cases, when the patient is "irretrievably inaccessible to human care," Paul Ramsey says quite correctly that it is "a matter of complete indifference whether death gains the victory over the patient in such impenetrable solitude by direct or indirect action."

One reason that clearly justifies maintaining this kind of life is to make possible the donation of organs since the human body thus sustained is considered to be the best possible tissue bank. Death, however, should first be declared and the law should be pushed to update itself to allow for this. Probably our society is not prepared to grant that the organs in such instances should, by eminent domain, be allocated to those who need them. Through cultural conditioning and taboo we are still disposed to prefer the "rights" of the dead to the needs of the living in this regard, and few people even carry donor cards now legal and available in all states. Thus the immoral waste of tissue that could be "a gift of life."

Our second case concerns the conscious patient whose life is artificially supported. This is a broad category and could be stretched to include those who attribute their perdurance to superabundant dosages of Vitamin C and E. We are thinking, rather, of those who are alive thanks to dialysis or an iron lung. As medicine advances this category will expand.

In assessing the moral right of such patients to die, one might be tempted to repair to the facile distinction between ordinary and extraordinary means. By the papal definition of ordinary means ("grave burden for oneself or another") both dialysis and the iron lung are extraordinary. Thus the patients involved would appear to be under no obligation to continue in this kind of therapy and would seem free to discontinue the treatment and die.

We must, however, avoid the simplistic allurements of a one-rubric ethics. There is more involved here than the patient and the extraordinary means. We must (as does the papal teaching cited) speak also of the patient's obligations of social justice and charity, remembering that an individualistic, asocial ethics has also infected medical morality.

We have all heard much of the redemptive power of suffering. Unfortunately, the pieties of the past often interpreted this in accord with primitive myths of vindication which implied that pain and bloodshed of themselves had an atoning power. These myths, replete as they may be with sado-masochistic elements, have had an enormous influence both on Christology and Christian asceticism, attributing a positive *per se* value to suffering. Such a value suffering does not have, but rather assumes its meaning, value, or disvalue from the concrete circumstances of the sufferer.

In this sense there *can* be a redemptive value to suffering. Take the example of a kidney patient who has not had a successful transplant. Could he not see his plight as a vocation? His disease can become a rostrum and he could become part of something that is needed . . . a lobby of the dying and the gravely ill. To be more specific, this nation which has borne him is also ill. Its priorities are out of joint. It has, by morbid selection, chosen for itself what Richard Barnet calls "an economy of death." Its heart is so askew that only if dialysis could be shown to have *military* significance would it receive the government and private funding it desperately needs. Many illnesses could now be contained if the nation had given the zeal and the budget to life that it now gives to death.

The gravely ill with their unique credentials could do more here than they realize. By working creatively with politicians, national health organizations, medical and legal societies, news media, writers, etc.—and in ways as yet unthought of which their healthy imaginations will bring forth—the lobby of the dying and the gravely ill could become a healing force in society. Man's nature is still so barbarous that the well attend more to the well than to the sick. Great obligations, therefore, devolve upon the sick. The well need the sick more than the well know.

Still, concerning persons in the category here discussed, if their artificially supported life becomes unbearable, they have a right to discontinue treatment and we owe them in justice and in charity the direct or indirect means to leave this life with the dignity, comfort and speed which they desire.

What then of the conscious, terminal patient whose life systems are functioning naturally? May he in any circumstance take direct, positive action to shorten his dying trajectory or is the natural course of his disease ethically normative in an absolute sense? Those who would deny the moral right to direct acceleration of the death process will adduce an "absolute" principle such as the unconditional inviolability of innocent human life. This is taken to mean that innocent life may in no circumstance be terminated by direct, positive action.

John Milhaven, in an important 1966 *Theological Studies* article, notes the tendency simply to proclaim this inviolability principle and work from there in right-to-life cases. He cites the failure of those who do this to indicate "the reasons that prove there is an absolute inviolability, holding under all circumstances." And he asks the fair and telling question: "How do we know this?" In point of fact, I submit that we do not know this because it is not knowable.

Those who use the principle accept no burden of proof and offer only a fideistic assertion. They present the principle without proof, as self-evident. In ethics, however, only the most generic propositions are self-evident, such as "do good and avoid evil"; "to each his own." The *evidence* emerges from the mere understanding of the terms. Such statements are universally true because they lack particularizing and complicating content.

Practical principles such as the one on direct termination of innocent life must be proved and the proof must come from wherever moral meaning is found. That is, it must come from a knowledge of the morally significant empirical data, the consequences, the existent alternatives, the unique circumstances of person, place and time, etc. Moral meaning is found not just in principles but in all the concrete circumstances that constitute the reality of a person's situation, and the principles themselves are rooted in empirical experience and must be constantly rewashed in an empirical bath to check their abiding validity.

To say that something is morally right or wrong in all possible circumstances implies a divine knowledge of all possible circumstances and their moral meaning. To say that something is universally good or bad regardless of circumstances is non-sense, for it is to say that something is *really* good or bad regardless of the *reality-constituting* circumstances.

An attempt can be made to base the "no direct killing of innocent life principle" on the reality of expected intolerable consequences. This is the cracked dike argument. *Après moi le déluge!* If X is allowed, then Y and Z and everything else will be allowed. This is a kind of ethical domino theory which has the deficiencies of any domino theory. It ignores the real meaning of the real differences between X, Y and Z. Good ethics is based on reality and makes real distinctions where there are real differences. It is, furthermore, fallacious to say that if an exception is allowed, it will be difficult to draw the line and therefore no exception should be allowed. It has been said quite rightly that ethics like art is precisely a matter of knowing where to draw lines.

Therefore, with regard to the principle in question, we can say that its absoluteness is, at the very least, doubtful. And then in accord with the hallowed moral axiom *ubi dubium ibi libertas* (where there is doubt there is liberty) we can proclaim moral freedom to terminate life directly in certain cases. It would perhaps be better to put it in terms of Aristotelian-Thomistic moral theory. This principle, like every practical moral principle, is valid most of the time *(in pluribus)* but in a particular instance *(in aliquo particulari)* it may not be applicable.

To apply this whole discussion to the case of conscious terminal patients, it can be said that in certain cases, direct positive intervention to bring on death may be morally permissible. The decision, of course, should not ignore issues of social responsibility and opportunity alluded to in the preceding case. The patient must consider also his cultural and legal context, the mind-set of insurance companies, and the ability of others to cope with the voluntary aspects of his death. He must also beware lest he is yielding to societal pressures to measure human dignity in terms of utility or to create the illusion that sickness and death are unreal.

To repeat therefore, direct action to bring on death in the situation described here may be moral. The absolutist stance opposed to this conclusion must assume the burden of proof—an impossible burden, I believe.

Finally, to self-killing in a non-medical context. It is estimated that more than eighty Americans a day kill themselves. Jacques Choron calculates that between six and seven million living Americans have attempted suicide and that 25 percent of these will try again and many will succeed. From a moral viewpoint it may be said that an enormous majority of these cases represent unmitigated tragedy. Most suicides flow from a loss of the vital ingredients of human life, hope and a supportive loving community. Studies show the suicide as a lonely, desperate person. His act is *The Cry for Help*, the title of a book on attempted suicide by Farberow and Schneidman. Eighty percent of all suicides signal their intentions in advance, apparently by way of final, desperate pleading. Contrary to the rationalizing myth, most suicides are not psychotic. Alcoholism and drug use are, of course, not unrelated to the suicidal syndrome.

The incidence of suicide in certain groups is revealing. Suicide rates among blacks and Indians in this nation are rated as of epidemic proportions. Two to three times more women than men attempt suicide. Suicide increases in socially disorganized communities and the rate of successful suicide among divorced persons is remarkably high. There can be no doubt that most suicides are an indictment of the surviving community which failed to give the possibility of life to their suicidal victims.

The prime moral reaction to suicide should be to attack the causes that yield such bitter fruit. Those of Christian persuasion should be in the forefront here. For Christians, the loss of hope is apostasy and to contribute to the loss of hope in others is the elementary Christian sin. Dietrich Bonhoeffer judged suicide severely from the perspective of Christian faith: "It is because there is a living God that suicide is wrongful as a sin of lack of faith." He did, however, realize that suicide is not univocal. Though usually akin to murder, "it would be very short-sighted simply to equate every form of self-killing with murder." Some suicides could be highly motivated and Bonhoeffer was inclined to suspend judgment with regard to these because "here we have reached the limits of human knowledge." Unless we do ethics by taboo, however, we must do more than suspend judgment. We must also discuss the possibility of objectively moral suicide. The discussion here relies on the points argued in the preceding cases.

Some moralists have not suspended favorable judgment on all suicides. Some medievals, weighing the suicides of such as Samson and virgin saints who killed themselves to avoid violation, concluded that the Holy Spirit had inspired their actions. (This, of course, implied the unraised question of whether the Holy Spirit could inspire other suicides.) Henry Davis, in his *Moral and Pastoral Theology*, leaned on the abused distinction between direct and indirect and concluded that a sexually threatened maiden "may leap from a great height to certain death, for her act has two effects, the first of which is to escape from violation, the second, her death, which is not directly wished but only permitted. The distinction between the jump and the fall is obvious. In the case, the maid wishes the jump and puts up with the fall." Davis' distinction is not obvious to today's moralists who would find the maiden's ability to dissect her intentionality even more remarkable than her passion for material chastity.

Some modern moralists have defended the suicide of a spy who, when captured, could be induced by chemicals or by torture to reveal damaging data. Less attended to are the social witness type suicides related, for example, to the early phase of the Vietnam war. In assessing the objective morality of the suicide of Roger LaPorte which first signaled to many in Communist China and Indochina the depth of anti-war feeling in America, our judgment should *at the least* show the kind of reserve that Bonhoeffer brings to limit situations. Of the possibility of other moral suicides it must be said that there are strong presumptions against them, arising from the experience of grounded hope, from the number of alternatives open to imaginative man and from the effects on the bereaved. However, no one is wise enough to say that those presumptions could not be overridden unless, of course, that someone is privy to knowledge of all possible circumstances. Realistic ethics requires more modesty than that. Thus, the possibility of objectively moral self-killing is an open question, and it may not be excluded that direct self-killing may be a good moral action, in spite of the strong presumptions against it.

In sum, then, death has lost its medical and moral simplicity. We know it now as a process, not a moment, and we have the means to extend or shorten that process. In the older ethics of dying, begged questions reigned unchallenged. The ordinary/extraordinary means rubric is still useful, but not self-sufficient. The borders between ordinary and extraordinary blur and shift, and years ago moralist Gerald Kelly, S.J., pointed out that even ordinary means are not always obligatory. The contention that life could be terminated indirectly and by omission but never directly by commission, was not proved—and that is a serious omission.

As Stanford law school Dean, Bayless Manning, has said: "The topic as a whole is still subterranean, and decisions are predominantly being made by thousands of doctors in millions of different situations and by undefined, particularized, *ad hoc* criteria." One partial solution to this would be a happily financed, well managed, hard-working, yearly study-meeting which would bring together doctors, lawyers, moralists of every stripe, insurance experts, nurses, social workers, morticians, sociologists, gravely ill persons, clergymen, journalists, etc., to discuss the current state of dying. The results each year should be energetically publicized in learned journals and in all news media, since death education is needed at every level. (Perhaps some mortal and affluent readers will let their treasures be where their hearts are in this regard.)

Hopefully, a healthier attitude toward death will emerge in our culture. We all could learn from a Donegal Irishman whose death a few years ago was a testimonial to culture wisdom. While on his deathbed, he was visited by friends who knew, as did he, that this would be their final visit. The dying man ordered his son to bring whisky for the guests. With this done, the son asked his father if he, too, would indulge. "Oh, no," replied the father with a gentle frown, "I don't want to be meeting the Lord with the smell of the drink on my breath." A few hours later, he died. If we were as at home with death as he, our deliberations on the subject might be more wise.

12 The Withdrawal of Treatment: The Costs and Benefits of Guidelines

THOMAS A. SHANNON

Problems associated with the withdrawal of individuals from life support systems have caused numerous policy problems. This article reviews four sets of guidelines that have been issued by hospitals and medical associations. The author concludes by indicating a variety of unresolved problems in intensive care that are not totally dealt with by the guidelines: competency, the management of the patient and staff, ensuring of conformity to guidelines, problems of funding, as well as the basic question of the necessity of such guidelines at all.

Thomas A. Shannon is an Associate Professor of Social Ethics at the Worcester Polytechnic Institute, and an Associate Professor of Medical Ethics at The University of Massachusetts Medical School

The Karen Ann Quinlan case, and the multileveled tragedies that followed in its wake, focused national and international attention on a problem partially caused by the very technology designed to resolve it: the treatment and care of individuals in a terminal illness or in irreversible coma. Technological advances have made it possible either to maintain physical life or to prolong dying almost endlessly. Unfortunately these advances have not been as effective in providing a cure or restoring such patients to an adequate level of health. Given this situation, many are questioning the wisdom of continuing treatment when no reasonable benefits are to be expected, while others question the ethics of not doing all that is possible.

Because of this ethical and medical dilemma, and as a result of a suggestion by the New Jersey Supreme Court, some institutions are turning to the use of "Ethics Committees" (Teel, 1975) whose responsibility is to determine whether there is a reasonable hope of the individual's ever returning to a cognitive, sapient state and, thus, to help decide whether to employ life-prolonging technology. The acceptance of such committees as the appropriate mechanism for dealing with the comatose patient is apparent from the number of guidelines for the evaluation of such patients that have recently been published—guidelines that include recommendations for terminating treatment if medically appropriate.

The purpose of this article is to review the guidelines proposed by Beth Israel Hospital, the Massachusetts General Hospital, Mount Sinai Hospital, and several New Jersey medical associations. While not wishing to suggest that these are the only guidelines available, I have chosen these particular documents for examination because they have all been recently published, they discuss a variety of issues, and they raise certain ethical problems. Consequently, they serve as a good point of departure for discussing the issues and problems surrounding such committees.

THE BETH ISRAEL GUIDELINES

The Beth Israel document (Rabkin, 1976) is the most general, dealing as it does with patient-physician decision making, and not exclusively with the intensive care unit. The document evolved from the Law and Ethics Working Group of a faculty seminar on an Analysis of Health and Medical Practice—an activity of the Center for the Analysis of Health Practice of the Harvard School of Public Health. Although two attorneys and one physician authored the article describing the guidelines, neither the names nor the professional status of the other participants is mentioned. At present this proposal is not actual policy at Beth Israel, although consideration is being given to using it as such.

The proposed Beth Israel guidelines are based on three general principles: (1) that the general policy of hospitals is to act affirmatively to preserve the life of all patients, including patients who suffer from irreversible, terminal illness; (2) that hospitals are to respect the competent patient's informed acceptance or rejection of treatment; and (3) that regardless of the hospital's prolife policy, the right of a patient to decline medical procedures must be respected.

A variety of approaches are proposed to help deal with problems that emerge as a terminal illness progresses. "When it appears that a patient is irreversibly and irreparably ill, and death seems imminent, the question of the appropriateness of cardiopulmonary resuscitation in the event of sudden cessation of vital functions may be considered by the patient's physician, if not already raised by the patient, to avoid an unnecessary abuse of the patient's presumed reliance on the physician and hospital for the continuation of life-supporting care." (Rabkin, 1976) Implied in the statement is the belief that although the patient comes to the hospital for treatment and consents to the procedure, it is nonetheless appropriate to test this consent at different stages of an illness to attempt to balance the interests, rights, and duties of both the hospital and the patient.

When the illness appears to be irreversible, then a medical evaluation and a diagnosis is made by the responsible physician. But any decision to withhold resuscitation becomes effective only upon the informed choice and consent of the competent patient. In this context, competence refers to the understanding of the relevant risks and alternatives, as well as a deliberate choice of the course to be pursued. The physician is cautioned to ensure that consent does not come from a temporary distortion because of pain or medication. Also, if the physician feels that the patient is not able to deal with making such a decision at this time, resuscitative efforts must be employed. Once the patient chooses, his or her choice may not be overridden by the staff or the family. Permission to inform the family must also be obtained from the patient and his or her instructions must be followed.

If the patient is incompetent, the final decision for orders not to resuscitate must be based on a concern for the patient's point of view; the focus for such a decision must be the clinical interest of the patient. No other factors can be considered because this would violate a fundamental prolife policy of the hospital. A final condition for the issuance of orders not to resuscitate an incompetent patient is the approval of the same family members who are required to consent to a postmortem examination.

The Beth Israel document stresses that orders not to resuscitate a patient do not imply any intent to diminish appropriate medical and nursing attention needed by the patient. It emphasizes that every effort must be made to provide a level of comfort and reassurance which is appropriate to the patient's state of consciousness and emotional condition, regardless of any designation of incompetence. The document concludes with a notice that any request not to resuscitate that originates from members of the family is not to be viewed as the patient's choice. Thus the attending physician and any committee involved in the decision-making process must not simply concur with the family's request, but should go through the procedures suggested above.

Several positive elements in the proposal need to be emphasized. First, it is an attempt to respond to the host of problems caused both by the complexities associated with a terminal illness and by the limits of technological intervention. Second, it implies that economic considerations not be the sole criterion by which decisions to terminate resuscitation measures are to be evaluated. Third, the document attempts to deal with the potential legal problems raised by the decision to terminate treatment. This is not done crassly to avoid a malpractice suit, but is a responsible effort to address problems that have been raised in the past. Fourth, the patient is the center of attention in that the patient is recognized as the ultimate decision maker. Finally, the document is respectful of the patient's decision, even to the point of arguing that if the physician finds the ". . . medical program as ordered by the patient so inconsistent with his own medical judgment as to be incompatible with his continuing as the responsible physician, he may attempt to transfer the care of the patient to another physician more sympathetic to the patient's desires." (Rabkin, 1976) This is also exemplified by the statement that the competent patient's decision may not be overridden by the family and that they are to be informed of it only after receiving the patient's permission to do so.

Two major problems, however, are created by the stipulation that physicians may unilaterally order resuscitation for those competent patients who they feel are psychologically incapable of participating in critical treatment decisions. The first problem revolves around establishing the criteria for evaluating whether a patient is able to cope with such decision making. The only criterion mentioned in the Beth Israel document is reasonable and humane medical practice, which, unfortunately, is vague and not subject to review. Consequently, this can open the way to possible abuse of the patient's *presumed* reliance on the hospital and physician for continued life support care. The second problem is that since neither the patient nor the family has given consent, the resuscitation effort constitutes treatment without consent and therefore is an assault on the patient. While one would not like to see critically ill patients treated cruelly or unfeelingly in the name of honesty and autonomy, neither is it appropriate to deny patients opportunities for giving or refusing consent in the name of humanity or reasonable medical practice. Indeed, such is not even *medical* practice; it is the unwarranted imposition of a possibly alien set of values on another.

Another problem arises from the use of the terms "medical judgment" and "clinical interest" of the patient. On the one hand, we must recognize the valued role of the physician in making a diagnosis and prognosis. On the other hand, there can be a variety of therapies to achieve the same end. Some of these may be more or less effective, more or less risky, more or less painful, more or less costly, or have more or fewer side effects. The choice of a treatment is not totally a medical decision for it also involves the patient's preferences, values, or religious beliefs. What the physician perceives to be the patient's best medical interest may indeed not be in the patient's best interest generally. Conflicts of interest, as well as significant value clashes, can occur. The possibility of such conflicts cannot be casually dismissed or glossed over by technical jargon without raising the problem of treatment without consent in yet another guise.

Another problem arises with the ad hoc committee which determines "whether the patient's death is so certain and so imminent that resuscitation in the event of sudden cessation of vital function would serve no purpose." (Rabkin, 1976) The problem is not that there is such a committee, but with its membership and status. The committee is to be composed of the attending physician and nurses and one staff member not previously in the patient's care. Such membership can easily lead to a biased discussion of the patient. The majority of the members may be locked into a preference for a certain therapy and thus may not discuss the full range of options. Also, the vested interests of the various professions may preclude an unbiased discussion of the situation. If the patient is competent and is given the opportunity to make decisions, such a committee is unnecessary. Also, in the case of a comatose patient, it is unclear who has the responsibility for convening such a committee. Nor is it clear, in this case, that a committee decision not to resuscitate can take precedence over the substitute consent of the next of kin. In the document, approval of the committee's decision by the next of kin is required. Such approval is obviously necessary; but it is unclear why the family's decision is *only* that of approval. Given a prognosis, it seems proper to allow the comatose patient's next of kin to make the decision since they may possibly know the patient's values, interests, and preferences better than the medical staff. Although problems may possibly arise here also, since the legal status of the next of kin is uncertain in this situation, this approach seems preferable to attempting to disguise a value judgment as a medical one.

THE MOUNT SINAI GUIDELINES

Other guidelines for dealing specifically with patients in intensive care units have been established. One example, that has been in practice for several years, is the informal patient-care classification system in use at the Falk Surgical-Respiratory Intensive Care Unit of the New York Mount Sinai Hospital. (Kirchner, 1976) Four levels of care have been established: (1) all-out therapeutic effort; (2) continuation of all-out effort, but with 24 hour reevaluation of the patient's prognosis; (3) conservative, passive care, with no heroic measures to prolong life; and (4) all therapy and life-support assistance discontinued for brain-dead patients (based on the Harvard Ad Hoc Committee's definition of death).

This classification system was born out of staff frustration arising from the use of heroic measures which seemed merely to prolong dying and from the lack of a mechanism for reaching a consensus on how to treat critically ill patients. The system, however, is an informal one and has never been presented formally to the hospital administration. Also the treatment category number never appears on the patient's chart—only on the charge nurse's record sheet. Finally the classification system is not routinely explained to attending physicians. Each patient's status is reviewed every morning on unit rounds and if the staff decides a change in classification is indicated, this is discussed with the attending physician. If the attending physician should want a change, this is discussed at afternoon rounds and, if possible, a decision is made that same day. The head of the intensive care unit has final responsibility for decision making, but if any members of the team insist that maximal therapy be continued, it is.

Two major problems emerge with the practice described here. First, the team does not involve the family in the decision-making process. "Rather," according to the director of the intensive care unit, "it's a matter of asking them to accept the doctor's decision." (Kirchner, 1976) The team keeps the family informed about the prognosis and the therapy being used; no change in therapy is made without the family's knowledge; and if the family wants to keep a terminally ill patient alive as long as possible, the team does so. However, the family is basically presented with a *fait accompli.* Since no routine explanation of the classification is given to the physician, one can hardly expect that such an explanation is given the family. Also, the decision to terminate treatment is a value judgment based on a medical prognosis, but this distinction is fatally blurred by the decision-making methodology.

The second and more critical problem is revealed in three sentences from the article reporting this system. "Patients with every chance of survival often beg the staff to stop treatment, but, naturally, nobody takes them at their word. 'All they really want is relief from constant pain and anxiety,' says Diane Adler. 'We give them more pain relievers and spend more time talking with them.' " (Kirchner, 1976) The problem words here are "naturally" and "really wants." Why is it *natural* that no one listens to the patients' requests?

Is it because they present an alternative value system or is it because their desires contradict what the staff thinks is best for them? Also, since no one listens to patients, how does anyone know what they really want? Patients indeed may want relief from pain and anxiety, but they may also want something else and they have the right to be heard and have their competent decisions carried out. Despite the argument that patients are usually too sick to participate in the decision-making process, the exclusion of competent patients from this process results in treatment without consent. Such treatment is legally an assault, as well as a questionable ethical practice.

In the intensive care unit, treatment decisions must often be made quickly and hence guidelines such as these could be helpful in assessing the available options. However, the purposeful elimination of the competent patient and the family from the decision-making process is a serious defect in these guidelines.

THE MASSACHUSETTS GENERAL HOSPITAL GUIDELINES

A subcommittee of the Critical Care Committee of the Massachusetts General Hospital in Boston has developed a set of guidelines similar to those at Mount Sinai. (Critical Care Committee, 1976) Although the four levels of patient classification are practically identical, several significant improvements have been introduced. First, the guidelines have been accepted by the hospital's General Executive Committee, the highest medical-policy committee in the hospital, thus giving the criteria official recognition. Second, there is an Optimal Care Committee available to all intensive care units and other physicians who request advice in the management of therapies. This committee provides the opportunity for the case to be reviewed by someone who is not directly involved in it and thus has the potential for widening the basis of decision making. Third, the guidelines require that "appropriate notes specifically describing the therapeutic plan should be made on the patient's record." (Critical Care Committee, 1976)

In spite of these improvements, the Massachusetts General document contains three critical problems. The primary one is that the attending physician makes the final decision about treatment regardless of what the Optimal Care Committee or staff may say. This raises two other major problems. The first deals with consent, which is not discussed in the document. If the patient is competent, any decision about treatment is properly the patient's; if the patient is incompetent, the authority to give proxy consent should fall to the next of kin. The other problem is that the centering of authority in the physician confuses medical and value judgments in therapy. The physician is the proper one to make a prognosis, but a variety of values enters into the implementation or rejection of the therapy and these values may not be the ones the physician would choose. Such confusion gives unfortunate support to the fiction that medicine is value free and further weakens efforts to obtain the consent of the patient or the next of kin.

A further shortcoming stems from the fact that the guidelines are management oriented. While not wrong—and in fact quite necessary—this orientation nonetheless leads to an emphasis on the procedures by which information is communicated between the physician and staff and the procedures by which the physician's decision is implemented. This makes the patient even more marginal to the decision-making process and even more a totally passive recipient of the medical care delivery system. An exaggerated emphasis on the management of the patient to achieve consensus on his or her treatment can lead to effectively neglecting the interests of the patient.

The third deficiency is the failure to provide clearly defined criteria for reaching treatment decisions. For example, when a patient is reclassified as Class C (selective limitation of therapeutic measures), he or she is to be transferred out of the intensive care unit and "certain procedures may cease to be justifiable and become contraindicated." (Critical Care Committee, 1976) The reclassification decision is to be based on the needs of the patient and only after required comfort measures become manageable in a nonintensive care setting. Unfortunately, the criteria for evaluating the patient's needs and the appropriateness of various procedures are not discussed. The assumption is that they are medical, value free, and presumably self-evident. While it may be quite proper to limit treatment selectively, this is not totally a medical decision and ought to be based on more than vague references to patient needs. The failure to designate who is to make this sort of decision and the lack of clear criteria for making it are serious lacunae.

In summary, the omission of a discussion on consent, the centering of authority on the attending physician, and the lack of criteria for decision making severely weaken the usefulness of the Massachusetts General Hospital guidelines.

New Jersey Medical Associations' Guidelines

Another orientation to some of these problems is provided by guidelines designed by three medical societies in New Jersey. (State Medical Society, 1977) These guidelines are a specific attempt to implement the New Jersey Supreme Court's ruling in the Quinlan case that an "Ethics Committee" be used to evaluate the attending physician's judgment that no hope of restoration to cognitive, sapient life exists. If such a committee agrees with the physician, life-supporting measures may be removed without criminal or civil liability. This committee, designated as a prognosis committee by the guidelines, is selected by the responsible governing authority of the hospital or other facility. Membership is to include representatives of standard medical disciplines plus any specialist who may be needed for a particular case. At least two nonstaff members should be on the committee; a physician of the family's choice may consult with the committee and attend its deliberations. The attending physicians are specifically exluded from the committee. The committee may be activated by either the attending physician or the patient's family or guardian. The committee is to review the patient's record, seek other medical information as necessary, and examine the patient. After a clear consensus is reached, the committee shall report whether there is a reasonable possibility of the patient's ever emerging from the comatose state to a cognitive, sapient state. This recommendation shall be part of the patient's medical record.

Of the several advantages of these guidelines, the first is the clear recognition that the committee's task is to make prognoses, not decisions regarding termination of treatment. The mandate of the committee is narrowly defined and this helps to clarify the different aspects of the decision to terminate treatment. Second, the attending physician may not be part of the committee. This both removes potential conflicts of interest and provides for independent review and evaluation. Third, the guidelines are clear on when and how the committee is to be activated. This procedural point is missing in the other documents even though it is crucial for the initiation of a review.

The major problem with these guidelines is best illustrated by the following excerpt from the guidelines: "The attending physician, guided by the committee's decision and with the concurrence of the family, may then proceed with the appropriate course of action and, if indicated, shall personally withdraw life support systems." (State Medical Committee, 1977) This provision reduces the role of the family or guardian to one of consultation rather than primary responsibility. The early part of this same paragraph recognizes that the committee's responsibility in evaluating a prognosis is limited to the application of specialized medical knowledge to the case. The above quoted, concluding sentence confuses the distinction between medical and nonmedical decisions and assumes that the attending physician is the appropriate one to make both types of decisions. Such generalization of expertise is basically fallacious and also serves to take the responsibility of decision making away from the family or guardian where, according to the New Jersey Supreme Court, it belongs. (In Re Karen Quinlan, 1976) Thus, while the guidelines follow one *suggestion* of the court, they ignore an important *requirement* of the court.

UNRESOLVED PROBLEMS IN INTENSIVE CARE

The medical and ethical dilemmas raised in the treatment of the terminally ill are serious and difficult. It seems that no matter which way one moves, problems emerge which further frustrate all parties. Therefore the remainder of this article is devoted to discerning the thematic connections among these problems and discussing their implications for the development and use of guidelines for critical care.

The most critical issue is that of patient competency. It is clear, legally and ethically, that the competent patient has the right to refuse any treatment. Given this, many of the provisions in the guidelines discussed above are useless because they assume that treatment decisions are within the scope of the physician's authority. This preempting of the patient's right needs to be more seriously addressed by the medical profession than it is here. For if the patient is competent and makes his or her decision, the guidelines are not needed.

The incompetent patient presents a variation on the consent for treatment theme. The central problem is determining who can give consent for the incompetent patient. Traditionally, parents have been allowed to give consent for minors. With respect to the incompetent adult, a legally appointed guardian of the person may make decisions regarding the individual's physical status and bodily integrity. Usually a spouse or other next of kin can make these decisions, although the legal status of their authority is somewhat unclear. A mechanism for circumventing the proxy consent question has been introduced in several states where statutes now allow competent adults to issue legally binding instructions regarding the treatment to be given them should they become incompetent. In the Quinlan decision, the New Jersey Supreme Court argued that third parties (i.e., guardians or next of kin) can exercise the incompetent patient's right of privacy, a right which includes the prerogative of terminating treatment. It is quite unclear why the guidelines exclude these people from the decision-making process. For the most part, the guidelines designate the physician as the proper decision maker with respect to the termination of treatment, a decision which does not turn solely on medical factors. Yet the physician is typically not the parent, spouse, next of kin, or legally appointed guardian. Since the rationale for this situation is not given, one can only infer that the working assumption is that the one best qualified to make the technical decision is also the one best

qualified to make the ethical or value judgment. This patently fallacious generalization of expertise is common to all of the guidelines
and represents their most critical weakness.

The second theme concerns the management of the patient and
staff. In such a critical area as the intensive care unit, where decisions
with extensive consequences must often be made quickly, it is most
important to have clear procedures for decision making. Confusion
and conflict serve no good purpose generally and, in an intensive care
unit, they can literally be fatal. Yet with respect to guidelines for
strengthening this process, the question of whom they are to benefit
must be raised. Are they to enhance the treatment and care of the patient? Or, are they to reduce the tension and guilt associated with
such decisions by artificially creating a show of unity and consensus
behind which hide all sorts of unresolved problems among the staff?
It should be kept in mind that the staff is there for the patient, not
vice versa. The Massachusetts General Hospital document in particular focuses on the benefits accruing to the attending physician in
terms of increased support from and reduced conflict within the staff.
But such a management goal is achieved only at the cost of ignoring
the patient. Guidelines that attempt to incorporate different points of
view have many procedural problems; those that resolve problems by
disregarding critical elements have many legal and ethical problems.
Unfortunately, some elements of the guidelines discussed here tend
toward the latter.

The third major issue (which is not addressed by any of the
guidelines discussed) is the question of who will ensure that the offi-
cially accepted guidelines will be followed. Unfortunately, it is much
easier to develop guidelines demonstrating one's legal and ethical
concern for the patient than it is to follow through on implementing
the guidelines and dealing with all of the problems and tensions they
inevitably raise. While wishing neither to burden anyone with further committee responsibility nor to establish yet another committee,
I must note that it is imperative that steps be taken to ensure that patients and families are informed of treatment guidelines, that the staff
follows them, and that changes in policy or interpretations of policy
are brought to the highest authority in the hospital for approval.

The fourth theme is that of economics. Although only the Massachusetts General Hospital guidelines explicitly reject economics as
the sole criterion for decision making, this rejection is implicit in the
other guidelines as well. In these days of the increasing use of cost-
benefit and cost-effectiveness analysis in health policy, as well as es-
calating costs for medical care, such a rejection provides a solid measure of comfort. It indicates that the primary concern is the patient

and his or her welfare. Yet, one can wonder if other types of quasi-economical decisions, such as the allocation of scarce resources (e.g., ICU beds), are not being made continually. If they are, the criteria for such decisions should be clearly stated and fully explained to the patient, staff, and family. The rejection of economic criteria narrowly defined as ability to pay seems to be a reasonable element in any guidelines and one that might be fairly easily agreed upon. The resource allocation decision, however, presents ethical, economic, and medical problems and will not lend itself to easy consensus. Nonetheless, this element of decision making must be clearly incorporated into treatment guidelines because of its far-reaching implications.

Finally, one can raise the ultimate question of the necessity of such guidelines at all. The right of the competent patient to refuse any treatment is well established. (Annas, 1977) At best the guidelines are superfluous in this situation. They have strategic value in that they reinforce this basic legal and ethical right; they give nothing new to the patient, however.

Regarding the incompetent patient, other issues arise with respect to who shall make the decision to terminate treatment. The first is that this is not entirely a medical decision. Although it is initiated by medical judgment, it is fundamentally a value, religious, or ethical decision. Hence, the physician *qua* physician should not be the primary decision maker; rather, the guardian or next of kin should be given this role since they have been charged with serving the interests of their wards. This was emphasized in the Quinlan case where the argument was made that third parties can exercise the right of privacy on behalf of their wards. Also, as "living wills" become more legally acceptable, the patient may already have made his or her decision with respect to treatment, thus removing the need for guidelines specifying the decision makers.

With respect to the content of the decision made by guardians, Veatch makes two helpful suggestions. The first is that a refusal of life-saving treatment is unreasonable if the treatment would restore reasonably normal health to the patient. The second is that it is not unreasonable to reject death-prolonging treatment. (Veatch, 1976) This view implies that treatment guidelines should be primarily concerned with establishing a mechanism by which guardians can obtain accurate long-range prognoses. The orientation of the guidelines, however, seems to be that of giving the physician the criteria for decisions regarding treatment. The New Jersey guidelines are somewhat of an exception for they recognize that the function of the committee is to clarify and/or confirm a prognosis. But, even here, the physician is the one to determine whether treatment should be continued. The implication is that the guardian or next of kin only consents to the decision of the physician. Unfortunately, the correct process is the reverse.

It is my contention that treatment guidelines are useful only insofar as they serve to clarify the prognosis and likely outcomes of various treatments so that the guardian or next of kin can make an informed decision. In their present form, then, the guidelines falsely serve to legitimize the physician as both the chief decision maker and the one who specifies the content of the decision. When this authority is restored to those to whom it rightfully belongs—the guardians and/or next of kin—then the guidelines will serve the more proper function of defining the medical status of the patient and the outcomes of various therapies.

The growing development and availability of sophisticated instrumentation in the intensive care unit has brought us to a keen awareness of the limitations of medical technology. Because such machines can serve to prolong dying as often as they serve to restore health, hospitals have developed guidelines to help in the decision-making process. As long as such guidelines serve as a means of generating information for competent patients, guardians, or next of kin to make the best decision they can, the guidelines are useful. But if they serve to validate the status quo of removing such decisions from the responsible parties and reserving it to the physician or staff, they overstep their own proper role and work against the patient and his or her interests and values.

REFERENCES

1. Annas, GJ (1975). *The Rights of Hospital Patients.* New York: Avon Books (79–91).

2. Critical Care Committee (1976). Optimum Care for Hopelessly Ill Patients. *New England Journal of Medicine* 295(7): 362–364, 12 August 1976.

3. In Re Karen Quinlan: An Alleged Incompetent (1976) 355 A 2d 647.

4. Kirchner, M (1976). How Far to Go Prolonging Life: One Hospital's System. *Medical Economics:* 12 July 1976.

5. Rabkin, MT; Gillerman, G; and Rice, NR (1976). Orders Not To Resuscitate. *New England Journal of Medicine* 295(7): 364–366, 12 August 1976.

6. State Medical Society (1977). Guidelines for Health Care Facilities to Implement Procedures Concerning the Care of Comatose Non-Cognitive Patients. January 1977.

7. Teel, K (1975). The Physician's Dilemma: A Doctor's View: What the Law Should Be. *Baylor Law Review* 27(1): 6–9.

8. Veatch, RM (1976). *Death, Dying, and the Biological Revolution: Our Last Quest for Responsibility.* New Haven: Yale University Press, (162).

Death and Dying/Further Reading

Annas, George J. "Law and the Life Sciences: The Incompetent's Right to Die: The Case of Joseph Saikewicz." *Hastings Center Report* 8 (February 1978), 21–23.

Aries, Philippe. *Western Attitudes toward Death: From the Middle Ages to the Present.* Translated by Patricia M. Ranum. Baltimore: Johns Hopkins University Press, 1974.

Callahan, Daniel. "On Defining a 'Natural Death.' " *Hastings Center Report* 7 (June 1977), 32–37.

Cantor, Norman L. "A Patient's Decision to Decline Life-Saving Medical Treatment: Bodily Integrity vs. the Preservation of Life." *Rutgers Law Review* 26 (Winter, 1972), 228–64.

Imbus, Sharon H., and Zawacki, Bruce E. "Autonomy for Burned Patients When Survival is Unprecedented." *New England Journal of Medicine* 297 (August 11, 1977), 038–11.

Mack, Arien, Editor. *Death in the American Experience.* New York: Schocken Books, 1973.

Maguire, Daniel C. *Death by Choice.* New York: Doubleday, 1974.

Vanderpool, Harold Y. "The Ethics of Terminal Care." *Journal of the American Medical Association* 239 (February 27, 1978), 850–52.

Veatch, Robert M. *Death, Dying, and the Biological Revolution.* New Haven: Yale University Press, 1976.

Weir, Robert F., Editor. *Ethical Issues in Death and Dying.* New York: Columbia University Press, 1977.

PART FOUR

RESEARCH AND HUMAN EXPERIMENTATION

Hans Jonas
David D. Rutstein
Sissela Bok
Robert J. Levine

13 Philosophical Reflections on Experimenting with Human Subjects

HANS JONAS, PH.D.

Jonas' article provides us with a comprehensive review of fundamental ethical issues in human experimentation. Part of the value of the article is his discussion of this topic within the context of the polarity between the individual and society. This fundamental theme is at the heart of the experimentation debate, and Jonas provides a basic description of the major issues. From this perspective, he also examines the motives either for participation in experiments or those used in the recruitment of volunteers. Jonas concludes by presenting basic rules for experimentation. The article points out general themes within the debate as well as providing specific resolutions of some of the important dilemmas.

Dr. Hans Jonas, Ph.D., is the Alvin Johnson Professor of Philosophy in the Graduate Faculty of The New School for Social Research, New York City.

When I was first asked to comment "philosophically" on the subject of human experimentation, I had all the hesitation natural to a layman in the face of matters on which experts of the highest competence have had their say and still carry on their dialogue. As I familiarized myself with the material,[1] any initial feeling of moral rectitude that might have facilitated my task quickly dissipated before the awesome complexity of the problem, and a state of great humility took its place. Nevertheless, because the subject is obscure by its nature and involves fundamental, transtechnical issues, any attempt at clarification can be of use, even without novelty. Even if the philosophical reflection should in the end achieve no more than the realization that in the dialectics of this area we must sin and fall into guilt, this insight may not be without its own gains.

The Peculiarity of Human Experimentation

Experimentation was originally sanctioned by natural science. There it is performed on inanimate objects, and this raises no moral problems. But as soon as animate, feeling beings become the subjects of experiment, as they do in the life sciences and especially in medical research, this innocence of the search for knowledge is lost and questions of conscience arise. The depth to which moral and religious sensibilities can become aroused is shown by the vivisection issue. Human experimentation must sharpen the issue as it involves ultimate questions of personal dignity and sacrosanctity. One difference between the human experiments and the physical is this: The physical experiment employs small-scale, artificially devised substitutes for that about which knowledge is to be obtained, and the experimenter extrapolates from these models and simulated conditions to nature at large. Something deputizes for the "real thing"—balls rolling down an inclined plane for sun and planets, electric discharges from a condenser for real lightning, and so on. For the most part, no such substitution is possible in the biologial sphere. We must operate on the original itself, the real thing in the fullest sense, and perhaps affect it irreversibly. No simulacrum can take its place. Especially in the human sphere, experimentation loses entirely the advantage of the clear division between vicarious model and true object. Up to a point, animals may fulfill the proxy role of the classical physical experiment. But in the end man himself must furnish knowledge about himself, and the comfortable separation of noncommittal experiment and definitive action vanishes. An experiment in education affects the lives of its subjects, perhaps a whole generation of schoolchildren. Human experimentation for whatever purpose is always *also* a responsible, nonexperimental, definitive dealing with the subject himself. And not even the noblest purpose abrogates the obligations this involves.

Can both that purpose and this obligation be satisfied? If not, what would be a just compromise? Which side should give way to the other? The question is inherently philosophical as it concerns not merely pragmatic difficulties and their arbitration, but a genuine conflict of values involving principles of a high order. On principle, it is felt, human beings *ought not* to be dealt with in that way (the "guinea pig" protest); on the other hand, such dealings are increasingly urged on us by considerations, in turn appealing to principle, that claim to override those objections. Such a claim must be carefully assessed, especially when it is swept along by a mighty tide. Putting the matter thus, we have already made one important assumption rooted in our "Western" culture tradition: The prohibitive rule is, to that way of thinking, the primary and axiomatic one; the permissive counter-rule, as qualifying the first, is secondary and stands in need of justification. We must justify the infringement of a primary inviolability, which needs no justification itself; and the justification of its infringement must be by values and needs of a dignity commensurate with those to be sacrificed.

"Individual Versus Society" as the Conceptual Framework

The setting for the conflict most consistently invoked in the literature is the polarity of individual versus society—the possible tension between the individual good and the common good, between private and public welfare. Thus, W. Wolfensberger speaks of "the tension between the long-range interests of society, science, and progress, on one hand, and the rights of the individual on the other."[2] Walsh McDermott says: "In essence, this is a problem of the rights of the individual versus the rights of society."[3] Somewhere I found the "social contract" invoked in support of claims that science may make on individuals in the matter of experimentation. I have grave doubts about the adequacy of this frame of reference, but I will go along with it part of the way. It does apply to some extent, and it has the advantage of being familiar. We concede, as a matter of course, to the common good some pragmatically determined measure of precedence over the individual good. In terms of rights, we let some of the basic rights of the individual be overruled by the acknowledged rights of society—as a matter of right and moral justness and not of mere force or dire necessity (much as such necessity may be adduced in defense of that right). But in making that concession, we require a careful clarification of what the needs, interests, and rights of society are, for society—as distinct from any plurality of individuals—is an abstract and as such is subject to our definition, while the individual is the primary concrete, prior to all definition, and his basic good is more or less known. Thus, the unknown in our problem is the so-called common or public good and its potentially superior claims, to which the individual good must or might sometimes be sacrificed, in circumstances that in turn must also be counted among the unknowns of our questions. Note that in putting the matter in this way—that is, in asking about the right of society to individual sacrifice—the consent of the sacrificial subject is no necessary part of the basic question.

"Consent," however, is the other most consistently emphasized and examined concept in discussions of this issue. This attention betrays a feeling that the "social" angle is not fully satisfactory. If society has a right, its exercise is not contingent on volunteering. On the other hand, if volunteering is fully genuine, no public right to the volunteered act need be construed. There is a difference between the moral or emotional appeal of a cause that elicits volunteering and a right that demands compliance—for example, with particular reference to the social sphere, between the *moral claim* of a common good and society's *right* to that good and to the means of its realization. A moral claim cannot be met without consent; a right can do without it. Where consent is present anyway, the distinction may become immaterial. But the awareness of the many ambiguities besetting the "consent" actually available and used in medical research prompts recourse to the idea of a public right conceived independently of (and valid prior to) consent; and, vice versa, the awareness of the problematic nature of such a right makes even its advocates still insist on the idea of consent with all its ambiguities: An uneasy situation exists for both sides.

Nor does it help much to replace the language of "rights" by that of "interests" and then argue the sheer cumulative weight of the interests of the many over against those of the few or the single individual. "Interests" range all the way from the most marginal and optional to the most vital and imperative, and only those sanctioned by particular importance and merit will be admitted to count in such a calculus— which simply brings us back to the question of right or moral claim. Moreover, the appeal to numbers is dangerous. Is the number of those afflicted with a particular disease great enough to warrant violating the interests of the nonafflicted? Since the number of the latter is usually so much greater, the argument can actually turn around to the contention that the cumulative weight of interest is on *their* side. Finally, it may well be the case that the individual's interest in his own inviolability is itself a public interest such that its publicly condoned violation, irrespective of numbers, violates the interest of all. In that case, its protection in *each* instance would be a paramount interest, and the comparison of numbers will not avail.

These are some of the difficulties hidden in the conceptual framework indicated by the terms "society-individual," "interest," and "rights." But we also spoke of a moral call, and this points to another dimension—not indeed divorced from the societal sphere, but transcending it. And there is something even beyond that: true sacrifice from highest devotion, for which there are no laws or rules except that it must be absolutely free. "No one has the right to choose martyrs for science" was a statement repeatedly quoted in the November, 1967, *Daedalus* conference. But no scientist can be prevented from making himself a martyr for his science. At all times, dedicated explorers, thinkers, and artists have immolated themselves on the altar of their vocation, and creative genius most often pays the price of happiness, health, and life for its own consummation. But no one, not even society, has the shred of a right to expect and ask these things. They come to the rest of us as a *gratia gratis data*.

The Sacrificial Theme

Yet we must face the somber truth that the *ultima ratio* of communal life is and has always been the compulsory, vicarious sacrifice of individual lives. The primordial sacrificial situation is that of outright human sacrifices in early communities. These were not acts of blood-lust or gleeful savagery; they were the solemn execution of a supreme, sacral necessity. One of the fellowship of men had to die so that all could live, the earth be fertile, the cycle of nature renewed. The victim often was not a captured enemy, but a select member of the group: "The king must die." If there was cruelty here, it was not that of men, but that of the gods, or rather of the stern order of things, which was believed to exact that price for the bounty of life. To assure it for the community, and to assure it ever again, the awesome *quid pro quo* had to be paid ever again.

Far be it from me, and far should it be from us, to belittle from the height of our enlightened knowledge the majesty of the underlying conception. The particular *causal* views that prompted our ancestors have long since been relegated to the realm of superstition. But in moments of national danger we still send the flower of our young manhood to offer their lives for the continued life of the community, and if it is a just war, we see them go forth as consecrated and strangely ennobled by a sacrificial role. Nor do we make their going forth depend on their own will and consent, much as we may desire and foster these: We conscript them according to law. We conscript the best and feel morally disturbed if the draft, either by design or in effect, works so that mainly the disadvantaged, socially less useful, more expendable, make up those whose lives are to buy ours. No rational persuasion of the pragmatic necessity here at work can do away with the feeling, mixed of gratitude and guilt, that the sphere of the sacred is touched with the vicarious offering of life for life. Quite apart from these dramatic occasions, there is, it appears, a persistent and constitutive aspect of human immolation to the very being and prospering of human society—an immolation in terms of life and happiness, imposed or voluntary, of few for many. What Goethe has said of the rise of Christianity may well apply to the nature of civilization in general: "*Opfer fallen hier, / Weder Lamm noch Stier, / Aber Menschenopfer unerhoert.*"[4] We can never rest comfortably in the belief that the soil from which our satisfactions sprout is not watered with the blood of martyrs. But a troubled conscience compels us, the undeserving beneficiaries, to ask: Who is to be martyred? in the service of what cause? and by whose choice?

Not for a moment do I wish to suggest that medical experimentation on human subjects, sick or healthy, is to be likened to primeval human sacrifices. Yet something sacrificial is involved in the selective abrogation of personal inviolability and the ritualized exposure to gratuitous risk of health and life, justified by a presumed greater, social good. My examples from the sphere of stark sacrifice were intended to sharpen the issues implied in that context and to set them off clearly from the kinds of obligations and constraints imposed on the citizen in the normal course of things or generally demanded of the individual in exchange for the advantages of civil society.

The "Social Contract" Theme

The first thing to say in such a setting-off is that the sacrificial area is not covered by what is called the "social contract." This fiction of political theory, premised on the primacy of the individual, was designed to supply a rationale for the *limitation* of individual freedom and power required for the existence of the body politic, whose existence in turn is for the benefit of the individuals. The principle of these limitations is that their *general* observance profits all, and that therefore the individual observer, assuring this general observance for his part, profits by it himself. I observe property rights because their general observance assures my own; I observe traffic rules because their general observance assures my own safety; and so on. The obligations here are mutual and general; no one is singled out for special sacrifice. For the most part, *qua* limitations of my liberty, the laws thus deducible from the hypothetical "social contract" enjoin me from certain actions rather than obligate me to positive actions (as did the laws of feudal society). Even where the latter is the case, as in the duty to pay taxes, the rationale is that I am myself a beneficiary of the services financed through these payments. Even the contributions levied by the welfare state, though not originally contemplated in the liberal version of the social contract theory, can be interpreted as a personal insurance policy of one sort or another—be it against the contingency of my own indigence, the dangers of disaffection from the laws in consequence of widespread unrelieved destitution, or the disadvantages of a diminished consumer market. Thus, by some stretch, such contributions can still be subsumed under the principle of enlightened self-interest. But no complete abrogation of self-interest at any time is in the terms of the social contract, and so pure sacrifice falls outside it. Under the putative terms of the contract alone, I cannot be required to die for the public good. (Thomas Hobbes made this forcibly clear.) Even short of this extreme, we like to think that nobody is entirely and one-sidedly the victim in any of the renunciations exacted under normal circumstances by society "in the general interest"—that is, for the benefit of others. "Under normal circumstances," as we shall see, is a necessary qualification. Moreover, the "contract" can legitimize claims only on our overt public actions and not on our invisible private being. Our powers, not our persons, are beholden to the commonweal. In one important respect, it is true, public interest and control do extend to the private sphere by general consent: in the compulsory education of our children. Even there, the assumption is that the learning and what is learned, apart from all future social usefulness, are also for the benefit of the individual in his own being. We would not tolerate education to degenerate into the conditioning of useful robots for the social machine.

Both restrictions of public claim in behalf of the "common good" —that concerning one-sided sacrifice and that concerning the private sphere—are valid only, let us remember, on the premise of the primacy of the individual, upon which the whole idea of the "social contract" rests. This primacy is itself a metaphysical axiom or option peculiar to our Western tradition, and the whittling away of this axiom would threaten the tradition's whole foundation. In passing, I may remark that systems adopting the alternative primacy of the community as their axiom are naturally less bound by the restrictions we postulate. Whereas we reject the idea of "expendables" and regard those not useful or even recalcitrant to the social purpose as a burden that society must carry (since their individual claim to existence is as absolute as that of the most useful), a truly totalitarian regime, Communist or other, may deem it right for the collective to rid itself of such encumbrances or to make them forcibly serve some social end by conscripting their persons (and there are effective combinations of both). We do not normally—that is, in nonemergency conditions—give the state the right to conscript labor, while we do give it the right to "conscript" money, for money is detachable from the person as labor is not. Even less than forced labor do we countenance forced risk, injury, and indignity.

But in time of war our society itself supersedes the nice balance of the social contract with an almost absolute precedence of public necessities over individual rights. In this and similar emergencies, the sacrosanctity of the individual is abrogated, and what for all practical purposes amounts to a near-totalitarian, quasi-Communist state of affairs is *temporarily* permitted to prevail. In such situations, the community is conceded the right to make calls on its members, or certain of its members, entirely different in magnitude and kind from the calls normally allowed. It is deemed right that a part of the population bears a disproportionate burden of risk of a disproportionate gravity; and it is deemed right that the rest of the community accepts this sacrifice, whether voluntary or enforced, and reaps its benefits—difficult as we find it to justify this acceptance and thus benefit by any normal ethical categories. We justify it transethically, as it were, by the supreme collective emergency, formalized, for example, by the declaration of a state of war.

Medical experimentation on human subjects falls somewhere between this overpowering case and the normal transactions of the social contract. On the one hand, no comparable extreme issue of social survival is (by and large) at stake. And no comparable extreme sacrifice or forseeable risk is (by and large) asked. On the other hand, what is asked goes decidedly beyond, even runs counter to, what it is otherwise deemed fair to let the individual sign over of his person to the benefit of the "common good." Indeed, our sensitivity to the kind of intrusion and use involved is such that only an end of transcendent value or overriding urgency can make it arguable and possibly acceptable in our eyes.

Health as a Public Good

The cause invoked is health and, in its more critical aspect, life itself—clearly superlative goods that the physician serves directly by curing and the researcher indirectly by the knowledge gained through his experiments. There is no question about the good served nor about the evil fought—disease and premature death. But a good to whom and an evil to whom? Here the issue tends to become somewhat clouded. In the attempt to give experimentation the proper dignity (on the problematic view that a value becomes greater by being "social" instead of merely individual), the health in question or the disease in question is somehow predicated of the social whole, as if it were society that, in the persons of its members, enjoyed the one and suffered the other. For the purposes of our problem, public interest can then be pitted against private interest, the common good against the individual good. Indeed, I have found health called a national resource, which of course it is, but surely not in the first place.

In trying to resolve some of the complexities and ambiguities lurking in these conceptualizations, I have pondered a particular statement, made in the form of a question, which I found in the *Proceedings* of the November *Dædalus* conference: "Can society afford to discard the tissues and organs of the hopelessly unconscious patient when they could be used to restore the otherwise hopelessly ill, but still salvageable individual?" And somewhat later: "A strong case can be made that society can ill afford to discard the tissues and organs of the hopelessly unconscious patient; they are greatly needed for study and experimental trial to help those who can be salvaged."[5] I hasten to add that any suspicion of callousness that the "commodity" language of these statements may suggest is immediately dispelled by the name of the speaker, Dr. Henry K. Beecher, for whose humanity and moral sensibility there can be nothing but admiration. But the use, in all innocence, of this language gives food for thought. Let me, for a moment, take the question literally. "Discarding" implies proprietary rights—nobody can discard what does not belong to him in the first place. Does society then own my body? "Salvaging" implies the same and, moreover, a use-value to the owner. Is the life-extension of certain individuals then a public interest? "Affording" implies a critically vital level of such an interest—that is, of the loss or gain involved. And "society" itself—what is it? When does a need, an aim, an obligation become social? Let us reflect on some of these terms.

What Society Can Afford

"Can Society afford . . .?" Afford what? To let people die intact, thereby withholding something from other people who desperately need it, who in consequence will have to die too? These other, unfortunate people indeed cannot afford not to have a kidney, heart, or other organ of the dying patient, on which they depend for an extension of their lease on life; but does that give them a right to it? Does it oblige society to procure it for them? What is it that *society* can or cannot afford—leaving aside for the moment the question of what it has a *right* to? It surely can afford to lose members through death; more than that, it is built on the balance of death and birth decreed by the order of life. This is too general, of course, for our question, but perhaps it is well to remember. The specific question seems to be whether society can afford to let some people die whose death might be deferred by particular means if these were authorized by society. Again, if it is merely a question of what society can or cannot afford, rather than of what it ought or ought not to do, the answer must be: Of course, it can. If cancer, heart disease, and other organic, noncontagious ills, especially those tending to strike the old more than the young, continue to exact their toll at the normal rate of incidence (including the toll of private anguish and misery), society can go on flourishing in every way.

Here, by contrast, are some examples of what, in sober truth, so-

Ur society literary cannot afford (handwritten margin note)

ciety cannot afford. It cannot afford to let an epidemic rage un-
checked; a persistent excess of deaths over births, but neither too
great an excess of births over deaths; too low an average life-expectancy even if demographically balanced by fertility, but neither too
great a longevity with the necessitated correlative dearth of youth in
the social body; a debilitating state of general health; and things of
this kind. These are plain cases where the whole condition of society is
critically affected, and the public interest can make its imperative
claims. The Black Death of the Middle Ages was a *public* calamity of
the acute kind; the life-sapping ravages of endemic malaria or sleeping
sickness in certain areas are a public calamity of the chronic kind. A
society as a whole can truly not "afford" such situations, and they
may call for extraordinary remedies, including, perhaps, the invasion
of private sacrosanctities.

This is not entirely a matter of numbers and numerical ratios. Society, in a subtler sense, cannot "afford" a single miscarriage of justice, a single inequity in the dispensation of its laws, the violation of
the rights of even the tiniest minority, because these undermine the
moral basis on which society's existence rests. Nor can it, for a similar
reason, afford the absence or atrophy in its midst of compassion and
of the effort to alleviate suffering—be it widespread or rare—one form
of which is the effort to conquer disease of any kind, whether "socially" significant (by reason of number) or not. And in short, society
cannot afford the absence among its members of *virtue* with its readi-
ness to sacrifice beyond defined duty. Since its presence—that is to
say, that of personal idealism—is a matter of grace and not of decree,
we have the paradox that society depends for its existence on intangi-
bles of nothing less than a religious order, for which it can hope, but
which it cannot enforce. All the more must it protect this most precious capital from abuse.

For what objectives connected with the medico-biological sphere
should this reserve be drawn upon—for example, in the form of accepting, soliciting, perhaps even imposing the submission of human
subjects to experimentation? We postulate that this must be not just
a worthy cause, as any promotion of the health of anybody doubtless-
ly is, but a cause qualifying for transcendent social sanction. Here one
thinks first of those cases critically affecting the whole condition,
present and future, of the community. Something equivalent to what
in the political sphere is called "clear and present danger" may be invoked and a state of emergency proclaimed, thereby suspending certain otherwise inviolable prohibitions and taboos. We may observe
that averting a disaster always carries greater weight than promoting a

good. Extraordinary danger excuses extraordinary means. This covers human experimentation, which we would like to count, as far as possible, among the extraordinary rather than the ordinary means of serving the common good under public auspices. Naturally, since foresight and responsibility for the future are of the essence of institutional society, averting disaster extends into long-term prevention, although the lesser urgency will warrant less sweeping licenses.

Society and the Cause of Progress

Much weaker is the case where it is a matter not of saving but of improving society. Much of medical research falls into this category. A permanent death rate from heart failure or cancer does not threaten society. So long as certain statistical ratios are maintained, the incidence of disease and of disease-induced mortality is not (in the strict sense) a "social" misfortune. I hasten to add that it is not therefore less of a human misfortune, and the call for relief issuing with silent eloquence from each victim and all potential victims is of no lesser dignity. But it is misleading to equate the fundamentally human response to it with what is owed to society: It is owed by man to man—and it is thereby owed by society to the individuals as soon as the adequate ministering to these concerns outgrows (as it progressively does) the scope of private spontaneity and is made a public mandate. It is thus that society assumes responsibility for medical care, research, old age, and innumerable other things not originally of the public realm (in the original "social contract"), and they become duties toward "society" (rather than directly toward one's fellow man) by the fact that they are socially operated.

Indeed, we expect from organized society no longer mere protection against harm and the securing of the conditions of our preservation, but active and constant improvement in all the domains of life: the waging of the battle against nature, the enhancement of the human estate—in short, the promotion of progress. This is an expansive goal, one far surpassing the disaster norm of our previous reflections. It lacks the urgency of the latter, but has the nobility of the free, forward thrust. It surely is worth sacrifices. It is not at all a question of what society can afford, but of what it is committed to, beyond all necessity, by our mandate. Its trusteeship has become an established, ongoing, institutionalized business of the body politic. As eager beneficiaries of its gains, we now owe to "society," as its chief agent, our individual contribution toward its *continued pursuit.* Maintaining the existing level requires no more than the orthodox means of taxation and enforcement of professional standards that raise no problems. The more optional goal of pushing forward is also more exacting. We have this syndrome: Progress is by our choosing an acknowledged interest of society, in which we have a stake in various degrees; science is a necessary instrument of progress; research is a necessary instrument of science; and in medical science experimentation on human subjects is a necessary instrument of research: Therefore, human experimentation has come to be a societal interest.

The destination of research is essentially melioristic. It does not serve the preservation of the existing good from which I profit myself and to which I am obligated. Unless the present state is intolerable, the melioristic goal is in a sense gratuitous, and not only from the vantage point of the present. Our descendants have a right to be left an unplundered planet; they do not have a right to new miracle cures. We have sinned against them if by our doing we have destroyed their inheritance—which we are doing at full blast; we have not sinned against them if by the time they come around arthritis has not yet been conquered (unless by sheer neglect). And generally, in the matter of progress, as humanity had no claim on a Newton, a Michelangelo, or a St. Francis to appear, and no right to the blessings of their unscheduled deeds, so progress, with all our methodical labor for it, cannot be budgeted in advance and its fruits received as a due. Its coming-about at all and its turning out for good (of which we can never be sure) must rather be regarded as something akin to grace.

The Melioristic Goal, Medical Research, and Individual Duty

Nowhere is the melioristic goal more inherent than in medicine. To the physician, it is not gratuitous. He is committed to curing and thus to improving the power to cure. Gratuitous we called it (outside disaster conditions) as a *social* goal, but noble at the same time. Both the nobility and the gratuitousness must influence the manner in which self-sacrifice for it is elicited and even its free offer accepted. Freedom is certainly the first condition to be observed here. The surrender of one's body to medical experimentation is entirely outside the enforceable "social contract."

Or can it be construed to fall within its terms—namely, as repayment for benefits from past experimentation that I have enjoyed myself? But I am indebted for these benefits not to society, but to the past "martyrs," to whom society is indebted itself, and society has no right to call in my personal debt by way of adding new to its own. Moreover, gratitude is not an enforceable social obligation; it anyway does not mean that I must emulate the deed. Most of all, if it was wrong to exact such sacrifice in the first place, it does not become right to exact it again with the plea of the profit it has brought me. If, however, it was not exacted, but entirely free, as it ought to have been, then it should remain so, and its precedence must not be used as a social pressure on others for doing the same under the sign of duty.

Indeed, we must look outside the sphere of the social contract, outside the whole realm of public rights and duties, for the motivations and norms by which we can expect ever again the upwelling of a will to give what nobody—neither society, nor fellow man, nor posterity—is entitled to. There are such dimensions in man with transsocial wellsprings of conduct, and I have already pointed to the paradox, or mystery, that society cannot prosper without them, that it must draw on them, but cannot command them.

What about the moral law as such a transcendent motivation of conduct? It goes considerably beyond the public law of the social contract. The latter, we saw, is founded on the rule of enlightened self-interest: *Do ut des*—I give so that I be given to. The law of individual conscience asks more. Under the Golden Rule, for example, I am required to give as I wish to be given to under like circumstances, but not in order that I be given to and not in expectation of return. Reciprocity, essential to the social law, is not a condition of the moral law. One subtle "expectation" and "self-interest," but of the moral order itself, may even then be in my mind: I prefer the environment of a moral society and can expect to contribute to the general morality by my own example. But even if I should always be the dupe, the Golden Rule holds. (If the social law breaks faith with me, I am released from its claim.)

Moral Law and Transmoral Dedication

Can I, then, be called upon to offer myself for medical experimentation in the name of the moral law? *Prima facie*, the Golden Rule seems to apply. I should wish, were I dying of a disease, that enough volunteers in the past had provided enough knowledge through the gift of their bodies that I could now be saved. I should wish, were I desperately in need of a transplant, that the dying patient next door had agreed to a definition of death by which his organs would become available to me in the freshest possible condition. I surely should also wish, were I drowning, that somebody would risk his life, even sacrifice his life, for mine.

But the last example reminds us that only the negative form of the Golden Rule ("Do not do unto others what you do not want done unto yourself") is fully prescriptive. The positive form ("Do unto others as you would wish them to do unto you"), in whose compass our issue falls, points into an infinite, open horizon where prescriptive force soon ceases. We may well say of somebody that he ought to have come to the succor of B, to have shared with him in his need, and the like. But we may not say that he ought to have given his life for him. To have done so would be praiseworthy; not to have done so is not blameworthy. It cannot be asked of him; if he fails to do so, he reneges on no duty. But *he* may say of himself, and only he, that he ought to have given his life. *This* "ought" is strictly between him and himself, or between him and God; no outside party—fellow man or society—can appropriate its voice. It can humbly receive the supererogatory gifts from the free enactment of it.

We must, in other words, distinguish between moral obligation and the much larger sphere of moral value. (This, incidentally, shows up the error in the widely held view of value theory that the higher a value, the stronger its claim and the greater the duty to realize it. The highest are in a region beyond duty and claim.) The ethical dimension far exceeds that of the moral law and reaches into the sublime solitude of dedication and ultimate commitment, away from all reckoning and rule—in short, into the sphere of the *holy*. From there alone can the offer of self-sacrifice genuinely spring, and this—its source—must be honored religiously. How? The first duty here falling on the research community, when it enlists and uses this source, is the safeguarding of true authenticity and spontaneity.

The "Conscription" of Consent

But here we must realize that the mere issuing of the appeal, the calling for volunteers, with the moral and social pressures it inevitably generates, amounts even under the most meticulous rules of consent to a sort of *conscripting*. And some soliciting is necessarily involved. This was in part meant by the earlier remark that in this area sin and guilt can perhaps not be wholly avoided. And this is why "consent," surely a non-negotiable minimum requirement, is not the full answer to the problem. Granting then that soliciting and therefore some degree of conscripting are part of the situation, who may conscript and who may be conscripted? Or less harshly expressed: Who should issue appeals and to whom?

The naturally qualified issuer of the appeal is the research scientist himself, collectively the main carrier of the impulse and the only one with the technical competence to judge. But his being very much an interested party (with vested interests, indeed, not purely in the public good, but in the scientific enterprise as such, in "his" project, and even in his career) makes him also suspect. The ineradicable dialectic of this situation—a delicate incompatibility problem—calls for particular controls by the research community and by public authority that we need not discuss. They can mitigate, but not eliminate the problem. We have to live with the ambiguity, the treacherous impurity of everything human.

Self-Recruitment of the Research Community

To whom should the appeal be addressed? The natural issuer of the call is also the first natural addressee; the physician-researcher himself and the scientific confraternity at large. With such a coincidence—indeed, the noble tradition with which the whole business of human experimentation started—almost all of the associated legal, ethical, and metaphysical problems vanish. If it is full, autonomous identification of the subject with the purpose that is required for the dignifying of his serving as a subject—here it is; if strongest motivation—here it is; if fullest understanding—here it is; if freest decision—here it is; if greatest integration with the person's total, chosen pursuit—here it is. With self-solicitation, the issue of consent in all its insoluble equivocality is bypassed *per se*. Not even the condition that the particular purpose be truly important and the project reasonably promising, which must hold in any solicitation of others, need be satisfied here. By himself, the scientist is free to obey his obsession, to play his hunch, to wager on chance, to follow the lure of ambition. It is all part of the "divine madness" that somehow animates the ceaseless pressing against frontiers. For the rest of society, which has a deep-seated disposition to look with reverence and awe upon the guardians of the mysteries of life, the profession assumes with this proof of its devotion the role of a self-chosen, consecrated fraternity, not unlike the monastic orders of the past; and this would come nearest to the actual, religious origins of the art of healing.

It would be the ideal, but not a real solution to keep the issue of human experimentation within the research community itself. Neither in numbers nor in variety of material would its potential suffice for the many-pronged, systematic, continual attack on disease into which the lonely exploits of the early investigators have grown. Statistical requirements alone make their voracious demands; were it not for what I have called the essentially "gratuitous" nature of the whole enterprise of progress, as against the mandatory respect for invasion-proof selfhood, the simplest answer would be to keep the whole population enrolled, and let the lot, or an equivalent of draft boards, decide which of each category will at any one time be called up for "service." It is not difficult to picture societies with whose philosophy this would be consonant. We are agreed that ours is not one such and should not become one. The specter of it is indeed among the threatening utopias on our own horizon from which we should recoil, and of whose advent by imperceptible steps we must beware. How then can our mandatory faith be honored when the recruitment for experimentation goes outside the scientific community, as it must in honoring another commitment of no mean dignity? We simply repeat the former question: To whom should the call be addressed?

"Identification" as the Principle of Recruitment in General

If the properties we adduced as the particular qualifications of the members of the scientific fraternity itself are taken as general criteria of selection, then one should look for additional subjects where a maximum of identification, understanding, and spontaneity can be expected—that is, among the most highly motivated, the most highly educated, and the least "captive" members of the community. From this naturally scarce resource, a descending order of permissibility leads to greater abundance and ease of supply, whose use should become proportionately more hesitant as the exculpating criteria are relaxed. An inversion of normal "market" behavior is demanded here —namely, to accept the lowest quotation last (and excused only by the greatest pressure of need), to pay the highest price first.

As such a rule of selection is bound to be rather hard on the number-hungry research industry, it will be asked: Why all the fuss? At this point we had better spell out some of the things we have been tacitly presupposing all the time. What is wrong with making a person an experimental subject is not so much that we make him thereby a means (which happens in social contexts of all kinds), as that we make him a thing—a passive thing merely to be acted on, and passive not even for real action, but for token action whose token object he is. His being is reduced to that of a mere token or "sample." This is different from even the most exploitative situations of social life; there the business is real, not fictitious. The subject, however much abused, remains an agent and thus a "subject" in the other sense of the word. The soldier's case, referred to earlier, is instructive: Subject to most unilateral discipline, forced to risk mutilation and death, conscripted without, perhaps against, his will—he is still conscripted with his capacities to act, to hold his own or fail in situations, to meet real challenges for real stakes. Though a mere "number" to the High Command, he is not a token and not a thing. (Imagine what he would say if it turned out that the war was a game staged to sample observations on his endurance, courage, or cowardice.)

These compensations of personhood are denied to the subject of experimentation, who is acted upon for an extraneous end without being engaged in a real relation where he would be the counterpoint to the other or to circumstance. Mere "consent" (mostly amounting to no more than permission) does not right this reification. The "wrong" of it can only be made "right" by such authentic identification with the cause that it is the subject's as well as the researcher's cause—whereby his role in its service is not just permitted by him, but *willed*. That sovereign will of his which embraces the end as his own restores his personhood to the otherwise depersonalizing context. To be valid it must be autonomous and informed. The latter condition can, outside the research community, only be fulfilled by degrees; but the higher the degree of the understanding regarding the purpose and the technique, the more valid becomes the endorsement of the will. A margin of mere trust inevitably remains. Ultimately, the appeal for volunteers should seek this free and generous endorsement, the appropriation of the research purpose into the person's own scheme of ends. Thus, the appeal is in truth addressed to the one, mysterious, and sacred source of any such generosity of the will—"devotion," whose forms and objects of commitment are various and may invest different motivations in different individuals. The following, for instance, may be responsive to the "call" we are discussing: compassion with human suffering, zeal for humanity, reverence for the Golden Rule, enthusiasm for progress, homage to the cause of knowledge, even longing for sacrificial justification (do not call that "masochism," please). On all these, I say, it is defensible and right to draw when the research objective is worthy enough: and it is a prime duty of the research community (especially in view of what we called the "margin of trust") to see that this sacred source is never abused for frivolous ends. For a less than adequate cause, not even the freest, unsolicited offer should be accepted.

The Rule of the "Descending Order" and Its Counter-Utility Sense

We have laid down what must seem to be a forbidding rule. Having faith in the transcendent potential of man, I do not fear that the "source" will ever fail a society that does not destroy it—and only such a one is worthy of the blessings of progress. But "elitistic" the rule is (as is the enterprise of progress itself), and elites are by nature small. The combined attribute of motivation and information, plus the absence of external pressures, tends to be socially so circumscribed that strict adherence to the rule might numerically starve the research process. This is why I spoke of a descending order of permissibility, which is itself permissive, but where the realization that it is a *descending* order is not without pragmatic import. Departing from the august norm, the appeal must needs shift from idealism to docility, from high-mindedness to compliance, from judgment to trust. Consent spreads over the whole spectrum. I will not go into the casuistics of this penumbral area. I merely indicate the principle of the order of preference: The poorer in knowledge, motivation, and freedom of decision (and that, alas, means the more readily available in terms of numbers and possible manipulation), the more sparingly and indeed reluctantly should the reservoir be used, and the more compelling must therefore become the countervailing justification.

Let us note that this is the opposite of a social utility standard, the reverse of the order by "availability and expendability": The most valuable and scarcest, the least expendable elements of the social organism, are to be the first candidates for risk and sacrifice. It is the standard of *noblesse oblige*; and with all its counter-utility and seeming "wastefulness," we feel a rightness about it and perhaps even a higher "utility," for the soul of the community lives by this spirit.[6] It is also the opposite of what the day-to-day interests of research clamor for, and for the scientific community to honor it will mean that it will have to fight a strong temptation to go by routine to the readiest sources of supply—the suggestible, the ignorant, the dependent, the "captive" in various senses.[7] I do not believe that heightened resistance here must cripple research, which cannot be permitted; but it may slow it down by the smaller numbers fed into experimentation in consequence. This price—a possibly slower rate of progress—may have to be paid for the preservation of the most precious capital of higher communal life.

Experimentation on Patients

So far we have been speaking on the tacit assumption that the subjects of experimentation are recruited from among the healthy. To the question "Who is conscriptable?" the spontaneous answer is: Least and last of all the sick—the most available source as they are under treatment and observation anyway. That the afflicted should not be called upon to bear additional burden and risk, that they are society's special trust and the physician's particular trust—these are elementary responses of our moral sense. Yet the very destination of medical research, the conquest of disease, requires at the crucial stage trial and verification on precisely the sufferers from the disease, and their total exemption would defeat the purpose itself. In acknowledging this inescapable necessity, we enter the most sensitive area of the whole complex, the one most keenly felt and most searchingly discussed by the practitioners themselves. This issue touches the heart of the doctor-patient relation, putting its most solemn obligations to the test. Some of the oldest verities of this area should be recalled.

The Fundamental Privilege of the Sick

In the course of treatment, the physician is obligated to the patient and to no one else. He is not the agent of society, nor of the interests of medical science, the patient's family, the patient's co-sufferers, or future sufferers from the same disease. The patient alone counts when he is under the physician's care. By the simple law of bilateral contract (analogous, for example, to the relation of lawyer to client and its "conflict of interest" rule), he is bound not to let any other interest interfere with that of the patient in being cured. But manifestly more sublime norms than contractual ones are involved. We may speak of a sacred trust; strictly by its terms, the doctor is, as it were, alone with his patient and God.

There is one normal exception to this—that is, to the doctor's not being the agent of society vis-à-vis the patient, but the trustee of his interests alone—the quarantining of the contagious sick. This is plainly not for the patient's interest, but for that of others threatened by him. (In vaccination, we have a combination of both: protection of the individual and others.) But preventing the patient from causing harm to others is not the same as exploiting him for the advantage of others. And there is, of course, the abnormal exception of collective catastrophe, the analogue to a state of war. The physician who desperately battles a raging epidemic is under a unique dispensation that suspends in a nonspecifiable way some of the strictures of normal practice, including possibly those against experimental liberties with his patients. No rules can be devised for the waiving of rules in extremities. And as with the famous shipwreck examples of ethical theory, the less said about it the better. But what is allowable there and may later be passed over in forgiving silence cannot serve as a precedent. We are concerned with non-extreme, non-emergency conditions where the voice of principle can be heard and claims can be adjudicated free from duress. We have conceded that there are such claims, and that if there is to be medical advance at all, not even the superlative privilege of the suffering and the sick can be kept wholly intact from the intrusion of its needs. About this least palatable, most disquieting part of our subject, I have to offer only groping, inconclusive remarks.

The Principle of "Identification" Applied to Patients

On the whole, the same principles would seem to hold here as are found to hold with "normal subjects": motivation, identification, understanding on the part of the subject. But it is clear that these conditions are peculiarly difficult to satisfy with regard to a patient. His physical state, psychic preoccupation, dependent relation to the doctor, the submissive attitude induced by treatment—everything connected with his condition and situation makes the sick person inherently less of a sovereign person than the healthy one. Spontaneity of self-offering has almost to be ruled out; consent is marred by lower resistance or captive circumstance, and so on. In fact, all the factors that make the patient, as a category, particularly accessible and welcome for experimentation at the same time compromise the quality of the responding affirmation that must morally redeem the making use of them. This, in addition to the primacy of the physician's duty, puts a heightened onus on the physician-researcher to limit his undue power to the most important and defensible research objectives and, of course, to keep persuasion at a minimum.

Still, with all the disabilities noted, there is scope among patients for observing the rule of the "descending order of permissibility" that we have laid down for normal subjects, in vexing inversion of the utility order of quantitative abundance and qualitative "expendability." By the principle of this order, those patients who most identify with and are cognizant of the cause of research—members of the medical profession (who after all are sometimes patients themselves)—come first; the highly motivated and educated, also least dependent, among the lay patients come next; and so on down the line. An added consideration here is seriousness of condition, which again operates in inverse proportion. Here the profession must fight the tempting sophistry that the hopeless case is expendable (because in prospect already expended) and therefore especially usable; and generally the attitude that the poorer the chances of the patient the more justifiable his recruitment for experimentation (other than for his own benefit). The opposite is true.

Nondisclosure as a Borderline Case

Then there is the case where ignorance of the subject, sometimes even of the experimenter, is of the essence of the experiment (the "double blind"-control group-placebo syndrome). It is said to be a necessary element of the scientific process. Whatever may be said about its ethics in regard to normal subjects, especially volunteers, it is an outright betrayal of trust in regard to the patient who believes that he is receiving treatment. Only supreme importance of the objective can exonerate it, without making it less of a transgression. The patient is definitely wronged even when not harmed. And ethics apart, the practice of such deception holds the danger of undermining the faith in the bona fides of treatment, the beneficial intent of the physician—the very basis of the doctor-patient relationship. In every respect, it follows that concealed experiment on patients—that is, experiment under the guise of treatment—should be the rarest exception, at best, if it cannot be wholly avoided.

This has still the merit of a borderline problem. This is not true of the other case of necessary ignorance of the subject—that of the unconscious patient. Drafting him for nontherapeutic experiments is simply and unqualifiedly impermissible; progress or not, he must never be used, on the inflexible principle that utter helplessness demands utter protection.

When preparing this paper, I filled pages with a casuistics of this harrowing field, but then scratched out most of it, realizing my dilettante status. The shadings are endless, and only the physician-researcher can discern them properly as the cases arise. Into his lap the decision is thrown. The philosophical rule, once it has admitted into itself the idea of a sliding scale, cannot really specify its own application. It can only impress on the practitioner a general maxim or attitude for the exercise of his judgment and conscience in the concrete occasions of his work. In our case, I am afraid, it means making life more difficult for him.

It will also be noted that, somewhat at variance with the emphasis in the literature, I have not dwelt on the elements of "risk" and very little on that of "consent." Discussion of the first is beyond the layman's competence; the emphasis on the second has been lessened because of its equivocal character. It is a truism to say that one should strive to minimize the risk and to maximize the consent. The more demanding concept of "identification," which I have used, includes "consent" in its maximal or authentic form, and the assumption of risk is its privilege.

No Experiments on Patients Unrelated to Their Own Disease

Although my ponderings have, on the whole, yielded points of view rather than definite prescriptions, premises rather than conclusions, they have led me to a few unequivocal yeses and noes. The first is the emphatic rule that patients should be experimented upon, if at all, *only* with reference to *their* disease. Never should there be added to the gratuitousness of the experiment as such the gratuitousness of service to an unrelated cause. This follows simply from what we have found to be the *only* excuse for infracting the special exemptions of the sick at all—namely, that the scientific war on disease cannot accomplish its goal without drawing the sufferers from disease into the investigative process. If under this excuse they become subjects of experiment, they do so *because*, and only because, of *their* disease.

This is the fundamental and self-sufficient consideration. That the patient cannot possibly benefit from the unrelated experiment therapeutically, while he might from experiments related to his condition, is also true, but lies beyond the problem area of pure experiment. Anyway, I am discussing nontherapeutic experimentation only, where *ex hypothesi* the patient does not benefit. Experiment as part of therapy—that is, directed toward helping the subject himself—is a different matter altogether and raises its own problems, but hardly philosophical ones. As long as a doctor can say, even if only in his own thought: "There is no known cure for your condition (or: You have responded to none); but there is promise in a new treatment still under investigation, not quite tested yet as to effectiveness and safety; you will be taking a chance, but all things considered, I judge it in your best interest to let me try it on you"—as long as he can speak thus, he speaks as the patient's physician and may err, but does not transform the patient into a subject of experimentation. Introduction of an untried therapy into the treatment where the tried ones have failed is not "experimentation on the patient."

Generally, there is something "experimental" (because tentative) about every individual treatment, beginning with the diagnosis itself; and he would be a poor doctor who would not learn from every case for the benefit of future cases, and a poor member of the profession who would not make any new insights gained from his treatments available to the profession at large. Thus, knowledge may be advanced in the treatment of any patient, and the interest of the medical art and all sufferers from the same affliction as well as the patient may be served if something happens to be learned from his case. But this gain to knowledge and future therapy is incidental to the *bona fide* service to the present patient. He has the right to expect that the doctor does nothing to him just in order to learn.

In that case, the doctor's imaginary speech would run, for instance, like this: "There is nothing more I can do for you. But you can do something for me. Speaking no longer as your physician but on behalf of medical science, we could learn a great deal about future cases of this kind if you would permit me to perform certain experiments on you. It is understood that you yourself would not benefit from any knowledge we might gain; but future patients would." This statement would express the purely experimental situation, assumedly here with the subject's concurrence and with all cards on the table. In Alexander Bickel's words: "It is a different situation when the doctor is no longer trying to make [the patient] well, but is trying to find out how to make others well in the future."[8]

But even in the second case of the nontherapeutic experiment where the patient does not benefit, the patient's own disease is enlisted in the cause of fighting that disease, even if only in others. It is yet another thing to say or think: "Since you are here—in the hospital with its facilities—under our care and observation, away from your job (or, perhaps, doomed), we wish to profit from your being available for some other research of great interest we are presently engaged in." From the standpoint of merely medical ethics, which has only to consider risk, consent, and the worth of the objective, there may be no cardinal difference between this case and the last one. I hope that my medical audience will not think I am making too fine a point when I say that from the standpoint of the subject and his dignity there is a cardinal difference that crosses the line between the permissible and the impermissible, and this by the same principle of "identification" I have been invoking all along. Whatever the rights and wrongs of any experimentation on any patient—in the one case, at least that residue of identification is left him that it is his own affliction by which he can contribute to the conquest of that affliction. his own kind of suffering which he helps to alleviate in others; and so in a sense it is his own cause. It is totally indefensible to rob the unfortunate of this intimacy with the purpose and make his misfortune a convenience for the furtherance of alien concerns. The observance of this rule is essential, I think, to attenuate at least the wrong that nontherapeutic experimenting on patients commits in any case.

On the Redefinition of Death

My other emphatic verdict concerns the question of the redefinition of death—acknowledging "irreversible coma as a new definition for death."[9] I wish not to be misunderstood. As long as it is merely a question of when it is permitted to cease the artificial prolongation of certain functions (like heartbeat) traditionally regarded as signs of life, I do not see anything ominous in the notion of "brain death." Indeed, a new definition of death is not even necessary to legitimize the same result if one adopts the position of the Roman Catholic Church, which here for once is eminently reasonable—namely that "when deep unconsciousness is judged to be permanent, extraordinary means to maintain life are not obligatory. They can be terminated and the patient allowed to die."[10] Given a clearly defined negative condition of the brain, the physician is allowed to allow the patient to die his own death by *any* definition, which of itself will lead through the gamut of all possible definitions. But a disquietingly contradictory purpose is combined with this purpose in the quest for a new definition of death, in the will to *advance* the moment of declaring him dead: Permission not to turn off the respirator, but, on the contrary, to keep it on and thereby maintain the body in a state of what would have been "life" by the older definition (but is only a "simulacrum" of life by the new) —so as to get at his organs and tissues under the ideal conditions of what would previously have been "vivisection."[11]

Now this, whether done for research or transplant purposes,

seems to me to overstep what the definition can warrant. Surely it is one thing when to cease delaying death, but another when to start doing violence to the body; one thing when to desist from protracting the process of dying, but another when to regard that process as complete and thereby the body as a cadaver free for inflicting on it what would be torture and death to any living body. For the first purpose, we need not know the exact borderline with absolute certainty between life and death—we leave it to nature to cross it wherever it is, or to traverse the whole spectrum if there is not just one line. All we need to know is that coma is irreversible. For the second purpose we must know the borderline; and to use any definition short of the maximal for perpetrating on a *possibly* penultimate state what only the ultimate state can permit is to arrogate a knowledge which, I think, we cannot possibly have. *Since we do not know the exact borderline between life and death,* nothing less than the maximum definition of death will do—brain death plus heart death plus any other indication that may be pertinent—before final violence is allowed to be done.

It would follow then, for this layman at least, that the use of the definition should itself be defined, and this in a restrictive sense. When only permanent coma can be gained with the artificial sustaining of functions, by all means turn off the respirator, the stimulator, any sustaining artifice, and let the patient die; but let him die all the way. Do not, instead, arrest the process and start using him as a mine while, with your own help and cunning, he is still kept this side of what may in truth be the final line. Who is to say that a shock, a final trauma, is not administered to a sensitivity diffusely situated elsewhere than in the brain and still vulnerable to suffering? a sensitivity that we ourselves have been keeping alive? No fiat of definition can settle this question.[12] But I wish to emphasize that the question of possible suffering (easily brushed aside by a sufficient show of reassuring expert consensus) is merely a subsidiary and not the real point of my argument; this, to reiterate, turns on the indeterminacy of the boundaries between *life and death,* not between sensitivity and insensitivity, and bids us to lean toward a maximal rather than a minimal determination of death in an area of basic uncertainty.

There is also this to consider: The patient must be absolutely sure that his doctor does not become his executioner, and that no definition authorizes him ever to become one. His right to this certainty is absolute, and so is his right to his own body with all its organs. Absolute respect for these rights violates no one else's rights, for no one has a right to another's body. Speaking in still another, religious vein: The expiring moments should be watched over with piety and be safe from exploitation.

I strongly feel, therefore, that it should be made quite clear that the proposed new definition of death is to authorize *only* the one and *not* the other of the two opposing things: only to break off a sustaining intervention and let things take their course, not to keep up the sustaining intervention for a final intervention of the most destructive kind.

There would now have to be said something about nonmedical experiments on human subjects, notably psychological and genetic, of which I have not lost sight. But having overextended my limits of space by the most generous interpretation, I must leave this for another occasion. Let me only say in conclusion that if some of the practical implications of my reasonings are felt to work out toward a slower rate of progress, this should not cause too great dismay. Let us not forget that progress is an optional goal, not an unconditional commitment, and that its tempo in particular, compulsive as it may become, has nothing sacred about it. Let us also remember that a slower progress in the conquest of disease would not threaten society, grievous as it is to those who have to deplore that their particular disease be not yet conquered, but that society would indeed be threatened by the erosion of those moral values whose loss, possibly caused by too ruthless a pursuit of scientific progress, would make its most dazzling triumphs not worth having. Let us finally remember that it cannot be the aim of progress to abolish the lot of mortality. Of some ill or other, each of us will die. Our mortal condition is upon us with its harshness but also its wisdom—because without it there would not be the eternally renewed promise of the freshness, immediacy, and eagerness of youth; nor, without it, would there be for any of us the incentive to number our days and make them count. With all our striving to wrest from our mortality what we can, we should bear its burden with patience and dignity.

REFERENCES

1. G. E. W. Wolstenholme and Maeve O'Connor (eds.), *CIBA Foundation Symposium, Ethics in Medical Progress: With Special Reference to Transplantation* (Boston, 1966); "The Changing Mores of Biomedical Research," *Annals of Internal Medicine* (Supplement 7), Vol. 67, No. 3 (Philadelphia, September, 1967); *Proceedings of the Conference on the Ethical Aspects of Experimentation on Human Subjects*, November 3-4, 1967 (Boston, Massachusetts; hereafter called *Proceedings*); H. K. Beecher, "Some Guiding Principles for Clinical Investigation," *Journal of the American Medical Association*, Vol. 195 (March 28, 1966), pp. 1135-36. H. K. Beecher, "Consent in Clinical Experimentation: Myth and Reality," *Journal of the American Medical Association*, Vol. 195 (January 3, 1966), pp. 34-35; P. A.

Freund, "Ethical Problems in Human Experimentation," *New England Journal of Medicine*, Vol. 273 (September 23, 1965), pp. 687-92; P. A. Freund, "Is the Law Ready for Human Experimentation?", *American Psychologist*, Vol. 22 (1967), pp. 394-99; W. Wolfensberger, "Ethical Issues in Research with Human Subjects," *World Science*, Vol. 155 (January 6, 1967), pp. 47-51; See also a series of five articles by Drs. Schoen, McGrath, and Kennedy, "Principles of Medical Ethics," which appeared from August to December in Volume 23 of *Arizona Medicine*. The most recent entry in the growing literature is E. Fuller Torrey (ed.), *Ethical Issues in Medicine* (New York, 1968), in which the chapter "Ethical Problems in Human Experimentation" by Otto E. Guttentag should be especially noted.

2. Wolfensberger, "Ethical Issues in Research with Human Subjects," p. 48.

3. *Proceedings*, p. 29.

4. *Die Braut von Korinth:* "Victims do fall here,/Neither lamb nor steer,/Nay, but human offerings untold."

5. *Proceedings*, pp. 50-51.

6. Socially, everyone is expendable relatively—that is, in different degrees; religiously, no one is expendable absolutely: The "image of God" is in all. If it can be enhanced, then not by any one being expended, but by someone expending himself.

7. This refers to captives of circumstance, not of justice. Prison inmates are with respect to our problem in a special class. If we hold to some idea of guilt, and to the supposition that our judicial system is not entirely at fault, they may be held to stand in a special debt to society, and their offer to serve—from whatever motive—may be accepted with a minimum of qualms as a means of reparation.

8. *Proceedings*, p. 33. To spell out the difference between the two cases: In the first case, the patient himself is meant to be the beneficiary of the experiment, and directly so; the "subject" of the experiment is at the same time its object, its end. It is performed not for gaining knowledge, but for helping him—and helping him in the *act* of performing it, even if by its results it also contributes to a broader testing process currently under way. It is in fact part of the treatment itself and an "experiment" only in the loose sense of being untried and highly tentative. But whatever the degree of uncertainty, the motivating anticipation (the wager, if you like) is for success, and success here means the subject's own good. To a pure experiment, by contrast, undertaken to gain knowledge, the difference of success and failure is not germane, only that of conclusiveness and inconclusiveness. The "negative" result has as much to teach as the "positive." Also, the true experiment is an act distinct from the uses later made of the findings. And, most important, the subject experimented on is distinct from the eventual beneficiaries of those findings: He lets himself be used as a means toward an end external to himself (even if he should at some later time happen to be among the beneficiaries himself). With respect to his own present needs and his own good, the act is gratuitous.

9. "A Definition of Irreversible Coma," Report of the *Ad Hoc* Committee of Harvard Medical School to Examine the Definition of Brain Death, *Journal of the American Medical Association*, Vol. 205, No. 6 (August 5, 1968), pp. 337-40.

10. As rendered by Dr. Beecher in *Proceedings*, p. 50.

11. The Report of the *Ad Hoc* Committee no more than indicates this possibility with the second of the "two reasons why there is need for a definition": "(2) Obsolete criteria for the definition of death can lead to controversy in obtaining organs for

transplantation." The first reason is relief from the burden of indefinitely drawn out coma. The report wisely confines its recommendations on application to what falls under this first reason—namely, turning off the respirator—and remains silent on the possible use of the definition under the second reason. But when "the patient is declared dead on the basis of these criteria," the road to the other use has theoretically been opened and will be taken (if I remember rightly, it has even been taken once, in a much debated case in England), unless it is blocked by a special barrier in good time. The above is my feeble attempt to help doing so.

12. Only a Cartesian view of the "animal machine," which I somehow see lingering here, could set the mind at rest, as in historical fact it did at its time in the matter of vivisection: But its truth is surely not established by definition.

14 *The Ethical Design of Human Experiments*

DAVID D. RUTSTEIN, M.D.

Beginning with the premise that it is ethical to experiment on humans in carefully controlled conditions, Rutstein goes on to the specific problem of how to design scientific experiments that will both produce the information needed and yet be ethical. This focus helps us to think our way through the many dilemmas inherent in this situation. The presentation is enhanced by its balanced blending of both ethical reasoning and scientific analysis. Rutstein clearly states his premise and arguments and thus provides the opportunity to engage in a genuine ethical and scientific dialogue.

Dr. David D. Rutstein, M.D., is Ridly Watts Professor of Preventive Medicine at Harvard Medical School.

This analysis of the ethical considerations governing human experiments is based on the assumption that it is ethical under carefully controlled conditions to study on human beings mechanisms of health and disease and to test new drugs, biological products, procedures, methods, and instruments that give promise of improving the health of human beings, of preventing or treating their diseases, or postponing their untimely deaths. Without such an assumption, there can be no systematic method of medical advance. Progress would have to depend on the surreptitious, illegal, or unsupervised research and testing of new modes of prevention and treatment of disease. The ethical standards of such irregular activities would certainly be at a far lower level than can be guaranteed when the testing of new methods of treatment is openly practiced.

Proceeding on that assumption, how can one design experiments upon human beings that will yield the desired scientific information and yet avoid or keep ethical contraindications to a minimum? This question is asked in the belief that in the design of the experiment itself many ethical dilemmas may be resolved. Attention must be given to the ways an experiment can be designed to maintain its scientific validity, meet ethical requirements, and yet yield the necessary new knowledge.

Let us concentrate on laying out new guidelines that might lead to the solution of ethical problems rather than on focusing our attention on the difficulties that these problems present. The ethical requirements that have created the most difficulty are obtaining informed consent from the potential subject; the need for the subject to derive a health benefit from the experiment; and keeping the risk to the subject as small as possible. Such questions are important and relevant because ethical considerations are paramount when experiments are to be performed on human subjects. It is the thesis of this essay that in the design of a human experiment it is mandatory to select those experimental conditions, subjects, and methods of measurement that impose the fewest ethical constraints. Such an approach will not cause the ethical problems of human experiments to disappear. If a definitive attempt is made, during the planning stages of an experiment on human beings, to keep the ethical as well as scientific criteria in mind, it is possible often to perform the necessary research to yield the desired information.

Scientifically Unsound Studies Are Unethical

It may be accepted as a *maxim that a poorly or improperly designed study involving human subjects*—one that could not possibly yield scientific facts (that is, reproducible observations) relevant to the question under study—*is by definition unethical.* Moreover, when a study is in itself scientifically invalid, all other ethical considerations become irrelevant. There is no point in obtaining "informed consent" to perform a useless study. A worthless study cannot possibly benefit anyone, least of all the experimental subject himself. Any risk to the patient, however small, cannot be justified. In essence, the scientific validity of a study on human beings is in itself an ethical principle.

How, then, can the experimental human subject be protected from incompetent investigators so that he will not become a victim in feckless studies? There are *two lines of* defense. The research committee of a medical school, institution, or hospital must be concerned with the ethical principles as well as the scientific validity of the proposals placed before them. To perform this task effectively, every committee must have among its membership a biostatistician to insure scientific validity and an expert (for whom there is as yet no name) who is concerned with the ethical aspects of human experimentation. The biostatistician can assist the committee in evaluating the scientific quality of the proposed investigation and make recommendations for improvement of the scientific aspects of the study design. *Experiments on human beings must not be performed without a carefully drawn protocol, which in turn can best be prepared in consultation with experts in study design.* In the same way, experts in the ethical aspects of human experimentation should assist the committee in passing on the ethical issues of proposed studies and in recommending modifications that might make the studies ethically acceptable.

The second line of defense can be provided by editors and editorial boards of journals that publish scientific reports of human experiments. If higher scientific standards of publication were established and adhered to, it would soon become clear to investigators that there would be small likelihood that improperly designed studies would be published. Automatically, many human subjects would be protected against participation in unsound and unethical medical research. When there has been a clear-cut violation of ethical principles, scientific reports of human studies should be refused publication. The reason for such refusal should be clearly stated.

For appropriate evaluation of manuscripts, therefore, the membership of editorial boards should include biostatisticians and experts in the ethical aspects of human experimentation to provide advice to the board and to the editor in their respective fields.

Anticipating Ethical Problems

Whenever possible, it is necessary to anticipate serious ethical problems in human experimentation in order to explore ways in which they can be avoided or kept to a minimum in the design of the experiment. The experiments in human heart transplantation are a case in point, particularly in the selection of the heart donor.

Death is not a simultaneous, instantaneous event for all of the organs of the body. Some organs die earlier than others—the brain being the most vulnerable. It is evident that the heart must be "alive" and free of disease at the time of transplantation if it is to be useful to the recipient. The selection of the heart donor, therefore, cannot be based on his "total death" in the usual sense—that is, a lack of any spontaneous activity and the complete absence of cerebral, cardiac, and pulmonary activity and of spinal reflex function.

As experience with this operation is accumulating, everyone concerned is becoming increasingly aware that to comply with both the scientific and ethical constraints, donor selection must be the responsibility of a specially constituted committee in hospitals where transplantation is performed and must not have among its members any member of the transplantation team. Within such a protective structure, the eligibility of the donor will in effect depend primarily on his having complete and irreversible cessation of cerebral function while the heart remains as normal as possible. Indeed, there may be clinical or electrocardiographic evidence at the time of donor selection that the heart continues to beat.

These criteria of eligibility represent a revolution in our cultural concept of death—and it has occurred by default. The dramatic nature of the heart transplant operation has obscured the underlying change in our ethical concept of deciding when an individual is dead—that is, when a physician may act as if there is no longer any need for treating the patient "to save his life." Until recently, if the heart were still beating, treatment would continue. It is remarkable that this major ethical change has occurred right before our eyes, and that this change is more and more widely accepted with little public discussion of its significance.

This new definition of heart donor eligibility that substitutes "irreversible brain damage" for "total death" raises more questions than it answers. Does acceptance of this concept mean that it is no longer necessary to treat, for example, the senile patient who would meet such criteria? How do eligible donors differ in principle from totally feeble-minded individuals? What are the implications for the inheritance of property if the heart of an intestate donor is kept beating with a pacemaker while the search for a recipient goes on and the donor's wife dies during the interval? Does this new definition of death for the heart donor open up new channels of criminal activity that will lead to the burking of patients to increase the supply of eligible donors?

Let us examine the nature of this revolutionary change in our ethics and pursue its implications to their logical conclusion. Substituting "irreversible brain damage" for "complete absence of any living manifestations in any organ" as essential in the diagnosis of "death" forces us to examine the meaning of the phrase "irreversible brain damage." The presently recommended definition of "irreversible brain damage" demands a complete absence of all manifestations of brain function, all the way from the higher levels of cortical activity down through the centers governing the emotions, sensations, automatic functions, and muscular control and including the spinal reflexes with, however, two special exceptions—the centers controlling respiration and circulation. These centers are excluded for practical and not for ethical reasons. In severely ill patients, the function of these centers can be taken over by machines that may not be stopped until the diagnosis of death is made on other counts. Moreover, for successful heart transplantation, the heart must be "alive" if it is to benefit the recipient.

Why do we insist on the absence of all activity of all of the other subcortical centers if we are willing in our diagnosis of irreversible brain damage to disregard these two centers of essentially automatic function? Again the reason is a practical one. We do not yet know how to make a firm diagnosis of irreversible brain damage limited to the higher cortical centers. We therefore turn in our ignorance to the requirement that all nervous activity be absent (with the exceptions noted) as the basis for the new diagnosis of "death." Again we are confronted with a practical and not a conceptual or ethical reason.

It is clear that in accepting the new definition of "irreversible brain damage" as equivalent to death in man, we are really concerned only with irreversible damage of higher cortical centers. We are saying that in man "life" exists only when he is aware of and can respond to his environment or, if he cannot, that he may recover to a point where he will react consciously within his environment. It remains then for intensive research to be conducted so that physicians will be able to identify specifically the irreversible loss of activity of higher cortical centers and distiguish it from lower reflex function. If, as a result, a reliable diagnosis of irreversible damage of higher cortical centers could be made and this concept generally accepted, the new diagnosis of death would be concerned only with the permanent loss of those functions that distinguish man from other animals. Eventually, many of the great problems imposed on society with its growing senile population might be overcome. In the meantime, however, it should be clear to all of us that such a concept has already been accepted in principle by those who would replace "irreversible brain damage" for "total absence of the functions of all organs" as satisfactory for the diagnosis of death in man.

It would have been better if some of these questions had been explored before the first heart transplantation was performed. Heart donor eligibility is becoming so complex that in the present stage of diagnosis the ethical problem of donor eligibility might best be avoided by the development of a suitable mechanical heart.

Asking the Right Question

The design of an experiment depends at first on the question asked by the investigator. Some questions are in themselves unethical. One cannot ask whether plague bacilli are more virulent in human beings when injected into the bloodstream than when they are sprayed into the throat. One may obtain hints as to the answer to such a question by epidemiologic comparison of the spread of pneumonic plague (spread from the lungs into the air) and bubonic plague (spread by insect bite). Anecdotal information on the spread of plague can also be obtained through the study of laboratory accidents. But a deliberate experiment to answer this question cannot be performed.

The human experiments performed by the Nazis during World

War II horrified the world because they were designed to answer unethical questions. "How long can a human being survive in ice cold water?" will, it is to be hoped, never again be a question to be answered by a scientific experiment. Thus, as a first step in the design of any human experiment, we must first be sure that the question itself is an ethical one.

Moreover, an unethical experiment can sometimes be converted into an ethical one by rephrasing the question. In drug testing, for example, it is not ethical to design an experiment to answer the question: "Is treatment of the disease with the new drug more effective than no treatment at all?" In answering such a question, the patients in the control group would literally have to receive "no treatment" and that is completely unacceptable. Instead, if the patients in the control group are given the best possible current treatment of the disease, we may now ask an ethical question: "Is treatment with the new drug more effective than the generally accepted treatment for this particular disease?"

We faced this problem in the design of the United States-United Kingdom Cooperative Rheumatic Fever Study, which was concerned with measuring the relative effectiveness of cortisone and ACTH in the treatment of that disease.[1] We could not give rheumatic fever patients in the control group "no treatment." We would have had to go so far as to prohibit bed rest, which in itself may be helpful to rheumatic fever patients, because patients in bed have a slower heart rate. Instead, we asked the question: "Is treatment with ACTH or cortisone better, worse, or the same as the best generally accepted drug treatment for this disease?"

Our control group, in addition to all the other non-specific treatments which the treated groups also received, were given large doses of aspirin—the generally accepted drug treatment of the time. A question that compares the new treatment with the most effective treatment of the time is not only ethical, but it is also the most practical question. If the new treatment is to replace the generally accepted treatment is must be demonstrated clearly to be better.

With a question framed in that way, one may obtain consent from the patient by explaining that he will receive either the best treatment of the time or the new drug. It is made clear that, although promising in animal and other experiments, the new drug has not yet been shown in human experiments to be better, worse, or the same as the generally accepted treatment. Most patients will accept these alternatives. The investigator himself would be reassured that he has done the best for his patient's health and safety, while evaluating a more promising remedy for his disease.

When designing a human experiment, the question under study

must not be so trivial as not to justify any risk to the human subject. One may not ask whether large doses of a cortico-steroid agent would remove freckles. Nor may there be an excessive risk when compared with the possible benefit of a successful experiment. One may not test in humans a "sure cure" for the common cold which causes paralysis in experimental animals. Thus, in selecting a question for human experimentation, the expectation of benefit to the subject and to mankind must clearly far exceed the risk to the human subject.

Ethical considerations that prohibit certain human experiments are similar in their effects as are scientific constraints on the design of experiments. At the moment, much research on infectious hepatitis, a serious human disease, is impossible because there is no method for isolating infectious hepatitis virus. No laboratory animal has been found which is susceptible to it and no other procedure for its isolation has been developed. This scientific constraint is serious because without isolation of the virus a vaccine cannot be made for the protection of susceptible human beings. Human beings are susceptible and theoretically could be used for the growth of large quantities of virus needed for vaccine manufacture, but now we face the ethical constraint. It is not ethical to use human subjects for the growth of a virus for any purpose. Here, then, we have an example of a scientific constraint that in turn creates an ethical constraint, both of which interfere with the conduct of experiments important to life and health. When asking a question that might be answered by human experimentation, both the scientific and ethical demands must carefully be taken into account. Unless both are satisfied, the experiment cannot be performed.

The Ethics of Controlled Human Experimentation

Controls are essential in such human experiments as the testing of a drug or a surgical procedure. In order to evaluate new therapy, it is necessary to identify those additional benefits of the new remedy which exceed the improvement that might be expected in the course of the natural history of the disease. To be sure, in a disease such as human rabies, which is practically 100 per cent fatal, controlled observations are not needed because any recovery of treated patients is an obvious benefit. Acute leukemia and virulent tumors such as reticulum-cell sarcoma are other examples of diseases whose natural history is one of immutable progression to death and where benefit can be identified without a controlled experiment.

Most diseases do not fall into such a clear-cut category. Even diseases such as cancer of the breast have such a variable course that one is not certain to this moment if surgery prolongs the life of the patient suffering from this disease. The variability of the disease from patient to patient makes it difficult, if not impossible, to evaluate the additional effectiveness of the surgical remedy without a controlled study.

Controlled studies also keep the investigator from leaving the world of reality. The enthusiastic research worker often concentrates on whether the new treatment seems to work. Psychologically, he is apt to pay less attention to possible harmful effects of the new treatment. The result of this attitude is documented repeatedly by the myriad of treatments that make the headlines and promise miraculous cures on the front pages of our best newspapers, only to be completely discarded a few years later. This phenomenon is not without its harmful effects. The definite, albeit limited, benefits of established treatments are often cast aside in favor of the dramatic new, but as yet unproven method of treatment for a human disease. For example, before the advent of the sulfa drugs and antibiotics, there were fairly effective procedures that alleviated and at times cured urinary-tract infection. When these new therapeutic agents became available, some physicians concentrated on intensive therapy with one of the new agents and often felt that it was no longer necessary to practice many of the important details of treatment that had been given in the past. Now, after several decades, there is a gradual return to a more balanced regimen of treatment that places each of the antibiotics in proper perspective in the total treatment of urinary-tract infection. In such situations, a controlled study that is properly performed permits a clean comparison of the helpful and harmful effects of the new treatment and of the older established method of therapy. If, in addition, circumstances permit the ethical use of a placebo control, information can also be obtained about the natural history of the disease.

One might ask whether it is unethical not to perform a controlled human experiment. The Pasteur experiment with rabies vaccine is classic. Controlled human experimentation was completely unknown when Pasteur first tested his new vaccine on those Russian *muzhiks* who were bitten by rabid wolves. All were given the vaccine and all recovered. The result was so dramatic and so electrifying that further experimentation seemed unnecessary. When it was later learned, however, that the chances were relatively low of developing this uniformly fatal disease—human rabies—even after a bite from a known rabid animal, and that the vaccine itself causes paralysis which is not infrequently fatal, it became important to determine whether the vaccine really does more good than harm.

But it became impossible to do a controlled experiment on rabies vaccine. After the general acceptance of the treatment, if a controlled experiment were to be performed, and if a subject in the control group developed rabies, the experimenter might not only be sued for mal-practice, but might even be deemed criminally culpable for not having given the patient the "accepted treatment of the time." This same situ-ation is now developing in the estimation of the benefits of heart transplantation and in measuring the value of intensive-care units in the treatment of heart attacks from coronary disease.

In hospitals where heart transplantation is performed, there are many more eligible recipients than donors. It would be relatively easy to randomize the procedure and measure the effectiveness of this new operation. Whenever a heart from a human donor became available, a random selection could be made among all of the eligible recipients. Those not selected would then comprise a control group whose course and outcome could be compared with those of the recipients of a transplanted heart.

The same situation obtains in the treatment of heart attacks from coronary disease in intensive-care units. There is as yet no published control study demonstrating the effectiveness of intensive care in the treatment of acute myocardial infarction. Once again, a control study is possible because in any one center the numbers eligible for care may far exceed the available facilities. A randomly allocated control study would provide the precise information that is needed to estimate the value of this procedure.[2] Instead, we are already beginning to hear ex-cathedra statements of the effectiveness of intensive-care units that are unsupported by the required scientific evidence.

Even up to the present moment, many patients suffer severe dis-comfort or are permanently harmed from treatments whose validity has been based on uncontrolled observations. Thousands of hyperten-sive patients in the 1930's and '40's were subjected to extensive sur-gery for the removal of their thoracolumbar sympathetic nervous sys-tem in the belief that the progress of the disease would be arrested. The treatment is no longer used.

Would this problem have been resolved had controlled studies been performed by means of random allocation in which half of the patients would have had sham operations? In an analogous situation when uncontrolled evidence suggested that the internal mammary ar-tery operation might be helpful in the treatment of angina pectoris, Dr. Henry Beecher recommended a study in which the patients in the control groups would be given a sham operation.[3] He indicated that a far smaller total number of human subjects would have been needed and a definitive answer could be obtained by such a procedure. Al-though scientifically sound, I do not believe that it is ethical to per-form sham operations on human subjects because of the operative risk and the lack of potential benefit to the patient. Instead, controlled studies could have been performed with the randomly allocated control patients being given the best medical treatment of the time together with a period of bed rest similar to that of the surgical convalescent.

The history of diseases of unknown etiology is replete with serially accepted and discarded fads of treatment. Peptic ulcer is an example. The Sippy rigid alkali and milk diet became the Meulengracht meat diet and in time became a bland diet with enough alkali to relieve the patient's symptoms. The short-circuiting surgical operation of gastro-enterostomy changed to one for removal of a large portion of the acid-secreting part of the stomach—partial gastrectomy—and then to the less traumatic procedure of removing the nerve supply to the stomach and upper intestine—vagotomy. Along the way, a procedure for freezing the stomach was introduced, then discarded, not because the fad had worn itself out, but because it was obviously harmful. All these treatments might or might not have been accompanied with psychiatric therapy. To this day, the treatment of peptic ulcer, as is the case of many diseases of unknown etiology, is an art with little solid scientific support.

In essence, a new treatment of a disease may be better, worse, or the same as the generally accepted one. The controlled clinical trial has not only been effective in rejecting useless treatments, but perhaps even more helpful in recognizing a harmful therapy, such as the anticoagulant treatment of cerebral thrombosis. That treatment was earning growing acceptance until, in a controlled clinical trial, it was recognized that cerebral hemorrhage was a more frequent complication in the group treated with anticoagulants than among the patients in the control group.[4] The trial was terminated.

One may conclude that if the question under study is an ethical one, and if the design of the study is sound, taking both the scientific and the ethical constraints into consideration, controlled studies when indicated impose fewer ethical problems than uncontrolled human experiments.

"Benefiting" the Subject

Years ago, after I had administered the first dose of streptomycin to a human patient, there was a need to obtain "normal" values for the absorption and excretion of this new antibiotic.[5] Because of the toxicity (subsequently eliminated) of the early lots of the antibiotics, it did not seem ethical to test normal subjects who could not conceivably benefit from the procedure.

A satisfactory compromise was reached. Streptomycin in early laboratory experiments gave promise of being effective in the treatment of infections due to gram-negative enteric bacilli—one of which is the typhoid bacillus. A typhoid carrier—that is, an individual who has recovered from typhoid fever, but continues to carry this pathogenic bacillus in his gall bladder, intestine, or genito-urinary tract—is relatively "normal" so far as the aims of our experiment were concerned. Moreover, he could have conceivably benefited from the experiment by elimination of his carrier state.

As a result of strict public health controls, the typhoid carrier is the pariah or leper of modern society. Everyone who knows him avoids him. Invited guests will not enter his home, and he will not be invited elsewhere. Indeed, the entire family suffers. Because of the great desire to be relieved of this burden, many typhoid carriers willingly volunteered, and the experiment on the absorption and excretion of streptomycin was performed on them. Unfortunately, the eradication of their carrier state was temporary, and it returned after drug therapy was stopped. But there was a possible benefit to the subject that could only be ascertained by experiment. The necessary human data on the absorption and excretion of streptomycin were collected.

A Proposed Design for Testing New Drugs

We will explore the hypothesis that a system of drug testing can be developed that will satisfy scientific requirements, meet a higher standard of ethical principles than now obtains, and yet release a new drug for general use more rapidly than is now feasible.

Now that the Federal Food and Drug Administration has imposed and implemented rigorous scientific standards for drug testing, there are frequent complaints that the procedure itself may be unethical in that it is so costly, time-consuming, and demanding of highly qualified investigators that the benefit to the public of the new therapeutic agent is unnecessarily delayed. And, yet, the critics are properly loath to recommend a return to a system that might permit a toxic drug, such as Thalidomide, to be sold in the open market. The horns of the dilemma are clearly visible—potential benefit on the one hand, and potential harm, on the other.

In outlining a plan to resolve this dilemma, let us first explore the existing process of rigorous testing through which a new drug becomes available for widespread use. From time to time, evidence is presented that a new medicament may be useful in the prevention or treatment of a particular disease. At that point, it is tested in animals for toxicity and for therapeutic efficacy in an appropriate model (for example, an animal disease or a laboratory test for inhibiting the growth of or eradicating a pathogenic micro-organism). If such efforts are successful—that is, if the drug continues to demonstrate therapeutic usefulness without serious toxicity—a decision finally is made to explore the use of the new drug in a few carefully selected patients.

Ideally, this next stage is carried out by a few investigators who have had great experience with the disease under study and who are most likely to detect in relatively few patients variations produced by the drug from fluctuations in the natural history of the illness. Let us assume that this hurdle has been surmounted and that the drug is deemed apparently effective, relatively safe, and ready for a large controlled clinical test. Assume also that an estimate has been made of the number of patients needed for a precise evaluation of the drug were its effectiveness to continue to be the same.

A controlled clinical trial is then instituted. An appropriate question is asked and a protocol of experiment is developed to satisfy scientific and ethical requirements. Patients are randomly admitted to treatment and control groups. Measurements of effectiveness and toxicity are made, recorded, and analyzed sequentially. If the results are satisfactory, a final decision is reached concerning the widespread availability of the new therapeutic agent. The drug may then be released and then monitored for continued effectiveness and for rare manifestations of severe toxicity.

At the present time, this process—performed in accordance with an ethical and scientifically precise plan—postpones the widespread use of an effective drug, involves too many competent investigators for too long a period of time, and through repetition in many countries throughout the world imposes the hazards of testing the same drug on an unnecessarily large number of human subjects. The process raises ethical questions because it is unethical to expose to unnecessary risk more human subjects than are needed to ascertain the scientific fact. Fortunately, when the situation is carefully analyzed, many of these difficulties may be overcome without sacrificing scientific standards or ethical principles.

The time required for testing can be reduced, and there can be an earlier release of the drug, if the trial is performed collaboratively, instead of haphazardly, by a group of investigators in different hospitals using a common protocol of experiment. In a coordinating center, the data are analyzed sequentially to be sure that the dose is maximal, the method of administration effective, and toxicity minimal. When satisfactory results are obtained, the drug could be released immediately for general use.

The collaborative clinical trial avoids one of the great ethical problems of drug testing. In any one hospital, when evidence that a new drug may be effective begins to become manifest, there are increasing pressures on the investigator to release it for general use before a statistically significant sample of patients has been studied.

At that point, in a collaborative study, the number of cases in all of the centers in the study is usually large enough to provide a solid estimate of the efficacy of the agent as well as of the nature and the degree of its toxicity.

A collaborative program by itself in any one country will not necessarily reduce the number of human subjects who are exposed to the risk of testing a particular drug. The number of human subjects can be reduced, however, by international agreements among developed countries that adhere to drug testing standards similar to our own.[6] A well-designed international collaborative and cooperative study in which the new drug could be tested on a specified but much smaller number of human subjects could suffice for all. At present, in many advanced countries a valid sample of treated and control patients is collected, and with relatively little exchange of information, the same experiments are repeated in many countries. The present process wastes professional skills and medical resources, unnecessarily exposes too many human subjects to the risk of drug testing, and is thereby unethical.

Experience in the United States-United Kingdom Cooperative Rheumatic Fever Study has demonstrated that international cooperation in drug testing can be successful.[7] From that experience, it is urged that an international agency, such as the World Health Organization, or an international pharmaceutical manufacturers' association explore the possibility of rapid, efficient international drug testing.

There are, of course, political and economic objections, including conflicts in patent policy, protection of trade secrets, and disturbance of international drug marketing agreements. These should be faced squarely to see whether or not a solution can be found so that the benefits of international testing might be reaped. One obvious benefit is much earlier marketability of the drug.

Another benefit for pharmaceutical companies adhering to the agreement is protection against damage suits resulting from relatively uncommon toxicity of the drug. For example, if it is determined by the international agency that to assure effectiveness and lack of serious toxicity a drug should be tested on a specific number of cases and the requirement is met by the manufacturer, it should be *prima facie* evidence that he cannot be responsible for severe drug reactions that occur less frequently than the number of cases treated. Thus, if it is estimated from preliminary testing that five hundred cases should be included in the clinical trial, and this requirement is satisfied and the drug released, a manufacturer could not reasonably be expected to be responsible for a rare, severe reaction that might occur in only one of two thousand patients.

In summary, then, an international drug testing program, including a monitoring system for following the continued effectiveness and the rare manifestations of toxicity of the drug, would decrease the number of human subjects at risk and bring the benefits of the new agent to the world in a shorter period of time.

Modern Design of Human Experiments

The advent of the electronic computer has increased the efficiency of biomedical research and, in turn, has had its ethical implications. In the days of collecting laboratory measurements on the smoked drum, it was easier to collect data than to analyze them. Indeed, final interpretation of experiments was often delayed for months as data were collated and analyzed by hand tabulations, often made by the investigator himself in odd moments between the pressing needs of laboratory duties and teaching. Rarely was the analysis of data given priority over his other activities. At the top of the list was the next experiment, the gross results of which were eagerly anticipated while the detailed analysis was again postponed.

As a result of these traditional limitations on data handling and processing, and with the increasing complexity of our understanding of biological systems, there has been a growing tendency to design experiments in simplified systems where at any one time a few variables can be measured with great precision. This method of research has yielded a great deal of generally applicable biological knowledge. But because the systems of the human body are so complex and the new simplified approach to research so remote from them, the research results have become less and less applicable to the solution of problems of human health and disease. Indeed, the clinical problem that may have originally inspired the research program may often be forgotten as the scientist concentrates on further and further detailed study of his simplified system. As a corollary, when such clinical investigations became more remote from human subjects, fewer ethical problems were created.

With the advent of the electronic computer, the underlying situation was completely reversed. The revolution in data handling and processing has now made it much easier to analyze and interpret than to collect scientific data. The analysis of scientific data, if the experiment's design is sound, should now take relatively little time and effort. Indeed, the immediate availability of an analysis of the data of an experiment should permit the scientist to concentrate on the significance of the experimental results. The scientist now has the time to take into consideration the results of this last experiment so as to plan better the next one.

More importantly, clinical experimentation no longer need be limited to the study of a few variables at any one time in simplified systems. Experiments can presently be designed to study the complex interrelationships of many variables at the same time. For example, instead of studying the salt and water metabolism of the kidney as if it were completely independent of all the other functions of this organ, it is now possible to study total organ function. Indeed, computer handling of data makes it feasible to study total body functions. Furthermore, with on-line data collection and analysis in real time, it is feasible, for example, to build into a physiologic experiment many contingent measurements depending upon what happens in the earlier stages of the experiment. With such study design, medical research can deal more directly with complex systems in the human subject. Moreover, because the research system is closer to that of the human being, the experimental results should be more easily applicable to the improvement of health and the prevention and treatment of human disease.

This modern method of research—with its more complicated design and more intensive and thorough study of each human subject—is uncovering new ethical questions: How long can one safely run a particular experiment on a patient with a certain disease? How much blood may be collected for research purposes over a specified period of time and how frequently may the experiment be repeated? Will a particularly long continued intensive experiment interfere with the best treatment of the patient? It is clear that more comprehensive clinical experiments requiring more intensive scientific planning will also demand more careful attention to the protection of the human subject. Moreover, now that more complete experiments can be performed, human subjects must not be "wasted" in trivial experiments. This is not to say that simple but complete and penetrating experiments should not be performed. It is a plea for more meticulous planning based on modern technology, computer facilities, and biostatistical consultation to yield more applicable experimental results without increasing the risk to the human subject. It will force us to ask the question: "Is it ethical to perform *limited experiments* if, with more careful planning and with no increased risk to the patient, much more valuable information could be collected of more immediate applicability to patients, including the subject himself?"

The design of medical experiments may also under certain circumstances help to surmount existing ethical problems. It is often difficult to justify ethically a human experiment concerned with the study of normal metabolism or physiology or with the changes produced by a particular illness, because benefits to the subject are likely to be limited. To be sure, the more profound study of the patient's condition implicit in the research measurements may yield information that permits better treatment of his disease. But we can do better than that. A properly designed experiment with on-line computer analysis of properly programmed data makes feasible the study of a physiologic, metabolic, or pathologic mechanism and simultaneously the evaluation of a therapeutic agent. For example, in the study of the mechanism of circulatory collapse in severe infections not amenable to antibiotic therapy, the testing of a drug or, perhaps in the future, the trial of a mechanical heart booster could be interwoven and both studies completed at the same time. Such design assures the experimenter that he could benefit the human subjects of his experiments.

Modern design of human experiments opens up new vistas in the understanding of complex biological systems with immediate promise of human benefit. Experiments so designed should yield more knowledge at the same risk to the human subject—and this is an ethical benefit. But these more complex experiments also uncover new ethical problems demanding increasing vigilance in the protection of the human subject.

The Role of Mathematical Theory

When it is impossible for ethical reasons to perform a given experiment to test a particular hypothesis, it may be useful to turn to the mathematician for the deduction of an alternative hypothesis which could be tested by experiment. As W. G. Cochran has pointed out:

A fruitful mathematical theory will predict the results of experiments not yet carried out—in some cases impossible to carry out. In the intensive studies of the Rhesus factor during the 1940's, Mendelian analysis predicted the existence of two genes not then discovered and the seriologic properties of two new antibodies. Epidemic theory can indicate by how much the probability of a major epidemic will be reduced by immunization of any given proportion of the susceptible population in a public-health program. In evolution and the study of inbreeding, the consequences of forces acting over many generations can be worked out.[8]

The increasing role of mathematics in the medical sciences should make this approach more and more feasible.

If we can agree that scientific medical research can continue to serve ethically as the basis for medical progress, there is an immediate need to re-examine the design of human experiments from both the scientific and ethical points of view; to reshape the design of human experiments and take advantage of new technology; to increase, improve, make more relevant the data collected in human experiments, and yet, at the same time, strengthen the ethical principles of medical research.

REFERENCES

1. Rheumatic Fever Working Party of the Medical Research Council of Great Britain and the Sub-committee of Principal Investigators of the American Council on Rheumatic Fever and Congenital Heart Disease, American Heart Association, "A Joint Report: The Treatment of Acute Rheumatic Fever in Children. A Cooperative Clinical Trial of ACTH, Cortisone, and Aspirin," *Circulation*, Vol. 11 (1955), pp. 343-77; *British Medical Journal*, Vol. 1 (1955), pp. 555-74.

2. Since this manuscript was presented and submitted, it has been learned that two control studies are now under way in the United Kingdom.

3. H. K. Beecher, "Surgery as Placebo—A Quantitative Study of Bias," *Journal of the American Medical Association*, Vol. 176 (1961), pp. 1102-1107.

4. A. B. Hill, J. Marshall, and D. A. Shaw, "A Controlled Clinical Trial of Long-Term Anticoagulant Therapy in Cerebrovascular Disease," *Quarterly Journal of Medicine*, Vol. 29, New Series (1960), pp. 597-609.

5. D. D. Rutstein, R. B. Stebbins, R. T. Cathcart, and R. M. Harvey, "The Absorption and Excretion of the Streptomycin in Human Chronic Typhoid Carriers, *Journal of Clinical Investigation*, Vol. 24 (1945), pp. 898-909.

6. Such an agreement should not be confused with the rumored surreptitious drug testing said to be conducted at times by a few United States drug manufacturers in countries not having standards so rigid as those of the Food and Drug Administration in the United States.

7. Rheumatic Fever Working Party of the Medical Research Council of Great Britain and the Sub-committee of Principal Investigators of the American Council on Rheumatic Fever and Congenital Heart Disease, American Heart Association, A Joint Report Prepared by D. D. Rutstein and E. Densen, "The Natural History of Rheumatic Fever and Rheumatic Heart Disease: Ten-Year Report of a Cooperative Clinical Trial of ACTH, Cortisone, and Aspirin," *Circulation*, Vol. 32, pp. 457-76; *British Medical Journal*, Vol. 2, pp. 607-615; *Canadian Medical Association Journal*, Vol. 93 (1965), pp. 519-31.

8. W. G. Cochran, "The Role of Mathematics in the Medical Sciences," *New England Journal of Medicine*, Vol. 265 (1961), p. 176.

15 The Ethics of Giving Placebos

Sissela Bok

This article explores one specific ethical dilemma in experimentation: placebos and their use in blind or double-blind studies. The ethical focus is on the meaning and role of informed consent in such experiments which necessarily imply a withholding of important knowledge from the subject. Bok carefully leads us through a variety of problems and draws up specific conclusions on the use of the placebo. This article provides an excellent test case for an examination of the scientific and ethical principles that are inherent in this type of human experimentation.

In 1971 a number of Mexican-American women applied to a family-planning clinic for contraceptives. Some of them were given oral contraceptives and other were given placebos, or dummy pills that looked like the real thing. Without knowing it the women were involved in an investigation of the side effects of various contraceptive pills. Those who were given placebos suffered from a predictable side effect: 10 of them became pregnant. Needless to say, the physician in charge did not assume financial responsibility for the babies. Nor did he indicate any concern about having bypassed the "informed consent" that is required in ethical experiments with human beings. He contented himself with the observation that if only the law had permitted it, he could have aborted the pregnant women!

The physician was not unusually thoughtless or hardhearted. The fact is that placebos are so widely prescribed for therapeutic reasons or administered to control groups in experiments, and are considered so harmless, that the fundamental issues they raise are seldom confronted. It appears to me, however, that physicians prescribing placebos cannot consider only the presumed benefit to an individual patient or to an experiment at a particular time. They must also take into account the potential risks, both to the patient or the experimental subject and to the medical profession. And the ethical dilemmas that are inherent in the various uses of placebos are central to such an estimate of possible benefits and risks.

The derivation of "placebo," from the Latin for "I shall please," gives the word a benevolent ring, somehow placing placebos beyond moral criticism and conjuring up images of hypochondriacs whose vague ailments are dispelled through adroit prescriptions of beneficent sugar pills. Physicians often give a humorous tinge to instructions for prescribing these substances, which helps to remove them from serious ethical concern. One authority wrote in a pharmacological journal that the placebo should be given a name previously unknown to the patient and preferably Latin and polysyllabic, and "it is wise if it be prescribed with some assurance and emphasis for psychotherapeutic effect. The older physicians each had his favorite placebic prescriptions—one chose tincture of Condurango, another the Fluidextract of *Cimicifuga nigra.*" After all, are not placebos far less dangerous than some genuine drugs? As another physician asked in a letter to *The Lancet*: "Whenever pain can be relieved with two milliliters of saline, why should we inject an opiate? Do anxieties or discomforts that are allayed with starch capsules require administration of a barbiturate, diazepam or propoxyphene?"

Before the 1960's placebos were commonly defined as just such pharmacologically inactive medications as salt water or starch, given primarily to satisfy patients that something is being done for them. It has only gradually become clear that any medical procedure has an implicit placebo effect and, whether it is active or inactive, can serve as a placebo whenever it has no specific effect on the condition for which it is prescribed. Nowadays fewer sugar pills are prescribed, but X rays, vitamin preparations, antibiotics and even surgery can function as placebos. Arthur K. Shapiro defines a placebo as "any therapy (or component of therapy) that is deliberately or knowingly used for its nonspecific, psychologic or psycho-physiologic effect, or that. . ., unknown to the patient or therapist, is without specific activity for the condition being treated."

Clearly the prescription of placebos is intentionally deceptive only when the physician himself knows they are without specific effect but keeps the patient in the dark. In considering the ethical issues attending deception with placebos I shall exclude the many procedures in which physicians have had—or still have—misplaced faith; that includes most of the treatments prescribed until this century and a great many still in use but of unproved or even disproved value.

Considering that in the past most therapies had little or no specific effect (yet sometimes succeeded thanks to faith on the part of healers and sufferers) and that we now have more effective remedies, it might be thought that the need to resort to placebos would have decreased. Improved treatment and diagnosis, however, have raised the expectations of patients and health professionals alike and consequently the incidence of reliance on placebos has risen. This is true of placebos given both in experiments and for therapeutic effect.

Modern techniques of experimentation with humans have vastly expanded the role of placebos as controls. New drugs, for example, are compared with placebos in order to distinguish the effects of the drug from chance events or effects associated with the mere administration of the drug. They can be tested in "blind" studies, in which the subjects do not know whether they are receiving the experimental drug or the placebo, and in "double-blind" studies, in which neither the subjects nor the investigators know.

Experiments involving humans are now subjected to increasingly careful safeguards for the people at risk, but it will be a long time before the practice of deceiving experimental subjects with respect to placebos is eradicated. In all the studies of the placebo effect that I surveyed in a study initiated as a fellow of the Interfaculty Program in Medical Ethics at Harvard University, only one indicated that those subjected to the experiment were informed that they would receive placebos; indeed, there was frequent mention of intentional deception. For example, a study titled "An Analysis of the Placebo Effect in Hospitalized Hypertensive Patients" reports that "six patients . . . were asked to accept hospitalization for approximately six weeks . . . to have their hypertension evaluated and to undertake a treatment with a new blood pressure drug. . . . No medication was given for the first five to seven days in the hospital. Placebo was then started."

As for therapeutic administration, there is no doubt that studies conducted in recent decades show placebos can be effective. Henry K. Beecher studied the effects of placebos on patients suffering from conditions including postoperative pain, angina pectoris and the common cold. He estimated that placebos achieved satisfactory relief for about 35 percent of the patients surveyed. Alan Leslie points out, moreover, that "some people are temperamentally impatient and demand results before they normally would be forthcoming. Occasionally, during a period of diagnostic observation or testing, a placebo will provide a gentle sop to their impatience and keep them under control while the important business is being conducted.

A number of other reasons are advanced to explain the continued practice of prescribing placebos. Physicians are acutely aware of the uncertainties of their profession and of how hard it is to give meaningful and correct answers to patients. They also know that disclosing uncertainty or a pessimistic prognosis can diminish benefits that depend on faith and the placebo effect. They dislike being the bearers of uncertain or bad news as much as anyone else. Sitting down to discuss an illness with a patient truthfully and sensitively may take much-needed time away from other patients. Finally, the patient who demands unneeded medication or operations may threaten to go to a more cooperative doctor or to resort to self-medication; such patient pressure is one of the most potent forces perpetuating and increasing the resort to placebos.

There are no conclusive figures for the extent to which placebos are prescribed, but clearly their use is widespread. Thorough studies have estimated that as many as 35 to 45 percent of all prescriptions are for substances that are incapable of having an effect on the condition for which they are prescribed. Kenneth L. Melmon and Howard F. Morrelli, in their textbook *Clinical Pharmacology*, cite a study of treatment for the common cold as indicating that 31 percent of the patients received a prescription for a broad-spectrum or medium-spectrum antibiotic, 22 percent received penicillin and 6 percent received sulfonamides—"none of which could possibly have any beneficial specific pharmacological effect on the viral infection per se." They point out further that thousands of doses of vitamin B-12 are administered every year "at considerable expense to patients without pernicious anemia," the only condition for which the vitamin is specifically indicated.

In view of all of this it is remarkable that medical textbooks provide little analysis of placebo treatment. In a sample of 19 popular recent textbooks in medicine, pediatrics, surgery, anesthesia, obstetrics and gynecology only three even mention placebos, and none of them deal with either the medical or the ethical dilemmas placebos present. Four out of six textbooks on pharmacology consider placebos, but with the exception of the book by Melmon and Morrelli they mention only the experimental role of placebos and are completely silent on ethical issues. Finally, four out of eight standard texts on psychiatry refer to placebos, again without ever mentioning ethical issues.

Yet little thought is required to see the dilemma placebos should pose for physicians. A placebo can provide a potent, although unreliable, weapon against suffering, but the very manner in which it can relieve suffering seems to depend on keeping the patient in the dark. The dilemma is an ethical one, reflecting contrary views about how human beings ought to deal with each other, an apparent conflict between helping patients and informing them about their condition.

This dilemma is pointed up by the concept of informed consent: the idea that the individual has the right to give prior consent to, and even to refuse, what is proposed to him in the way of medical care. The doctrine is recognized in proliferating "bills of rights" for patients. The one recommended by the American Hospital Association states, for example, that the patient has the right to complete, understandable information on his diagnosis, treatment and prognosis; the right to whatever information is needed so that he can give informed consent to any treatment; the right to refuse treatment to the extent permitted by law.

Few physicians appear to consider the implications of informed consent when they prescribe placebos, however. One reason is surely that the usefulness of a placebo may be destroyed if informed consent is sought, since its success is assumed to depend specifically on the patient's ignorance and suggestibility. Then too the substances employed as placebos have been considered so harmless, and at the same time so potentially beneficial, that it is easy to assume that the lack of consent cannot possibly matter. In any case health professionals in general have not considered the possibility that the prescription of a placebo is so intrinsically misleading as to make informed consent impossible.

Some authorities have argued that there need not be any deception at all. Placebos can be described in such a way that no outright verbal lie is required. For example: "I believe these pills may help you." Lawrence J. Henderson went so far as to maintain that "it is meaningless to speak of telling the truth, the whole truth and nothing but the truth to a patient . . . because it is . . . a sheer impossibility. . . . Since telling the truth is impossible, there can be no sharp distinction between what is false and what is true."

Can one really think of prescribing placebos as not being deceptive at all as long as the words were sufficiently vague? In order to answer this question it is necessary to consider the nature of deception. When someone intentionally deceives another person, he causes that person to believe what is false. Such deception may be verbal, in which case it is a lie, or it may be nonverbal, conveyed by gestures, false visual cues or the myriad other means human beings have devised for misleading one another. What is common to all intentional deception is the intent to deceive and the providing of misleading information, whether that information is verbal or nonverbal.

The statement that a placebo may help a patient is not a lie or even, in itself, deceitful. Yet the circumstances in which a placebo is prescribed introduce an element of deception. The setting in a doctor's office or hospital room, the impressive terminology, the mystique of the all-powerful physician prescribing a cure—all of these tend to give the patient faith in the remedy; they convey the impression that the treatment prescribed will have the ingredients necessary to improve the patient's condition. The actions of the physician are therefore deceptive even if the words are so general as not to be lies. Verbal deception may be more direct, but all kinds of deception can be equally misleading.

The view that merely withholding information is not deceptive is particularly inappropriate in the case of placebo prescriptions because information that is material and important is withheld. The crucial fact that the physician may not know what the patient's problems are is not communicated. Information concerning the prognosis is vague and information about the specific way in which the treatment may affect the condition is not provided. Henderson's view fails to make the distinction between such relevant information, which it is usually feasible to provide, and infinite details of decreasing importance, which to be sure can never be provided with any completeness. It also fails to distinguish between two ways in which the information reaching the patient may be altered: it may be withheld or it may be distorted. Often the two are mingled. Consider the intertwining of distortion, mystification and failure to inform in the following statement, made to unsuspecting recipients of placebos in an experiment performed in a psychiatric outpatient clinic: "You are to receive a test that all patients receive as part of their evaluation. The test medication is a nonspecific autonomous nervous system stimulant."

Even those who recognize that placebos are deceptive often dispel any misgivings with the thought that they involve no serious deception. Placebos are regarded as being analogous to the innocent white lies of everyday life, so trivial as to be quite outside the realm of ethical evaluation. Such liberties with language as telling someone that his necktie is beautiful or that a visit has been a pleasure, when neither statement reflects the speaker's honest opinion, are commonly accepted as being so trivial that to evaluate them morally would seem unduly fastidious and, from a utilitarian point of view, unjustified. Placebos are not trivial, however. Spending for them runs into millions of dollars. Patients incur greater risks of discomfort and harm than is commonly understood. Finally, any placebo uses that are in fact trivial and harmless in themselves may combine to form nontrivial practices, so that repeated reliance on placebos can do serious harm in the long run to the medical profession and the general public.

Consider first the cost to patients. A number of the procedures undertaken for their placebo effect are extremely costly in terms of available resources and of expense, discomfort and risk of harm to patients. Many temporarily successful new surgical procedures owe their success to the placebo effect alone. In such cases there is no intention to deceive the patient; physician and patient alike are deceived. On occasion, however, surgery is deliberately performed as a placebo measure. Children may undergo appendectomies or tonsillectomies that are known to be unnecessary simply to give the impression that powerful measures are being taken or because parents press for the operation. Hysterectomies and other operations may be performed on adults for analogous reasons. A great many diagnostic procedures that are known to be unnecessary are undertaken to give patients a sense that efforts are being made on their behalf. Some of these carry risks; many involve discomfort and the expenditure of time and money. The potential for damage by an active drug given as a placebo is similarly clear-cut. Calvin M. Kunin, T. Tupasi and W. Craig have described the ill effects—including death—suffered by hospital patients as a result of excessive prescription of antibiotics, more than half of which they found had been unneeded, inappropriately selected or given in incorrect dosages.

Even inactive placebos can have toxic effects in a substantial proportion of cases; nausea, dermatitis, hearing loss, headache, diarrhea and other symptoms have been cited. Stewart Wolf reported on a double-blind experiment to test the effects of the drug mephenesin and a placebo on disorders associated with anxiety and tension. Depending on the symptom studied, roughly 20 to 30 percent of the patients were better while taking the pills and 50 to 70 percent were unchanged, but 10 to 20 percent were worse—"whether the patient was taking mephenesin or placebo." A particularly serious possible side effect of even a harmless substance is dependency. In one case a psychotic patient was given placebo pills and told they were a "new major tranquilizer without any side effects." After four years she was taking 12 tablets a day and complaining of insomnia and anxiety. After the self-medication reached 25 pills a day and a crisis had occurred, the physician intervened, talked over the addictive problem (but not the deception) with the patient and succeeded in reducing the dose to two a day, a level that was still being maintained a year later. Other cases have been reported of patients' becoming addicted or habituated to these substances to the point of not being able to function without them, at times even requiring that they be stepped up to very high dosages.

Most obvious, of course, is the damage done when placebos are given in place of a well-established therapy that is clearly indicated for the patient's condition. The Mexican-American women I mentioned at the outset, for example, were actually harmed by being given placebo pills in the guise of contraceptive pills. In 1966 Beecher, in an article on the ethics of experiments with human subjects, documented a case in which 109 servicemen with streptococcal respiratory infections were given injections of a placebo instead of injections of penicillin, which was already known to prevent the development of rheumatic fever in such patients and which was being given to a larger group of patients. Two of the placebo subjects developed rheumatic fever and one developed an acute kidney infection, whereas such complications did not occur in the penicillin-treated group.

There have been a number of other experiments in which patients suffering from illnesses with known cures have been given placebos in order to study the course of the illness when it is untreated or to determine the precise effectiveness of the known therapy in another group of patients. Because of the very nature of their aims the investigators have failed to ask subjects for their informed consent. The subjects have tended to be those least able to object or defend themselves: members of minority groups, the poor, the institutionalized and the very young.

A final type of harm to patients given placebos stems not so much from the placebo itself as from the manipulation and deception that accompany its prescription. Inevitably some patients find out that they have been duped. They may then lose confidence in physicians and in bona fide medication, which they may need in the future. They may even resort on their own to more harmful drugs or other supposed cures. That is a danger associated with all deception: its discovery leads to a failure of trust when trust may be most needed. Alternatively, some people who do not discover the deception and are left believing that a placebic remedy works may continue to rely on it under the wrong circumstances. This is particularly true with respect to drugs, such as antibiotics, that are used sometimes for their specific action and sometimes as placebos. Many parents, for example, come to believe they must ask for the prescription of antibiotics every time their child has a fever.

The major costs associated with placebos may not be the costs to patients themselves that I have discussed up to this point. Rather they may be costs to new categories of patients in the future, to physicians who do not abuse placebo treatment and to society in general.

Deceptive practices, by their very nature, tend to escape the normal restraints of accountability and so can spread more easily. There are many instances in which an innocuous-seeming practice has grown to become a large-scale and more dangerous one; warnings against "the entering wedge" are often rhetorical devices but may sometimes be justified when there are great pressures to move along the undesirable path and when the safeguards against undesirable developments are insufficient. In this perspective there is reason for con-

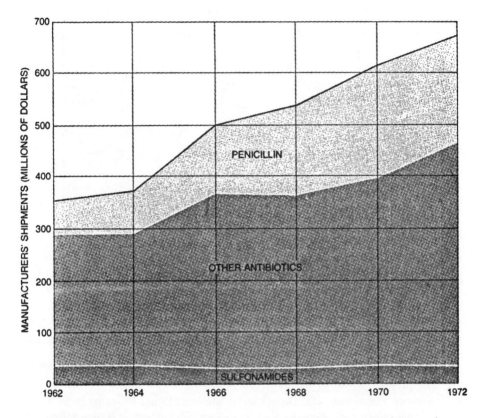

ANTIBIOTICS serve primarily as placebos when they are prescribed for minor virus diseases such as the common cold. Curves, based on Department of Commerce figures, show the value of manufacturers' shipments of systemic antibiotics and of sulfonamides, which are chemical anti-infective drugs. Among antibiotics, single penicillin preparations are shown separately. Antibiotic-sulfonamide combinations are included with other antibiotics.

cern about placebos. The safeguards are few or nonexistent against a practice that is secretive by its very nature. And there are ever stronger pressures—from drug companies, patients eager for cures and busy physicians—for more medication, whether it is needed or not. Given such pressures the use of placebos can spread along a number of dimensions.

The clearest danger lies in the gradual shift from pharmacologically inert placebos to more active ones. It is not always easy to distinguish completely inert substances from somewhat active ones and these in turn from more active ones. It may be hard to distinguish between a quantity of an active substance so low that it has little or no effect and quantities that have some effect. It is not always clear to physicians whether patients require an inert placebo or possibly a

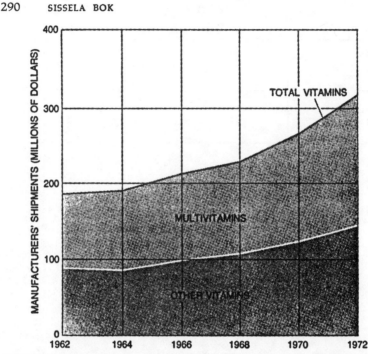

RISING SALES of vitamins (*left*) and of hematinics, or drugs that improve the quality of the blood (*right*), reflect their frequent prescription or purchase across the counter as placebos rather than for their specific effects. The Department of Commerce figures used here include proprietary preparations (sold across the counter) as well as "ethical" ones (sold only on a physician's prescription).

more active one, and there can be the temptation to resort to an active one just in case it might also have a specific effect. It is also much easier to deceive a patient with a medication that is known to be "real" and to have power. One recent textbook in medicine goes so far as to advocate the use of small doses of effective compounds as placebos rather than inert substances—because it is important for both the doctor and the patient to believe in the treatment! The fact that the dangers and side effects of active agents are not always known or considered important by the physician is yet another factor contributing to the shift from innocuous placebos to active ones.

Meanwhile the number of patients receiving placebos increases as more and more people seek and receive medical care and as their desire for instant, push-button alleviation of symptoms is stimulated by drug advertising and by rising expectations of what "science" can do. Reliance on placebic therapy in turn strengthens the belief that there really is a pill or some other kind of remedy for every ailment. As long ago as 1909 Richard C. Cabot wrote, in a perceptive paper on the subject of truth and deception in medicine: "The majority of placebos are given because we believe the patient . . . has learned to expect medi-

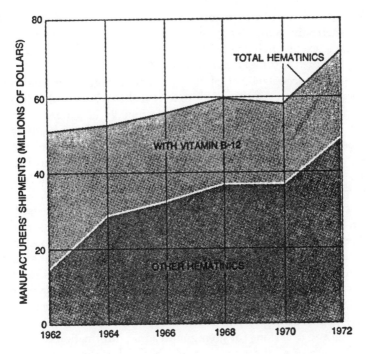

cine for every symptom, and without it he simply won't get well. True, but who taught him to expect a medicine for every symptom? He was not born with that expectation. . . . It is we physicians who are responsible for perpetuating false ideas about disease and its cure. . . . With every placebo that we give we do our part in perpetuating error, and harmful error at that."

A particularly troubling aspect of the spread of placebos is that it now affects so many children. Parents increasingly demand pills, such as powerful stimulants, to modify their children's behavior with a minimum of effort on their part; there are some children who may need such medication but many receive it without proper diagnosis. As I have mentioned, parents demand antibiotics even when told they are unnecessary, and physicians may give in to the demands. In these cases the very meaning of "placebo" has shifted subtly from "I shall please the patient" to "I shall please the patient's parents."

Deception by placebo can also spread from therapy and diagnosis to experimental applications. Although placebos can be given non-deceptively in experimentation, someone who is accustomed to prescribing placebos therapeutically without consent may not take the precaution of obtaining such consent when he undertakes an experiment on human subjects. Yet therapeutic deception is at least thought to be for the patient's own good, whereas experimental deception may not benefit the subject and may actually harm him; even the paternalistic excuse that the investigator is deceiving the patient for his own good then becomes inapplicable.

Finally, acceptance of placebos can encourage other kinds of deception in medicine such as failure to reveal to a patient the risks connected with an operation, or lying to terminally ill patients. Medicine lends itself with particular ease to deception for benevolent reasons because physicians are so clearly more knowledgeable than their patients and the patients are so often in a weakened or even irrational state. As Melvin Levine has put it, "the medical profession has practiced as if the truth is, in fact, a kind of therapeutic instrument [that] . . . can be altered or given in small doses . . . [or] not used at all when deemed detrimental to the patients. . . . Many physicians have utilized truth distortion as a kind of anesthetic to promote comfort and ease treatment." Such practices are presumably for the good of patients. No matter how cogent and benevolent the reasons for resorting to deception may seem, when those reasons are considered in secret, without the consent of the doctored, they tend to be reinforced by less benevolent pressures, self-deception begins to blur nice distinctions and occasions for giving misleading information multiply.

Because of all these ways in which placebo usage can spread it is impossible to look at each incident of manipulation in isolation. There are no water-tight compartments in medicine. When the costs and benefits of any therapeutic, diagnostic or experimental procedure are weighed, not only the individual consequences but also the cumulative ones must be taken into account. Reports of deceptive practices inevitably filter out, and the resulting suspicion is heightened by the anxiety that threats to health always create. And so even the health professionals who do not mislead their patients are injured by those who do and the entire institution of medicine is threatened by practices lacking in candor, however harmless the results may appear to be in some individual cases.

What should be the profession's attitude with regard to placebos? In the case of most experimental applications there are ways of avoiding deception without abandoning placebo controls. Subjects can be informed of the nature of the experiment and of the fact that placebos will be administered; if they then consent to the experiment, the use of placebos cannot be considered surreptitious. Although the subjects in a blind or double-blind experiment will not know exactly when they are receiving placebos or even whether they are receiving them, the initial consent to the experimental design, including placebos, removes the ethical problems having to do with deception. If, on the other hand, there are experiments of such a nature that asking subjects for their informed consent to the use of placebos would invalidate the results or cause too many subjects to decline, then the experiment ought not to be performed and the desired knowledge should be sought by means of a different research design.

As for the diagnostic and therapeutic use of placebos, we must start with the presumption that it is undesirable. By and large, given the principle of informed consent as well as concern for human integrity, no measures that affect someone's health should be undertaken without explanation and permission. Placebos are not so trivial as to be unworthy of ethical evaluation; they carry a definite possibility of harm and discomfort to patients as well as high collective costs; as a result placebo prescriptions present a more serious inroad on patient decision making than has been appreciated up to now. Surreptitious diagnostic and therapeutic administration of placebos should therefore be ruled out whenever possible.

The prohibition should not be absolute, however. In some cases the balance of benefit over cost is so overwhelming that reasonable people would choose to be deceived. There is no clear formula that will quickly reveal in each case whether the benefits will greatly outweigh the possible harm. Much of the problem can be avoided if care is taken to avoid placebos if possible and to observe the following principles in the remaining cases: (1) Placebos should be used only after a careful diagnosis; (2) no active placebos should be employed, merely inert ones; (3) no outright lie should be told and questions should be answered honestly; (4) placebos should never be given to patients who have asked not to receive them; (5) placebos should never be used when other treatment is clearly called for or all possible alternatives have not been weighed.

If placebo medicine is to be thus limited, the information provided to both medical personnel and patients will have to change radically. Placebos, so often resorted to and yet so rarely mentioned, will have to be discussed from scientific as well as ethical points of view during medical training. Textbooks will have to confront the medical and ethical dilemmas analytically and exhaustively. Similarly, much education must be provided for the public. There must be greater stress on the autonomy of the patient and on his right to consent to treatment or to refuse treatment after being informed of its nature. Understanding of the normal courses of illnesses should be stressed, including the fact that most minor conditions clear up by themselves rather quickly. The great pressure patients exert for more medication must be countered by limitations on drug advertising and by information concerning the side effects and dangers of drugs.

I have tried to show that the benevolent deception exemplified by placebos is widespread, that it carries risks not usually taken into account, that it represents an inroad on informed consent, that it damages the institution of medicine and contributes to the erosion of confidence in medical personnel.

Honesty may not be the highest social value; at exceptional times, when survival is at stake, it may have to be set aside. To permit a widespread practice of deception, however, is to set the stage for abuses and growing mistrust. Augustine, considering the possibility of giving official sanction to white lies, pointed out that "little by little and bit by bit this will grow and by gradual accessions will slowly increase until it becomes such a mass of wicked lies that it will be utterly impossible to find any means of resisting such a plague grown to huge proportions through small additions."

16 Clarifying the Concepts of Research Ethics

ROBERT J. LEVINE

The way in which words are used by different groups of people has always caused a variety of problems; this is particularly true in ethics where a different understanding of a word can give a completely different outcome to an argument. One of the major contributions that the National Commission for the Protection of Human Subjects has made is the clarification of a variety of terms that are routinely used in discussing ethical issues in the conducting of research on human subjects. In this article, Levine discusses the Commission's definition of research and practice, its abandoning of the distinction between therapeutic and nontherapeutic research, the clarification of the concept of risk and the different purposes of obtaining informed consent and its relationship to a consent form.

Robert J. Levine, M.D., is Professor of Medicine at the Yale University School of Medicine

One of the most important achievements of the National Commission for the Protection of Human Subjects of Biomedical and Behavioral Research (the Commission) was to begin the process of correcting the conceptual and semantic errors that had undermined virtually all previous attempts to develop rational public policy on research involving human subjects. The fruits of this achievement are seen most clearly in the later reports of the Commission—on children,[1] on those institutionalized as mentally infirm,[2] and on Institutional Review Boards (IRBs).[3] Earlier reports, for example, on the fetus,[4] were flawed in several respects[5] because they were prepared before the Commission had completed its conceptual clarifications. Four of these clarifications are especially important. The Commission (1) developed satisfactory definitions of research and practice; (2) abandoned the distinction between therapeutic and nontherapeutic research; (3) clarified the concept of risk; and (4) identified the different purposes of informed consent and the consent form.

DISTINCTIONS BETWEEN RESEARCH AND PRACTICE

Distinguishing research from practice might not seem to present any problems. Yet, the legislative history of the Act that created the Commission reflects the fact that some physicians regarded this as a very important and exceedingly difficult task.[6] Jay Katz identified "drawing the line between research and accepted practice . . . (as) the most difficult and complex problem facing the Commission." And Thomas Chalmers asserted: "It is extremely hard to distinguish between clinical research and the practice of good medicine. Because episodes of illness and individual people are so variable, every physician is carrying out a small research project when he diagnoses and treats a patient."

Chalmers, of course, was only echoing the views of many distinguished physicians who had spoken to this issue earlier. For example, in *Experimentation with Human Subjects,* Hermann Blumgart stated: "Every time a physician administers a drug to a patient, he is in a sense performing an experiment."[7] To this Francis Moore added: "Every (surgical) operation of any type contains certain aspects of experimental work."[8] While these statements are true, they tend to cloud the real issues and make it more difficult to distinguish research from practice.

Bewildered by these statements, Congress directed the Commission to "consider . . . the boundaries between biomedical or behavioral research involving human subjects and the accepted and routine practice of medicine."[9] The Commission, after its deliberations, concluded in *The Belmont Report* that there were no overlapping boundaries. Rather, when used correctly, the terms "practice" and "research" describe mutually exclusive sets of activities.[10]

For the most part, the term "practice" refers to interventions that are designed solely to enhance the well-being of an individual patient or client and that have a reasonable expectation of success. The purpose of medical or behavioral practice is to provide diagnosis, preventive treatment or therapy to particular individuals. By contrast, the term "research" designates an activity designed to test a hypothesis, permit conclusions to be drawn and thereby to develop or contribute to generalizable knowledge (expressed, for example, in theories, principles and statements of relationships) (pp. 2–3).

The Commission further observed:

The distinction between research and practice is blurred partly because both often occur together (as in research designed to evaluate a therapy) and partly because notable departures from standard practice are often called "experimental" when the terms "experimental" and "research" are not carefully defined (p. 2).

In order to state its views clearly, the Commission avoided such terms as "experimentation" and "experimental"[11] unless it carefully defined the intended sense of the word.[11a] Further, it identified two classes of practices that are commonly confused with research and provided advice as to how to deal with them.

Nonvalidated practices. A class of procedures performed by physicians conforms to the definition of "practice" to the extent that these procedures are "designed solely to enhance the well-being of an individual patient or client." However, they may not have been tested sufficiently often or sufficiently well to meet the standard of having "a reasonable expectation of success." The Commission uses various terms to describe these procedures: "innovative therapies,"[12] "nonvalidated practices,"[13] and, most commonly, "interventions that hold out the prospect of direct benefit for the individual subjects."[14] The regulations of the Food and Drug Administration refer to "investigational" drugs or devices. In my opinion, the best designation for this class of procedures is "nonvalidated practices."[15] Novelty is not the attribute that defines this class of practices; rather it is the lack of suitable validation of the safety or efficacy of the practice.

In *The Belmont Report,* the Commission focuses upon innovations:

When a clinician departs in a significant way from standard or accepted practice, the innovation does not, in and of itself, constitute research. The fact that a procedure is "experimental," in the sense of new, untested or different, does not automatically place it in the category of research. Radically new procedures of this description *should,* however, be made the object of formal research at an early stage in order to determine whether they are safe and effective (emphasis supplied). Thus, it is the responsibility of medical practice committees, for example, to insist that a major innovation be incorporated into a formal research project (p. 3).

Thus, performing a procedure that is innovative, or, for some other reason, nonvalidated is not research; rather, it is a form of practice that ought to be made the object of formal research. The Commission concludes that when research is designed to evaluate a practice, the entire activity—research and practice components—should be reviewed by an IRB.[16] However, the harm-benefit analysis of the nonvalidated practice component should be conducted according to the standards of practice, not research, a point to which I shall return in the section on therapeutic versus nontherapeutic research.

Practice for the benefit of others. The Commission recognizes still another class of activities that conforms to its definition of practice in that it has "a reasonable expectation of success"; however, it is not designed *solely* to enhance the well-being of an individual patient. This class includes some interventions that are applied to one individual in order to enhance the well-being of another (for example, blood donation and organ transplants), and some others that have the dual purpose of enhancing the well-being of a particular individual and, at the same time, providing some benefit to others (for example, vaccination). While these activities often raise harm-benefit questions similar to those presented by research,[17] the Commission concluded that they need not be reviewed by an IRB.[18]

THERAPEUTIC VERSUS NONTHERAPEUTIC RESEARCH

It is not clear when the distinction between therapeutic and nontherapeutic research began to be made in discussions of the ethics and regulation of research. The Nuremberg Code (1947) draws no such distinction. The Declaration of Helsinki (1964) distinguishes nontherapeutic clinical research from clinical research combined with professional care. In the 1975 revision of this Declaration, "medical research combined with professional care" is designated "clinical research," while "nontherapeutic biomedical research involving human subjects" is designated "non-clinical biomedical research."

One major problem with this dichotomy is illustrated by placing one principle of the Helsinki Declaration developed for clinical research (II.6) in immediate proximity to one developed for nonclinical research (III.2):

II.6 The doctor can combine medical research with professional care, the objective being the acquisition of new medical knowledge only to the extent that medical research is justified by its potential diagnostic or therapeutic value for the patient. III.2 The subjects should be volunteers—either healthy persons or patients for whom the experimental design is not related to the patient's illness.

This classification has several unfortunate (and unintended) consequences, two of which are:

1. Many types of research cannot be defined as either therapeutic or nontherapeutic. Consider, for example, the placebo-controlled, "double-blind," drug trial, in which neither patient nor physician knows whether the drug or the placebo is being administered. Certainly, the administration of a placebo for research purposes is not "justified by its potential diagnostic or therapeutic value for the patient." Therefore, according to the principles, this is nontherapeutic, and those who receive the placebo must be "either healthy persons or patients for whom the experimental design is not related to the patient's illness." This, of couse, makes no sense.

2. A strict interpretation of the Declaration of Helsinki would lead to the conclusion that all rational research designed to explore the pathogenesis of a disease is to be forbidden. Since it cannot be justified as prescribed in principle II.6., it must be considered nontherapeutic and therefore done only on healthy persons or patients not having the disease one wishes to investigate. Again, this makes no sense.

By 1974, when the Commission was first convened, the distinction between therapeutic and nontherapeutic research had assumed a central position in most discussions of the ethics and regulation of research involving human subjects. These discussions generally reflected the assumption that there was a class of activities that could be defined as "therapeutic research" and that this was always done—at least in part—for the benefit of the subject. They further reflected the assumption that "nontherapeutic research," because it was not done for the benefit of the subject, was somehow more ethically suspect; that it required stronger justification to proceed; and that it required more stringent mechanisms designed to safeguard the rights and welfare of the subjects. In fact, there were some DHEW proposals to foreclose entire categories of "non-beneficial research" without regard to considerations of whether any risk was involved.[19]

The Commission's first report, *Research on the Fetus*, did not show much promise of correcting these problems; it presented the following definitions:

Research refers to a systematic collection of data or observations in accordance with a designed protocol.

Therapeutic research refers to research designed to improve the health condition of the research subject by prophylactic, diagnostic or treatment methods that depart from standard medical practice but hold out a reasonable expectation of success.

Nontherapeutic research refers to research not designed to improve the health condition of the . . . subject (p. 6).

Of course, there is no such thing as a "systematic collection of data or observations . . . designed to improve the health condition of a research subject . . . that departs from standard medical practice." Thus, the Commission developed recommendations for the conduct of a null set of activities.[20]

After its report on the fetus was completed, the Commission began the process of systematically addressing the general conceptual charges in its mandate. As it developed its definitions of research and medical practice, it identified a distinct class of activities called "nonvalidated practices." It is these practices that are "designed to improve the health condition of the research subject by prophylactic, diagnostic or treatment methods that depart from standard medical practice." These practices are therapeutic but they are not research. As the Commission concluded, research should be done to validate these practices—that is, by establishing their safety and efficacy.

In all subsequent reports, the Commission completely abandoned the language of therapeutic and nontherapeutic research and used instead the concept of nonvalidated practice, for example: "Research on practices, both innovative and accepted, which have the intent and reasonable probability of improving the health or well-being of the individual prisoner . . ."[21] The same concept is reflected in Recommendation 4 on children[22] and Recommendation 3 on those institutionalized as mentally infirm;[23] these Recommendations make it clear that the risks and benefits of therapeutic maneuvers are to be analyzed similarly, notwithstanding the status of a maneuver as either nonvalidated or standard (accepted): "The relation of anticipated benefit to such risk . . . (should be) at least as favorable to the subjects as that presented by available alternative approaches." The risks of research maneuvers (designed to benefit the collective) are perceived differently. If they are more than minimal, special justifications and procedural protections are required.

THE CONCEPT OF RISK

It is widely believed that being a research subject is a highly risky undertaking. This assumption is clearly reflected in the legislative history of the Act that established the Commission; it is further reflected in some of the Commission's early deliberations. For example, because the Commission considered the role of research subject a hazardous occupation, it called upon the philospher Marx Wartofsky to analyze the distinctions between this and other hazardous occupations.[24]

Biomedical researchers have contributed to this incorrect belief.[25] For example, it is often stated that accepting the role of subject nearly always entails the assumption of some risk without specifying the nature of the risk. To many members of the public and to many analysts who are not themselves researchers, the word "risk" seems to carry the implication of some possibly dreadful consequence; this is made to seem even more terrifying when it is acknowledged that, in some cases, the very nature of this dreadful consequence cannot be anticipated. And yet, when biomedical researchers discuss risk, they more commonly mean a possibility that something like a bruise might be observed after a venipuncture.[26]

Some recent data indicate that, in general, it is not especially hazardous to be a research subject. For example, John D. Arnold estimates the risks of physical or psychological harm to subjects in Phase I drug testing[27] are slightly greater than those of being an office secretary, one-seventh those of window washers, and one-ninth those of miners. Chris J. D. Zarafonetis and his coworkers found that, in Phase I drug testing among prisoners, a "clinically significant medical event" occurred once every 26.3 years of individual subject exposure.[28] In 805 protocols involving 29,162 prisoner volunteers over 614,534 days, there were 58 adverse drug reactions, none of which produced death or permanent disability. The only subject who died did so while receiving a placebo. Philippe Cardon and his colleagues, in a large-scale survey of investigators designed to determine the incidence of injuries to research subjects, found that in "nontherapeutic" research, the risk of being disabled either temporarily or permanently was substantially less than that of being similarly harmed in an accident.[29] None of the nearly 100,000 subjects of "nontherapeutic research" died. The apparent hazards of being a subject in "therapeutic research" were substantially greater. This reflects the fact that the subjects of "therapeutic research" tend to have diseases that produce disability or death. Although no direct comparisons were made, it appears that the frequency of either death or disability in subjects of "therapeutic research" was substantially lower than similar unfortunate outcomes in comparable medical settings involving no research.

Mere inconvenience and minimal risk. In considering the burdens imposed upon the human research subject, risk of physical or psychological injury ought to be distinguished from various phenomena that are more appropriately referred to as inconvenience, discomfort, and embarrassment; "mere inconvenience" is a general term for these phenomena.[30] Research presenting the burden of mere inconvenience is characterized as presenting no greater risk of consequential injury to the subject than that inherent in his or her particular life situation.

This class of research is by far the most common. In general, what researchers ask is that a prospective subject give up some time (for example, to reside in a clinical research center, to be observed in a physiology laboratory, or to fill out a questionnaire). Often there is a request to draw some blood or to collect urine or feces.

The term "mere inconvenience" is used in Recommendation 1 in the Commission's report on prisoners.[31] Other reports attempt to reflect a similar conceptualization by using the term "minimal risk."[32] For example:

> Minimal risk[33] is the probability and magnitude of physical or psychological harm that is normally encountered in the daily lives, or in the routine medical or psychological examination, of healthy children.[33a]

Incorporation of this concept into the Commission's recommendations has had a very important effect. Like HEW, the Commission has recognized that stringent procedures—such as review at the national level by an Ethics Advisory Board (EAB)—may be necessary to protect subjects from harm. However, because HEW viewed virtually all research subjects as being "at risk," without distinguishing harm from inconvenience, it proposed bureaucratic procedural protections for all research on certain classes of persons. The favorable effect of the Commission's conceptual clarification is to bring the benefit of these procedural protections to the small minority of research proposals in which some meaningful purpose might be served. For example, proposals to do research involving children must be reviewed by the EAB only if more than minor increments above minimal risk are presented by interventions that do not "hold out the prospect of direct benefit for the individual subjects."[34]

INFORMED CONSENT

HEW regulations require that for all grants and contracts supporting research, development, and related activities in which human subjects are involved, an IRB must determine whether these subjects will be placed at risk. If risk is involved the IRB must assure that "legally effective informed consent will be obtained by adequate and appropriate methods." HEW interpretations of the meaning of "at risk" create a requirement for informed consent in virtually all of these activities. A careful reading of HEW regulations indicates that informed consent must always be documented on a consent form.[35]

Various critics have argued that these requirements are not merely a waste of time, but they also cause many investigators to disregard significant aspects of obtaining informed consent and its documenta-tion.[36] In addition, prospective "subjects" are commonly burdened by the necessity of making decisions that are made to seem more conse-quential than they really are.[37]

The Commission has recommended changes in the regulations that respond to these criticisms. These recommendations not only re-flect the conceptual clarifications I have discussed but also elucidate the function of informed consent and the very different purposes of its documentation on a consent form. Informed consent is designed to show respect for the research subjects' right to self-determination—that is, to make free choices. Researchers tell the subjects what they wish to do and the subjects decide whether they wish to become in-volved. For example, by telling prospective subjects of the risks in-volved in procedures, researchers provide them with an opportunity to protect themselves by deciding whether the potential benefits are worth the risks.

The purpose of documenting consent on a consent form is entire-ly different. It is designed to protect the investigator and the institu-tion against legal liability.[38] The retention of a signed consent form tends to give the advantage to the investigator in any adversary pro-ceeding. In fact, signed consent forms may be detrimental to subjects' interests even in the absence of litigation; their availability in institu-tional records may lead to violations of privacy and confidentiality.[39]

Recognizing these facts, the Commission recommends that there need be no written documentation of consent if the IRB determines either: "(1) The existence of signed consent forms would place sub-jects at risk, or (2) the research presents no more than minimal risk and involves no procedures for which written consent is normally re-quired" (p. 21).

The Commission goes even further by pointing out some circum-stances in which informed consent itself is unnecessary. These in-clude research activities in which:

(1) The subjects' interests are determined to be adequately pro-tected in studies of documents, records or pathological speci-mens and the importance of the research justifies such invasion of the subjects' privacy, or (2) in studies of public behavior where the research presents no more than minimal risk, is unlikely to cause embarrassment, and has scientific merit (p. 21).

Finally, it should be noted that the Commission has abandoned the use of the word "consent," except in situations in which an indi-vidual can provide "legally effective consent" on his or her own behalf. Respect for persons such as children or others who have either immature or impaired capacities for rational self-determination is manifested by negotiating for "assent" or, at least, by recognizing and respecting a "deliberate objection." As a corollary to this decision, the Commission does not use such terms as "proxy consent" or "consent of a legally authorized representative"; the parent or guardian of such a person is asked to give "permission." This terminology was introduced in the Commission's recommendations on children and on those institutionalized as mentally infirm; it is tentatively adopted in HEW proposed regulations on these two classes of subjects.

Conclusions

In its later reports the Commission recommended policies that are far more rational than those developed by any of its predecessors. To illustrate this I have compared these recommendations with earlier HEW final and proposed regulations, the Declaration of Helsinki, and an early report of the Commission. The Commission's achievement can be partially attributed to the identification and correction of several conceptual and semantic errors that had undermined previous attempts to develop suitable standards for the ethical conduct of research involving human subjects.

Still, many authors continue to use these incorrect concepts in their current writings in this field. For example, several commentators have used the language of therapeutic and nontherapeutic research in their analyses of the Commission's reports on prisoners, children, and those institutionalized as mentally infirm. This indicates that they do not understand the recommendations that they are analyzing. Such analyses, if taken seriously, could contribute to the development of a regressive public policy. As Confucius said:[40]

If language is not used rightly,
then what is said is not what is meant.
If what is said is not what is meant,
then that which ought to be done is left undone;
if it remains undone, morals and art will be corrupted;
if morals and art are corrupted, justice will go awry;
and if justice goes awry, the people will stand about
in helpless confusion.

REFERENCES

1. Commission, *Report and Recommendations: Research Involving Children.* DHEW Publication No. (OS) 77-0004, Washington, 1977.

2. Commission, *Report and Recommendations: Research Involving Those Institutionalized as Mentally Infirm.* DHEW Publication No. (OS) 78-0006, Washington, 1978.

3. Commission, *Report and Recommendations: Institutional Review Boards.* DHEW Publication No. (OS) 78-0008, Washington, 1978.

4. Commission, *Report and Recommendations: Research on the Fetus.* DHEW Publication No. (OS) 76-127, Washington, 1975.

5. Robert J. Levine, "The Impact on Fetal Research of the Report of the National Commission for the Protection of Human Subjects of Biomedical and Behavioral Research," *Villanova Law Review* 22 (1977), 367–82.

6. E. M. Kay, "Legislative History of Title II—Protection of Human Subjects of Biomedical and Behavioral Research—of the National Research Act: P.L. 93–348." Prepared for the National Commission, 1975, pp. 16–18.

7. Hermann L. Blumgart, "The Medical Framework for Viewing the Problems of Human Experimentation," in *Experimentation with Human Subjects,* Paul A. Freund, ed. (New York: George Braziller, 1970), pp. 39–65.

8. Francis D. Moore: "Therapeutic Innovation: Ethical Boundaries in the Initial Clinical Trials of New Drugs and Surgical Procedures," in *Experimentation with Human Subjects,* pp. 358–78.

9. Public Law 93–348: The National Research Act.

10. Commission, *The Belmont Report: Ethical Principles and Guidelines for the Protection of Human Subjects of Research.* DHEW Publication No. (OS) 78-0012, Washington, 1978.

11. To experiment means to test something or to try something out. In another sense, an experiment is a tentative procedure, especially one adopted in uncertainty as to whether it will bring about the desired purposes or results. Much of the practice of diagnosis and therapy is experimental in nature. For example, a physician tries out a drug to see if it brings about the desired result in the patient. If it does not, the physician either increases the dose, changes to another therapy, or adds a new therapeutic modality to the first drug. All of this "experimentation" is done in the interests of enhancing the well-being of the patient.

    When experimentation is conducted for purposes of developing generalizable knowledge, it is regarded as research. One of the problems presented by much research designed to test the safety or efficacy of drugs is that this activity is much less experimental than the practice of medicine. It must, in general, conform to the specifications of a protocol. Thus, the individualized dosage adjustments and changes in therapeutic modalities are less likely to occur in the context of a clinical trial than they are in the practice of medicine. This deprivation of the "experimentation" ordinarily done to enhance the well-being of a patient is one of the burdens imposed on the patient-subject in a randomized clinical trial.

11a. Robert J. Levine and Karen Lebacqz: "Some Ethical Considerations in Clinical Trials," *Clinical Pharmacology and Therapeutics* 25 (1979), 728–41.

12. Early drafts of *The Belmont Report.*

13. Commission, *Report and Recommendations: Research Involving Prisoners.* DHEW Publication No. (OS) 76–131, Washington, 1976.

14. Commission, *Report and Recommendations: Research Involving Children* and *Report and Recommendations: Research Involving Those Institutionalized as Mentally Infirm.*

15. See Levine, "The Impact on Fetal Research of the Report of the National Commission. . . . ," pp. 380–82.

16. Commission, *The Belmont Report,* p. 4.

17. Robert J. Levine: "On the Relevance of Ethical Principles and Guidelines Developed for Research to Health Services Conducted or Supported by the Secretary, DHEW" in: *Appendix to the Commission's Report: Ethical Guidelines for the Delivery of Health Services by DHEW.* DHEW Publication No. (OS) 78-0011, Washington, 1978, pp. 2.14–2.16.

18. While the concept of "practice for the benefit of others" is reflected in *The Belmont Report* (p. 3), the term is not used. A more complete description of this class of practices and the problems it presents may be found in Levine, "On the Relevance of Ethical Principles. . . . ," pp. 2.11–2.13.

19. DHEW: Proposed Policy, Federal Register 39 (No. 165) (August 23, 1974), 30648–57.

20. I do not wish to be excessively critical of the Commission's first report. The Commission was required by Congress to place fetal research first on its agenda. Considering the time constraints imposed by Congress, the report is remarkably good.

21. Commission, *Report and Recommendations: Research Involving Prisoners,* p. 15.

22. Commission, *Report and Recommendations: Research Involving Children,* pp. 5–6.

23. Commission, *Report and Recommendations Involving Those Institutionalized as Mentally Infirm,* pp. 11–13.

24. Marx W. Wartofsky: "On Doing It for Money," in *Appendix to Reference No. 13.* DHEW Publication No. (OS) 76-132, Washington, 1976, pp. 3.1–3.24.

25. Robert J. Levine: "Nondevelopmental Research on Human Subjects: The Impact of the Recommendations of the National Commission for the Protection of Human Subjects of Biomedical and Behavioral Research," *Federation Proceedings* 36 (1977) 2359–64; and Robert J. Levine, "Commentary: Terminological Inexactitude," in *Legal and Ethical Issues in Human Research and Treatment: Psychopharmacologic Considerations,* Donald M. Gallant and Robert Force, eds. (New York: Spectrum Publications, 1978), pp. 85–98.

26. A comprehensive discussion of the origins and consequences of this belief is beyond the scope of this essay. Further details may be found in the articles cited in reference 25.

27. J. D. Arnold: "Alternatives to the Use of Prisoners in Research in the United States," in *Appendix to reference 13,* pp. 8.1–8.18.

28. C. J. D. Zarafonetis et al, "Clinically Significant Adverse Effects in a Phase 1 Testing Program," *Clinical Pharmacology and Therapeutics* 24 (1978) 127–32.

29. Philippe V. Cardon, F. William Dommel, and Robert R. Trumble: "Injuries to Research Subjects: A Survey of Investigators," *New England Journal of Medicine* 295 (1976) 650–54.

30. Robert J. Levine, "Appropriate Guidelines for the Selection of Human Subjects for Participation in Biomedical and Behavioral Research," in *Appendix I to reference 10,* DHEW Publication No. (OS) 78-0013, pp. 4.8–4.10.

31. Commission, *Report and Recommendations: Research Involving Prisoners,* p. 14.

32. Commission, *Report and Recommendations: Research Involving Children,* p. xx.

33. The Commission's choice of the term "minimal risk" troubles me. It is one of the unusual instances that the Commission decided to stipulate a definition for words in common usage. The use of stipulated definitions in regulations commonly creates confusion. Citations of regulations using this term will be misinterpreted by the unsophisticated reader, unless the full stipulated definition is reproduced in each citation.

The unsophisticated reader might think that the intended meaning is that presented in standard dictionaries; "minimal" means least possible. The term "risk" is even more problematic. For examples: could minimal risk mean an infinitesimal chance of a substantial harm? A substantial chance of an infinitesimal harm? In this case the confusion will be compounded by the fact that the term is used in the report on children (p. xx) with one stipulated definition, in the report on those institutionalized as mentally infirm with a different definition (p. 8), and in the report on the fetus (p. 73) with no definition.

33a. Robert J. Levine: "The Role of Assessment of Risk-Benefit Criteria in the Determination of the Appropriateness of Research Involving Human Subjects," in *Appendix I to reference No. 10,* pp. 2.1–2.59.

34. Commission: *Report and Recommendations: Research Involving Children,* p. 10.

35. Robert J. Levine: "The Nature and Definition of Informed Consent in Various Research Settings," in *Appendix I to reference No. 10,* pp. 3.1–3.91.

36. Franz J. Ingelfinger, "The Unethical in Medical Ethics," *Annals of Internal Medicine,* 83 (1975), 264–69.

37. Angela R. Holder and Robert J. Levine: "Informed Consent for Research on Specimens Obtained at Autopsy or Surgery: A Case Study in the Overprotection of Human Subjects," *Clinical Research* 24 (1976), 68–77.

38. Levine, "On the Relevance of Ethical Principles, . . . ," pp. 2.17–2.19.

39. Karen Lebacqz and Robert J. Levine, "Respect for Persons and Informed Consent to Participate in Research," *Clinical Research* 254 (1977), 101–07.

40. Lois DeBakey, "Literacy: Mirror of Society," *Journal of Technical Writing and Communication* 8 (1978), 297–319.

Research and
Human Experimentation/Further Reading

Barber, Bernard, et al. *Research on Human Subjects: Problems of Social Control In Medical Experimentation.* New York: Russell Sage Foundation, 1973.

Capron, Alexander Morgan. "Informed Consent in Catastrophic Disease Research and Treatment." *University of Pennsylvania Law Review* 123 (December, 1974), 340–438.

Childress, James F. "Compensating Injured Research Subjects: I. The Moral Argument." *Hastings Center Report* 6 (December, 1976), 21–27.

Eisenberg, Leon. "The Social Imperatives of Medical Research." *Science* 198 (December, 1977), 1105–10.

"The Freedom of Inquiry and Subjects." *American Journal of Psychiatry* 134 (August, 1977), 891–913.

Fried, Charles. *Medical Experimentation: Personal Integrity and Social Policy. Volume 5 of Clinical Studies.* Edited by A. G. Bearn, D. A. K. Black, and H. H. Hiatt. New York: Elsevier, 1974.

Katz, Jay, Edited with Alexander A. Capron and Eleanor Swift Glass. *Experimentation With Human Beings.* New York: Russell Sage Foundation, 1972.

Rivlin, Alice M., and Timpane, P. Michael, Editors. *Ethical and Legal Issues of Social Experimentation.* Washington, D.C.: The Brookings Institution, 1975.

Roth, Loren H., et al. "Tests of Competency to Consent to Treatment." *American Journal of Psychiatry* 134 (March 1977), 279–84.

Walters, LeRoy. "Some Ethical Issues in Research Involving Human Subjects." *Perspectives in Biology and Medicine.* 20 (Winter, 1977), 193–211.

PART FIVE

ETHICAL DILEMMAS IN OBTAINING INFORMED CONSENT

Thomas A. Shannon
Bernard Barber
Karen Lebacqz and
Robert J. Levine
Alan Soble

17 The Problem of Interests and Loyalties: Ethical Dilemmas in Obtaining Informed Consent

THOMAS A. SHANNON

This article discusses the problem of the conflict of interests and loyalties that occur when a professional seeks to negotiate informed consent with a research subject. Shannon proposes an approach to professions that will see them in a broader social context so that the tension between interests and loyalties can be diminished and the research subject better protected.

Thomas A. Shannon is an Associate Professor of Social Ethics at the Worcester Polytechnic Institute, and an Associate Professor of Medical Ethics at The University of Massachusetts Medical School

In all professions, the problem of the conflict of interests and loyalties inevitably arises. Eventually all persons must at some time and in some way resolve this dilemma. This common problem of such a conflict is of special importance to the various health professions, especially in the area of research and experimentation. Here such a dilemma will always involve at least two elements—the model of the profession and the research subject. Both of these issues must be taken into account in resolving the potential conflict of interests and loyalties in this situation.

I. THE PROBLEM

For purposes of clarification, a few working definitions are in order to help set a context for the discussion. As such these suggested definitions are not meant to be exhaustive; they are proposed as indicators to provide a partial background for the proposed analysis.

"Interest" refers to a relation of being concerned with respect to some advantage or disadvantage. When we speak of interests in general, we are referring to concerns, needs, or goals of ours in which we have something at stake. There is an involvement on a personal level with the subject under discussion. The problem of interests becomes compounded when one's interests become vested. Parsons defines these vested interests as follows:

> . . . the interests in maintaining the gratification involved in an established system of role expectation . . .[1]

Here the notion of interest becomes linked with an institution and its multi-faceted relations with other institutions. It then becomes much more difficult to sort out specific personal interests because they become merged with the interests inherent in institutional workings and relations.

"Loyalty" refers to a set of normative values or ideals of the person or institution which serves as the basis of ultimate commitment. Loyalties are derived from membership in a normative class and are, in many ways, a set of commonly shared values. Loyalties differ from interests in that loyalties refer to one's normative ideals rather than to one's actual concerns. Hopefully one's interests and loyalties would be in harmony, but this is not necessarily the case.

In defining "informed consent" one must carefully avoid omitting a significant detail or including too many details. Perhaps a better approach is to indicate some common themes underlying the concept of informed consent. Obviously information is one of these for the intent is that the subject be informed in a meaningful way of what the experiment is and what may occur during its course. Included in the act of informing is a discussion of benefits and risks. The element of consent assures that the subject is voluntarily agreeing to participate in the research. This would also imply the right to terminate participation in an experiment at any point. Ramsey suggests that another needed element is prudence: ". . . practical wisdom in the appraisal of cases and specific situations."[2] This is to emphasize that the obtaining and giving of informed consent are never simplistic applications of a definition, but rather the perception of the situation by both investigator and subject from which emerge loyalty and partnership.[3]

Another important element of informed consent, but one that is difficult to include in a definition, is the reality of power. It is obvious that the researcher has a tremendous amount of power over the subject. This may be due to the researcher's institutional affiliation, professional title, connection with a hospital or university, specialized knowledge, or the class difference that exists between the researcher and the subject. Since subjects are often drawn from lower classes, there is either an actual or perceived power difference between the two.[4] Informed consent can give the subject some countervailing power within the experimental situation. At a minimum level, the obligation and practice of obtaining informed consent say a researcher cannot do whatever he or she wishes and help the subject learn that he or she has rights that must be respected, thus equalizing a potentially unequal situation.[5]

With these themes and Ramsey's caveat in mind, the following DHEW definition of informed consent will be used in the rest of the paper.

'Informed consent' means the knowing consent of an individual or his legally authorized representative, so situated as to be able to exercise free power of choice without undue inducement, or any element of force, fraud, deceit, duress, or other forms of constraint or coercion.[6]

Regardless of how we articulate or of how we exemplify it, all agree that progress in medicine and the abolition of disease are positive goods and health a significant value. Implicit in these values, however, is the necessity for research and experimentation on human subjects. Chemotherapy, for example, is one critical element in the protection and preservation of health.[7] However, drugs ultimately need to be tested on humans, for animal trials or mathematical models simply do not indicate fully how a drug may react on a human subject. Consequently, at some stage in the development of a new drug, human experimentation becomes a sine qua non which we recognize and accept as a critical part of health maintenance.

But, such an acceptance reveals a serious problem: the need to protect the human subject. Most would subscribe to a general claim that the human is a bearer of certain rights and dignity which are an intrinsic part of one's personal nature. And, as such, most would also generally affirm that there are limits to what one can do to a consenting or unconsenting subject. The necessity of human experimentation to help promote the value of health is qualified by the personhood of the subject and there are religious and ethical arguments that prevent the researcher from doing whatever he or she might want to do, even though the subject has consented.

The general consequence of such a situation is the recognition of a need to protect human subjects. This is stated and argued for in a variety of ways and codes of ethics, but the general thrust is clear: the personhood of a subject in some way places limits on how that subject may be treated. We affirm the value of health and medical advances, but not to the exclusion or negation of the value of the person.

II. THE ETHICAL PROBLEM

The need for the protection of human subjects is recognized by the various professions that have to do with health care, experimentation, or other related areas such as anthropology, psychology, or sociology. Although it is not necessary to analyze all of these codes of ethics,[8] some should be presented to give the flavor of their orientation.

Possibly the strongest of such codes is the Nuremberg Code of 1946. Coming in the wake of the Nazi war trials, the code particularly emphasized voluntary consent, the necessity of avoiding unnecessary pain or mental suffering, and a balancing of risks in proportion to the humanitarian importance of the experiment.[9] The 1954 World Medical Association Code continued to emphasize the obtaining of consent and informing of a subject, but not as ponderously or sternly as the Nuremberg Code. The World Medical Association Code states:

> The paramount factor in experimentation on human beings is the responsibility of the research worker and not the willingness of the person submitting to the experiment.[10]

This can have two possible readings. A cynical one would suggest that the rights of the subjects have just been placed on the back burner. But the statement also suggests the ethical demand for the responsible use of power on the subject by the researcher, for indeed during the course of the experiment the subject must trust the researcher to be faithful to the information given and promises made at the beginning of the experiment. Thus there is a subtle emphasis on the continuing loyalty of the researcher to the subject.

An interesting dilemma is posed by the 1962 Army regulations on the use of volunteer subjects. The statement reads, in part:

> He must have sufficient understanding of the implications of his participation to enable him to make an informed decision, so far as such knowledge does not compromise the experiment. He will be told as much of the nature, duration, and purpose of the experiment, the method and means by which it is to be conduct-

ed, and the inconveniences and hazards to be expected, as will not invalidate the results. He will be fully informed of the effect upon his health or person which may possibly come from his participation in the experiment.[11]

These guidelines recognize several areas of possible conflicts—but do not appear to resolve them in favor of the subject. The rights of the subject seem to be subordinated to the outcome of the experiment. Also, there appears to be an unresolved conflict between informing the subject of possible risks to health and withholding information that may invalidate the experiment.

The 1964 Declaration of Helsinki is a strong statement on the protection of human subjects. An advance over the Nuremberg Code is the section on nontherapeutic clinical research which spells out a different orientation from therapeutic research, or as the Declaration states it, clinical research combined with professional care. The distinction is an important one for different values emerge in each situation which cannot be resolved in the same ways. What may be extremely risk-ladened in a nontherapeutic situation, may be the only possible alternative in a therapeutic situation. The decision to enter into such a risk-ladened therapeutic situation cannot be judged by criteria used to evaluate a nontherapeutic one.

The 1971 "Opinions and Report on the Judicial Council" of the American Medical Association deals with the ethical regulation of experimentation under Section II of its Code of Ethics which has to do with the improvement of medical knowledge and skills. The guidelines list three general principles: the subject must give voluntary consent; the dangers of each experiment must be investigated by prior animal experimentation; and experimentation must be performed under proper medical protection and management.[12] These are followed by a separate discussion of the ethical requirements for clinical investigation, either for treatment or for the accumulation of scientific knowledge. Consent and protection of the subject from effects of the experiment are common to both situations. The guidelines stress the responsibility of the investigator throughout the procedure. They seem to present no innate ethical dilemmas, but neither do they explicitly deal with the problem of what should be done, in the nontherapeutic context, when knowledge of possible results may either invalidate or weaken the experiment.

The 1973 Principles of Medical Ethics as applied to Psychiatry[13] do not specifically mention experimentation. The special focus of this Code of Ethics is the problem and dilemma of confidentiality. The psychiatrists are referred to the AMA Code of Ethics for a general statement of ethical regulations dealing with research. An explicit statement on confidentiality as it specifically applies to some types of psychological studies and investigations—other than the admission that "... materials used in teaching and writing must be adequately disguised in order to preserve the anonymity of the individual involved"[14]—could be helpful in terms of specifying the requirements of the investigation and the range of rights of the subject in this situation.

The Code of Ethics of the American Sociological Association offers two general principles which relate to research:

> 3. Every person is entitled to the right of privacy and dignity of treatment. The sociologists must respect these rights.
> 4. All research should avoid causing personal harm to subjects.[15]

The code continues, in paragraph five, to discuss confidentiality, surely an important topic in sociological investigation. The basic problem here is not the general principles, which are correct enough, but rather with their vagueness. The meaning of the terms are left ambiguous enough to be able to cover a wide variety of situations and therefore possibly skirt the ethical dilemmas. "Right of privacy" and "personal harm" are important concepts, but their significance and usefulness diminish in direct proportion to their vagueness.

The 1973 statement "Ethical Principles in the Conduct of Research with Human Participants" by the American Psychological Association[16] presents some interesting value dilemmas. Paragraphs 3, 4, and 5 all deal with the relationship among informed consent, the obligation of disclosure, and the responsibility of the investigator to protect the subject's dignity. Paragraph 4 states the dilemma most clearly:

> 4. Openness and honesty are essential characteristics of the relationship between investigator and research participant. When the methodological requirements of a study necessitate concealment or deception, the investigator is required to insure the participant's understanding of the reasons for this action and to restore the quality of the relationship with the investigator.[17]

Two basic problems appear at once. First, the document begs the question of whether concealment and deception are necessary; it glibly assumes they are and proceeds from there. Second, the document does not specify when the participant is to be informed of the "necessary" deception. If the informing is before deception, there may be no ethical problems involved; if the informing is post-factum, there are ethical problems involved—the violation of the rights of the subject and negation of the values of openness and honesty. This situation is problematic at least and to my mind is an unsatisfactory resolution of the problem.

Paragraph 5 presents a similar problem in that it requires the respect for the subject's freedom of choice, but simultaneously says the researcher may sometimes have to limit this freedom. The researcher, then, must take special care to protect the dignity of the subject. In addition to the repeated problem of question-begging, a new dimension enters in here: paternalism. Obviously the researcher is the one who is making the decision and is the one who decides what is best for the subject in the light of the needs of the experiment. This is a serious and unresolved problem in this code of ethics.

As a result of greater sensitivity to ethical issues inherent in the research situation, a growing concern with the rights of the person, and disclosure of several problematic, if not totally unethical research situations and experiments, the Federal Government is in the process of generating guidelines for the protection of human subjects in experimental situations. A federal commission was established for this purpose and has recently submitted recommendations for guidelines on fetal research which have been approved and promulgated by DHEW. What is almost more important than whatever final guidelines will emerge, is the national concern and debate over the problem of the protection of human subjects. One cannot predict the full effect and lasting significance of such debate, but a major step forward has been taken that will provide the basis for further developments.

Of special importance in the proposed DHEW guidelines of 30 May 1974 is the statement that all research proposals which place a subject at risk must indicate that informed consent has been obtained from that subject and cannot in any way be waived and, in fact, must be documented. This basically means that all research proposals, submitted for DHEW funding, which place subjects at risk must have a section on the ethics of the research in the protocol itself. This implies that procedures which might have ethical dilemmas must be faced and resolved.

A definition of informed consent that the proposed DHEW guidelines offer has been quoted above. The definition is supplemented by statements which affirm that full explanation of the material must be given, a description of risks and benefits must be offered, appropriate alternatives must be disclosed, a declaration of freedom must be included, and the ability to withdraw from the experiment at any time must be clearly stated.[18] One critique raised against these guidelines is the exclusive use of the risk-benefit calculus in evaluating some of the ethical dilemmas of experimentation. Another problem may involve the need to separate more clearly therapeutic from nontherapeutic experimentation—the distinction could be made clearer. Also, there is need for the composition of the ethical review board to be specified lest an unfair power advantage be built into such boards. The final word has yet to be said concerning these guidelines; but as they stand, they indicate basic areas of concerns and situations where value dilemmas may occur.

All professionals have an interest in their occupations. They wish to perform well, they want their work to be important, and they want it to be recognized and well-received. They too must provide for the necessities of life and, in a highly competitive market, the competition for a limited amount of money could tend to bring out high levels of creativity in research proposal and design—and possibly questionable means of obtaining results.[19]

These all contribute to creating a set of personal interests for a professional, goals or concerns in which he or she is personally engaged. These personal interests may link up with social interests and become magnified. If, for instance, science has a high priority within a society, the energies and finances of that society go towards promoting science and scientific endeavors. Interest groups gather to promote special projects or to help establish priorities. Interests become merged and fused; personal interests may become public interests and public interests may be manipulated to attain one's private interests.

Many charges of scientists' having manipulated various interests have been leveled; within the last decade science has been the whipping boy for many social and political problems. Yet, in the main, people esteem both science and scientists (including physicians and other researchers). H. R. Niebuhr suggests several basic reasons for this esteem: (1) Science has provided good things for us; (2) Science makes predictions that come true; and (3) Scientists have been faithful to us.[20] These reasons are augmented by the feeling that science has used its knowledge:

> ... for the benefit of the whole human society and for each individual in it as though humanity and the individual had a value not derived from their relations to a nation or a caste, or some other special value-center.[21]

Even though this basic confidence is present and operative, some problems appear. The very value of science itself may serve as the problem, for vested interests may develop which may shift values and priorities. Parson's suggests that:

> Collectivity-orientation on the other hand converts this 'propensity' (of loyalty) into an institutional obligation of the role expectation. Then whether the actor 'feels like it' or not, he is obliged to act in certain ways and risk the application of negative sanctions if he does not.[22]

This implies that pressures brought to bear by institutions—i.e., universities, research organizations, or profesional societies—may interject their value preferences between the scientist and his or her work. This is not to suggest that such pressures are necessarily illegitimate or in any way a perversion of genuine research. It is to suggest, however, that other interests may enter into the picture and may shift one's perception and priorities.

> Insofar as one plays an institutionalized role in interaction with other institutionalized roles, the alternatives for action are presented here in terms of the institutional definition of the situation.[23]

H. R. Niebuhr suggests that potential problems exist when science begins to accept uncritically narrow or closed value frameworks such as nationalism or truth as an exclusive end in itself to the detriment of other values such as justice and equality. Such conflict of interest has the potential to weaken the faith and trust we have in science.[24]

There are also a number of potential personal conflicts of interest which should be mentioned. All are aware of the decreasing amount of money available for research, the tightening of the job market, and other pressures in the academic community. Such pressures may cause a person to shift interests or focus on the protection of one's interest by engaging in research whose purpose or methods may be questionable. The extent to which this actually happens is one question; the other question—the potential for it to happen given the nature of the case—cannot be naively ignored. We all know that results are important, and when our interests are at stake, values which are normally important to us may fall by the wayside in the ensuing conflict.[25]

In addition to the fact of professional interests, there is also the fact of professional loyalties. A professional tends to have a certain standard for action. This may be derived from the code of ethics of a professional society, or from a strong set of convictions of what ought to be done in fulfilling one's role. These codes of moral values give the professional a guide by which actions, priorities, and interests may be evaluated.

H. R. Niebuhr suggests that there are elements within science itself that may help create a set of loyalties that can guide one's actions. The first of these is what he calls a morality of enterprise which implies that one's commitment is to the service one performs or the cause which one serves. This type of commitment to the larger goods of one's profession can help to override a short-term focusing on what could be exclusively private interests. Second, there is the element of self-examination and self-criticism which is an attitude of openness towards one's research and its design, procedure, and evaluation. This social dimension enhances one's loyalty to values such as honesty and integrity. Such openness, which is expected in research, again thwarts a possible tendency to base one's research on vested interests. The third and fourth dimensions of morality in science deal with truthtelling within both the scientific community itself and the human community at large. Minimally this would imply a willingness to have errors brought to light, but it also implies a willingness to share knowledge gained and to examine the purpose for which it can be used.[26]

These different elements within the nature of the scientific endeavor reinforce the loyalties of the researcher to certain values and ways of proceeding. They set forth certain standards and goals. In doing this they provide an ethical foundation for self-evaluation and criticism.

In many ways, professional codes of ethics are specifications of these values which are inherent in scientific procedures. They function as types of middle axioms for judging particular cases. In doing this, they focus the professional's attention on value issues and value claims so that loyalties and interests can be judged. Some elements in various codes of ethics—particularly but not exclusively, those dealing with biomedical research—are also concerned with values that go beyond the researcher, but which are obviously related to research. Such values would be concerned with the dignity and rights of the person who is the subject of the experiment. Although one can easily and correctly argue that values such as respect for the person, truthtelling, and respect for privacy are part of the scientific endeavor itself, it is important that such value claims are also grounded outside the profession. This clearly separates the issue of loyalties and interests. When a professional's loyalties transcend a situation at hand, a conflict of interests and loyalties can show up more clearly and has the potential for a more equitable resolution. What happens, then, is that values impinge on the professional both from within and without the profession itself. This gives a broader base of support to one's fidelity to loyalties and has a potential to weaken narrow interest-based claims. A summary statement of the issue is provided by H. R. Niebuhr:

> ... does not the issue lie between those whose good is the collective representation of a special group and those who trust in and are loyal to the collective representation of mankind as a whole?[27]

In a recent article,[28] B. B. Page suggests that professions open to individuals are not the property of the individuals in the profession but rather of the society. Page argues that this is so for two reasons: (1) "... any individual in *any* society—even a counterculture commune—acquires recognition, relevance, and even meaning primarily in terms of his or her relationships to that society, its culture and institutions, and its other members."[29]; (2) "... professions acquire recognition and relevance primarily in terms of the needs, conditions, and traditions of particular societies and their members."[30]

What this means is that membership in any profession cannot be thought of as separate from one's membership in society. In fact, membership in a profession is a way of specifying social obligations and responsibilities. Thus, rather than reinforcing a rather tenuous separation of the individual and society, the fact of membership in a profession is a way of integrating the individual and society, in that the professional has specific skills to deal with social needs or problems.

Such an understanding of a practice of a profession would go far in helping to resolve, or at least place in a more significant context, the dilemmas that can exist between interests and loyalties. First, the profession is not perceived as being in a conflict situation (us against them). Rather, it is one institution among others seeking to contribute to the common good. It obviously has its own priorities and its own orientations, but what it has to offer is a contribution or service to society. Its primary focus is not exclusively its own preservation, but the preservation and well-being of the society of which it is a member and which it serves. Secondly, the professional is not simply a member of a profession. He or she is also a member of various other civic or voluntary associations. As such, a professional is, to use Michael Walzer's phrase, a "pluralist citizen,"[31] one who shares in ruling and being ruled precisely because of one's multiple membership in various associations. Each of these associations, to a greater or lesser degree, proposes various values or ideals. The internalization of these values is one additional means of the professional's developing a set of loyalties which ultimately may transcend the interests of any single association. Through this process, interests, endemic to only one group, and loyalties, which transcend any one group, are more easily separated and evaluated. Such a separation of interests and loyalties does not guarantee more ethical decisions but does contain a strong potential for forcing professional interests—vested or not—into the clear light of day to be evaluated either on their own merits or hopefully, in the light of one's transcendent loyalties.

The quality of one's ethical decisions may be in direct relationship to one's degree of specialization. If a professional becomes increasingly focused on only one issue or orientation, then there is a corresponding danger that everything will be evaluated in terms of that one issue and its interest. This situation obviously places a professional in opposition to society, for society consists of more than one issue, need, or service. If, on the other hand, the professional—with his or her needed and valued specialization—has the ability to participate at least in an inner dialogue from a variety of value positions and social viewpoints which stem from membership in various associations, then the issue may be resolved more in terms of loyalties to personal or social values, rather than narrow interest-based reasons. Such a professional will continually be in the process of evaluating issues and programs from a variety of viewpoints with the positive result that none of these will *a priori* have the upper hand. In this situation, loyalties and transcendent values have at least an equal chance to compete against powerful interest-based reasoning.

As applied specifically to the research or experimental situation, such a framework may be of ethical significance in that it forces the researcher to view his or her work in a wider social perspective, within the context of a variety of loyalties in addition to the specific interests involved. The significance of such a framework is that it casts the ethical dilemmas in research in a context much broader than one determined solely by professional or other interests. Second, if a researcher is in this wider context, he or she will have a variety of viewpoints from which to evaluate ethical dilemmas. One will not automatically exclude a set of issues because they do not relate to or bear upon one's own personal or professional interests. Involvement in a variety of associations can also have the beneficial effect of forcing us to entertain a variety of values. And it is this pause for reflection that can induce a greater degree of ethical sensitivity in the research situation. Third, this type of context may cause the process of decision making to be extended. While this is not necessarily a virtue, neither is it intrinsically a vice. In areas where significant value issues are at stake, sometimes an extended period of time is necessary if the full range of values is to be considered. Such a process helps to ensure that interests will not be the sole element in decision making. Interests are significant and need representation; but if they are the only elements that are represented, then a variety of needs and values are always automatically excluded. Such a situation may well produce more short-term experimental benefits; but one also needs to look at the long run. Decisions that incorporate a variety of loyalties and values have the potential to expand one's horizon and to be more critical in the necessary evaluation of consequences, benefits, and risks.

III. CONCLUSION

One of the major purposes of this paper is to suggest a new model for understanding a profession. Traditionally, professions have been viewed as specialized groups, quite often far removed from the public and its needs, and concerned primarily with their own interests and goals. Oftentimes, however, this can lead members of the profession into an adversary relation with society (i.e., society may not need or want what the profession has to offer) or into a paternalistic relation with society (the profession's knowing what is best for society). What this often means in practice is that a profession becomes another interest group, pursuing its own interests without much concern until these interests are challenged and the profession is forced to defend them.

The model suggested here would attempt to broaden the concept of a profession by incorporating wider loyalties into its structure. These wider loyalties, which could come from a professional code of ethics, the values or goals of civic, religious, or other voluntary organizations of which one is a member, or one's personal code of ethics, can help the professional realize that he or she cannot define social participation or social responsibility simply from one perspective, i.e., the profession. Other dimensions of life, although perhaps not as dominant as the professional, need to be incorporated into one's modus vivendi so that a pattern of personal and social coherence may emerge. The significance of this model is that it attempts to open up the concept of a profession from within so that the profession may be more easily perceived and defined in relation to and as a part of society—rather than being in an adversary relation. This has the major advantage of helping professionals to understand themselves as an organic part of a social whole in which a variety of nonprofessional loyalties must be harmonized with professional ones.

This wider understanding of a profession is not intended to reject needed specialization, the valued services offered by professions, or the legitimate interests of a profession. Rather it is suggesting that such professional services need to be defined to heighten the relation of the profession to society and to a wider loyalty base. Such an orientation will not automatically ensure that more or better ethical decisions will be made. What this model will do, however, is to introduce into this very concept and model of a profession value dilemmas that cannot be easily brushed aside. For if membership in a profession can highlight other social values and can help in the personal integration of alternate points of view, then the professional will be forced to become more conscious of his or her multi-faceted relations with the larger society and the responsibilities that such relations bring with it. By suggesting that the model of a profession be broadened from within through the incorporation of a wider loyalty base rather than accepting a model which encourages specialization together with free-market style competition for power, influence and the attaining of self-defined interests, there is the possibility that the professional will be forced to perceive his or her self in a new light and in a new, more integrated relation with society.

As applied to the informed consent issue, this framework, simply stated, will help insure that values other than the interests involved in the research itself—its methodologies, its claims, and its potential for social good—be brought to bear on the rights and values of the subject of the experiment. If the professional conducting the experiment has been involved in a wide variety of activities, he or she may have been in situations in which values and loyalties have been proposed from other contexts or perspectives. These values and loyalties will hopefully be integrated within the professional person to produce a moral whole and to serve as a means of testing and evaluating specific interests, which might automatically predispose him or her to act, if not routinely, at least uncritically. Also the paternalism which is quite inherent in a traditional profession may be broken down or weakened. For if the professional is forced to perceive the research subject as a fellow citizen, a degree of equality may be restored to the situation. This in turn will, of course, weaken the power of the professional over the subject. But this is precisely the function of informed consent and the suggested model of a profession may help reestablish the significant role informed consent ought to play in experimentation.

In the issue of informed consent, such a new professional context may, at the least, give the subject an even break; at the most, it will ensure that he or she will be treated with the full dignity due a human being. Since health is a value and since ultimately human subjects are needed for experiments to test products to make or keep individuals healthy (physically or psychologically), it is incumbent upon professionals to separate clearly narrow, profession-based interests from the value inherent in the research situation so that the ethical dilemmas may be clearly stated and the major issues set forth before a final decision is made. Such a model of a profession would enhance the value and significance of the profession itself. But it would do this by incorporating the profession into a wide range of social interests and loyalties rather than by understanding the profession as separate from the community because of narrowly defined interests and loyalties.

REFERENCES

1. Parsons T: *The Social System.* The Free Press of Glencoe, 1964, p. 492.

2. Ramsey R: *The Patient as Person.* New Haven, Yale University Press, 1972, p. 3.

3. Ramsey, op. cit., cf. pp. 5–6.

4. Cf. Herbert Kelman, "The Rights of the Subject in Social Research: An Analysis in Terms of Relative Power and Legitimacy." *American Psychologist* 27:989–1016, Nov., 1972.

5. I am grateful to Dr. Dan McGee of Baylor University for suggesting this theme to me.

6. Protection of Human Subjects, DHEW. Federal Register, Vol. 39, No. 105, 30 May 1974, p. 18917.

7. Fuchs VR: *Who Shall Live?* New York, Basic Books, Inc., 1974, p. 30.

8. For an excellent representative sampling of such codes, cf. Henry K. Beecher, *Research and the Individual.* Boston, Little Brown and Co., 1970, pp. 217–309.

9. Ibid., pp. 227–28.

10. Ibid., p. 240.

11. Ibid., pp. 252–53.

12. Opinions and Reports of the Judicial Council. American Medical Association, Chicago, 1971, p. 10.

13. "The Principles of Medical Ethics With Annotations Especially Applicable to Psychiatry." Reprint from *American Journal of Psychiatry,* September, 1973.

14. Loc. cit.

15. Code of Ethics. American Sociological Association. Reprint from the *American Sociological Association.*

16. *American Psychologist,* January, 1973, pp. 79–80.

17. Loc. cit.

18. *Federal Register,* op. cit., p. 18917.

19. Barber B, Lally JJ, Maharahka JL, et al: "The Structures of Scientific Competition and Reward and its Consequences for Ethical Practice in Biomedical Research." Paper presented at the 67th annual meeting of the American Sociological Association, 1972. Cf. also, by the same authors, *Research on Human Subjects: Problems of Social Control in Medical Experimentation.* New York, Russell Sage Foundation. 1973.

20. Niebuhr HR: *Radical Monotheism and Western Culture,* New York, Harper and Brothers, 1960, pp. 79–80.

21. Ibid., p. 81.

22. Parsons, op. cit., p. 98.

23. Parsons T: *Essays in Sociological Theory.* New York, The Free Press, 1967, p. 145.

24. Niebuhr, op. cit., pp. 82–83.

25. Barber, et al., op. cit.

26. Ibid., pp. 132–135.

27. Ibid., p. 88.

28. Page BB: "Who Owns the Professions?" *Hastings Center Report,* October, 1975, pp. 7–8.

29. Ibid., p. 7.

30. Loc. cit.

31. Walzer M: *Obligations: Essays on Disobedience, War and Citizenship.* Cambridge, Harvard University Press, 1970, p. 218.

18 *The Ethics of Experimentation with Human Subjects*

Bernard Barber

This article is a summary of two studies conducted on the attitudes and practices of investigators who used human subjects in research. This study provided empirical data for arguing that there is reason for some concern about how subjects are treated in research protocols. Barber argues that such abuses can be traced to defects in the training of physicians and in the screening and monitoring of research by review committees as well as a function of the fundamental tension between investigation and therapy.

Bernard Barber is Professor of Sociology at Barnard College

The power, scope and funding of biomedical research have expanded enormously in the past 40 years. So also, inevitably, has clinical research with human subjects. That expansion has led in the past decade to widespread reflection on what is increasingly perceived as a new social problem: the abuse of human subjects of medical experimentation. In particular it is alleged that human subjects are not always protected from undue risk and do not always have the opportunity to voluntarily give their adequately informed consent to participation in experiments.

A social problem is defined in part by the concern it arouses, and this one has clearly aroused concern. Members of the medical profession itself led the way, with increasing numbers of journal articles, books and seminars on the issues. The public has become aroused, largely through popular accounts of dramatic incidents—genuine scandals in certain cases—involving the violation of the dignity and rights of patients. And the Federal Government has moved to protect human subjects, potential or actual. Beginning in 1966 the National

Institutes of Health, the Food and Drug Administration and the Department of Health, Education, and Welfare have issued increasingly detailed regulations governing experimentation with human subjects in projects they support, which means in most of the biomedical research done in the country. In 1974 a National Commission for the Protection of Human Subjects of Biomedical and Behavioral Research was established to advise the Department of Health, Education, and Welfare, and it is to be replaced by a long-term National Advisory Council that is to deal with the same issues.

The regulations, commissions and councils and the very fact of interference in medical activities by outsiders are viewed by many investigators as being onerous and even dangerous. On the other hand, many outsiders believe far more social control is required. The debate on the issue has been conducted without much reference to objective evidence. In 1970 our Research Group on Human Experimentation undertook two studies of investigators' attitudes and practices. On the basis of our results I would argue that there is indeed inadequate ethical concern among biomedical investigators, that it is reflected in excessively risky procedures and that better internal and external controls are essential.

There are two major reasons for the general recognition that experimentation with humans is a subject for concern, one of which I alluded to at the outset: the increased power, scope and funding of biomedical research. The other reason is a change in values: increased emphasis on equality, participation and the challenging of arbitrary authority.

It is easy to forget how new scientific medicine is. The revolutionary advances based on knowledge of physiology and biochemistry have come in the past 40 years, and they came from research. The basic work could be done with test-tube preparations and laboratory animals, but eventually human subjects had to be involved. Man is "the final test site," as Henry K. Beecher, a pioneer among physicians concerned about the ethics of research, once put it. Unfortunately there are no statistics on the number of people who are subjects in medical experiments or even on how many projects involve human subjects; the National Institutes of Health keeps records according to area of research (a disease or a physiological process, for example) rather than according to species of experimental subject; the NIH can say only that recently about a third of the projects it approves involve human subjects. It is clear, however, that the number of human subjects is larger than it used to be and that some small but significant minority of those subjects are involved in risky experiments. If more people have been put at more risk, then there is a rational basis for concern about the satisfactory balancing of risks and benefits, about adequate protection from unnecessary risk and about some groups being put at more risk than other groups.

Over and beyond this utilitarian basis for the new social concern with medical experimentation is the value factor, which arises from recent social changes. All over the world individuals have been demanding more equality of treatment and the right to be informed about and to participate in decisions affecting them and have been challenging the right of experts to make those decisions unilaterally. People who define themselves as being unequal, underprivileged or exploited are demanding better treatment and better protection, whether it is underdeveloped countries as against developed ones, blacks as against whites, women as against men, young as against old, patients as against doctors—or subjects as against investigators. This moral revolution of rising value-expectations has combined with the revolution in medicine to focus attention on the ethics of experimentation with human subjects.

Public awareness of the problem is too much the result of headlined scandals, but the scandals do illustrate some of the possible abuses. In the 1960's two respected cancer investigators who were studying the immune response to malignancies injected live cancer cells into a number of geriatric patients at the Jewish Hospital and Medical Center of Brooklyn without first obtaining the patients' informed consent. A few years later a leading virologist conducted an experiment at Willowbrook, a New York State institution for the severely retarded. Reasoning that a serious liver infection, hepatitis, was in effect endemic in the hospital anyway, he deliberately exposed some children to hepatitis virus in an attempt to achieve controlled conditions for testing a vaccine. The accusation was that the children's parents were not given enough information on which to base informed consent, and that in some cases consent was given perfunctorily by administrators of the institution.

More recently there was the exposure by the press of the ongoing syphilis experiment in Tuskegee, Ala. Since the 1930's a group of black subjects with syphilis had been kept under observation in an effort to study the course of the disease. That was not considered wrong in the 1930's, when the known treatments for the disease were only marginally effective, but by 1945 penicillin had become available as a safe and extremely effective cure for syphilis. Yet somehow the experiment was continued, and presumably some men died of the disease who could have been cured.

How significant are such scandals? We do not know, because no one has been doing the kind of social bookkeeping about numbers of

45. A researcher plans to study bone metabolism in children suffering from a serious bone disease. He intends to determine the degree of appropriation of calcium into the bone by using radioactive calcium. In order to make an adequate comparison, he intends to use some healthy children as controls, and he plans to obtain the consent of the parents of both groups of children after explaining to them the nature and purposes of the investigation and the short and long-term risks to their children. Evidence from animals and earlier studies in humans indicates that the size of the radioactive dose to be administered here would only very slightly (say, by 5–10 chances in a million) increase the probability of the subjects involved contracting leukemia or experiencing other problems in the long run. While there are no definitive data as yet on the incidence of leukemia in children, a number of doctors and statistical sources indicate that the rate is about 250/million in persons under 18 years of age. Assume for the purpose of this question that the incidence of the bone disease being discussed is about the same as that for leukemia in children under 18 years of age. The investigation, if successful, would add greatly to medical knowledge regarding this particular bone disease, but the administration of the radioactive calcium would not be of immediate therapeutic benefit for either group of children. The results of the investigation may, however, eventually benefit the group of children suffering from the bone disease. Please assume for the purposes of this question that there is no other method that would produce the data the researcher desires. The researcher is known to be highly competent in this area.

45A. Hypothetically assuming that you constitute an institutional review "committee of one," and that the proposed investigation has never been done before, please check the *lowest* probability that *you* would consider acceptable for *your* approval of the proposed investigation. (Check only *one*)

( ) 1. If the chances are 1 in 10 that the investigation will lead to an important medical discovery.

( ) 2. If the chances are 3 in 10 that the investigation will lead to an important medical discovery.

( ) 3. If the chances are 5 in 10 that the investigation will lead to an important medical discovery.

( ) 4. If the chances are 7 in 10 that the investigation will lead to an important medical discovery.

( ) 5. If the chances are 9 in 10 that the investigation will lead to an important medical discovery.

( ) 6. Place a check here if you feel that, as the proposal stands, the researcher should not attempt the investigation, no matter what the probability that an important medical discovery will result. (*IF YOU CHECKED HERE,* please explain): _____

45B. Which of the above responses comes closest to what you feel the *existing institutional review committee* in your institution would make? _____ (Please write in the number of the response.)

45C. Which of the above responses comes closest to what you feel the *majority of the researchers* in your institution would make, acting in their role as researcher rather than as a "committee of one"? _____ (Please write in the number of the response.)

HYPOTHETICAL EXPERIMENT *described here was one of six experiments submitted to investigators and administrators in hospitals and other research centers in a mailed questionnaire. In each case respondents were asked whether, under specified conditions, they would approve of the experiment. This proposal involved giving radioactive calcium to children with a bone disease and to a control group and measuring its uptake by bone.*

subjects, degree of risk, adequacy of consent and efficacy of protective mechanisms that would yield an overall view of experimentation with human beings and that might contradict the more extreme allegations of abuse elicited by the publicized scandals. In the absence of such intensive record keeping it remains for social research to fill the gap by sampling the total range of experimentation with human subjects. To that end our group conducted first a national mail survey of nearly 300 biomedical research institutions and then an intensive interview study of 350 individual investigators at two institutions.

Our national survey questionnaire was answered by 293 teaching and nonteaching hospitals and other research institutions that, our analysis showed, constituted a nationally representative sample of all such institutions. Those who filled out the questionnaire were generally themselves active researchers and members of their institution's review committee, set up to pass on research proposals. We asked the investigators to give us their response to six simulated proposals such as those that might come before a review committee. The proposals were detailed research protocols designed to measure the degree of the investigators' concern about informed consent and their willingness to approve of studies involving various levels of risk. We could be confident that the protocols were "hypothetical-actual" rather than "hypothetical-fantastic" because we constructed them with careful attention to the research literature, checked them with specialists and pretested them with a dozen chiefs of research at medical centers, who found them to be convincingly real.

One protocol described a study of chromosome breakage in users of hallucinogenic drugs; blood samples (for chromosomes) and urine samples (for evidence of drug use) were to be taken, at no risk but also without notification of the experimental purpose, from students routinely visiting the university health center. Another protocol proposed that the thymus gland, which is a component of the immune system, be removed unnecessarily from a random sample of children undergoing heart surgery; the objective was to learn the effect of the thymectomy on the survival of an experimental skin graft made at the same time. The other protocols dealt with a random test of alternative treatments for a congenital heart defect in children; with an evaluation of the efficacy of a new drug for severe depression (placebos were given to some patients); with a study of lung function in patients kept under unnecessarily prolonged anesthesia after undergoing a routine hernia repair, and with an investigation of the effect of radioactive calcium on bone metabolism in children.

---

44. It has been shown that the thymus has an important bearing on the development and maintenance of immunity. For this reason the researcher proposes an investigation to determine the effect of thymus removal on the survival of tissue transplants, a very timely and important problem. In a sample of children and adolescents admitted for surgery to correct congenital heart lesions, he would randomly select an experimental group for thymectomy. Though the thymectomy will prolong the heart surgery by a few minutes, there is otherwise extremely little additional surgical risk from this procedure. At the conclusion of each heart operation, a full-thickness skin graft, approximately one cm. in diameter and obtained from an unrelated adult donor, would be sutured in place on the chest wall of both the experimental and control groups. He would then compare the survival of the skin grafts in each of the groups. It has been shown in a number of investigations of neonatal rats and other animals that those whose thymus had been removed were much less likely to reject skin grafts. The possible long-term immunological problems that might result are as yet not completely known, but a number of studies in animals indicate significant immunological deficiencies after thymectomy. Studies done in humans with myasthenia gravis, some of whom had undergone thymectomy, have not definitively demonstrated that the immunological abnormalities discovered in these patients were the result of thymectomies. To quote one authority: "There were no immunologic abnormalities that could be attributed to the effect of thymectomy *per se.*"

The research will result in no therapeutic benefits for the patients involved. The researcher plans to obtain the consent of his potential patient-volunteers and/or their parents after explaining the procedures involved in the investigation as well as the possible short-term surgical and long-term immunological hazards for the subjects.

REMOVAL OF THYMUS GLAND *during heart surgery was the experimental procedure proposed in another protocol in the questionnaire. Respondents were asked if they would approve of the experiment, given various probabilities that it would show thymectomy "considerably increases the probability of tissue-transplant survival in children and adolescents."*

The answers to the thymectomy, anesthesia and radioactive-calcium protocols in particular gave us measures of the respondents' attitudes toward the balancing of risks and benefits. A clear pattern emerged. In the case of the high-risk thymectomy, for example, 72 percent of the respondents said the project should not be approved no matter how high the probability was that it would establish the efficacy of thymectomy in promoting transplant survival. On the other hand, 28 percent of the respondents said they would approve the experiment; 6 percent said they would approve it even if the chance of significant results was no better than one in 10. Similarly, 54 percent were against doing the calcium study at all—but 14 percent said they would approve it even if the odds were only one in 10 that it would lead to an important medical discovery. Our basic finding was that whereas the majority of the investigators were what we called "strict" with regard to balancing risks against benefits, a significant minority were "permissive," that is, they were much more willing to accept an unsatisfactory risk-benefit ratio.

The same general pattern of a strict majority and a permissive minority emerged from our second study, in which we interviewed 350 investigators actively engaged in research with human subjects. The investigators were at institutions to which we gave the synthetic names University Hospital and Research Center and Community and Teaching Hospital. The institutions were picked (by a technique known as cluster analysis) as being representative of two kinds of medical center that do considerable amounts of research. The interviewees told us about 424 different studies involving human subjects, and for each study they estimated the risk for subjects, the potential benefit for subjects, the potential benefit for future patients and the potential scientific importance of the study. It was reassuring to find that the investigators considered that only 56 percent of the clinical investigations graded for risk and benefits involved any risk for the subjects. We went on, however, to cross-tabulate the estimated risks and benefits [see Chart IV], and we concluded that in 18 percent of the studies the risk was not adequately counterbalanced by the benefits. We called those studies the "less favorable" ones, and we proceeded to classify them further according to their potential benefits for other patients or for medical science. Even when these compensating justifications were taken into account, tabulation revealed a "least favorable" category of studies in which the poor immediate risk-benefit ratio was not compensated for by possible future benefits. These "least favorable" investigations constituted 8 percent of the investigations in our analysis.

The concept of informed consent is a troublesome one. The in-

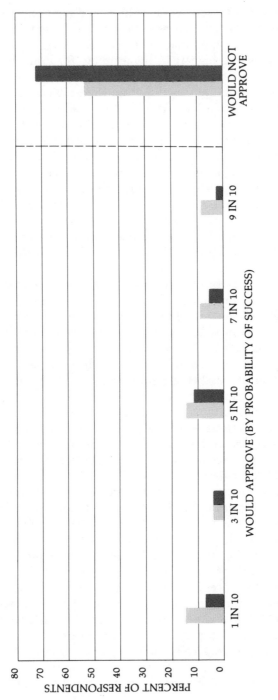

REACTION OF RESPONDENTS to the two hypothetical-experiment questions illustrated on the preceding pages is shown: the calcium study *(light gray bars)* and the experimental removal of the thymus gland for a skin-transplant study *(dark gray)*. "As the proposal stands," 54 percent of the respondents would refuse to approve the calcium study and 72 percent would refuse to approve the thymectomy, regardless of the probability of success. Substantial minorities were much more "permissive," however.

vestigator wants to have enough subjects and is afraid of scaring them off. Patients are likely to be concerned about their own condition, may feel powerless with respect to the physician or hospital and often have difficulty understanding medical language or concepts. Even established medical procedures can have somewhat unpredictable consequences, so that physicians feel there is a limit to how completely "informed" a patient can be. The fact remains that regulations of Government funding agencies and most institutions now require that the human subject of an experiment (or his guardian, in the case of small children and mentally incompetent patients) understand that something is being done (or some treatment is being withheld) for reasons other than immediate therapeutic ones; the subject or guardian must be informed of any risks and must give consent voluntarily.

With regard to informed consent, our questionnaires and interviews again revealed a minority with "permissive" views and practices, although that minority was smaller than it was for unfavorable risk-benefit ratios. For example, 23 percent of the questionnaire respondents said they would approve the chromosome-break proposal, which presented the informed-consent issue clearly and in effect by itself. The situation was more complex in the heart-defect protocol. Here other dubious elements competed with the fact that the investigator would not inform the parents that his decision whether or not to operate would be a random one, not based on therapeutic considerations. Only 12 percent of our respondents said they would approve of the study without requiring any revisions, but only 65 percent specifically mentioned the lack of informed consent as a problem.

The best available research evidence on informed consent comes from a study conducted by Bradford H. Gray, who was then a graduate student at Yale University, at a distinguished university hospital and research center (not the one in our interview study). With the consent of the responsible investigator, Gray interviewed 51 women who were the subjects in a study of the effects of a new labor-inducing drug. Although the women had signed a consent form, often in the hectic course of the admitting procedure or in the labor room itself, 20 of them (39 percent) learned only from Gray's interview, which was held after the drug infusion had been started or even after the delivery, that they were the subjects of research. Among those who did know, most of them did not understand at least one aspect of the study: that there might be hazards, that it was a double-blind experiment, that they would be subjected to special monitoring and test procedures or that they were not required to participate; four of the women said they would have refused to participate if they had known there was any choice. Many of the women had been referred for the study by their private physician, but instead of being informed that an experimental drug was to be administered they were told that it would be a "new" drug; they trusted their doctor and assumed that "new" meant "better."

How does it happen that the treatment of human subjects is sometimes less than ethical, even in some of the most respected university-hospital centers? We think the abuses can be traced to defects in the training of physicians and in the screening and monitoring of research by review committees, and also to a fundamental tension between investigation and therapy. We have data bearing on each of these causative factors.

It is in medical school that the profession's central and most serious concerns are presumably given time and place and that its basic knowledge and values are instilled. Yet the evidence from our interviews shows that there is not much training in research ethics in medical school. Of the more than 300 investigators who responded to

Chart IV

THERAPEUTIC BENEFIT FOR SUBJECTS	RISK		
	NONE	VERY LITTLE	SOME MODERATE OR LARGE
MINOR, LITTLE OR NONE	11	14	2
SOME	14	12	2
GREAT	10	19	7

RISKS AND BENEFITS *were cross-tabulated for some 400 current research projects reported by investigators in two hospitals. Studies falling on or below the diagonal were considered to have risks for subjects that were more or less counterbalanced by benefits for subjects. (In 9 percent of the studies respondents reported no risk and were not asked about benefits.) Studies above the diagonal (colored boxes) were classified as "less favorable" for their subjects: they contained risks for subjects and, according to the investigators, offered relatively low benefits. These cases, 18 percent of the total, were further subdivided (in a table not reproduced here) according to benefit for others or for science. Studies that were low in those justifications (8 percent of total) were called "least favorable."*

questions in this area, only 13 percent reported they had been exposed in medical school to part of a course, a seminar or even a single lecture devoted to the ethical issues involved in experimentation with human subjects; only one respondent said he had taken an entire course dealing with the issues. Another 13 percent reported that the subject had come to their attention when, as students, they did practice procedures on one another; for 24 percent it was in the course of experiments with animals; 34 percent remembered discussion of ethical issues in specific research projects. One or more of these learning experiences were reported by 43 percent of the respondents—but the remaining 57 percent reported not a single such experience. The figures were about the same whether the investigators were graduates of elite U.S. medical schools, other U.S. schools or foreign schools. The figures were a little better, however, for those who had graduated since 1950 than for older investigators.

What little ethics training there is is apparently not very effective: the investigators who reported having learned something about research ethics were only slightly less permissive in response to protocols presenting the risk-benefit issue than those who reported no such experiences. It would appear that both the amount and the quality of medical-school training in the ethics of research could be improved. In this connection it is worth remembering that the many physicians who are not engaged in investigation at all also need some background in experimentation ethics, if only so they can evaluate requests that they direct their patients toward a colleague's research project.

Scientific "peer review" is a keystone of scientific inquiry, operating implicitly in many ways and explicitly in the case of professional journals, grant-awarding committees and many institutional reviewing boards such as the "tissue committees" that assess the results of surgery in hospitals. Ethical peer review of experimentation with human beings should be the counterpart of scientific peer review, but until the mid-1960's such activity received limited support among biomedical researchers. Even after 1966, when the NIH mandated ethical peer review for all its grantees, effective review did not become universal. Our questionnaire went to hospitals and other research centers that had filed with the NIH formal assurances that the required institutional review committee had been established, but 10 percent of the respondents said their institution's committee reviewed only proposals for outside funds and 5 percent reported that only formal proposals to the NIH were reviewed. The two institutions in our interview study were among the 85 percent that stated they were reviewing all research proposals, and yet 8 percent of our interviewees volunteered the information that at least one of their own investigations with human subjects had not been reviewed.

How effective are the review committees in handling the protocols that do come before them? Our questionnaire respondents told us that in 34 percent of the institutions the committees had never required any revisions, rejected any proposals or had any proposals withdrawn in anticipation of rejection for ethical reasons; 31 percent reported revisions, 32 percent outright rejections and 19 percent withdrawals. Either some of these committees have very few ethical problems coming before them or they are ineffective. Gray's study in an institution with an active and strong committee suggests that they are ineffective rather than underworked. The committee whose performance he examined found relatively few proposals that did not need some kind of modification, and he thinks "a record of few actions by committees is an indication that their members are indifferent or that their standards are loose."

The peer-review groups seemed weak in other ways. In some institutions there was no face-to-face discussion among the reviewers. Only 22 percent of the committees had members from outside the institution, something that was then recommended and has since been mandated by the Department of Health, Education, and Welfare. In practically none of the institutions was there continuous monitoring of studies that were approved, although this was even then required by Government regulations. In general ethical peer review is hampered by the fact that each committee operates in isolation and must consider every new issue on its own and without benefit of precedent. A case-reporting system, such as operates in the law, would make that unnecessary and would promote both equity among institutions and high standards. The major weakness in the system is the lack of keen interest in and support of the review committees on the part of most working biomedical investigators. Research is their business; research is their mission and predominant interest, not applied ethics or active advocacy of patients' rights.

Most biomedical investigators are, however, interested in taking care of patients and making them well. As a result medical institutions and individual investigators operate today with two powerful sets of values and goals. On the one hand there is the pursuit and advancement of scientific knowledge. On the other there is the provision of humane and effective therapy for patients. Through a broad range of complex interactions these two sets of values and goals are harmonious, even complementary and mutually reinforcing. Occasionally, however, scientific research and humane therapy can be in conflict. When that happens, there is sometimes a tendency to choose

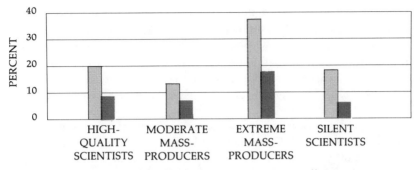

PRESSURE TO PRODUCE *leads to "permissiveness." Investigators were classed as "high-quality scientists" (most cited), "moderate mass-producers" (many papers, few citations), "extreme mass-producers" (many papers, no citations) or "silent scientists" (few papers and citations). Extreme mass-producers were twice as likely as high-quality scientists to have a role in one of the less favorable (light gray) or least favorable (dark gray) studies.*

the pursuit of knowledge at the expense of the ethical treatment of patients. An irreducible minimum of conflict may be inevitable. The ethical task now is to come as close as possible to that minimum—and to resolve unavoidable conflict in favor of humane therapy.

There is evidence that the enhanced excitement attending scientific achievement and the rewards bestowed on it in recent decades have skewed the decision-making process in many cases of conflict. As our data show, the medical schools have been largely indifferent to training their students in the ethics of research. Moreover, their record in peer review has been inferior to that of other institutions. Answers to our questionnaire showed they were less likely than other research centers to have set up a review committee before the NIH required one, less likely to have one that met the first NIH guidelines in 1966, less likely to have a committee that reviews all clinical research and less likely to include on their committee medical or nonmedical members from outside the institution. Medical schools, the Association of American Medical Colleges and professional associations of clinical investigators have been much quicker to seek research funds or to protest funding cuts than to organize seriously for the purpose of studying the ethics of research and making policy in that area.

The same emphasis on the pursuit of knowledge rather than on ethics is apparent among individual biomedical investigators. Ethical concern for the subjects of their research is not a major factor when they select their collaborators; at least it is not often mentioned as a characteristic they look for in collaborators. Scientific ability is a major concern. When we asked our 350 interview subjects, "What three characteristics do you most want to know about another researcher before entering into a collaborative relationship with him?" 86 percent of the respondents mentioned scientific ability, 45 percent mentioned motivation to work hard and 43 percent mentioned personality. Only 6 percent of them listed anything we could classify as "ethical concern for research subjects."

The tension between investigation and ethical concern is perhaps best illustrated by indications that the struggle for scientific priority and recognition exerts pressure on ethical considerations. Our data show that the social structure of competition and reward is one of the sources of permissive behavior in experimentation with human subjects; the relatively unsuccessful scientist, striving for recognition, was most likely to be permissive both in his approval of hypothetical protocols and in his own investigative work. We divided our respondents into four categories based on the number of papers they had published and the number of times their work had been cited by other workers; the frequency of citation has been shown to be a good measure of scientific excellence. We called the most-cited investigators the "high quality" scientists and those who had published a great deal but were never cited the "extreme mass-producer" scientists. It was the extreme mass-producers who were most often engaged in investigations with less favorable risk-benefit ratios, who approved of the protocols with poorer risk-benefit ratios and who least often expressed awareness of the importance of consent. Caught up in the socially structured competitive system of science, unsuccessful in it but still pursuing the prize of peer recognition, they appear to be more likely to overvalue scientific work as against humane therapy.

It is not only the mass-producers, contending for recognition among peers in their discipline, who are apt to be more permissive. We also weighed the rank achieved by each worker within his own institution against various measures of his effectiveness compared with that of his colleagues. We found that the "underrewarded" investigators tended to be the more permissive. There is also a quite different kind of medical investigator who we think is likely to be pushed toward permissive practices by scientific competition: some of the professionally esteemed, highly successful medical scientists who are engaged in intense competition for priority and recognition in well-publicized areas of research. There are not many of those people, and they did not emerge in our sample, although some workers who refused to be interviewed may belong in that category. In the absence of real data we can only point to such evidence as published discussions concerning the worldwide heart-transplant competition of a few years ago, which raised questions about the premature exposure of human subjects to what were then still experimental procedures.

Given the fact that there are ethical defects in current medical-research standards and practices, do the resulting abuses strike particularly, as is often alleged, at certain social groups: at the poor, at children and at institutionalized patients (prisoners in particular)?

The evidence from our interviews with 350 investigators indicates that the poorer patients in hospitals are indeed at a disadvantage as subjects of research. For each of the 424 studies our respondents reported, they told us whether fewer than 50 percent, between 50 and 75 percent or more than 75 percent of the subjects were ward or clinic patients (as opposed to patients in private or semiprivate rooms and under the care of their own physician). We found first of all that ward and clinic patients were more likely to be subjects of experiments. Moreover, when we examined the cases we had previously identified as having "less favorable" and "least favorable" risk-benefit ratios, we found that both categories were almost twice as likely to involve subjects more than three-quarters of whom were ward and clinic patients as the studies with the more favorable ratios were.

The ward and clinic patients are, of course, vulnerable to that kind of discrimination. They can most readily be channeled into an experimental group by admitting physicians and clerks without interference from a personal physician. They tend to be less knowledgeable about hospitals, more readily intimidated and less likely to understand what they are told about an experimental project, and therefore less likely to be able to withhold their consent or to give genuinely informed consent. In sum, they are the least likely to be able to protect themselves.

Many institutionalized patients are poor and perhaps incompetent, and they may feel completely dependent on the institution's administrators and physicians. Prisoners are a special case: they are institutionalized in an implicitly coercive situation, so that genuinely informed consent may be a logical impossibility. On the other hand, a prison population is by definition a good source of experimental and control subjects living under controllable conditions, and there have been instances where prison studies have been conducted humanely, with good scientific results and apparently with good effect on the prisoners' morale. Experimentation with prisoners is nevertheless subject to grave abuses. Last summer the head of the Food and Drug Administration told a Senate committee that a review of experimentation in 19 prisons revealed abuses ranging from unprofessional supervision of drug tests to inadequate medical care and follow-up treatment.

Children constitute still another special group. Small children

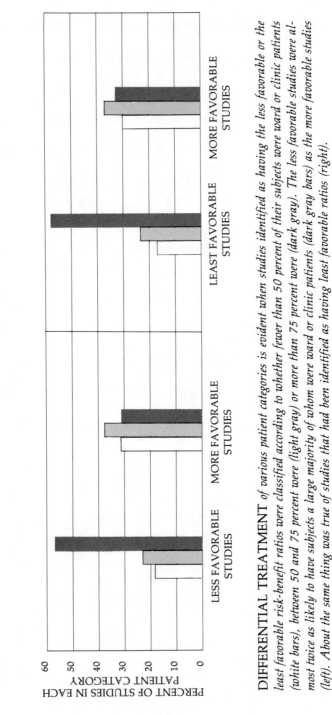

DIFFERENTIAL TREATMENT *of various patient categories is evident when studies identified as having the less favorable or the least favorable risk-benefit ratios were classified according to whether fewer than 50 percent of their subjects were ward or clinic patients (white bars), between 50 and 75 percent were (light gray) or more than 75 percent were (dark gray). The less favorable studies were almost twice as likely to have subjects a large majority of whom were ward or clinic patients (dark gray bars) as the more favorable studies (left). About the same thing was true of studies that had been identified as having least favorable ratios (right).*

cannot give consent for their own participation in experiments; older children, who could, are often not asked. As the Willowbrook incident demonstrated, parents are not always adequately protective of their children's interests. In the case of institutionalized patients, prisoners and children, new regulations of the Department of Health, Education, and Welfare call for special protective committees and procedures. These will only be effective, however, in a context of better ethical training for investigators and more effective peer review.

The ethical problems that attend medical research with human subjects are representative of an entire class of problems created by the impact of professionals and professional power on the general public and on public policy. In the area of research with human subjects the medical investigators are not alone; there is a tendency in other fields too for humane concerns to be left at the laboratory door. Psychologists and sociologists have often been accused of circumventing the requirement for consent and of applying unethical manipulative techniques in their investigations of human behavior, and neither profession has welcomed scrutiny from outsiders or restrictive regulation. The issue goes beyond research ethics, however. Many professions now command knowledge that has great potential usefulness for human welfare but bestows power that can be abused. Because professional power is largely based on knowledge that has not yet diffused to the general public it must to a considerable degree be self-regulated, but because professional power is of such major public consequence it must also be subject to significant public control. The medical-research profession does not have a proud record of self-regulation or acceptance of public controls.

19 Respect for Persons and Informed Consent To Participate in Research

KAREN LEBACQZ AND ROBERT J. LEVINE

This article investigates the foundations of and requirements for informed consent. Basing the requirement on the ethical principle of respect for persons, the authors argue that major disagreements over the requirement come from two different interpretations of the principle: 1) it requires that we protect another from harm and 2) it requires that we respect another's right to self-determination.

Karen Lebacqz is an Associate Professor of Ethics at the Pacific School of Religion
Robert J. Levine is Professor of Medicine at the Yale University School of Medicine

The first sentence of the Nuremberg Code (1947), "The voluntary consent of the human subject is absolutely essential," signals the centrality of the consent requirement in research using human subjects. Prior to Nuremberg, statements of medical and other professional organizations made no mention of the necessity for consent. Subsequently, the tendency to focus on "informed consent" has been reinforced by public outcry over the inadequacy of consent in certain landmark cases: *eg,* Willowbrook,[1, at p. 1007] Jewish Chronic Disease Hospital,[1, at p. 9] Tea Room Trade,[1, at p. 325] and most recently, Tuskegee.[2] Indeed, the issue of informed consent has so dominated recent discussion of the ethics of research that one might be led to think erroneously that other issues (*eg,* research design, selection of subjects) are either less important or more satisfactorily resolved.

The purpose of this essay is to explore the foundations of and requirements for informed consent. The requirement is best derived from the ethical principle of *respect for persons;* we contend that the major disagreements over the requirement arise from two conflicting interpretations of the principle: (1) it requires that we protect another from harm, and (2) it requires that we respect another's right to self-determination—*ie*, to be left alone or to make free choices. We shall provide examples of how these differing views have been expressed in the development of the legal requirements for informed consent to biomedical research.[a]

RESPECT FOR PERSONS

The principle of respect for persons was stated formally by Immanuel Kant: "So act as to treat humanity, whether in thine own person or in that of any other, in every case as an end withal, never as a means only."[5, at p. 441] However, what it means to treat a person as an *end* and not only as a *means* to another's end may be variously interpreted.

One interpretation of respect for persons is that it requires that we protect others from harm. This interpretation would be in accord with the Hippocratic admonition ". . . to help, or at least, to do no harm." The consent requirement can be justified as a form of helping, or at least not harming, the patient: seeking consent provides a mechanism for ascertaining what the patient might consider a "benefit" and whether the patient considers the anticipated benefits worth the taking of the risks.

However, a focus on benefits and harms does not establish a strict requirement for informed consent; physicians could avoid seeking consent in any situation in which they consider that informing the patient would be more harmful than beneficial. Such reasoning is commonly used to legitimate incomplete disclosure (*infra*).

Another interpretation of respect for persons is that it requires us to leave them alone—even to the point of allowing them to choose activities that might be harmful (*eg*, parachuting). Authors who share this interpretation speak of respecting a person's autonomy, self-determination, liberty, and so on. As stated by Justice Cardozo[1, at p. 526]: ". . . Every human being of adult years and sound mind has a right to determine what shall be done with his (*sic*) own body. . . ." In this interpretation of the principle, the purpose of informed consent is to assure the person's right to choose; the requirement for it cannot be overridden by concern for the possible harm that might be done by informing the person.[b]

AUTONOMY VERSUS PROTECTION

The tensions between protecting persons from harm and respecting their autonomy are reflected in two rather distinct approaches to interpret respect for persons to require the fostering of a covenantal relationship in which subjects are protected by giving truly *informed* consent. Benjamin Freedman,[8] on the other hand, construes the principle as requiring primarily and almost exclusively a respect for the individual's *freedom* to choose, regardless of whether the choice is informed.

For Jonas and Ramsey, the use of persons as means to another's end, as in research, is justified only if they so identify with the purposes of research that they *will* them as their own. Only such absolute identification rectifies the "sacrifice" of the individual for the collective good. Researcher and subject become "co-adventurers." The consent requirement affirms a basic covenantal bond between subject and researcher, and ensures that the person is not used simply as a means to an end.

While the focus on the will of persons suggests a primary interest in their freedom of choice, Jonas and Ramsey also argue that in order to establish a proper covenant (to become truly "co-adventurers"), the subject's consent must be *informed*: it must reflect a genuine appreciation of the purposes and especially of the risks of the research. The less one understands the risks and identifies with the purposes of research, the less valid is one's consent and the less desirable one's participation. Hence, for Jonas and Ramsey respect for the individual's autonomy is always strongly tempered by a concern to protect subjects from harm through ensuring that their consent is informed.

Therefore, the ideal subjects are those maximally able to identify with the goals of research and to understand the risk—namely, researchers themselves![6, at p. 17] Recognizing the need for a larger subject population than this ideal would permit, Jonas proposes a "descending order of permissibility" for the recruitment ("conscription") of volunteers. Therefore, he would be reluctant to allow the use of such vulnerable subjects as the sick.

While neither Jonas nor Ramsey focuses exclusively on patients as subjects, their treatment of informed consent appears to be influenced largely by the medical practice model. The stress on covenantal bonds and on the duty to do no harm is analogous to traditional assumptions about the physician-patient relationship. Research is seen as a violation of covenant because the physician-researcher no longer has the good of the patient-subject at heart and the subject is asked to sacrifice for the good of others.

This interpretation reflects certain assumptions which can be challenged:

1. The medical practice model may not be appropriate for many types of research: *eg,* social and behavioral research.[9, at p. 17]

Even in biomedical research—particularly basic research—there is often no physician-patient relationship between investigator and subject; thus one may question the relevance of discussing either the creation or the violation of these particular covenantal bonds.[c]

2. Research is not generally so risky (so much of a "sacrifice") as Jonas and Ramsey assume.[d]

3. This "ideal" model is probably impracticable.[15]

The most important challenge to the Jonas-Ramsey construction of the requirements for informed consent lies in an alternative interpretation of the principle of respect for persons. Freedman[8] interprets the principle to include a respect for the validity of an expression of will that is not "informed," even if the price of not being "informed" is not being able to "protect" oneself. Overprotection is seen as a form of dehumanization and lack of respect. For example, to classify persons as incompetent in order to "protect" them from their own judgments is the worst form of abuse.

In this interpretation, respect for persons means primarily to let persons alone and respect their expressions of autonomy. Protecting the welfare of persons must not be bought at the price of repudiating their autonomy. "Fully informed consent" is not only unattainable, but in most cases undesirable! Rather, contends Freedman, we should strive for "valid" consent, which entails ". . . only the imparting of that information which the patient/subject requires in order to make a responsible decision." The two fundamental requirements are that the choice be responsible[e] and that it be voluntary. A person may *choose* not to be fully informed, and a decision based on less information (or different information) than another person might consider essential is not to be regarded as a sign of irresponsibility. Thus Freedman opens up the possibility of a "valid yet ignorant consent." Information and protection must not be forced upon potential subjects.

REQUIREMENTS FOR INFORMED CONSENT

According to the Nuremberg Code, to consent to participate in research one must: (a) have the "legal capacity" to give consent, (b) be "so situated as to be able to exercise free power of choice," (c) have "sufficient knowledge" on which to decide, and (d) have "sufficient . . . comprehension" to make an "enlightened" decision. Most commentators agree that compromise of any one of these conditions jeopardizes the ethical acceptability of research. However, they differ on what constitutes a compromise, in part owing to different interpretations of what is required by the underlying principle of respect for persons.

"Free Power of Choice"

The Nuremberg Code proscribes ". . . any element of force . . . or other ulterior form of constraint or coercion . . ." in obtaining consent. Any flagrant coercion—*eg,* as when competent, comprehending persons are forced to submit to research against their expressed wills—clearly renders consent invalid.

Yet, more subtle or indirect "constraints" or "coercions" may obtain when prospective subjects are highly dependent, impoverished, or needy, as exemplified by persons who are confined involuntarily in institutions. Some commentators argue that consent to participate in research is not sufficiently voluntary when it is given (a) to procure financial reward in situations offering few alternatives for remuneration; (b) to seek release from an institution either by evidencing "good behavior" or by ameliorating the condition for which one was confined; or (c) to please physicians or authorities upon whom one's continued welfare depends.[16]

In contrast, Cornell West[17] argues that such indirect constraints (or inner motivations) do not invalidate consent by making it involuntary. True coercion, he asserts, consists in a *threat*—*eg,* a threat to put prisoners in "the hole" unless they cooperate. Consent given under such circumstances would not be voluntary. But offers of reward—*eg,* better living conditions or financial remuneration—are not, strictly speaking, "coercion" and do not render consent involuntary. While West agrees with other commentators that rewards should not be so high as to constitute "undue" or "unfair" inducement and that advantage should not be taken of the confined circumstances of prisoners, mental patients, and other institutionalized persons, his argument nonetheless provides grounds for rejecting an overprotection of confined persons on the basis that they cannot give valid consent. Parenthetically, we note that many prisoners protest such overprotection and do not consider it an appropriate expression of respect for them as persons.

Competence and Comprehension

The Nuremberg Code requires both "legal capacity" to consent and "sufficient understanding" to reach an "enlightened" decision, without specifying what is meant by either. Contemporary interpretations tend to link the two, defining competence in terms of comprehension—*eg,* the ability to evaluate relevant information,[18, at p. 55] to understand the consequences of action,[19, at p. 183] or to reach a decision for rational reasons.[3, at p. 203; 5, at p. 445] However, Joseph Goldstein[20, at p. 15] charges that this link is "pernicious," since refusal to participate in research might be judged "irrational" and used as grounds for declaring a person incompetent. He argues that the purpose of the consent requirement is to guarantee the exercise of free choice, not to judge the rationality of the choice. Here again we see the tension between the desire to protect the individual (by ensuring the "rationality" of choice) and the requirement to respect that individual's freedom of choice.

This tension is further highlighted in debates about the validity of "proxy consent" to do research on subjects who lack the legal capacity to consent. Arguing strictly from the interpretation that respect for persons requires that we let them alone, Ramsey claims that the use of a nonconsenting subject (eg. a child) is wrong whether or not there is risk, simply because it involves an "unconsented touching." "Wrongful touching" is rectified only when it is for the good of the individual, because then the person is treated as an end as well as a means. Hence, proxy consent may be given for nonconsenting subjects only when the research includes therapeutic interventions related to the subject's own recovery.[7, at p. 11]

But in a strict interpretation of leaving alone, the unconsented touching of a competent adult is wrong even if it benefits that person. Why, then, should benefit justify such touching for a child (or other subject unable to give consent)? Richard McCormick proposes that the validity of such interventions rests on the presumption that the person, if capable, *would* consent to therapy. This presumption in turn derives from a person's obligation to seek therapy—an obligation which people possess simply as human beings.[18, at p. 9] Because people have an obligation to seek their own well-being, we presume they *would* consent if they *could* and thus presume also that "proxy consent" for therapeutic interventions will not violate respect for them as persons.

By analogy, people have other obligations, as members of a moral community, to which one would presume their consent; this presumption justifies "proxy consent." One such obligation is to contribute to the general welfare when to do so requires little or no sacrifice. Hence, McCormick concludes that nonconsenting subjects may be used in research not directly related to their own benefit so long as the research fulfills an important social need and involves no discernible risk. In McCormick's view, respecting persons includes recognizing that they are members of a moral community with attendant obligations.

To this, Ramsey counters that children, at least, are not adults with a full range of duties and obligations. Therefore, they have no obligation to contribute to the general welfare and respect for them requires that they be protected from harm and from unconsented touching.

While Freedman[8] adopts Ramsey's premise that a child is not a moral being in the same sense as an adult, his analysis yields a different conclusion. Precisely because children are *not* autonomous they have no right to be let alone. Instead they have a right to custody. Thus the only relevant moral issue is the risk involved in research; the child must be protected from harm. Freedman, therefore, agrees with McCormick that children may be used in research unrelated to their therapy provided it presents them no discernible risk.

The inevitable conflict between the goals of promoting autonomy and providing protection are thus demonstrated in disagreements over both the standards of competence and the use of incompetent subjects.

Disclosure of Information

The Nuremberg Code requires that the subject be told "the nature, duration, and purpose of the experiment; the method and means by which it is to be conducted; all inconveniences and hazards reasonably to be expected; and the effects upon his (*sic*) health or person which may possibly come. . . ." There is no universal agreement on what constitutes "sufficient knowledge" to give an "informed" consent. Many commentators disagree over what sorts of information should be provided.[8] Those who agree on the need for disclosure of information of a particular sort—*eg*, the risks—often disagree on the extent of the information that must be provided. The Nuremberg Code requires explication of hazards "reasonably" to be expected. Does this include an infinitesimal chance of a substantial harm? A substantial chance of an infinitesimal harm?

Disagreements over particulars arise in part from disagreements about underlying standards: is disclosure to be determined by (1) general medical practice or opinion, (2) the requirements of a "reasonable person," or (3) the idiosyncratic judgment of the individual? While the legal trend may be shifting from the first to the second,[24, at p. 25] it may be argued that only the third is truly compatible with the requirement of respect for the autonomy of the individual person.[3, at pp. 31-37;23, at pp. 25-28]

Yet even those who adopt the third standard disagree as to its implications. As noted earlier, Freedman holds that the idiosyncratic judgment of the individual is overriding—to the point that the prospective subject can choose to have less information than a "reasonable" person might require. Veatch, however, argues that anyone refusing to accept as much information as would be expected of a "reasonable person" should not be accepted as a subject.[23, at p. 29]

Professional codes (eg, AMA code) and federal regulations[22] reflect the traditional medical maximum of "do no harm" and the legal doctrine of *therapeutic privilege* according to which a physician may withhold information when in his/her judgment disclosure is either infeasible or potentially harmful.[3, at p. 35] Recent critics have argued that invoking the doctrine of therapeutic privilege to justify withholding of information from a prospective subject in order to assure cooperation in a research project is almost never appropriate; it gives the investigator entirely too much license to serve vested interests by withholding information[h] that might be material to the decision of a prospective subject.[3, at p. 36]

THE CONSENT FORM

Considerations of informed consent in federal regulations and by members of Institutional Review Boards tend to focus on the composition and disposition of consent forms—ie, the *documentation* of the negotiations for informed consent.[11, at pp. 52-65] While the purpose of the *negotiations* for consent is to make operational the requirements of the principle of respect for persons, the purposes of the documentation are different. The most important function of meticulous and formal documentation of informed consent is to protect the interests of the investigator and the institution.[26, at p. 18] The net effect of the documentation may, in fact, be harmful to the interests of the subject: The retention of a signed consent form tends to give the advantage to the investigator in any adversary proceeding; moreover, the availability of such documents in institutional records may lead to violations of privacy and confidentiality.[27, at pp. 74-86] Indeed, Albert Reiss suggests that a change in the current system of documentation might better serve the interests of subjects: Where investigators now retain consent forms signed by subjects, Reiss would prefer that subjects retain statements of responsibility signed by investigators.[27]

CONCLUSIONS AND IMPLICATIONS

The consent requirement is derived from the ethical principle of respect for persons. Most major disagreements over the requirements arise from two conflicting interpretations of the principle: (1) it requires that we protect another from harm, and (2) it requires that we respect another's right to self-determination—ie, to be left alone or to make free choices. Tensions between these conflicting interpretations are reflected in debates over who is sufficiently free to give valid consent, whether subjects must make rational decisions, whether consent is valid if it is not informed, and so on.

The recent trend in the evolution of federal regulation of research reflects an emphasis on the protection interpretation of the principle. This is manifest by, among other things, increasing requirements for formal and meticulous documentation of the negotiations for informed consent (mistakenly thought to provide protection for subjects) as well as proposals to establish consent committees and to assign monitoring functions to Institutional Review Boards. The development of excessive or inappropriate mechanisms for the protection of subjects is contrary to the interpretation that the principle of respect for persons requires that we respect another's right to self-determination.

It will be necessary to clarify further the principle of respect for persons and its interpretation in the requirement to give consent to participate in research. Because we see overprotection as a form of disrespect for persons, we favor an approach to consent that emphasizes the autonomy of prospective subjects.

REFERENCES AND NOTES

1. Katz J: Experimentation with Human Beings, New York, Russell Sage Foundation, 1972.

2. Final Report of the Tuskegee Syphilis Study, Ad Hoc Advisory Panel, Washington, DC, DHEW, Apr 1973.

3. Annas GJ, Glantz LH, Katz BF: The law of informed consent to human experimentation. Prepared for The National Commission for the Protection of Human Subjects of Biomedical and Behavioral Research, Jun 1976.

4. Fried C: Medical Experimentation, Personal Integrity and Social Policy. Amsterdam, North-Holland Publishing Company, 1974.

5. Macklin R, Sherwin S: Experimenting on human subjects, philosophical perspectives. Case Western Reserve Law Rev 25:434–471, 1975.

6. Jonas H: Philosophical reflections on experimenting with human subjects. In Freund PA (ed): Experimentation with Human Subjects. New York, George Braziller, 1970, pp. 1–31.

7. Ramsey P: The Patient as Person. New Haven, Yale University Press, 1970.

8. Freedman B: A moral theory of informed consent. Hastings Center Rep 5 (No. 4):32–39, 1975.

9. Baumrind D: Nature and definition of informed consent in research involving deception. Prepared for The National Commission for the Protection of Human Subjects of Biomedical and Behavioral Research, Jan, 1976.

10. Feinstein AR: Medical ethics and the architecture of clinical research. Clin Pharmacol Therapeut 15:316–334, 1974.

11. Levine RJ: The nature and definition of informed consent in various research settings. Prepared for The National Commission for the Protection of Human Subjects of Biomedical and Behavioral Research, Dec, 1975.

12. Levine RJ: Appropriate guidelines for the selection of human subjects for participation in biomedical and behavioral research. Prepared for The National Commission for the Protection of Human Subjects of Biomedical and Behavioral Research, Feb, 1976.

13. Arnold JD: Alternatives to the use of prisoners in research in the United States. Prepared for The National Commission for the Protection of Human Subjects of Biomedical and Behavioral Research, Mar, 1976.

14. Cardon PV, Dommel FW, Trumble RR: Injuries to research subjects. N Eng J Med 295:650–654, 1976.

15. Ingelfinger FJ: Informed (but uneducated) consent. N Eng J Med 287:465–466, 1972.

16. Branson R: Philosophical perspectives on experimentation with prisoners. Prepared for The National Commission for the Protection of Human Subjects of Biomedical and Behavioral Research, Feb, 1976.

17. West CR: Philosophical perspective on the participation of prisoners in experimental research. Prepared for The National Commission for the Protection of Human Subjects of Biomedical and Behavioral Research, Jan, 1976.

18. Shuman SI: The emotional, medical and legal reasons for the special concern about psychosurgery. In Ayd FJ Jr (ed): Medical, Moral and Legal Issues in Mental Health Care. Baltimore, Williams and Wilkins, 1974.

19. Katz J: Human rights and human experimentation. In Protection of Human Rights in the Light of Scientific and Technological Progress in Biology and Medicine (Proc 8th Round Table of Council for International Organizations of Medical Sciences, 1973), Geneva, World Health Organization, 1974.

20. Goldstein J: On the right of the institutionalized mentally infirm to consent to or refuse to participate as subjects in biomedical and behavioral research. Prepared for The National Commission for the Protection of Human Subjects of Biomedical and Behavioral Research, Feb, 1976.

21. McCormick RA: Proxy consent in the experimentation situation. Perspect Biol Med 18:2–20, 1974.

22. DHEW: Protection of human subjects: technical amendments. Fed Reg 40(50): 11854–11858, Mar 13, 1975.

23. Veatch RM: Three theories of informed consent: philosophical foundations and policy implications. Prepared for The National Commission for the Protection of Human Subjects of Biomedical and Behavioral Research, Feb, 1976.

24. Curran WJ: Ethical issues in short term and long term psychiatric research. In Ayd FJ Jr (ed): Medical, Moral and Legal Issues in Mental Health Care. Baltimore, William and Wilkins, 1974.

25. Ashley BM: Ethics of experimenting with persons. In Schoolar JC, Gaitz CM (eds): Research and the Psychiatric Patient. New York, Brunner and Mazel, 1975.

26. Levine RJ: On the relevance of ethical principles and guidelines developed for research to health services conducted or supported by the Secretary, DHEW. Prepared for The National Commission for the Protection of Human Subjects of Biomedical and Behavioral Research, May, 1976.

27. Reiss AJ: Selected issues in informed consent and confidentiality with special reference to behavioral-social science research-inquiry. Prepared for the National Commission for the Protection of Human Subjects of Biomedical and Behavioral Research, Feb, 1976.

a. It should be recognized that the legal grounding for the requirement for consent to research[3, at pp. 29-37;4, at p. 18-25] is based on the outcome of litigation of disputes arising almost exclusively in the context of the practice of medicine. There is very nearly no case law upon which legal standards for consent to research, as distinguished from practice, might be defined (there is one Canadian case: *Halushka v. University of Saskatchewan*).

b. The law defines, in general, the circumstances under which a patient, or by extension, a research subject, may recover damages for having been wronged or harmed as a consequence of failure to negotiate adequate consent.[3, at p. 29-37] Traditionally, failure to negotiate adequate consent was treated as a *battery* action. In accord with the view that respect for persons requires us to leave them alone, the law of battery makes it wrong *a priori* to touch, treat, or do research upon a person without the person's consent. Whether or not harm befalls the person is irrelevant; it is the "unconsented-to touching" that is wrong.

The modern trend in malpractice litigation is to treat cases based upon failure to obtain proper consent as *negligence* rather than battery actions. The negligence doctrine combines elements of patient benefit and self-determination. To bring a negligence action, a patient/subject must prove: that the physician had a *duty* toward the patient; that the duty was *breached;* that *damage* occurred to the patient; and that the damage was *caused* by the breach. In contrast to battery actions, negligence actions remove as a basis for the requirement for consent the simple notion that "unconsented-to touching" is a wrong—rather, such touching is wrong (actionable) only if it is negligent and results in harm; otherwise, the patient/subject cannot recover damages.

Under both battery and negligence doctrines, consent is invalid if any information is withheld that might be considered material to the decision to give consent.

c. Whether or not negotiations for informed consent to research should be conducted according to different standards than consent to practice is a controversial matter. Feinstein[10] has observed that it is our custom to adhere to a "double standard": "An act that receives no special concern when performed as part of clinical practice may become a major ethical or legal issue if done as part of a formally designed investigation." In his view there is less need for formality in the negotiations for informed consent to a relationship where the interests of research and practice are conjoined— eg, as in research conducted by a physician-investigator having the aim of demonstrating the safety and/or efficacy of a non-validated therapeutic maneuver—than when the only purpose of the investigator-subject relationship is to perform research. Capron,[1, at p. 574] on the other hand, has asserted: "Higher requirements for informed consent should be imposed in therapy than in investigation, particularly when an element of honest experimentation is joined with therapy." Levine[11, at pp. 41-42] concluded that patients are entitled to the same degree of thoroughness of negotiations for informed consent as are subjects. However, patients may be offered the opportunity to delegate decision-making authority to a physician while subjects should rarely be offered this option. The most important distinction is that the prospective subject should be informed that in research, as contrasted with practice, the subject will be at least in part a means and perhaps only a means to another's end.

d. Most research does not present risk of physical or psychological harm; rather, it calls upon the subject to assume a burden that might more appropriately be named inconvenience.[12, at p. 8-11] The researcher calls upon subjects to give of their time, to perform tasks that might be tedious or embarrassing, or to experience transitory pains or other discomfort. The risk of physical or psychological harm from Phase I drug testing, for example, has been estimated as slightly greater than that of being an office secretary, $1/7$ that of window washers and $1/9$ that of miners.[13, at p. 18] For biomedical research generally, the following estimates have been reported[14]: In basic research (called "nontherapeutic" by the authors) the risk of being disabled either temporarily or permanently is substantially less than that of being similarly harmed by an accident. The risks of dying or becoming disabled in what the authors call "therapeutic research" are much higher; however, they seem much less than the risks of similar unfortunate outcomes for comparable patients in other medical settings involving no research.

e. A full elaboration of Freedman's use of the word *responsible* is beyond the scope of this essay. Interested readers are referred to his paper.[8]

f. This argument is based on the erroneous assumption—which Ramsey shares with many other authors—that research can be dichotomized into two distinct sets: "therapeutic" and "nontherapeutic." Authors using this classification usually argue that, since the purpose of "therapeutic research" is to benefit the subject, it is more easily justified than "nontherapeutic research." These arguments fail because they do not recognize that activities commonly referred to as "therapeutic research" generally include maneuvers designed to benefit persons other than the subject—*eg*, randomization is designed to develop generalizable knowledge about the safety and efficacy of therapies. The many problems that have been created by the categorization of research as "therapeutic" and "nontherapeutic" are elaborated in two essays in a forthcoming issue of the *Villanova Law Review:* (1) Lebacqz, K.: Some reflections on *Report and Recommendations: Research on the Fetus;* (2) Levine, R.J.: The impact on fetal research of the report of The National Commission for the Protection of Human Subjects of Biomedical and Behavioral Research.

g. Codes developed subsequent to Nuremberg have modified its requirements in various ways: disclosure of the purpose of research is often ignored, and other stipulations such as disclosure of the availability of alternative modes of treatment are added. Federal regulations[22] require (1) a fair explanation of procedures; (2) disclosure of risks; (3) explanation of benefits; (4) description of alternatives; (5) an offer to answer questions; and (6) a statement that the subject may withdraw at any time.

While these requirements have the force of law, they are by no means exhaustive of possible standards for disclosure. To them one might add: a statement of overall purpose; a clear invitation to participate in research, distinguishing maneuvers required for research purposes from those necessary for therapy; an explanation of why the particular person is invited (selected); a suggestion that the prospective subject might wish to discuss the research with another person; and an explanation (when appropriate) of the fact that the risks are unknown (much biomedical research is designated to determine the nature, probability and magnitude of harms that might be produced by a therapeutic maneuver).[11] Veatch[23] would add the names of members of any review boards that had approved the research, an explanation of who is responsible for harm done, and an explanation of the right, if any, to continue receiving treatments found useful.

h. Similarly, in research activities contingent upon subjects' lack of awareness of purposes or procedures it has been thought permissible to either withhold information or practice deliberate deception provided harms are minimized and subjects "debriefed" (given a full explanation) afterwards. Baumrind[9] opposes deceptive practices arguing not only that they violate the principle of respect for persons but also that eventually they will invalidate research on scientific grounds. Various proposals have been made to minimize the need for and harmful effects of deceptive practices: subjects might be invited to consent to incomplete disclosure with a promise of full disclosure at the termination of the research[11, at pp. 30-31]; subjects might be told as much as possible and asked to consent for specified limits of time and risk[25, at p. 19]; and consent based on full disclosure might be negotiated with mock or "surrogate" populations.[12, at pp. 26-29]

20 Deception and Informed Consent in Research

ALAN SOBLE

The practice of deceiving some subjects in a social science or other research protocol seems to contradict one of the very core elements of the concept of informed consent: informing the subject. In this article, Soble surveys a variety of approaches which argue either for not obtaining consent prior to participation in a protocol involving deception or provides strategies by which some form of consent can be obtained by the subject prior to participation in the experiment. This survey article provides a good guide for thinking one's way through the many problematic issues surrounding deception in research.

Alan Soble, Ph.D., is with the Department of Philosophy, University of New Orleans

The principle of informed consent generally includes two necessary conditions for the proper treatment of human subjects in experimentation. The first condition, which has been widely discussed, is that the consent be obtained from subjects who agree to participate *voluntarily*, where voluntarism is understood negatively as the absence of coercion. The second condition, which is less often discussed, is that the consent must be *informed*. The Articles of the Nuremberg Tribunal and The Declaration of Helsinki both state that the subjects must be told the duration, methods, possible risks, and the purpose or aim of the experiment. The most recent HEW regulations agree: informed consent has not been obtained if there has been any element of deceit or fraud. These guidelines reflect our ordinary moral view that deception is morally unacceptable.

During the past quarter-century the size of the scientific research establishment has vastly increased. Medical, sociological, and psychological research is being carried out at our universities and other institutions at a rapid rate. The success of this effort, measured in terms of the amount of knowledge gained, has been well documented. In some of this research, however, the human beings serving as subjects are deceived as to the purpose of the experiment. In social psychology, for example, the incidence of the use of deceptive research designs has been estimated to be as high as 38 percent,[1] and even though deception is less common in medical research, many examples are available.[2] One immediate response is to say that "the experiment ought not to be performed and the desired knowledge should be sought by means of a different research design."[3] But this response overlooks the crucial point that certain bits of knowledge cannot, for logical reasons alone, be obtained without the use of deception. The testing of some hypotheses, within both psychology and medicine, requires that the subjects not be informed of the purpose or aim of the experiment being conducted.

We are faced then with a moral dilemma. Since the search for knowledge is at least morally permissible (if not, to a certain extent, morally obligatory), and since the use of deception is morally unacceptable (at least on a *prima facie* basis), in some situations both moral desiderata cannot be satisfied. And it is not clear which moral value ought to be sacrificed for the sake of the other.

I am aware of a handful of proposed solutions for this dilemma. First, we can maintain that subjects ought to be told the full purpose of an experiment for which they have volunteered. In this view no experiments logically requiring deception are permissible. Second, we can say that the subject's knowing the purpose of an experiment is not a central element of informed consent and therefore that experiments using deception are always permissible, as long as they satisfy the other conditions of the principle of informed consent. These are the two extreme solutions.[4]

The other positions are more complicated. According to one proposal, ineliminable deception is permissible *only* when there are substantial paternalistic reasons for withholding the purpose of the experiment from the subjects. According to another proposal, deception is permissible *only* when there are firm utilitarian reasons for doing so. This view, the standard argument for the use of deception, claims that the knowledge to be gained from deceptive experiments is so valuable to society that it is only a minor defect that persons must be deceived in the process. Finally, a number of strategies have been recently proposed to resolve the dilemma. These include the method of *ex post facto* consent (getting approval of the subjects retroactively), the method of presumptive consent (getting approval from a group of mock subjects and inferring that the real subjects would have consented), the method of prior general consent (in effect, getting consent to deceptive procedures well before the experiment is actually conducted), and a method that combines prior general consent and proxy consent. Before I discuss each of these ways of resolving the dilemma, let me examine briefly the major presupposition that underlies the dilemma.

ARE DECEPTIVE DESIGNS NECESSARY?

Many types of experiments seem to require that the subjects not be told the purpose of the study, and in some cases that they be induced to hold false beliefs about the nature of the experiment during the experiment itself. Experiments, for example, that are designed to yield information about the influence of expectation or other psychological factors on the psychoactivity of drugs or on physiological processes would be ruined if subjects were told that what was being studied was their "mental" contribution to drug effects. Often subjects are not merely ignorant as to whether they have received the drug being tested or a placebo, but rather are told that they will receive one of these but in fact receive the other. Similarly, experiments designed to test for the existence of psychological phenomena such as obedience and trustworthiness seem necessarily to involve the use of deception.[5] Telling subjects that what is being studied is, for example, the extent to which they conform with the judgments of persons who are really cohorts of the experimenter, will destroy the attempt to discover the extent of conformity.

But how does one go about proving that deception is required in order to obtain certain bits of knowledge? In some cases, of course, it is quite easy. If what we want to know is something like "the effect of LSD-25 on the behavior of a group of unsuspecting enlisted men," it is quite obvious that the subjects must be deceived in order to assure that they are unsuspecting. This is an easy case because the statement of the relevant hypothesis being tested will include reference to deceived persons. But the hypothesis that persons will tend to judge in conformity with the judgments of persons in their immediate vicinity does not contain a reference to deceived subjects. It *seems* obvious that knowledge about conformity requires that we deceive subjects, but how can this intuition be supported?

Certainly, there is a way of proving that a given case of the use of deception was *not* required: all we need to do is to construct an experiment that is designed to yield the same information but that does not involve any deception. But the failure to find such an alternative nondeceptive research design does not prove that the deception *was* required by the nature of the knowledge being sought. The failure may only show how unimaginative we are in constructing research designs. (This is very ironic. Some deceptive research designs are extraordinarily ingenious.[6]) At least for this reason we ought not to take lightly the claim that deceptive research is ultimately justifiable because deceptive designs are necessary.

Knowledge that can be obtained only by using a deceptive research design must be contrasted with knowledge that can be obtained without deception but that can be obtained more efficiently by employing deception. Deception that is motivated out of a need to secure enough subjects, or deception that is pragmatically useful in terms of conserving time, effort, and expense, is not generally deception that is required purely on account of the nature of the knowledge being sought.[7] In these cases, of course, the dilemma I outlined earlier does not arise. But we have to be careful, for there is the danger that if we do allow deception because it is logically necessary for the testing of specific hypotheses, then it becomes slightly more plausible to argue that deception that saves the experimenter (and society) time, effort, and money also should be permitted. One major fault of the paternalistic and utilitarian solutions to the dilemma is that they also tend to justify deception that is only pragmatically, and not logically, required.

Before discussing the various proposals, I would like to comment on two related issues. First, the dilemma as I have stated it involves the acceptability of deceiving subjects who know, at least, that they are subjects in an experiment. What they do not know is precisely what experiment they are subjects in. Experiments done especially within sociology, however, involve deception in which the subjects do not even know that they are subjects in the experiment (covert observation, for example).[8] I will not discuss the issue of the morality of this practice here, for in covert observation there is apparently not simply a violation of the "informed" condition of the principle of informed consent, but also a violation of the "voluntary" condition. Discussion of this issue would take us too far from the resolution of the dilemma.

Second, my discussion is meant only to examine the acceptability of deception in thoroughly experimental situations, and is not meant to bear upon the use of deception by physicians and others in situations that are purely therapeutic. Therefore, my conclusion on the acceptability of deceptive techniques in experimentation does not necessarily apply to the morality of placebo therapy and the practice of lying to patients in the course of treating disease or disability.

The Paternalistic Defense of Deception

A defender of research employing ineliminable deception might try to justify the deception by relying on an argument like this: the deceptive procedures employed in these experiments can be viewed as being therapeutic for the subjects, and since there are many contexts in which the principle of informed consent is temporarily abandoned for the sake of persons who need therapy (for example, unconscious adults requiring emergency treatment), it ought to be acceptable temporarily to ignore the principle in these experiments.[9]

The argument, however, does not provide an adequate way of resolving the dilemma. First, it is not a global justification of the use of ineliminable deception, for it would only justify a small percentage of the experiments in question, those in which some real benefit to the subjects could be demonstrated. But there are more serious problems with the argument. It assumes much too quickly that the experimenter who plans to deceive subjects can know that the subjects will agree that the deception is in fact beneficial for them. Even if it is true, however, that the subjects do agree that the deception is beneficial for them (by, for example, exposing to them certain psychological traits they have but would rather not have), this does not mean that the deception was also therapeutic. Possessing certain psychological traits may not be beneficial for a person, but possessing these traits does not constitute being unhealthy, and therefore procedures that tend to expose and to remove these traits cannot be called "therapeutic." But even if it makes some sense to say that the removal of certain psychological traits is therapeutic, whether a procedure that removes these traits is therapeutic will depend on the context in which the procedure is carried out. Persons presenting themselves at a physician's office or at a clinic acknowledge that the context is a therapeutic one, but this acknowledgement is absent when persons volunteer for experiments.

Finally, the paternalistic argument carries the danger of justifying not only ineliminable deception but also deception motivated out of a concern to conserve time, effort, and expense. The paternalistic argument can be extended to something like this: deception that conserves the experimenter's time and effort, which enables the experimenter to carry out the research less expensively, ultimately is beneficial for each of the *individual* subjects, who are of course taxpayers. (The paternalistic argument might also say that the money saved could be reallocated and used in, for example, cancer research. The deception then can be tied to a therapeutic intent.) But an argument like this would justify more deception than we would find comfortable.

The Utilitarian Defense of Deception

A utilitarian justification of the use of ineliminable deception is far superior to a paternalistic one because it does not have to blur the distinction between therapy and experiments. It is also more plausible because it argues that the deception is acceptable because it promises to benefit in many cases the whole of society and not merely the individuals who participate as subjects. Simply put, the utilitarian argues that experiments utilizing ineliminable deception and that do *not* cause any other harm to subjects are acceptable because the knowledge gained from them is socially valuable. When balancing the needs of society and a desire of individual subjects not to be deceived, experiments that do contribute substantially to our knowledge are justified.

Like the paternalistic argument, the utilitarian argument does not provide a global justification of experiments employing ineliminable deception. If the deceptive procedure is accompanied by the possibility of grave harm to the subjects, then the needs of society no longer overshadow the needs of the subjects. Or if the deceptive procedure is part of an experiment that is designed to yield only trivial knowledge, there is no longer a justification for the experiment. Even so, the utilitarian argument has the potential for justifying most of the experiments recently carried out that employ deception. But I do not find the argument to be a convincing one.

One weakness of the utilitarian argument can be seen by examining what the argument has to prove. Certainly we can agree that the scientific research establishment as a whole is to be justified on utilitarian grounds. The reason we spend so much time, effort, and money on research is that the research as a whole is bound to have beneficial results for society and the individuals who make up society. But the utilitarian argument has to prove that because the research establishment itself is justified on utilitarian grounds, anything else done within that establishment is also justified by utilitarian considerations. In particular, the utilitarian argument has the burden of showing that not only the principle of informed consent is ultimately based on utilitarian considerations, but also that exceptions to the principle (for example, the use of ineliminable deception) are also grounded in utilitarian considerations.

This argument may be very difficult to prove. For example, one who relied on a notion of "rules" developed by John Rawls[10] might say that, yes, the principle of informed consent can be justified on utilitarian grounds, and yes, the research establishment can be similarly justified. But this Rawlsian would go on to claim that a system of experimentation on human subjects that publicly included a rule permitting violations of the principle of informed consent (by allowing ineliminable deception) would not be justified on utilitarian grounds, given certain sociological, psychological, economic, and ideological properties of this society. For example, public outrage at the use of deception could undermine the status of the experimental scientists and eventually result in a curtailment of research and a subsequent decrease in the knowledge that the research establishment provides. The use of deception then from a utilitarian point of view would be counterproductive. The heavy burden of the utilitarian is thus to show that it is unreasonable to believe that the proposed modification of the principle of informed consent would be counterproductive in this way.

Alternatively, the utilitarian can drop the requirement that a rule permitting violations of the principle of informed consent must be public. This modification might eliminate the possibility that deception is counterproductive. But I think that this alternative is inadequate, for at least three reasons. First, it relies on the assumption, which is likely false, that this deception (the failure to make a rule public) is especially immune from discovery. Indeed, it could turn out that this deception, when exposed, is more counterproductive than the deception permitted by the public rule in certain experiments. Second, it solves the dilemma about the use of deception in experiments by introducing deception at another level, and therefore in a sense just begs the question. Third, in allowing rules to be nonpublic, it seems to violate one of the so-called logical or conceptual requirements of morality.

The utilitarian argument is also not convincing because it ignores the history or the genesis of the principle of informed consent.[11] The whole point of the principle (including its prohibition of deception) is precisely to protect individual rights against exactly these sorts of claims of social need or benefit. Those who designed and those who now support the principle freely admit that there might be utilitarian reasons for not always obeying the principle, but they announce that the individual has a sphere of autonomy that cannot be sacrificed or invaded for the good of society. To try to justify experiments employing ineliminable deception on utilitarian grounds would be to deny the intent and significance of a principle that has only recently appeared in our history, that took much effort to develop and apply, and that represents one of the major advances of modern society. In a word, the utilitarian argument proposes that we undo the moral progress we have made in this century.

Finally, like the paternalistic argument, the utilitarian argument would justify too much deception. If experiments employing ineliminable deception are acceptable for utilitarian reasons, then what about deception motivated by pragmatic considerations? Indeed, the utilitarian justification of merely pragmatic deception seems stronger than the corresponding paternalistic justification of pragmatic deception. Again we end up allowing more deception than we find comfortable, and the utilitarian argument therefore does not provide a safe way of resolving the dilemma.[12]

EX POST FACTO CONSENT: ANOTHER DEFENSE

Stanley Milgram has tried to justify the ineliminable deception used in his obedience experiments this way:

> Misinformation . . . [and] illusion . . . are justified for one reason only; they are, in the end, accepted and endorsed by those who are exposed to them. . . . *The central moral justification for allowing a procedure of the sort used in my experiments is that it is judged acceptable by those who have taken part in it.*[13]

The thrust of this argument is that the principle of informed consent can be modified to allow for consent being obtained after a procedure has been carried out on subjects, rather than before, in order to permit the successful execution of deceptive techniques vital to the knowledge being sought. I want to argue that such a modification is not acceptable.

Consider first the possibility that after the experiment is over, some of the subjects do not agree that they should have been deceived. We already know that violations of the principle of informed consent, when the principle is understood in its usual way as requiring consent prior to an experiment, have been increasingly met by claims of subjects that they deserve compensation and that punitive measures be directed at the experimenters. But an experimenter who wants to rely on *ex post facto* consent is in a rather shaky position. If any subject withholds consent afterwards, this subject was a participant in an experimental procedure to which he or she *never* consented. Their failure to agree means that such subjects never should have been exposed to the experimental procedures. They would have, I think, a good argument for compensation. The experimenter who relies on *ex post facto* consent therefore faces practical problems that endanger the continuance of the research, professional standing, and perhaps financial status.[14] But even if compensation is exacted, this does not mean that it was morally correct that the procedures were carried out. The 1.3 percent of Milgram's subjects who expressed disapproval afterwards were morally wronged, and it is no defense to say that 98.7 percent found the deception acceptable.

Consider now the situation in which all the subjects do approve afterwards. Is it safe to say that this postexperimental approval counts as *bona fide* consent? Steven Patten has argued that the approval of Milgram's subjects was not really consent.[15] We know, says Patten, from Milgram's study that persons are submissive to authority; this gives us reason to think that when the subjects approved afterwards they were merely obeying (once again) and not really consenting. Although Patten has a point here, I think his argument is too strong. If the subjects in Milgram's experiments are representative of people in general, then *no* experiments at all on human subjects would be permissible. For if we take Patten's point seriously, it means that we ought not to believe any subject who shows up for an experiment; the subject's showing up is just an act of obedience and real consent cannot be obtained. From Patten's argument, then, we can only draw the conclusion that Milgram's study is only as objectionable as any other study. It does not provide a way of singling out deceptive experiments as especially objectionable.

The approval given by the subjects afterwards might not really be consent because the experimental procedures themselves elicit the approval of the subjects or make it difficult for them not to agree afterwards. Consent, whether it be prior or retroactive, ought to be *independent* approval of the experimental procedures. (One might want to interpret the requirement of independence simply as the requirement that consent be given voluntarily.) I don't want to argue that *ex post facto* consent is always nonindependent, but rather that many experimental procedures do contain ingredients that elicit the subsequent approval of the subjects. Because obtaining consent prior to an experiment is the best way to ensure that consent is independent, we ought not to allow deceptive procedures that can be given approval only, if at all, in retrospect.

In the case of Milgram's studies, it is plausible that persons who have been exposed (to themselves and, quasi-publicly, to the experimenters) as obedient and as unfaithful to their own moral beliefs will be embarrassed and shamed by this exposure and will attempt to alleviate their unpleasant position by agreeing afterwards that the experimenters were correct to have used deception. Even those subjects who were not obedient during the course of the experiment have good reasons for giving approval afterwards. To disapprove of the deception would be to undermine their fine performance. Thus for both obedient and defiant subjects the nature of the experiment provides powerful psychological reasons (self-respect, exculpation, self-righteousness) for giving approval afterwards.[16]

What is astounding is that Milgram recognizes the influence of just these kinds of psychological factors on other features of his experiment but not on the credibility of retrospective approval. Commenting on the fact that 3.8 percent of the obedient subjects *later* said that they were certain that the learner was not receiving real shocks, Milgram writes:

> Even now I am not willing to dismiss those subjects because it is not clear that their rejection of the technical illusion was a cause of their obedience or a consequence of it. Cognitive processes may serve to rationalize behavior. . . . [S]ome subjects may have come to this position as a post facto explanation. It cost them nothing and would go a long way toward preserving their positive self-conception.[17]

If the doctrine of *ex post facto* consent is to be taken seriously, surely the burden of proof is on the one who wants to justify the use of deception in research to demonstrate convincingly that retrospective approval has in no way been manufactured by the experimental procedures. It is necessary to eliminate the possibility that the procedures for which consent is requested do not themselves elicit that consent.[18]

PRESUMPTIVE CONSENT: A PROPOSAL

Robert Veatch has proposed a method for resolving the dilemma and thereby for permitting ineliminable deception:

> In those rare, special cases where knowledge of the purpose would destroy the experiment . . . it might be acceptable to ask a group of mock subjects drawn from the same experimental population if they would consent to participate in the experiment knowing its purpose. If there is substantial agreement (say, 95 percent), then it seems reasonable to conclude that most real subjects would have agreed to participate even if they had had the information that would destroy the experiment's validity.[19]

It seems to me that Veatch's proposal is just as controversial as the dilemma it was intended to resolve. His method relies on our assuming that real subjects *would* consent on the basis of what other persons *do* consent to. Although in some cases (for example, when a person in need of treatment is temporarily unconscious) we allow that the next-of-kin consent to therapy assuming that the patient would consent, the experimental situation is too far removed from these emergency cases for this kind of hypothetical reasoning to be compelling.

An objection that was raised to the doctrine of *ex post facto* consent can be raised in this context also. Why does Veatch settle for only 95 percent agreement from the mock subjects? This figure suggests that

Veatch is willing to expose 5 percent of the subject population to procedures to which they would not have consented, had they known
the information that they in fact do not know. If we were to use his
method, we ought to set the level of mock subject agreement at 100
percent. In a typical nondeceptive experiment, there are some persons
who, having heard the terms of the experiment, decide not to participate. This percentage can be weeded out at the start. But in an experiment involving ineliminable deception and governed by presumptive
consent, if the agreement of the mock subjects is only 95 percent,
then those persons who would have been weeded out are not going to
be weeded out. (A side thought: regarding the "voluntary" element of
the principle of informed consent, would we allow an experiment in
which we have evidence that only 95 percent of the subjects *would*
have consented?)

Of course, if the criterion of mock subject agreement is set as
high as 100 percent then perhaps in practice Veatch's proposal will
mean that very few deceptive experiments will be carried out. This
possibility does not fit very well with Veatch's apparent sympathy
with the attempt to gain knowledge with deceptive and otherwise
nonharmful experiments. Furthermore, even if all the mock subjects
give approval, this in no way guarantees that all the real subjects
would have consented had they known the purpose of the experiment. Even when the criterion is set as high as it can be, there is still a
possibility that some subjects will be exposed to procedures to which
they would not have consented.

Note also that Veatch's proposal has a loose tie with Milgram's
suggestion of *ex post facto* consent. If we were to use a group of real
subjects in a deceptive experiment on the grounds that mock subjects
have given their approval, we are using real subjects on the strength
of a claim that is somewhat testable. It would be nice to know, afterwards, whether the percentage of mock subjects who agreed was a reliable indicator of the percentage of real subjects who would have
agreed. In order to obtain this information we must seek the approval
of the real subjects *ex post facto*. But it is possible that this check on the
accuracy of the prediction, however, may very well be contaminated
by the influence of the experimental procedures themselves.[20]

PRIOR GENERAL CONSENT: STILL ANOTHER METHOD

Milgram has proposed another method designed to satisfy the
principle of informed consent and at the same time make deceptive
experiments possible:

[Prior general consent] is a form of consent that would be based
on subjects' knowing the general types of procedures used in

psychological investigations, but without their knowing what specific manipulations would be employed in the particular experiment in which they would take part. The first step would be to create a pool of volunteers to serve in psychology experiments. Before volunteering to join the pool people would be told explicitly that sometimes subjects are misinformed about the purposes of an experiment. . . . Only persons who had indicated a willingness to participate in experiments involving deception . . . would, in the course of the year, be recruited for experiments that involved [deception].[21]

This proposal fails to resolve the dilemma because it creates a new dilemma: experiments based upon this proposal *either* will yield no useful information at all *or* will require that additional experiments be performed which do violate the principle of informed consent (see fn. 20). If an experiment relying on this technique for recruitment yields no useful information, then the technique has not preserved one of the original goals we had: to secure valuable information. And if experiments relying on this technique require (for their validity, as I argue below) that further experiments be done which do involve violations of the principle of informed consent, then the technique has not preserved our other goal.

The first question is whether we could have reason to believe that any deception eventually carried out on this subject pool really worked and really provided us with the information we were seeking. At first glance, if people know or think that the procedures to which they have consented involve deception, then they will be more suspicious of the experimental protocol and may very well not be "tricked" in the necessary way. Even when subjects are *not* told in advance that the experiment involves deception, there is always some doubt as to whether the illusion was successful and whether the experiment has generated any useful knowledge. But if the technique of prior general consent is used, the subjects know in advance that they might be exposed to deception, and this knowledge makes the success of the illusion even more problematic. In order to show that the knowledge gained in experiments relying on prior general consent is useful, experimenters will have to demonstrate that subjects' foreknowledge of the deception did not interfere with the success of the illusion. And, as far as I can tell, to establish this kind of fact one must resort to deceptive procedures.[22] If this is so, prior general consent does not solve the dilemma. One might say here that even though it is true that only those subjects who generally consented to deception are used in deceptive experiments, this pool of subjects does not have to know this fact about itself. Withholding *this* information from the subjects would certainly bypass the problems I just mentioned but at the cost of replacing one act of deception with another.

The second question is whether the information gained in an experiment relying on this technique is useful because the only subjects who are exposed to deceptive procedures are those who express a willingness to be exposed to deception. Application of this technique, that is, restricts the nature of the subject pool and may possibly insert a bias into the characteristics of the research population. This complaint is raised often in the context of sex research; the information obtained by studying only those who volunteer for sex experiments may be misleading because the research population is lacking other kinds of persons. If deceptive procedures are carried out only upon those who are willing to undergo deception, then psychological studies may be misleading because they have excluded from the research population those people who were not willing to undergo deception and who may have different personality structures or profiles than willing subjects. Again, in order to show that such a bias is not present, it is quite likely that deceptive experiments *not* relying on prior general consent will have to be carried out. At the very least, the burden of proof is on the experimenter who relies on prior general consent to establish that the knowledge is not contaminated by either of the two factors I mentioned. And in establishing this, the experimenter must not conduct experiments that violate the principle of informed consent.

Prior General Consent and Proxy Consent

I have so far rejected all but one of the more complicated ways of resolving the dilemma. In addition to the method that I am about to describe, then, the only positions left are the two extreme views. According to one, no experiments involving ineliminable deception are permissible; according to the other, all such experiments are permissible. This latter alternative wants to decrease substantially the significance of the "informed" condition of the principle of informed consent. But there is very little that can be said in favor of doing so. I have already suggested that paternalistic and utilitarian arguments for exceptions to the prohibition on the use of deception are inadequate. But paternalistic and utilitarian reasons are the only ones we could have for decreasing the significance of the "informed" condition. The second extreme solution, then, is in practice no different from the solutions proposed by the paternalist or by the utilitarian. There are simply no other arguments to use in defending the second extreme solution.

There are of course perfectly good reasons for accepting the first extreme solution. Experiments without deception respect those individuals who have already volunteered to be subjects at least in part for the sake of other people. Conversely, experiments with deception show disrespect for these persons who have willingly undertaken the risks of an experiment so that other persons might benefit. Deceiving an experimental subject who has volunteered is an acute expression of ingratitude. And it deserves the scorn that we ordinarily give to the person who passes through the cafeteria line twice but pays only once. There is, however one final method that seems to satisfy our requirements; it allows some ineliminable deception, and so preserves the search for knowledge, without (1) expressing ingratitude to the subjects and (2) undermining the epistemological status of the data collected during the experiment. In this method prior general consent is combined with proxy consent.[23]

I suggest that we make the method of prior general consent applicable to the whole realm of experimental science employing human subjects. If the method of prior general consent is employed for any and every subject pool, the likelihood that forewarning of deception will disrupt the experimental illusion is greatly decreased. In this method, furthermore, the experimental bias introduced in Milgram's proposal (only those subjects who consented in general to deception would be used in deceptive experiments) is overcome in the following way. Subjects are *not* told that only those who approve of deception will be used in experiments utilizing deception; rather, all subjects are candidates for participating in deceptive experiments. But the usual objection to doing this is vitiated by the use of an additional procedure: proxy consent. Each subject in the pool designates some relative or friend as one who will inspect the experiment in which the subject might participate. This relative or friend is empowered by the subject to reject or accept experiments on the basis of whether they posed too much risk, employed deception that was too devious, or was aimed at providing knowledge that might be misused. The proxy makes these judgments from the point of view of the subject who has empowered him or her to do so. It is important to note that combining the method of prior general consent with that of proxy consent combines what is acceptable from both Milgram's and Veatch's proposals. From Milgram's it takes the idea that consent to deception is compatible with the principle of informed consent; from Veatch's proposal it takes the idea that we can resolve the dilemma by consulting persons other than the subjects themselves. But the method of proxy consent used as a conjunct to prior general consent has an obvious advantage to Veatch's proposal: the necessity of having to argue from the approval of mock subjects to the hypothetical approval of real subjects is eliminated by consulting persons empowered by the real subjects to give consent for them.

A procedure employing both prior general consent (as standard for all subject pools) and proxy consent is very far removed from what exists at the present: the use of deception in experiments without the protection for subjects of either prior general consent or proxy consent. For this reason many changes will have to be made in the structure of experimental science using human subjects; so many changes, in fact, that I suspect that the initial reaction of experimental scientists will be that the proposal is impractical, that it will create too many bureaucratic impediments to the conduct of research. Indeed, the experimental scientist could argue that the method, in solving the original dilemma, gives rise to a new dilemma. Either we employ the method of prior general/proxy consent (and abandon a large part of the research enterprise because the method is too costly in terms of time, effort, and money), or we retain the large bulk of the research enterprise (but employ a less ethically satisfying method of obtaining the approval of the subjects). My response to this argument would be to say that as long as we reject the paternalistic and utilitarian arguments for the use of ineliminable deception because those justifications could very well justify deception required only for pragmatic reasons, then we must also be prepared to embrace the relative inefficiency of the method of general/proxy consent. Pragmatic considerations, we had decided, are not compelling enough to warrant the less-than-full satisfaction of the principle of informed consent.

REFERENCES

1. See Donald Warwick, "Social Scientists Ought to Stop Lying," *Psychology Today* (February 1975), p. 105; and Jay Katz, *Experimentation with Human Beings* (New York: Russell Sage Foundation, 1972), pp. 323–433, esp. p. 358. A very recent example is the experiment done by Diane Ruble, "Premenstrual Symptoms: A Reinterpretation," *Science* 197 (1977), 291–292.

2. Three recent examples are: Robert Heaton, "Subject Expectancy and Environmental Factors as Determinants of Psychedelic Flashback Experience," *Journal of Nervous and Mental Disease* 161 (1975), 157–65; Monte S. Buchsbaum, Robert D. Coursey, and Dennis L. Murphy, "The Biochemical High-Risk Paradigm: Behavioral and Familial Correlates of Low Platelet Monoamine Oxidase Activity," *Science* 194 (1976), 339–41; and C. P. O'Brien, Thomas Testa, T. J. O'Brien, J. P. Brady, and Barbara Wells, "Conditioned Narcotic Withdrawal in Humans," *Science* 195 (1977), pp. 1000–02.

3. Sissela Bok, "The Ethics of Giving Placebos," *Scientific American* 231 (November 1974), 17–23.

4. Perry London says that "neither extreme position will do for those of us who are equally concerned with the need for valid scientific information and for the protection of human subjects" (*Psychology Today* [November 1977], p. 23), and he laments the fact that the current HEW guidelines "offer no clues as to when some amount of deception may be necessary and proper." The guidelines however do seem to be rather explicit in prohibiting "any element of force, fraud, deceit, duress . . ." (*Federal Register, 39*, No. 105, at section 46.3c).

5. For example: Seymour Feshbach and Robert Singer, *Television and Aggression* (San Francisco: Jossey-Bass, 1971); C. K. Hofling, E. Brotzman, S. Dalrymple, N. Graves, and C. M. Pierce, "An Experimental Study in Nurse-Physician Relationships," *Journal of Nervous and Mental Disease* 143 (1966), 171–80; Stanley Milgram, *Obedience to Authority* (New York: Harper & Row, 1974); L. Paige, "The Effects of Oral Contraceptives on Affective Fluctuations Associated With the Menstrual Cycle," *Psychosomatic Medicine* 33 (1971), 515–37. An interesting discussion and a number of references can be found in S. Wolf, "The Pharmacology of Placebos," *Pharmacological Review*, 11 (1959), 689-704.

6. For an experiment involving extensive deception of different types, see Stuart Valins, "Cognitive Effects of False Heart-Rate Feedback," *Journal of Personality and Social Psychology* 4 (1966) 400–08. Incidentally, Valin's comment on p. 401 is noteworthy. "Male introductory psychology students, whose course *requirements* included 6 hours of participation in experiments, *volunteered* for a psychophysiological experiment" (italics added).

7. One might want to interpret the San Antonio oral contraceptive study in this way. For discussion, see Robert Veatch, "Medical Ethics: Professional or Universal?" *Harvard Theological Review 65* (1972), 550, and his " 'Experimental' Pregnancy," *Hastings Center Report*, 1 (1971), 2–3, and the editors' note in Robert Hunt and John Arras, eds., *Ethical Issues in Modern Medicine* (Encino, Cal.: Mayfield, 1977), p. 266. Some deception in personality studies is carried out because the investigators fear, ironically, that the subjects would otherwise lie when answering questionnaires. Is this deception logically required by the knowledge being sought, or is it only pragmatically useful in obtaining that knowledge?

8. For example: D. Rosenhan, "On Being Sane in Insane Places," *Science* 179 (1973) 250–59; and Laud Humphreys, *Tearoom Trade. Impersonal Sex in Public Places* (Chicago: Aldine, 1975), enlarged edition. A general critique of Humphreys is presented by Donald Warwick, "Tearoom Trade: Means and Ends in Social Research," *Hastings Center Studies* 1 (1973), 27–38; and Murray Wax has recently discussed problems of consent in sociology, in "Fieldworkers and Research Subjects: Who Needs Protection?" *Hastings Center Report* 7, No. 4 (August 1977), 29–32.

9. See the remarks of some of Milgram's subjects after they had been debriefed, in *Obedience to Authority*, pp. 196 and 200, and the "Foreword" to *Tearoom Trade* by Lee Rainwater, pp. xiv–xv.

10. John Rawls, "Two Concepts of Rules," *Philosophical Review* 64 (1955), 3–32. For a similar position see H. L. A. Hart, "Prolegomenon to the Principles of Punishment," in *Punishment and Responsibility* (New York: Oxford University Press, 1968), pp. 1–27.

11. A concise history can be found in Alan Donagan, "Informed Consent in Therapy and Experimentation," *Journal of Medicine and Philosophy* 2 (1977), 307–29.

12. It might be argued that some studies employing ineliminable deception are permissible or even mandatory because they are designed to expose or produce facts

about the occurrence of harmful or immoral activities (for example, the studies done by Rosenhan and Hofling). But this utilitarian argument justifies only a very small percentage of the experiments involving ineliminable deception. Furthermore, there are dangers in this rationale, in that the argument not only comes close to justifying police entrapment but also suggests that policework is a legitimate activity for scientists.

13. *Obedience to Authority,* pp. 198–99. Milgram's position is repeated in his "Subject Reaction: The Neglected Factor in the Ethics of Experimentation," *Hastings Center Report* 7, No. 5 (August 1977). 21.

14. Murray Wax (in "Fieldworkers and Research Subjects," p. 32) does not mention compensation to subjects when *ex post facto* consent is not obtained, but only punitive measures taken against the experimenter by professional peers.

15. "The Case That Milgram Makes," *Philosophical Review* 86 (1977), 350–64.

16. The fact that Milgram's subjects were given supportive debriefing may also have influenced their retroactive approval. See A. K. Ring, et al., "Mode of Debriefing as a Factor Affecting Subjective Reaction to a Milgram-type Experiment—An Ethical Inquiry," in Katz, *Experimentation on Human Beings,* pp. 395–99.

17. *Obedience to Authority,* pp. 173–74. See also *ibid.,* p. 204.

18. An example from ordinary life will help explain my point. Consider a man who persists with amorous advances, as in attempted seduction, in spite of the fact that the woman is verbally and physically resisting. If later, having finally given in, she expresses approval, we cannot take her approval at face value because it has been induced by the procedure she has approved of. Her approval is not independent of the procedure being judged.

19. "Ethical Principles in Medical Experimentation," in *Ethical and Legal Issues of Social Experimentation,* A. M. Rivlin and P. M. Timpane, eds. (Washington: The Brookings Institution, 1975), p. 52. Milgram also makes this suggestion, in "Subject Reaction," p. 23.

20. This last objection to Veatch's proposal is similar to an objection raised by Milgram to the technique known as "role-playing." In role-playing, the subject is fully informed but is asked to go through the experimental protocol *as if* he or she didn't know the relevant information. Clearly, role-playing does not involve any violation of the principle of informed consent. But, as Milgram points out, "we must still perform the crucial experiment [with deception] to determine whether role-played behavior corresponds to nonrole-played behavior" ("Subject Reaction," p. 23). What the experimental scientist needs to do, then, is to show *without* the use of deception that role-played and nonrole-played behavior closely correspond, and this may be impossible to do.

21. "Subject Reaction," p. 23.

22. See, for example, L. J. Stricker, S. Messick, and D. N. Jackson, "Suspicion of Deception: Implications for Conformity Research," *Journal of Personality and Social Psychology* 5 (1967), pp. 379–89; and Z. Rubin and J. C. Moore, Jr., "Assessment of Subjects' Suspicions," *Journal of Personality and Social Psychology* 17 (1971) 163–70.

23. This method was suggested to me by Professor Richard T. Hull, Department of Philosophy, SUNY/Buffalo.

Ethical Dilemmas In
Obtaining Informed Consent/Further Reading

Callan, Dennis E. "Patients' Rights and Informed Consent: An Emergency Case For Hospitals?" *California Western Law Review* 12 (Winter, 1976), 406–28.

Freedman, Benjamin. "A Moral Theory of Informed Consent." *Hastings Center Report.* 5 (August, 1975), 32–39.

Kelman, Herbert C. "The Rights of the Subject in Social Research: An Analysis in Terms of Relative Power and Legitimacy." *American Psychologist* 27 (November, 1972), 989–1016.

Macklin, Ruth. "Consent, Coercion, and Conflicts of Rights." *Perspectives in Biology and Medicine* 20 (Spring, 1977), 360–71.

Morris, R. Curtis, et al. "Guidelines for Accepting Volunteers: Consent, Ethical Implications, and The Function of Peerly View." *Clinical Pharmacology and Therapeutics* 13 (September, 1972), 782–802.

Patten, Stephen C. "The Case That Milgram Makes." *The Philosophical Review* (July, 1977), 350–63.

The following reports have all been issued by the National Commission for the Protection of Human Subjects Biomedical and Behavioral Research, and are available from the United States Printing Office.

1. Disclosure of Research Information Under the Freedom of Information Act. DHEW (OS) 77-003.

2. Report and Recommendations: Psychosurgery. DHEW (OS) 77-001.

3. Report and Recommendations: Research Involving Children. DHEW (OS) 77-004.

4. Report and Recommendations: Research Involving Prisoners. DHEW (OS) 76-131.

5. Report and Recommendations: Research Involving Those Institutionalized as Mentally Infirm. DHEW (OS) 78-006.

6. Report and Recommendations: Research On The Fetus. DHEW (OS) 76-127.

7. Report and Recommendations: Institutional Review Boards. DHEW (OS) 78-008.

8. Report and Recommendations: Ethical Guidelines for the Delivery of Health Services by DHEW. DHEW (OS) 78-0010.

PART SIX

GENETIC ENGINEERING AND GENETIC POLICY

Leon R. Kass
Tabitha M. Powledge
and John Fletcher
Key Dismukes

21 The New Biology: What Price Relieving Man's Estate?

1971

Leon R. Kass, M.D.

The basic dilemma in the new biology, Kass suggests, is that now the engineers can be engineered. This potential for self-recreation or self-replication presents a variety of personal and social ethical problems. Kass approaches the problem by reviewing the new biomedical technologies and some of the ethical and social problems inherent in them. In his conclusion, Kass focuses on the potential problem of dehumanization and an empty concept of what is humanly good. By providing a blend of scientific and ethical data, Kass offers a good foundation for a discussion of these issues as well as presenting his own approach to the problems.

Recent advances in biology and medicine suggest that we may be rapidly acquiring the power to modify and control the capacities and activities of men by direct intervention and manipulation of their bodies and minds. Certain means are already in use or at hand, others await the solution of relatively minor technical problems, while yet others, those offering perhaps the most precise kind of control, depend upon further basic research. Biologists who have considered these matters disagree on the question of how much how soon, but all agree that the power for "human engineering," to borrow from the jargon, is coming and that it will probably have profound social consequences.

These developments have been viewed both with enthusiasm and with alarm; they are only just beginning to receive serious attention. Several biologists have undertaken to inform the public about the technical possibilities, present and future. Practitioners of social science "futurology" are attempting to predict and describe the likely social consequences of and public responses to the new technologies. Lawyers and legislators are exploring institutional innovations for assessing new technologies. All of these activities are based upon the hope that we can harness the new technology of man for the betterment of mankind.

Yet this commendable aspiration points to another set of questions, which are, in my view, sorely neglected—questions that inquire into the meaning of phrases such as the "betterment of mankind." A *full* understanding of the new technology of man requires an exploration of ends, values, standards. What ends will or should the new techniques serve? What values should guide society's adjustments? By what standards should the assessment agencies assess? Behind these questions lie others: what is a good man, what is a good life for man, what is a good community? This article is an attempt to provoke discussion of these neglected and important questions.

While these questions about ends and ultimate ends are never unimportant or irrelevant, they have rarely been more important or more relevant. That this is so can be seen once we recognize that we are dealing here with a group of technologies that are in a decisive respect unique: the object upon which they operate is man himself. The technologies of energy or food production, of communication, of manufacture, and of motion greatly alter the implements available to man and the conditions in which he uses them. In contrast, the biomedical technology works to change the user himself. To be sure, the printing press, the automobile, the television, and the jet airplane have greatly altered the conditions under which and the way in which men live; but men as biological beings have remained largely unchanged. They have been, and remain, able to accept or reject, to use and abuse these technologies; they choose, whether wisely or foolishly, the ends to which these technologies are means. Biomedical technology may make it possible to change the inherent capacity for choice itself. Indeed, both those who welcome and those who fear the advent of "human engineering" ground their hopes and fears in the same prospect: *that man can for the first time recreate himself.*

Engineering the engineer seems to differ in kind from engineering his engine. Some have argued, however, that biomedical engineering does not differ qualitatively from toilet training, education, and moral teachings—all of which are forms of so-called "social engineering," which has man as its object, and is used by one generation to mold the next. In reply, it must at least be said that the techniques which have hitherto been employed are feeble and inefficient when compared to those on the horizon. This quantitative difference rests in part on a qualitative difference in the means of intervention. The traditional influences operate by speech or by symbolic deeds. They pay tribute to man as the animal who lives by speech and who understands the meanings of actions. Also, their effects are, in general, reversible, or at least subject to attempts at reversal. Each person has greater or lesser power to accept or reject or abandon them. In contrast, biomedical engineering circumvents the human context of speech and meaning, bypasses choice, and goes directly to work to modify the human material itself. Moreover, the changes wrought may be irreversible.

In addition, there is an important practical reason for considering the biomedical technology apart from other technologies. The advances we shall examine are fruits of a large, humane project dedicated to the conquest of disease and the relief of human suffering. The biologist and physician, regardless of their private motives, are seen, with justification, to be the well-wishers and benefactors of mankind. Thus, in a time in which technological advance is more carefully scrutinized and increasingly criticized, biomedical developments are still viewed by most people as benefits largely without qualification. The price we pay for these developments is thus more likely to go unrecognized. For this reason, I shall consider only the dangers and costs of biomedical advance. As the benefits are well known, there is no need to dwell upon them here. My discussion is deliberately partial.

I begin with a survey of the pertinent technologies. Next, I will consider some of the basic ethical and social problems in the use of these technologies. Then, I will briefly raise some fundamental questions to which these problems point. Finally, I shall offer some very general reflections on what is to be done.

The Biomedical Technologies

The biomedical technologies can be usefully organized into three groups, according to their major purpose: (i) control of death and life, (ii) control of human potentialities, and (iii) control of human achievement. The corresponding technologies are (i) medicine, especially the arts of prolonging life and of controlling reproduction, (ii) genetic engineering, and (iii) neurological and psychological manipulation. I shall briefly summarize each group of techniques.

1) *Control of death and life*. Previous medical triumphs have greatly increased average life expectancy. Yet other developments, such as organ transplantation or replacement and research into aging, hold forth the promise of increasing not just the average, but also the maximum life expectancy. Indeed, medicine seems to be sharpening its tools to do battle with death itself, as if death were just one more disease.

More immediately and concretely, available techniques of prolonging life—respirators, cardiac pacemakers, artificial kidneys—are already in the lists against death. Ironically, the success of these devices in forestalling death has introduced confusion in determining that death has, in fact, occurred. The traditional signs of life—heartbeat and respiration—can now be maintained entirely by machines. Some physicians are now busily trying to devise so-called "new definitions of death," while others maintain that the technical advances show that death is not a concrete event at all, but rather a gradual process, like twilight, incapable of precise temporal localization.

The real challenge to death will come from research into aging and senescence, a field just entering puberty. Recent studies suggest that aging is a genetically controlled process, distinct from disease, but one that can be manipulated and altered by diet or drugs. Extrapolating from animal studies, some scientists have suggested that a decrease in the rate of aging might also be achieved simply by effecting a very small decrease in human body temperature. According to some estimates, by the year 2000 it may be technically possible to add from 20 to 40 useful years to the period of middle life.

Medicine's success in extending life is already a major cause of excessive population growth: death control points to birth control. Although we are already technically competent, new techniques for lowering fertility and chemical agents for inducing abortion will greatly enhance our powers over conception and gestation. Problems of definition have been raised here as well. The need to determine when individuals acquire enforceable legal rights gives society an interest in the definition of human life and of the time when it begins. These matters are too familiar to need elaboration.

Technologies to conquer infertility proceed alongside those to promote it. The first successful laboratory fertilization of human egg by human sperm was reported in 1969 (1). In 1970, British scientists learned how to grow human embryos in the laboratory up to at least the blastocyst stage [that is, to the age of 1 week (2)]. We may soon hear about the next stage, the successful reimplantation of such an embryo into a woman previously infertile because of oviduct disease. The development of an artificial placenta, now under investigation, will make possible full laboratory control of fertilization and gestation. In addition, sophisticated biochemical and cytological techniques of monitoring the "quality" of the fetus have been and are being developed and used. These developments not only give us more power over the generation of human life, but make it possible to manipulate and to modify the quality of the human material.

2) *Control of human potentialities.* Genetic engineering, when fully developed, will wield two powers not shared by ordinary medical practice. Medicine treats existing individuals and seeks to correct de-

viations from a norm of health. Genetic engineering, in contrast, will be able to make changes that can be transmitted to succeeding generations and will be able to create new capacities, and hence to establish new norms of health and fitness.

Nevertheless, one of the major interests in genetic manipulation is strictly medical: to develop treatments for individuals with inherited diseases. Genetic disease is prevalent and increasing, thanks partly to medical advances that enable those affected to survive and perpetuate their mutant genes. The hope is that normal copies of the appropriate gene, obtained biologically or synthesized chemically, can be introduced into defective individuals to correct their deficiencies. This *therapeutic* use of genetic technology appears to be far in the future. Moreover, there is some doubt that it will ever be practical, since the same end could be more easily achieved by transplanting cells or organs that could compensate for the missing or defective gene product.

Far less remote are technologies that could serve *eugenic* ends. Their development has been endorsed by those concerned about a general deterioration of the human gene pool and by others who believe that even an undeteriorated human gene pool needs upgrading. Artificial insemination with selected donors, the eugenic proposal of Herman Muller (3), has been possible for several years because of the perfection of methods for long-term storage of human spermatozoa. The successful maturation of human oocytes in the laboratory and their subsequent fertilization now make it possible to select donors of ova as well. But a far more suitable technique for eugenic purposes will soon be upon us—namely, nuclear transplantation, or cloning. Bypassing the lottery of sexual recombination, nuclear transplantation permits the asexual reproduction or copying of an already developed individual. The nucleus of a mature but unfertilized egg is replaced by a nucleus obtained from a specialized cell of an adult organism or embryo (for example, a cell from the intestines or the skin). The egg with its transplanted nucleus develops as if it had been fertilized and, barring complications, will give rise to a normal adult organism. Since almost all the hereditary material (DNA) of a cell is contained within its nucleus, the renucleated egg and the individual into which it develops are genetically identical to the adult organism that was the source of the donor nucleus. Cloning could be used to produce sets of unlimited numbers of genetically identical individuals, each set derived from a single parent. Cloning has been successful in amphibians and is now being tried in mice; its extension to man merely requires the solution of certain technical problems.

Production of man-animal chimeras by the introduction of selected nonhuman material into developing human embryos is also expected. Fusion of human and nonhuman cells in tissue culture has already been achieved.

Other, less direct means for influencing the gene pool are already available, thanks to our increasing ability to identify and diagnose genetic diseases. Genetic counselors can now detect biochemically and cytologically a variety of severe genetic defects (for example, Mongolism, Tay-Sachs disease) while the fetus is still in utero. Since treatments are at present largely unavailable, diagnosis is often followed by abortion of the affected fetus. In the future, more sensitive tests will also permit the detection of heterozygote carriers, the unaffected individuals who carry but a single dose of a given deleterious gene. The eradication of a given genetic disease might then be attempted by aborting all such carriers. In fact, it was recently suggested that the fairly common disease cystic fibrosis could be completely eliminated over the next 40 years by screening all pregnancies and aborting the 17,000,000 unaffected fetuses that will carry a single gene for this disease. Such zealots need to be reminded of the consequences should each geneticist be allowed an equal assault on his favorite genetic disorder, given that each human being is a carrier for some four to eight such recessive, lethal genetic diseases.

3) Control of human achievement. Although human achievement depends at least in part upon genetic endowment, heredity determines only the material upon which experience and education impose the form. The limits of many capacities and powers of an individual are indeed genetically determined, but the nurturing and perfection of these capacities depend upon other influences. Neurological and psychological manipulation hold forth the promise of controlling the development of human capacities, particularly those long considered most distinctively human: speech, thought, choice, emotion, memory, and imagination.

These techniques are now in a rather primitive state because we understand so little about the brain and mind. Nevertheless, we have already seen the use of electrical stimulation of the human brain to produce sensations of intense pleasure and to control rage, the use of brain surgery (for example, frontal lobotomy) for the relief of severe anxiety, and the use of aversive conditioning with electric shock to treat sexual perversion. Operant-conditioning techniques are widely used, apparently with success, in schools and mental hospitals. The use of so-called consciousness-expanding and hallucinogenic drugs is widespread, to say nothing of tranquilizers and stimulants. We are promised drugs to modify memory, intelligence, libido, and aggressiveness.

The following passages from a recent book by Yale neurophysiologist José Delgado—a book instructively entitled *Physical Control of the Mind: Toward a Psychocivilized Society*—should serve to make this discussion more concrete. In the early 1950's, it was discovered that, with electrodes placed in certain discrete regions of their brains, animals would repeatedly and indefatigably press levers to stimulate their own brains, with obvious resultant enjoyment. Even starving animals preferred stimulating these so-called pleasure centers to eating. Delgado comments on the electrical stimulation of a similar center in a human subject (4, p. 185).

[T]he patient reported a pleasant tingling sensation in the left side of her body 'from my face down to the bottom of my legs.' She started giggling and making funny comments, stating that she enjoyed the sensation 'very much.' Repetition of these stimulations made the patient more communicative and flirtatious, and she ended by openly expressing her desire to marry the therapist.

And one further quotation from Delgado (4, p. 88).

Leaving wires inside of a thinking brain may appear unpleasant or dangerous, but actually the many patients who have undergone this experience have not been concerned about the fact of being wired, nor have they felt any discomfort due to the presence of conductors in their heads. Some women have shown their feminine adaptability to circumstances by wearing attractive hats or wigs to conceal their electrical headgear, and many people have been able to enjoy a normal life as outpatients, returning to the clinic periodically for examination and stimulation. In a few cases in which contacts were located in pleasurable areas, patients have had the opportunity to stimulate their own brains by pressing the button of a portable instrument, and this procedure is reported to have therapeutic benefits.

It bears repeating that the sciences of neurophysiology and psychopharmacology are in their infancy. The techniques that are now available are crude, imprecise, weak, and unpredictable, compared to those that may flow from a more mature neurobiology.

Basic Ethical and Social Problems in the Use of Biomedical Technology

After this cursory review of the powers now and soon to be at our disposal, I turn to the questions concerning the use of these powers. First, we must recognize that questions of use of science and technology are always moral and political questions, never simply technical ones. All private or public decisions to develop or to use biomedical technology—and decisions *not* to do so—inevitably contain judgments about value. This is true even if the values guiding those decisions are not articulated or made clear, as indeed they often are not. Secondly, the value judgments cannot be derived from biomedical science. This is true even if scientists themselves make the decisions.

These important points are often overlooked for at least three reasons.

1) They are obscured by those who like to speak of "the control of nature by science." It is men who control, not that abstraction "science." Science may provide the means, but men choose the ends; the choice of ends comes from beyond science.

2) Introduction of new technologies often appears to be the result of no decision whatsoever, or of the culmination of decisions too small or unconscious to be recognized as such. What can be done is done. However, someone is deciding on the basis of some notions of desirability, no matter how self-serving or altruistic.

3) Desires to gain or keep money and power no doubt influence much of what happens, but these desires can also be formulated as reasons and then discussed and debated.

Insofar as our society has tried to deliberate about questions of use, how has it done so? Pragmatists that we are, we prefer a utilitarian calculus: we weigh "benefits" against "risks," and we weigh them for both the individual and "society." We often ignore the fact that the very definitions of "a benefit" and "a risk" are themselves based upon judgments about value. In the biomedical areas just reviewed, the benefits are considered to be self-evident: prolongation of life, control of fertility and of population size, treatment and prevention of genetic disease, the reduction of anxiety and aggressiveness, and the enhancement of memory, intelligence, and pleasure. The assessment of risk is, in general, simply pragmatic—will the technique work effectively and reliably, how much will it cost, will it do detectable bodily harm, and who will complain if we proceed with development? As these questions are familiar and congenial, there is no need to belabor them.

The very pragmatism that makes us sensitive to considerations of economic cost often blinds us to the larger social costs exacted by biomedical advances. For one thing, we seem to be unaware that we may not be able to maximize all the benefits, that several of the goals we are promoting conflict with each other. On the one hand, we seek to control population growth by lowering fertility; on the other hand, we develop techniques to enable every infertile woman to bear a child. On the one hand, we try to extend the lives of individuals with genetic disease; on the other, we wish to eliminate deleterious genes from the human population. I am not urging that we resolve these conflicts in favor of one side or the other, but simply that we recognize that such conflicts exist. Once we do, we are more likely to appreciate that most "progress" is heavily paid for in terms not generally included in the simply utilitarian calculus.

To become sensitive to the larger costs of biomedical progress, we must attend to several serious ethical and social questions. I will briefly discuss three of them: (i) questions of distributive justice; (ii) questions of the use and abuse of power, and (iii) questions of self-degradation and dehumanization.

Ethical / Social Questions

Distributive Justice

The introduction of any biomedical technology presents a new instance of an old problem—how to distribute scarce resources justly. We should assume that demand will usually exceed supply. Which people should receive a kidney transplant or an artificial heart? Who should get the benefits of genetic therapy or of brain stimulation? Is "first-come, first-served" the fairest principle? Or are certain people "more worthy," and if so, on what grounds?

just allocation

It is unlikely that we will arrive at answers to these questions in the form of deliberate decisions. More likely, the problem of distribution will continue to be decided ad hoc and locally. If so, the consequence will probably be a sharp increase in the already far too great inequality of medical care. The extreme case will be longevity, which will probably be, at first, obtainable only at great expense. Who is likely to be able to buy it? Do conscience and prudence permit us to enlarge the gap between rich and poor, especially with respect to something as fundamental as life itself?

Questions of distributive justice also arise in the earlier decisions to acquire new knowledge and to develop new techniques. Personnel and facilities for medical research and treatment are scarce resources. Is the development of a new technology the best use of the limited resources, given current circumstances? How should we balance efforts aimed at prevention against those aimed at cure, or either of these against efforts to redesign the species? How should we balance the delivery of available levels of care against further basic research? More fundamentally, how should we balance efforts in biology and medicine against efforts to eliminate poverty, pollution, urban decay, discrimination, and poor education? This last question about dis-

problem of prior decision

tribution is perhaps the most profound. We should reflect upon the social consequences of seducing many of our brightest young people to spend their lives locating the biochemical defects in rare genetic diseases, while our more serious problems go begging. The current squeeze on money for research provides us with an opportunity to rethink and reorder our priorities.

Problems of distributive justice are frequently mentioned and discussed, but they are hard to resolve in a rational manner. We find them especially difficult because of the enormous range of conflicting values and interests that characterizes our pluralistic society. We cannot agree—unfortunately, we often do not even try to agree—on standards for just distribution. Rather, decisions tend to be made largely out of a clash of competing interests. Thus, regrettably, the question of how to distribute justly often gets reduced to who shall decide how to distribute. The question about justice has led us to the question about power.

Use and Abuse of Power

We have difficulty recognizing the problems of the exercise of power in the biomedical enterprise because of our delight with the wondrous fruits it has yielded. This is ironic because the notion of power is absolutely central to the modern conception of science. The ancients conceived of science as the *understanding* of nature, pursued for its own sake. We moderns view science as power, as *control* over nature; the conquest of nature "for the relief of man's estate" was the charge issued by Francis Bacon, one of the leading architects of the modern scientific project (5).

Another source of difficulty is our fondness for speaking of the abstraction "Man." I suspect that we prefer to speak figuratively about "Man's power over Nature" because it obscures an unpleasant reality about human affairs. It is in fact particular men who wield power, not Man. What we really mean by "Man's power over Nature" is a power exercised by some men over other men, with a knowledge of nature as their instrument.

While applicable to technology in general, these reflections are especially pertinent to the technologies of human engineering, with which men deliberately exercise power over future generations. An excellent discussion of this question is found in *The Abolition of Man*, by C. S. Lewis (6).

It is, of course, a commonplace to complain that men have hitherto used badly, and against their fellows, the powers that science has given them. But that is not the point I am trying to make. I am not speaking of particular corruptions and abuses which an increase of moral virtue would cure: I am considering what the thing called "Man's power over Nature" must always and essentially be. . . .

In reality, of course, if any one age really attains, by eugenics and scientific education, the power to make its descendants what it pleases, all men who live after it are the patients of that power. They are weaker, not stronger: for though we may have put wonderful machines in their hands, we have pre-ordained how they are to use them. . . . The real picture is that of one dominant age . . . which resists all previous ages most successfully and dominates all subsequent ages most irresistibly, and thus is the real master of the human species. But even within this master generation (itself an infinitesimal minority of the species) the power will be exercised by a minority smaller still. Man's conquest of Nature, if the dreams of some scientific planners are realized, means the rule of a few hundreds of men over billions upon billions of men. There neither is nor can be any simple increase of power on Man's side. Each new power won *by* man is a power *over* man as well. Each advance leaves him weaker as well as stronger. In every victory, besides being the general who triumphs, he is also the prisoner who follows the triumphal car.

Please note that I am not yet speaking about the problem of the misuse or abuse of power. The point is rather that the power which grows is unavoidably the power of only some men, and that the number of powerful men decreases as power increases.

Specific problems of abuse and misuse of specific powers must not, however, be overlooked. Some have voiced the fear that the technologies of genetic engineering and behavior control, though developed for good purposes, will be put to evil uses. These fears are perhaps somewhat exaggerated, if only because biomedical technologies would add very little to our highly developed arsenal for mischief, destruction, and stultification. Nevertheless, any proposal for large-scale human engineering should make us wary. Consider a program of positive eugenics based upon the widespread practice of asexual reproduction. Who shall decide what constitutes a superior individual worthy of replication? Who shall decide which individuals may or must reproduce, and by which method? These are questions easily answered only for a tyrannical regime.

Concern about the use of power is equally necessary in the selection of means for desirable or agreed-upon ends. Consider the desired end of limiting population growth. An effective program of fertility control is likely to be coercive. Who should decide the choice of means? Will the program penalize "conscientious objectors"?

Serious problems arise simply from obtaining and disseminating information, as in the mass screening programs now being proposed for detection of genetic disease. For what kinds of disorders is compulsory screening justified? Who shall have access to the data obtained, and for what purposes? To whom does information about a person's genotype belong? In ordinary medical practice, the patient's privacy is protected by the doctor's adherence to the principle of confidentiality. What will protect his privacy under conditions of mass screening?

More than privacy is at stake if screening is undertaken to detect psychological or behavioral abnormalities. A recent proposal, tendered and supported high in government, called for the psychological testing of all 6-year-olds to detect future criminals and misfits. The proposal was rejected; current tests lack the requisite predictive powers. But will such a proposal be rejected if reliable tests become available? What if certain genetic disorders, diagnosable in childhood, can be shown to correlate with subsequent antisocial behavior? For what degree of correlation and for what kinds of behavior can mandatory screening be justified? What use should be made of the data? Might not the dissemination of the information itself undermine the individual's chance for a worthy life and contribute to his so-called antisocial tendencies?

Consider the seemingly harmless effort to redefine clinical death. If the need for organs for transplantation is the stimulus for redefining death, might not this concern influence the definition at the expense of the dying? One physician, in fact, refers in writing to the revised criteria for declaring a patient dead as a "new definition of heart donor eligibility" (7, p. 526).

Problems of abuse of power arise even in the acquisition of basic knowledge. The securing of a voluntary and informed consent is an abiding problem in the use of human subjects in experimentation. Gross coercion and deception are now rarely a problem; the pressures are generally subtle, often related to an intrinsic power imbalance in favor of the experimentalist.

A special problem arises in experiments on or manipulations of the unborn. Here it is impossible to obtain the consent of the human subject. If the purpose of the intervention is therapeutic—to correct a known genetic abnormality, for example—consent can reasonably be implied. But can anyone ethically consent to nontherapeutic interventions in which parents or scientists work their wills or their eugenic visions on the child-to-be? Would not such manipulation represent in itself an abuse of power, independent of consequences?

There are many clinical situations which already permit, if not invite, the manipulative or arbitrary use of powers provided by biomedical technology: obtaining organs for transplantation, refusing to let a person die with dignity, giving genetic counselling to a frightened couple, recommending eugenic sterilization for a mental retardate, ordering electric shock for a homosexual. In each situation, there is an opportunity to violate the will of the patient or subject. Such opportunities have generally existed in medical practice, but the dangers are becoming increasingly serious. With the growing complexity of the technologies, the technician gains in authority, since he alone can understand what he is doing. The patient's lack of knowledge makes him deferential and often inhibits him from speaking up when he feels threatened. Physicians *are* sometimes troubled by their increasing power, yet they feel they cannot avoid its exercise. "Reluctantly," one commented to me, "we shall have to play God." With what guidance and to what ends I shall consider later. For the moment, I merely ask: "By whose authority?"

While these questions about power are pertinent and important, they are in one sense misleading. They imply an inherent conflict of purpose between physician and patient, between scientist and citizen. The discussion conjures up images of master and slave, or oppressor and oppressed. Yet it must be remembered that conflict of purpose is largely absent, especially with regard to general goals. To be sure, the purposes of medical scientists are not always the same as those of the subjects experimented on. Nevertheless, basic sponsors and partisans of biomedical technology are precisely those upon whom the technology will operate. The will of the scientist and physician is happily married to (rather, is the offspring of) the desire of all of us for better health, longer life, and peace of mind.

Most future biomedical technologies will probably be welcomed, as have those of the past. Their use will require little or no coercion. Some developments, such as pills to improve memory, control mood, or induce pleasure, are likely to need no promotion. Thus, even if we should escape from the dangers of coercive manipulation, we shall still face large problems posed by the voluntary use of biomedical technology, problems to which I now turn.

Voluntary Self-Degradation and Dehumanization

Modern opinion is sensitive to problems of restriction of freedom and abuse of power. Indeed, many hold that a man can be injured only by violating his will. But this view is much too narrow. It fails to recognize the great dangers we shall face in the use of biomedical technology, dangers that stem from an excess of freedom, from the uninhibited exercises of will. In my view, our greatest problem will increasingly be one of voluntary self-degradation, or willing dehumanization.

Certain desired and perfected medical technologies have already had some dehumanizing consequences. Improved methods of resuscitation have made possible heroic efforts to "save" the severely ill and injured. Yet these efforts are sometimes only partly successful; they may succeed in salvaging individuals with severe brain damage, capable of only a less-than-human, vegetating existence. Such patients, increasingly found in the intensive care units of university hospitals, have been denied a death with dignity. Families are forced to suffer seeing their loved ones so reduced, and are made to bear the burdens of a protracted death watch.

Even the ordinary methods of treating disease and prolonging life have impoverished the context in which men die. Fewer and fewer people die in the familiar surroundings of home or in the company of family and friends. At that time of life when there is perhaps the greatest need for human warmth and comfort, the dying patient is kept company by cardiac pacemakers and defibrillators, respirators, aspirators, oxygenators, catheters, and his intravenous drip.

But the loneliness is not confined to the dying patient in the hospital bed. Consider the increasing number of old people who are still alive, thanks to medical progress. As a group, the elderly are the most alienated members of our society. Not yet ready for the world of the dead, not deemed fit for the world of the living, they are shunted aside. More and more of them spend the extra years medicine has given them in "homes for senior citizens," in chronic hospitals, in nursing homes—waiting for the end. We have learned how to increase their years, but we have not learned how to help them enjoy their days. And yet, we bravely and relentlessly push back the frontiers against death.

Paradoxically, even the young and vigorous may be suffering because of medicine's success in removing death from their personal experience. Those born since penicillin represent the first generation ever to grow up without the experience or fear of probable unexpected death at an early age. They look around and see that virtually all of their friends are alive. A thoughtful physician, Eric Cassell, has remarked on this in "Death and the physician" (8, p. 76):

> [W]hile the gift of time must surely be marked as a great blessing, the *perception* of time, as stretching out endlessly before us, is somewhat threatening. Many of us function best under deadlines, and tend to procrastinate when time limits are not set. . . . Thus, this unquestioned boon, the extension of life,

and the removal of the threat of premature death, carries with it an unexpected anxiety: the anxiety of an unlimited future.

In the young, the sense of limitless time has apparently imparted not a feeling of limitless opportunity, but increased stress and anxiety, in addition to the anxiety which results from other modern freedoms: personal mobility, a wide range of occupational choice, and independence from the limitations of class and familial patterns of work. . . . A certain aimlessness (often ringed around with great social consciousness) characterizes discussions about their own aspirations. The future is endless, and their inner demands seem minimal. Although it may appear uncharitable to say so, they seem to be acting in a way best described as "childish"—particularly in their lack of a time sense. They behave as though there were no tomorrow, or as though the time limits imposed by the biological facts of life had become so vague for them as to be nonexistent.

Consider next the coming power over reproduction and genotype. We endorse the project that will enable us to control numbers and to treat individuals with genetic disease. But our desires outrun these defensible goals. Many would welcome the chance to become parents without the inconvenience of pregnancy; others would wish to know in advance the characteristics of their offspring (sex, height, eye color, intelligence); still others would wish to design these characteristics to suit their tastes. Some scientists have called for the use of the new technologies to assure the "quality" of all new babies (9). As one obstetrician put it: "The business of obstetrics is to produce *optimum* babies." But the price to be paid for the "optimum baby" is the transfer of procreation from the home to the laboratory and its coincident transformation into manufacture. Increasing control over the product is purchased by the increasing depersonalization of the process. The complete depersonalization of procreation (possible with the development of an artificial placenta) shall be, in itself, seriously dehumanizing, no matter how optimum the product. It should not be forgotten that human procreation not only issues new human beings, but is itself a human activity.

Procreation is not simply an activity of the rational will. It is a more complete human activity precisely because it engages us bodily and spiritually, as well as rationally. Is there perhaps some wisdom in that mystery of nature which joins the pleasure of sex, the communication of love, and the desire for children in the very activity by which we continue the chain of human existence? Is not biological parenthood a built-in "mechanism," selected because it fosters and supports in parents an adequate concern for and commitment to their children? Would not the laboratory production of human beings no longer be *human* procreation? Could it keep human parenthood human?

The dehumanizing consequences of programmed reproduction extend beyond the mere acts and processes of life-giving. Transfer of procreation to the laboratory will no doubt weaken what is presently for many people the best remaining justification and support for the existence of marriage and the family. Sex is now comfortably at home outside of marriage; child-rearing is progressively being given over to the state, the schools, the mass media, and the child-care centers. Some have argued that the family, long the nursery of humanity, has outlived its usefulness. To be sure, laboratory and governmental alternatives might be designed for procreation and child-rearing, but at what cost?

This is not the place to conduct a full evaluation of the biological family. Nevertheless, some of its important virtues are, nowadays, too often overlooked. The family is rapidly becoming the only institution in an increasingly impersonal world where each person is loved not for what he does or makes, but simply because he is. The family is also the institution where most of us, both as children and as parents, acquire a sense of continuity with the past and a sense of commitment to the future. Without the family, we would have little incentive to take an interest in anything after our own deaths. These observations suggest that the elimination of the family would weaken ties to past and future, and would throw us, even more than we are now, to the mercy of an impersonal, lonely present.

Neurobiology and psychobiology probe most directly into the distinctively human. The technological fruit of these sciences is likely to be both more tempting than Eve's apple and more "catastrophic" in its result (10). One need only consider contemporary drug use to see what people are willing to risk or sacrifice for novel experiences, heightened perceptions, or just "kicks." The possibility of drug-induced, instant, and effortless gratification will be welcomed. Recall the possibilities of voluntary self-stimulation of the brain to reduce anxiety, to heighten pleasure, or to create visual and auditory sensations unavailable through the peripheral sense organs. Once these techniques are perfected and safe, is there much doubt that they will be desired, demanded, and used?

What ends will these techniques serve? Most likely, only the most elemental, those most tied to the bodily pleasures. What will happen to thought, to love, to friendship, to art, to judgment, to public-spiritedness in a society with a perfected technology of plea-

sure? What kinds of creatures will we become if we obtain our pleasure by drug or electrical stimulation without the usual kind of human
efforts and frustrations? What kind of society will we have?

We need only consult Aldous Huxley's prophetic novel *Brave
New World* for a likely answer to these questions. There we encounter
a society dedicated to homogeneity and stability, administered by
means of instant gratifications and peopled by creatures of human
shape but of stunted humanity. They consume, fornicate, take
"soma," and operate the machinery that makes it all possible. They do
not read, write, think, love, or govern themselves. Creativity and curi-
osity, reason and passion, exist only in a rudimentary and mutilated
form. In short, they are not men at all.

True, our techniques, like theirs, may in fact enable us to treat
schizophrenia, to alleviate anxiety, to curb aggressiveness. We, like
them, may indeed be able to save mankind from itself, but probably
only at the cost of its humanness. In the end, the price of relieving
man's estate might well be the abolition of man (11).

There are, of course, many other routes leading to the abolition of
man. There are many other and better known causes of dehumanization. Disease, starvation, mental retardation, slavery, and brutality—to
name just a few—have long prevented many, if not most, people from
living a fully human life. We should work to reduce and eventually to
eliminate these evils. But the existence of these evils should not prevent us from appreciating that the use of the technology of man,
uninformed by wisdom concerning proper human ends, and untem-
pered by an appropriate humility and awe, can unwittingly render us
all irreversibly less than human. For, unlike the man reduced by disease or slavery, the people dehumanized à la *Brave New World* are not
miserable, do not know that they are dehumanized, and, what is
worse, would not care if they knew. They are, indeed, happy slaves,
with a slavish happiness.

Some Fundamental Questions

The practical problems of distributing scarce resources, of curb-
ing the abuses of power, and of preventing voluntary dehumanization
point beyond themselves to some large, enduring, and most difficult
questions: the nature of justice and the good community, the nature
of man and the good for man. My appreciation of the profundity of
these questions and my own ignorance before them makes me hesitant
to say any more about them. Nevertheless, previous failures to find a
shortcut around them have led me to believe that these questions must
be faced if we are to have any hope of understanding where biology is
taking us. Therefore, I shall try to show in outline how I think some

of the larger questions arise from my discussion of dehumanization and self-degradation.

My remarks on dehumanization can hardly fail to arouse argument. It might be said, correctly, that to speak about dehumanization presupposes a concept of "the distinctively human." It might also be said, correctly, that to speak about wisdom concerning proper human ends presupposes that such ends do in fact exist and that they may be more or less accessible to human understanding, or at least to rational inquiry. It is true that neither presupposition is at home in modern thought.

The notion of the "distinctively human" has been seriously challenged by modern scientists. Darwinists hold that man is, at least in origin, tied to the subhuman; his seeming distinctiveness is an illusion or, at most, not very important. Biochemists and molecular biologists extend the challenge by blurring the distinction between the living and the nonliving. The laws of physics and chemistry are found to be valid and are held to be sufficient for explaining biological systems. Man is a collection of molecules, an accident on the stage of evolution, endowed by chance with the power to change himself, but only along determined lines.

Psychoanalysts have also debunked the "distinctly human." The essence of man is seen to be located in those drives he shares with other animals—pursuit of pleasure and avoidance of pain. The so-called "higher functions" are understood to be servants of the more elementary, the more base. Any distinctiveness or "dignity" that man has consists of his superior capacity for gratifying his animal needs.

The idea of "human good" fares no better. In the social sciences, historicists and existentialists have helped drive this question underground. The former hold all notions of human good to be culturally and historically bound, and hence mutable. The latter hold that values are subjective: each man makes his own, and ethics becomes simply the cataloging of personal tastes.

Such appear to be the prevailing opinions. Yet there is nothing novel about reductionism, hedonism, and relativism; these are doctrines with which Socrates contended. What is new is that these doctrines seem to be vindicated by scientific advance. Not only do the scientific notions of nature and of man flower into verifiable predictions, but they yield marvelous fruit. The technological triumphs are held to validate their scientific foundations. Here, perhaps, is the most pernicious result of technological progress—more dehumanizing than any actual manipulation or technique, present or future. We are witnessing the erosion, perhaps the final erosion, of the idea of man as something splendid or divine, and its replacement with a view that sees man, no less than nature, as simply more raw material for manipulation and homogenization. Hence, our peculiar moral crisis. We are in turbulent seas without a landmark precisely because we adhere more and more to a view of nature and of man which both gives us enormous power and, at the same time, denies all possibility of standards to guide its use. Though well-equipped, we know not who we are nor where we are going. We are left to the accidents of our hasty, biased, and ephemeral judgments.

Let us not fail to note a painful irony: our conquest of nature has made us the slaves of blind chance. We triumph over nature's unpredictabilities only to subject ourselves to the still greater unpredictability of our capricious wills and our fickle opinions. That we have a method is no proof against our madness. Thus, engineering the engineer as well as the engine, we race our train we know not where (12).

While the disastrous consequences of ethical nihilism are insufficient to refute it, they invite and make urgent a reinvestigation of the ancient and enduring questions of what is a proper life for a human being, what is a good community, and how are they achieved (13). We must not be deterred from these questions simply because the best minds in human history have failed to settle them. Should we not rather be encouraged by the fact that they considered them to be the most important questions?

As I have hinted before, our ethical dilemma is caused by the victory of modern natural science with its non-teleological view of man. We ought therefore to reexamine with great care the modern notions of nature and of man, which undermine those earlier notions that provide a basis for ethics. If we consult our common experience, we are likely to discover some grounds for believing that the questions about man and human good are far from closed. Our common experience suggests many difficulties for the modern "scientific view of man." For example, this view fails to account for the concern for justice and freedom that appears to be characteristic of all human societies (14). It also fails to account for or to explain the fact that men have speech and not merely voice, that men can choose and act and not merely move or react. It fails to explain why men engage in moral discourse, or, for that matter, why they speak at all. Finally, the "scientific view of man" cannot account for scientific inquiry itself, for why men seek to know. Might there not be something the matter with a knowledge of man that does not explain or take account of his most distinctive activities, aspirations, and concerns (15)?

Having gone this far, let me offer one suggestion as to where the difficulty might lie: in the modern understanding of knowledge. Since Bacon, as I have mentioned earlier, technology has increasingly come to be the basic justification for scientific inquiry. The end is power, not knowledge for its own sake. But power is not only the end. It is also an important *validation* of knowledge. One definitely knows that one knows only if one can make. Synthesis is held to be the ultimate proof of understanding (16). A more radical formulation holds that one knows only what one makes: knowing *equals* making.

Yet therein lies a difficulty. If truth be the power to change or to make the object studied, then of what do we have knowledge? If there are no fixed realities, but only material upon which we may work our wills, will not "science" be merely the "knowledge" of the transient and the manipulatable? We might indeed have knowledge of the laws by which things change and the rules for their manipulation, but no knowledge of the things themselves. Can such a view of "science" yield any knowledge about the nature of man, or indeed, about the nature of anything? Our questions appear to lead back to the most basic of questions: What does it mean to know? What is it that is knowable (17)?

We have seen that the practical problems point toward and make urgent certain enduring, fundamental questions. Yet while pursuing these questions, we cannot afford to neglect the practical problems as such. Let us not forget Delgado and the "psychocivilized society." The philosophical inquiry could be rendered moot by our blind, confident efforts to dissect and redesign ourselves. While awaiting a reconstruction of theory, we must act as best we can.

What Is To Be Done?

First, we sorely need to recover some humility in the face of our awesome powers. The arguments I have presented should make apparent the folly of arrogance, of the presumption that we are wise enough to remake ourselves. Because we lack wisdom, caution is our urgent need. Or to put it another way, in the absence of that "ultimate wisdom," we can be wise enough to know that we are not wise enough. When we lack sufficient wisdom to do, wisdom consists in not doing. Caution, restraint, delay, abstention are what this second-best (and, perhaps, only) wisdom dictates with respect to the technology for human engineering.

If we can recognize that biomedical advances carry significant social costs, we may be willing to adopt a less permissive, more critical stance toward new developments. We need to reexamine our prejudice not only that all biomedical innovation is progress, but also that it is inevitable. Precedent certainly favors the view that what can be done will be done, but is this necessarily so? Ought we not to be suspicious when technologists speak of coming developments as automatic, not subject to human control? Is there not something contradictory in the notion that we have the power to control all the untoward consequences of a technology, but lack the power to determine whether it should be developed in the first place?

What will be the likely consequences of the perpetuation of our permissive and fatalistic attitude toward human engineering? How will the large decisions be made? Technocratically and self-servingly, if our experience with previous technologies is any guide. Under conditions of laissez-faire, most technologists will pursue techniques, and most private industries will pursue profits. We are fortunate that, apart from the drug manufacturers, there are at present in the biomedical area few large industries that influence public policy. Once these appear, the voice of "the public interest" will have to shout very loudly to be heard above their whisperings in the halls of Congress. These reflections point to the need for institutional controls.

Scientists understandably balk at the notion of the regulation of science and technology. Censorship is ugly and often based upon ignorant fear; bureaucratic regulation is often stupid and inefficient. Yet there is something disingenuous about a scientist who professes concern about the social consequences of science, but who responds to every suggestion of regulation with one or both of the following: "No restrictions on scientific research," and "Technological progress should not be curtailed." Surely, to suggest that *certain* technologies ought to be regulated or forestalled is not to call for the halt of *all* technological progress (and says nothing at all about basic research). Each development should be considered on its own merits. Although the dangers of regulation cannot be dismissed, who, for example, would still object to efforts to obtain an effective, complete, global prohibition on the development, testing, and use of biological and nuclear weapons?

The proponents of laissez-faire ignore two fundamental points. They ignore the fact that not to regulate is as much a policy decision as the opposite, and that it merely postpones the time of regulation. Controls will eventually be called for—as they are now being demanded to end environmental pollution. If attempts are not made early to detect and diminish the social costs of biomedical advances by intelligent institutional regulation, the society is likely to react later with more sweeping, immoderate, and throttling controls.

The proponents of laissez-faire also ignore the fact that much of technology is already regulated. The federal government is already deep in research and development (for example, space, electronics, and weapons) and is the principal sponsor of biomedical research. One may well question the wisdom of the direction given, but one would be wrong in arguing that technology cannot survive social control. Clearly, the question is not control versus no control, but rather what kind of control, when, by whom, and for what purpose.

Means for achieving international regulation and control need to be devised. Biomedical technology can be no nation's monopoly. The need for international agreements and supervision can readily be understood if we consider the likely American response to the successful asexual reproduction of 10,000 Mao Tse-tungs.

To repeat, the basic short-term need is caution. Practically, this means that we should shift the burden of proof to the proponents of a new biomedical technology. Concepts of "risk" and "cost" need to be broadened to include some of the social and ethical consequences discussed earlier. The probable or possible harmful effects of the widespread use of a new technique should be anticipated and introduced as "costs" to be weighed in deciding about the first use. The regulatory institutions should be encouraged to exercise restraint and to formulate the grounds for saying "no." We must all get used to the idea that biomedical technology makes possible many things we should never do.

But caution is not enough. Nor are clever institutional arrangements. Institutions can be little better than the people who make them work. However worthy our intentions, we are deficient in understanding. In the long run, our hope can only lie in education: in a public educated about the meanings and limits of science and enlightened in its use of technology; in scientists better educated to understand the relationships between science and technology on the one hand, and ethics and politics on the other; in human beings who are as wise in the latter as they are clever in the former.

REFERENCES AND NOTES

1. R. G. Edwards, B. D. Bavister, P. C. Steptoe, Nature 221, 632 (1969).

2. R. G. Edwards, P. C. Steptoe, J. M. Purdy, ibid. 227, 1307 (1970).

3. H. J. Muller, Science 134, 643 (1961).

4. J. M. R. Delgado, Physical Control of the Mind: Toward a Psychocivilized Society (Harper & Row, New York, 1969).

5. F. Bacon, The Advancement of Learning, Book I, H. G. Dick, Ed. (Random House, New York, 1955), p. 193.

6. C. S. Lewis, The Abolition of Man (Macmillan, New York, 1965), pp. 69-71.

7. D. D. Rutstein, Daedalus (Spring 1969), p. 523.

8. E. J. Cassell, *Commentary* (June 1969), p. 73.

9. B. Glass, *Science* 171, 23 (1971).

10. It is, of course, a long-debated question as to whether the fall of Adam and Eve ought to be considered "catastrophic," or more precisely, whether the Hebrew tradition considered it so. I do not mean here to be taking sides in this quarrel by my use of the term "catastrophic," and, in fact, tend to line up on the negative side of the questions, as put above. Curiously, as Aldous Huxley's *Brave New World* [(Harper & Row, New York, 1969)] suggests, the implicit goal of the biomedical technology could well be said to be the reversal of the Fall and a return of man to the hedonic and immortal existence of the Garden of Eden. Yet I can point to at least two problems. First, the new Garden of Eden will probably have no gardens; the received, splendid world of nature will be buried beneath asphalt, concrete, and other human fabrications, a transformation that is already far along. (Recall that in *Brave New World* elaborate consumption-oriented, mechanical amusement parks—featuring, for example, centrifugal bumble-puppy—had supplanted wilderness and even ordinary gardens.) Second, the new inhabitant of the new "Garden" will have to be a creature for whom we have no precedent, a creature as difficult to imagine as to bring into existence. He will have to be simultaneously an innocent like Adam and a technological wizard who keeps the "Garden" running. (I am indebted to Dean Robert Goldwin, St. John's College, for this last insight.)

11. Some scientists naively believe that an engineered increase in human intelligence will steer us in the right direction. Surely we have learned by now that intelligence, whatever it is and however measured, is not synonymous with wisdom and that, if harnessed to the wrong ends, it can cleverly perpetrate great folly and evil. Given the activities in which many, if not most, of our best minds are now engaged, we should not simply rejoice in the prospect of enhancing IQ. On what would this increased intelligence operate? At best, the programming of further increases in IQ. It would design and operate techniques for prolonging life, for engineering reproduction, for delivering gratifications. With no gain in wisdom, our gain in intelligence can only enhance the rate of our dehumanization.

12. The philosopher Hans Jonas has made the identical point: "Thus the slow-working accidents of nature, which by the very patience of their small increments, large numbers, and gradual decisions, may well cease to be 'accident' in outcome, are to be replaced by the fast-working accidents of man's hasty and biased decisions, not exposed to the long test of the ages. His uncertain ideas are to set the goals of generations, with a certainty borrowed from the presumptive certainty of the means. The latter presumption is doubtful enough, but this doubtfulness becomes secondary to the prime question that arises when man indeed undertakes to 'make himself': in what image of his own devising shall he do so, even granted that he can be sure of the means? In fact, of course, he can be sure of neither, not of the end, nor of the means, once he enters the realm where he plays with the roots of life. Of one thing only can he be sure: of his power to move the foundations and to cause incalculable and irreversible consequences. Never was so much power coupled with so little guidance for its use." [*J. Cent. Conf. Amer. Rabbis* (January 1968), p. 27.] These remarks demonstrate that, contrary to popular belief, we are not even on the right road toward a rational understanding of and rational control over human nature and human life. It is indeed the height of irrationality triumphantly to pursue rationalized technique, while at the same time insisting that questions of ends, values, and purposes lie beyond rational discourse.

13. It is encouraging to note that these questions are seriously being raised in other quarters—for example, by persons concerned with the decay of cities or the pollution of nature. There is a growing dissatisfaction with ethical nihilism. In fact, its tenets are unwittingly abandoned, by even its staunchest adherents, in any discussion of "what to do." For example, in the biomedical area, everyone, including the most unre-

constructed and technocratic reductionist, finds himself speaking about the use of powers for "human betterment." He has wandered unawares onto ethical ground. One cannot speak of "human betterment" without considering what is meant by *the human* and by the related notion of *the good for man*. These questions can be avoided only by asserting that practical matters reduce to tastes and power, and by confessing that the use of the phrase "human betterment" is a deception to cloak one's own will to power. In other words, these questions can be avoided only by ceasing to discuss.

14. Consider, for example, the widespread acceptance, in the legal systems of very different societies and cultures, of the principle and the practice of third-party adjudication of disputes. And consider why, although many societies have practiced slavery, no slave-holder has preferred his own enslavement to his own freedom. It would seem that some notions of justice and freedom, as well as right and truthfulness, are constitutive for any society, and that a concern for these values may be a fundamental characteristic of "human nature."

15. Scientists may, of course, continue to believe in righteousness or justice or truth, but these beliefs are not grounded in their "scientific knowledge" of man. They rest instead upon the receding wisdom of an earlier age.

16. This belief, silently shared by many contemporary biologists, has recently been given the following clear expression: "One of the acid tests of understanding an object is the ability to put it together from its component parts. Ultimately, molecular biologists will attempt to subject their understanding of all structure and function to this sort of test by trying to synthesize a cell. It is of some interest to see how close we are to this goal." [P. Handler, Ed, *Biology and the Future of Man* (Oxford Univ. Press, New York, 1970), p. 55.]

17. When an earlier version of this article was presented publicly, it was criticized by one questioner as being "antiscientific." He suggested that my remarks "were the kind that gave science a bad name." He went on to argue that, far from being the enemy of morality, the pursuit of truth was itself a highly moral activity, perhaps the highest. The relation of science and morals is a long and difficult question with an illustrious history, and it deserves a more extensive discussion than space permits. However, because some readers may share the questioner's response, I offer a brief reply. First, on the matter of reputation, we should recall that the pursuit of truth may be in tension with keeping a good name (witness Oedipus, Socrates, Galileo, Spinoza, Solzhenitsyn). For most of human history, the pursuit of truth (including "science") was not a reputable activity among the many, and was, in fact, highly suspect. Even today, it is doubtful whether more than a few appreciate knowledge as an end in itself. Science has acquired a "good name" in recent times largely because of its technological fruit; it is therefore to be expected that a disenchantment with technology will reflect badly upon science. Second, my own attack has not been directed against science, but against the use of *some* technologies and, even more, against the unexamined belief—indeed, I would say, superstition—that all biomedical technology is an unmixed blessing. I share the questioner's belief that the pursuit of truth is a highly moral activity. In fact, I am inviting him and others to join in a pursuit of the truth about whether all these new technologies are really good for us. This is a question that merits and is susceptible of serious intellectual inquiry. Finally, we must ask whether what we call "science" has a monopoly on the pursuit of truth. What is "truth"? What is knowable, and what does it mean to know? Surely, these are also questions that can be examined. Unless we do so, we shall remain ignorant about what "science" is and about what it discovers. Yet "science"—that is, modern natural science—cannot begin to answer them; they are philosophical questions, the very ones I am trying to raise at this point in the text.

22 Guidelines for the Ethical, Social and Legal Issues in Prenatal Diagnosis

TABITHA M. POWLEDGE AND JOHN FLETCHER

This report from the Genetics Research Group of the Hastings Center proposes 18 guidelines which are relevant to ethical, social and legal issues in prenatal diagnosis. This interdisciplinary research group focused on nontechnical dimensions of technical and procedural problems, the use of the information after prenatal diagnosis has occurred, and problems of access to prenatal diagnosis.

John Fletcher is Director of Bioethics, National Institute of Health

Tabitha Powledge is an Associate for Genetics, Institute of Society, Ethics, and the Life Sciences

Medicine's ability to collect information about the fetus has increased dramatically in the last few years. A variety of technics—collectively known as prenatal diagnosis—have made it possible to learn much about a fetus's genetic and metabolic state, chromosomal constitution (including gender), bone structure and other information, and, moreover, to learn it earlier and earlier in gestation.

Many of the technics yield this information at a time that makes the selective abortion of a fetus legally possible. Selective abortion is morally unacceptable to many people. This report does not discuss this question in detail. It recognizes that selective abortion raises a variety of deeply troubling moral, social and legal questions that must always be kept in mind for the exercise of responsible decision making.

Since the release in the United States of a prospective study of mid-trimester amniocentesis in the fall of 1975 and its subsequent publication, the comparative safety and accuracy of this form of prenatal diagnosis has been widely accepted.[1] A study carried out in Canada and published shortly after the United States study disclosed similar findings.[2] As a result, it has been predicted that amniocentesis will soon become much more common, and ultimately part of standard obstetric practice, where indicated.

Rapid development of other forms of prenatal diagnosis is proceeding along several fronts. Ultrasound, often used as an adjunct of amniocentesis, is also used as a separate form of fetal diagnosis, not only for the detection of birth defects but also to establish gestational age, or to detect multiple pregnancy early.[3] This technic has been expanded to include real-time scanning, which yields a moving picture of the fetus.[4]

Another of the new procedures involves aspiration of blood from the placental vessels, making possible diagnosis of the hemoglobulinopathies and other blood and biochemical disorders.[5] In some centers this technic is accomplished by direct visualization of the fetus within the uterus by means of a tiny fiberoptic endoscope, also called the fetoscope.[6] This developing technology may also prove very helpful in the detection of a variety of other anomalies.

Early fetal assessment does not have to depend on methods that directly involve the uterus and its contents; the development of indirect diagnostic technics has also been proceeding rapidly. α-Fetoprotein levels in maternal serum can give information about birth defects, especially the neural-tube defects, on a mass basis.[7] This test is being widely used in the United Kingdom, where neural-tube defects are among the most common major malformations, and is also frequently employed in the United States and Canada. A pilot program in Nassau County, New York, has as its goal screening the serum of every pregnant woman in the county to identify those with high risk of fetal disease. Women with positive tests are then studied with more definitive diagnostic methods like ultrasonography and determination of α-fetoprotein concentrations in amniotic fluid.[8]

In short, various kinds of prenatal diagnoses are proliferating, and the array of technics is likely to expand even more in the future. None of them are without technical problems. Most are experimental and not yet standard practice. All of them also present other kinds of problems—moral, social, and legal as well—that, though often acknowledged, remain largely unresolved.

In early stages of the procedures' development, informal or ad hoc solutions to these problems must be attempted while experience is being acquired. But that is not a satisfactory long-term situation for society as a whole or for prenatal diagnosticians, many of whom are rightly uncomfortable cast in the unfamiliar and undesirable role of moral decision makers on behalf of society.

The Genetics Research Group of the Hastings Center, Institute of Society, Ethics and the Life Sciences, an interdisciplinary task force of people with academic training in law, medicine, philosophy, theology, biology, genetics and the social sciences, has been considering many of those questions for more than six years. Recently, this group decided to attempt a systematic exploration of them, with the intention of offering assistance to workers in this rapidly developing field. This document represents the result of that effort. Its purpose is to propose guidelines for the development and institutionalization of prenatal diagnostic programs and to help workers in this area provide the most favorable circumstances for thoughtful, informed, morally responsible decision making by parents.

NONTECHNICAL DIMENSIONS OF TECHNICAL AND PROCEDURAL PROBLEMS

The design, operation and validation of prenatal diagnostic technics, the organization and conduct of the programs themselves and the associated laboratory procedures involved have moral, social and legal dimensions often unrecognized or inadequately examined and discussed. This section considers these dimensions.

Prenatal diagnostic programs can be regarded as one type of screening program and should be organized in ways that are consonant with the principles of optimum screening programs. We emphasize "optimum" because most screening programs are conceded to be sometimes deficient in practice.[9] With the exception of prenatal diagnosis done for research purposes (dealt with in point 8, below), programs of prenatal diagnosis ought to meet at least the following criteria:

1. The programs should be designed to reach well defined groups of pregnant women known to be at risk.
2. Prenatal diagnosis should be undertaken only where high-quality laboratory work is available, even though, at the moment, lack of adequate laboratory services is an important barrier to the availability of amniocentesis. Though the current error rates are low, it is possible that frequency of error will increase as the technics become more widely applied by practitioners and laboratories new to the field, and by experienced workers burdened by a growing volume of tests. Laboratory tests must be standardized, and quality-control measures

implemented to assure accurate results. Inaccurate results have even heavier moral and legal consequences in prenatal diagnosis than in many screening procedures or other routine medical tests. To avoid the consequences, training programs for laboratory workers and a comprehensive prenatal-diagnosis laboratory monitoring and surveillance system should also be developed. Uncontrolled expansion of prenatal diagnosis and, particularly, inadequate laboratory work, would be a disastrous step for technics that can offer hope and reassurance to many families.

3. The lowest practical limits for false-positive and false-negative results should be established, and continuing attempts to improve test sensitivity should be made. In many other kinds of screening programs, a positive diagnosis is confirmed by additional tests, and a test that errs on the side of false-positive results is therefore acceptable. In most kinds of prenatal diagnosis, however, the test will be done only once, and a positive test result may lead to an abortion. False-negative results will lead to the undesired birth of an affected child. Error in either direction will cause pain and suffering in families, and may also open up the prenatal diagnostician to subsequent legal liability. Positive diagnoses should be confirmed, whether after abortion, stillbirth or live birth.

4. Efforts to follow up and evaluate all types of prenatal diagnosis procedures should be made and should be part of large programs from their outset, with the intention of determining the short-term and long-term safety for both mother and fetus. Reports of adverse psychologic responses by some parents when abortion is performed for a medical indication after prenatal diagnosis[10] raise troubling questions and need rapid validation or refutation, though care must also be taken to avoid unnecessary intrusion into family life. And though the short-term unfavorable health effects of amniocentesis and ultrasound appear to be minimal or nonexistent, we emphasize that very little is as yet known of their long-term consequences. Prudence dictates caution in the widespread application of any of these technics.

5. To assist prospective parents in the exercise of well informed and responsible decision making, they should be given adequate information about the procedure before prenatal diagnosis and adequate counseling after it. The professional should ascertain that the patient is aware of a number of extremely important factors relevant to a decision, such as the need for thoughtful consideration of the various options available, depending on the results of the diagnosis (abortion vs. possible al-

ternatives, taking into account the nature of the prenatally diagnosed condition and the present and future possibilities for prenatal and postnatal treatment); the possibility and meaning of inaccurate diagnoses ("falsely positive" or "falsely negative") and the consequences for the family in such an instance; the mechanics of the diagnostic intervention as well as the risks involved; the type of information that the procedure can and cannot provide; and the cost. Every effort should be made to include prospective fathers in the sessions.

6. The patient's privacy should be scrupulously protected ("patient" is here defined as the mother of the fetus, and, in most cases, the father as well). Great difficulties with this provision are to be expected, in view of the trend to genetic registries and similar medical-data banks, but those difficulties do not absolve the keepers of this information from their moral—and perhaps legal—obligation to protect patients. Central registries for epidemiologic data collection and needs assessment are justifiable, but individual patients should never be identifiable. Individual genetic disorders are, by definition, often found largely concentrated in particular racial and ethnic groups, but special care should be taken in the public presentation of these data not to perpetuate past instances of misinformation and stigmatization of particular groups.[11] It is desirable for physicians to follow their patients, but individually identifiable information should not be stored in centralized data banks. In the rare situation in which information obtained via prenatal diagnosis might be of direct benefit to collateral relatives, permission to disclose should be sought from the patients. If that permission is refused, professionals can face a conflict between their ethical and their legal obligations, and may need to justify disclosure from a legal standpoint.

7. The apparent trend toward performing amniocentesis in the office of the private physician, with the fluids sometimes being sent for assessment to commercial laboratories, should not be allowed to lead to a diminution of quality control in prenatal diagnosis. Nor does it relieve professionals of their obligation to maintain confidentiality, or to provide adequate counseling, information and follow-up study. Physicians should demonstrate competence in the technic—perhaps by way of special training—before applying it to their patients.

8. The distinction between attempts at prenatal diagnosis for research purposes and prenatal diagnosis as a relatively routine service for selected obstetric patients should be clear in the

minds of both the people performing the procedure and their patients. Research can be an important, rational and legitimate purpose of prenatal diagnosis, but researchers are under strong moral and legal obligation to make sure that their patients understand that, in these circumstances, they are experimental subjects as well as patients who may benefit from the procedure. The obligation to explain, so far as they are known, what the possible risks are, and to obtain the subject's comprehending consent to the procedure, is, if anything, greater for researchers than for ordinary physicians.

Use of Information after Prenatal Diagnosis

The desired and intended result of prenatal diagnosis is information about the presence or absence of a possible disease or defect in the fetus. In practice the test results are negative in more than 96 per cent of amniocentesis cases, thus providing these families with many months of relief from anxiety. In addition, pregnancies that would otherwise be terminated because the fetus is at a substantial risk of being abnormal can be carried to term. When diagnosis of the presence of disease or defect is made, parents and physicians use that information to make choices about subsequent action. The alternatives after such a diagnosis include abortion, treatment of the fetus in the rare cases in which prenatal therapy is or may be possible, bringing the pregnancy to term and providing the parents with whatever help is available for treating the condition after delivery, if possible, or bringing the pregnancy to term and arranging for the parents to give over their responsibilities by adoption, foster care or institutionalization. Each of these alternatives is intimately related to individual views on the ethics of abortion. Ethical considerations make it imperative to separate the fact of a positive diagnosis from the choice about the subsequent action. What parents and physicians should decide to do is not automatically dictated by the diagnosis, but ought to be shaped by their ethical and social views.

Most women who seek prenatal diagnosis want a child; prenatal diagnosis provides information that can enhance the freedom and responsibility of parents in reproductive decision making. Before development of prenatal diagnosis, parents and physicians were more at the mercy of outcomes unknown until delivery or after. However, the lack of a moral consensus on abortion also makes it inappropriate to suggest that women have a moral (or "medically indicated") obligation to undergo prenatal diagnosis.

One of the ultimate goals of prenatal diagnosis should be the treatment and eventual cure of disease in the fetus or infant. Many decades of work will be required to make substantial progress toward this goal, and, in the meantime, abortion in these cases is a limited and imperfect response when an effective therapy is lacking. Abortion is never therapeutic for the fetus, but we believe it can be morally justified for the relief of suffering and burden to family and society. These guidelines were developed in a moral framework favoring the protection of individual choice and the autonomy of parents, even when we disagree with their courses of action.

The profound conflict of values embodied in the abortion debate cannot be finally resolved in a set of guidelines that offer directions for choices that must always be made in the light of competing, and sometimes polar, claims. These guidelines cannot reconcile the views of those who believe that abortion is wrong virtually without exception with the views of those who exclude the welfare of the fetus completely from any argument about reproductive decisions. These guidelines attempt to respect the pluralism of values about abortion that exists in society, while offering concrete guidance in special cases that are cause for concern.

9. Prenatal diagnosis should not be denied to the woman who is at risk and desires the procedure, but has decided that she will not undergo an abortion no matter what the diagnosis. In the past, particularly before release of the collaborative studies, some researchers argued that such use of prenatal diagnosis offered possible risk to the unaffected fetus and was a waste of scarce resources; without abortion of an affected fetus, prenatal diagnosis was seen by them to have no purpose. By contrast, we regard the provision of information as an important and legitimate purpose of prenatal diagnosis. Depending on the disorder, certain families might find very useful a few months of planning and preparation for the birth of an affected child. In most cases, diagnosis of an unaffected fetus will provide months of relief from anxiety.

10. Counseling for the procedure should be noncoercive and respectful of parental views about abortion. Each family's life situation is special, and therefore, ideally, a counselor should have a tolerance for the moral ambiguities of abortion and the complexities of decision making.

11. In conditions for which both prenatal diagnosis and postnatal treatment are available, parents should be informed about both possibilities—including their relative disadvantages—and allowed to make their own decisions about what action to take. This is the situation that now prevails in the rare

metabolic disorder galactosemia. It has been argued that where treatment is available, the need for prenatal diagnosis vanishes. We believe that the medical profession should acknowledge that there are some families, particularly those who already have a child with a disorder, who will not want to bear another child who will need treatment, but who may wish to have additional, unaffected children. Prenatal diagnosis allows such families to have available wide choices about future childbearing, and it should not be denied them.

12. Prenatal ascertainment of sex in fetuses at risk for otherwise undiagnosable sex-linked disorders should be available to parents who want it. Since in such cases there will be at least a 50 per cent probability of aborting an unaffected fetus, which is an extremely unsatisfactory situation from both a moral and social perspective, the development of specific tests for the presence of such disorders should have a very high research priority.

13. Moral and legal strictures require that findings not be withheld from the parents, even when they are of disputed importance. This situation will become increasingly common as amniocentesis for Down's syndrome increases and chromosome anomalies of unknown or controversial consequence are discovered in the process. Prognosis for many children with these abnormalities is ambiguous, a well known example being the XYY condition.[12] We strongly recommend the development, perhaps under the auspices of an appropriate professional organization, of a series of fact sheets describing what is known about each of these conditions, to aid counselors who may have to give out this information. Counseling before prenatal diagnosis as part of the process of obtaining consent will alert couples to the possibility of such findings, giving them the choice of whether they want that information. Requests for both disclosure and nondisclosure should be honored.

14. Although we strongly oppose any movement aimed at making diagnosis of sex and selective abortion a part of ordinary medical practice and family planning, we recommend that no legal restrictions be placed on ascertainment of fetal sex. We think such restrictions would be ineffective and impossible to administer, would lead to subterfuge and, more important, would violate our objective of noninterference with parental choice, even when we disagree with that choice. Prenatal diagnosis is not now widely available for this purpose; indeed, amniocentesis is often refused to women who request it for that reason, largely on grounds that the procedure is expensive and possibly risky,

and should be reserved for grave medical conditions. Though we support the right of individual physicians to refuse to perform prenatal diagnosis for sex choice, we also recognize that in special situations, sex choice can appear to parents to be justifiable. We think most couples should not seek such information, however. Discouragement of this use of prenatal diagnosis, by pointing out that the risks and stresses of second-trimester abortions are not trivial, will mean that such cases will at least not be very great in number, though availability of earlier sex-ascertainment technics, now in development,[13] is likely to expand them considerably.

15. Standard medical practice on resuscitation, maintenance and treatment of newborns with severe abnormalities should apply in the event of a live birth after attempted abortion for a medical indication. Such measures may occasionally need to be considered because a positive prenatal diagnosis in most of its current forms is usually followed by a second-trimester abortion, occasionally close to the limits of viability for the fetus. Such infants should be regarded as members of the general class of gravely ill newborns, and decisions about them should be made on the same basis.

16. The existence of prenatal diagnosis should never be used as a justification for the withholding of support and services for those who are born with prenatally diagnosable diseases and whose parents want such support and services.

ACCESS

In the United States, women who undergo prenatal diagnosis belong largely to higher economic and social groups.[14] There are large numbers of women who are appropriate candidates for it and who might reasonably decide to use these technics, but for whom they have not been available. The last guidelines address that issue.

17. We endorse third-party payments, including Medicaid, for most kinds of prenatal diagnosis now available. In view of the data on economic and educational level of the typical prenatal-diagnosis patient, we regard the availability of prenatal diagnosis to pregnant women of lower educational and income levels as a very high priority, and subsidy, in one form or another, is critical to that availability.

18. The government, the medical profession, major foundations and voluntary health agencies could all do more, at

relatively low cost, to make the existence of prenatal diagnosis more widely known. This communication effort should be aimed at both professionals and consumers, but at a pace that does not create greater demand than existing services can safely meet.

We are indebted for the suggestions of the following people, though this acknowledgment is not necessarily meant to imply their endorsement of this report: Sherry Arnstein, M.S., National Center for Health Services Research; Stanley Bergen, Jr., M.D., College of Medicine and Dentistry of New Jersey; Arthur Caplan, Ph.D., Hastings Center; Joseph Fletcher, S.T.D., University of Virginia School of Medicine; Paul Freund, LL.B., Harvard Law School; Ruth Hanft, M.A., Department of Health, Education and Welfare; Leonard Isaacs, Ph.D., Michigan State University; Paul Ramsey, Ph.D., Princeton University; Philip Reilly, J.D., Yale University School of Medicine; Margaret Steinfels, M.A., Hastings Center; and Robert M. Veatch, Ph.D., Hastings Center.

REFERENCES

1. The NICHD National Registry for Amniocentesis Study Group: Midtrimester amniocentesis for prenatal diagnosis: safety and accuracy. JAMA 236:1471–1476, 1976

2. Simpson NE, Dallaire L, Miller JR, et al: Prenatal diagnosis of genetic disease in Canada: report of a collaborative study. Can Med Assoc J 115:739–748, 1976

3. Leopold GR, Asher WM: Ultrasound in obstetrics and gynecology. Radiol Clin North Am 12:127–146, 1974

4. Platt LD, Manning FA, Lemay M: Real-time B-scan-directed amniocentesis. Am J. Obstet Gynecol 130:700–703, 1978

5. Alter BP, Modell CB, Fairweather D, et al: Prenatal diagnosis of hemoglobinopathies: a review of 15 cases. N Engl J Med 295:1437–1443, 1976

6. Hobbins JC, Mahoney MJ: Fetoscopy in continuing pregnancies. Am J Obstet Gynecol 129:440–442, 1977

7. Wald NJ, Cuckle H, Brock DJH, et al: Maternal serum-alpha-fetoprotein measurement in antenatal screening for anencephaly and spina bifida in early pregnancy. Lancet 1:1323–1332, 1977

8. Macri JN, Weiss RR, Tillitt R, et al: Prenatal diagnosis of neural tube defects. JAMA 236:1251–1254, 1976

9. Wilson JMG, Jungner G: Principles and Practice of Screening for Disease, Geneva, World Health Organization, 1968

10. Blumberg BD, Golbus MS, Hanson KH: The psychological sequelae of abortion performed for a genetic indication. Am J Obstet Gynecol 122:799–808, 1975

11. Lappé M, Gustafson JM, Roblin R: Ethical and social issues in screening for genetic disease. N Engl J Med 286:1129–1132, 1972

12. Hamerton JL: Human population cytogenetics: dilemmas and problems. Am J Hum Genet 28:107–122, 1976

13. Pirani BBK, Pairaudeau N, Doran TA, et al: Amniotic fluid testosterone in the prenatal determination of fetal sex. Am J Obstet Gynecol 129:518–520, 1977

14. Bannerman RM, Gillick D, Van Coevering R, et al: Amniocentesis and educational attainment. N Engl J Med 297:449, 1977

23 *Moral Obligations and the Fallacies of "Genetic Control"*

Marc Lappé, Ph.D.

Lappé presents us with an analysis of several problems running through genetic engineering: eugenics, the problem of a genetic apocalypse, social and individual costs of genetic decisions and the problems in setting genetic policy. What is important in this discussion is the introduction of social and environmental data into an examination of ethical issues in genetics. Combined with this is a sensitivity to the reality of individual and social rights and obligations. The eugenics debate, which lies behind Lappé's presentation, helps us focus on the ways in which some members of the society are evaluated by others and the values presupposed by this.

Dr. Marc Lappé, Ph.D., is Associate for the Biological Sciences at the Institute of Society, Ethics, and the Life Sciences.

The sciences of molecular biology and human genetics emerged within my own lifetime. Partly as a consequence of the development of these two sciences, my generation was the first to become swept up in what we now recognize as "The Biological Revolution." What made genetics "revolutionary" is that it was transformed from a science whose content was discernible only by inference, to one which seemingly could be known with certainty: the discovery which made the unknown knowable was made the year I was born.

Chance Events and the Myth of Genetic Certainty

In 1943, Oswald Avery wrote his brother Roy to describe his findings about a physiological principle which appeared to be able to confer the properties of virulence to a bacterium. The excited tone of his letter reflected the utter incredulity that Avery must have felt upon learning the outcome of his experiments: a *chemical* had made it possible to induce predictable and hereditable changes in living cells. Genes were molecules! As such they were subject to human control and manipulation. Avery wrote: "This is something which has long been the dream of geneticists. . . . [Up until now] the mutations they induced . . . are always unpredictable and random and chance changes."[1]

Although Avery was mistaken in his assumption that this knowledge would allow us generally to control where and when mutations occur, he was correct in concluding that his discovery revolutionized our ability potentially to control what specific genetic information a cell contained or expressed. Thus, when he discovered the molecular basis for a "transforming principle," he simultaneously acquired the ability to effect genetic transformations. The phenomenon by which the acquisition of knowledge per se changes that which has become known (or affords the potential for such change) represents a subtle mechanism by which genetic information (as well as much other knowledge in science) escapes the moral scrutiny of its possessors. Hans Jonas perceptively observed:

> Effecting changes in nature as a means and as a result of knowing it are inextricably interlocked, and once this combination is at work it no longer matters whether the pragmatic destination of theory is expressly accepted . . . or not. The very process of attaining knowledge leads through manipulation of the things to be known, and this origin fits of itself the theoretical results for an application whose possibility is irresistible . . . whether or not it was contemplated in the first place.[2]

In Avery's case, he might well have foreseen that transformation could be used to confer virulence to normally nonpathogenic bacteria, but he certainly could not have anticipated that his principle, in conjunction with the later to be discovered "R" factors, would be used to make potent, antibiotic resistant biological warfare agents![3] But the prospect of nefarious application is *not* what makes genetic knowledge unique. Rather, its uniqueness lies in the manner in which "knowing" the genetics of something changes it.

For example, the simple act of acquiring prenatal genetic information about a fetus—whether or not he is carrying a particular gene, or if he will develop a genetically determined disease later in his life—automatically sets into motion a train of events which themselves change that individual's future. At the very moment you acquire a "bit" of genetic information about a fetus (or any person, for that matter), you have begun to define him in entirely novel terms. You tell him (and sometimes others) something about where he came from and who is responsible for what he is now. You project who he may or may not become in the future. You set certain limits on his potential. You say something about what his children will be like, and whether or not he will be encouraged or discouraged to think of himself as a parent. In this way the information you obtain changes both the individual who possesses it, and in turn the future of that information itself.

In addition to the potential for individual stigmatization, there is also sufficient ambiguity in genetic "facts" themselves to seriously question the judiciousness of massive operations designed to ascertain the genetic composition of whole populations. In contrast to the simplistic view of genetics in Avery's time, we now know that genetic information, by its very nature, tends to confound rational analysis. It is *redundant*, such that a flaw in replication or a mutational event need not irrevocably distort or destroy (as had previously been assumed) the information contained in the genetic material. It is *self-correcting*, containing enzymes whose sole function is to recognize damaged segments of DNA molecules, excise them, and faithfully reconstruct the whole (thereby compelling reconsideration of estimations of mutation rates and their causes). It is *heterogeneous*, with most seemingly "single" genes being in fact clusters of genes with related functions ("pleiotrophy") or products ("alleles," or "pseudalleles"), each gene having the potential property of producing different effects in different organs at different times in development (frustrating any simplistic analysis of whether or not a single gene or many is responsible for a given complex constellation of developmental defects).

These observations begin to explain why at the human level, for example, medical researchers have been at a loss to explain why some individuals who by all measurements have the defective genes for phenylketonuria[4] do *not* in fact show the physical stigmata of the condition. If, as appears likely, this genetic "defeat" (and perhaps the one responsible for the related condition galactosemia) is not an "all or none" phenomenon, but can actually be compensated for by the operation of other genes, all of our assumptions about the nature of such genes, *and* our moral decisions of what should be done in the event that an individual is discovered with them, have to be seriously reassessed. The fact that this reassessment is *not* currently going on reflects, I believe, an underlying cultural bias that affects our analyses of genetic problems. It is not just that we want simple answers to complex questions; it is that we would like to be able to *control* a material whose nature eludes our dominion.

Intolerance for Uncertainty and the Quest for Genetic Control

If genetic systems are so inherently difficult to understand, why do we feel impelled to seek to control them? The problem appears to be rooted in our Western psyche and philosophical assumptions about the use of knowledge. Avery's letter gives us a sense of the deep-seated aversion most Western scientists (and philosophers) feel towards the chance events that appear to govern genetic systems. (Recall Avery's mistaken assumption that he had discovered the means to *control* the class of events we call "random mutations," when in fact he had merely discovered an analogue for one specific mutational event.)

Joseph Fletcher, an ethicist, echoes this profound disquiet towards uncertainty in genetic systems when he states: "We cannot accept the 'invisible hand' of blind chance or random nature in genetics."[5] An implicit assumption in Fletcher's remarks is that the reduction of uncertainty is equivalent to progress, a view widely held in the West.[6] In genetic systems, the paradox is that progress (in this sense evolutionary progress) is accomplished *because of* genetic instability and susceptibility to chance events, not in spite of it.

James Crow, a renowned population geneticist, has described the operation of chance in sexual reproduction by pointing out that "In a sexual population, genotypes are formed and broken up by recombination every generation, and a particular genotype is therefore evanescent: what is transmitted to the next generation is a sample of genes, not a [whole] genotype."[7] It is difficult to reconcile evolutionary progress with this image alone, since Crow omits (by intention, I am sure) discussion of the mechanisms by which variation is introduced into sexual populations. Faced with the reality of incessant fluctuation and change of genetic systems, the Nobel laureate geneticist Joshua Lederberg asked at one point: "If a superior individual . . . is identified, why not copy it directly, rather than suffer all the risks of recombinational disruption, including those of sex? . . . Leave sexual reproduction for experimental purposes; when a suitable type is ascertained, take care to maintain it by clonal propagation."[8]

If from Avery's day scientists believed they had discovered the means to control the transmission of hereditary information, why does Joshua Lederberg believe that the only real means of control for man would be to clone him? The answer in part is that the kinds of control which were possible in bacteria thirty years ago remain an illusive quest for human organisms today. Not only is cloning a distant and limited prospect for man, but so is the much-vaunted genetic engineering which would precede it. Mammalian cells, unlike bacterial ones, appear to be extraordinarily resistant to the introduction of most forms of genetic information. Although reports have appeared indicating that bacterial viral genes will function after being introduced into human cells in tissue culture[9] (a feat proving difficult to replicate), enormous difficulties remain in attempting to use the same techniques actually to treat individuals with the genetic defect that the virus appears to correct. Another technique for correcting "defective genes" also appears to pose currently insuperable problems for human application. It entails fusing or "hybridizing" a cell lacking a particular gene with one containing the active equivalent.[10] This technique may prove to be limited to tissue-culture studies, since the number of cells needed to correct the same defect in a person would be astronomically large and the problem of immunologic acceptance of the cells a thorny one.

While tantalizing in the control that such techniques appear to promise for the future, there is a danger in their seductiveness in the present. In the first place, they obfuscate the need for solving current problems which do not need novel technical solutions, such as general health care. Secondly, they pose the threat of dehumanization that Jacques Ellul identifies with technique per se. Ellul observes that "When technique enters into every area of life, including the human, it ceases to be external to man and becomes his very substance. It is no longer face to face with man but is integrated with him, and it progressively absorbs him."[11] In the context of the above examples, Ellul would envision man's existence becoming dependent upon and inevitably indistinguishable from the vast array of artificially engineered genes and tissue-culture support systems needed to sustain him. More importantly, such techniques do not offer permanent solutions to human problems but merely transiently replace one technique (e.g., insulin for treating diabetes) with another (genetic engineering of Islets of Langerhans cells in the pancreas) for coping with man's medico-genetic dilemmas. Since none of these projected genetic techniques offer the prospect of the permanent change that can only be accomplished by changing the germ plasm itself, they offer only the illusion of changing man.

The "New" Eugenics and the "Old"

In presenting a scenario of genetically "engineering" man,[12] Lederberg and Fletcher believe that current knowledge of genetics mandates a new eugenics to meet pressing human needs. There are two points to be made about any such proposal: (1) Concern about effecting widespread genetic changes in a population is unwarranted, given existing demographic trends; but (2) the general *motivation* for proposing cloning or other engineering of man must be taken seriously, because it reveals a tacit approval by some of the best minds of the country for both the legitimacy and the need for introducing genetic controls.

To some geneticists, the recrudescence of a social concern for applied human genetics is mandated by an assumed or projected deterioration of the genetic quality of the species. They frankly admit that this concern must be properly construed as a "eugenic" one, but insist that it is based on hard facts. They maintain that their concern is not tainted with the racial connotation that irrational eugenicists had applied in the past. Nevertheless, both the basis for this concern—a progressive "genetic deterioration" of man—and the proposed remedy —a humane form of "genetic counseling" or at an extreme "negative eugenics"—actually are synonymous with the analyses of a hundred years ago.

While Galton is the name usually associated with the "eugenics" movement of the late 1800's, it is actually Darwin whose ideas have endured. Galton described the aim of eugenics[13] (a word he coined) in blatantly racist, class-society terms. Its purpose was "to give the more suitable races or strains of blood a better chance of prevailing speedily over the less suitable [races] than they otherwise would have had."[14] Darwin, not Galton, represented the more representative and "morally enlightened" tone of the eugenics movement:

> With savages, the weak in body or mind are soon eliminated; and those that survive commonly exhibit a vigorous state of health. We civilized men, on the other hand, do our utmost to check the process of elimination; we build asylums for the imbecile, the maimed, and the sick; we institute poor-laws; and our medical men exert their utmost skill to save the life of everyone to the last moment. There is reason to believe that vaccination has preserved thousands, who from a weak constitution would formerly have succumbed to small pox. Thus the weak members of civilized society propagate their kind. No one who has attended to the breeding of domestic animals will doubt that this must be highly injurious to the race of man. It is surprising how soon want of care, or care wrongly directed leads to the degeneration of a domesticated race; but excepting in the case of man himself, hardly anyone is so ignorant as to allow his worst animals to breed. . . .
>
> The aid which we feel impelled to give to the helpless is mainly an incidental result of the instinct of sympathy, which was originally acquired as part of the social instincts, but subsequently rendered, in the manner previously indicated, more tender and more widely diffused. Nor could we check our sympathy, even at the urging of hard reason, without deterioration in the noblest part of our nature . . . if we were to neglect the weak and helpless, it could only be for a contingent benefit, with

an overwhelming present evil. *We must therefore bear the un-doubtedly bad effects of the weak surviving and propagating their kind; but there appears to be at least one check in steady action, namely that the weaker and inferior members of society do not marry so freely as the sound; and this check might be in-definitely increased by the weak in body or mind refraining from marriage, though this is more to be hoped for than expected.*[15]

"Expecting" the weak to refrain from marriage may strike us as a quaint nineteenth-century idea; but it faithfully echoes some contemporary statements of the value of "quasi-coercive" genetic counseling. There are some today who no longer "hope" but "expect" the weak in body and mind to refrain from marriage or its genetic equivalent childbearing. A growing number of people use moral arguments to urge those who are genetically "handicapped" (and this may only mean individuals who *carry* but do not express aberrant genes) to fulfil their social responsibility by refraining from procreation.[16] This moral suasion is mistakenly based on the assumption that genetic deterioration of the species will be the inevitable consequence of the "unbridled" procreation of the unfit.

Moral Obligations in the Face of Genetic Realities

Darwin's focus on the moral dilemmas facing those who think they recognize a genetic basis for human suffering and feel impelled to act on this assumption has a contemporary ring. Theodosius Dobzhansky assessed the eugenic situation in 1961 in this Darwinian tradition: "We are then faced with a dilemma—if we enable the weak and the deformed to live and to propagate their kind, we face the prospect of a genetic twilight; but if we let them die or suffer when we can save them, we face the certainty of a moral twilight. How to escape this dilemma?"[17] Thus ten years ago, the moral problems were not posed in terms of the need for genetic improvement, but rather in terms of the need for societal protection against genetic deterioration. The genetic information which made such an analysis valid thirty or even ten years ago has been substantially amended today.

In the recent past, the chief proponent of the need for eugenic practice was Hermann Muller. In 1959 he stated: "If we fail to act now to eradicate genetic defects, the job of ministering to infirmities would come to consume all the energy that society could muster for it, leaving no surplus for general, cultural purposes."[18] Other, more contemporary authors have voiced similarly concerned if not alarmist views.[19]

While no one can conclusively refute the contention that *some-*

time in the future we may have to come to grips with an increased incidence of genetically disabling disorders, it would have been extremely difficult to have made the case, even in 1959, for our moral obligations to act to anticipate them. As Martin Golding, in a review of genetic responsibility to future generations, concluded: "We are thus raising a question about our moral obligation to the community of the remote future. I submit that this relationship is far from clear, certainly less clear than our moral obligations to communities of the present. . . ."[20]

What actually is the "threat" posed to future generations (or, for that matter, to our very own children) by the specter of genetic deterioration? Golding and others appear to believe that current trends in medical treatment and protection of the "genetically unfit" condemn the future to suffer the weight of our omissions. He states, for example, that "the tragedy of the situation may be that we will have to reckon with the fact that the amelioration of short-term evils . . . and the promotion of good for the remote future are mutually exclusive alternatives."[21]

Part of the fallacy of this form of pessimism is the assumption that genes and genes alone are the only means by which we project ourselves into the future. Certainly, most anthropologists, when faced with the question of the most important way in which we influence the future, would emphasize the primacy of *cultural* factors in establishing human societies through time, because purely genetic trends are highly uncertain in fluctuating and migrating human populations.

The Fallacy of a Genetic Apocalypse

The other part of the fallacy is the assumption that we actually do face a genetically deteriorating situation. In the ten years since Dobzhansky originally posed the dilemma of a "genetic twilight," we have acquired enough information to enable us to draw back from the vision of a genetic apocalypse. Imminent "genetic deterioration" of the species is, for all intents and purposes, a red herring. The officers of the American Eugenics Society acknowledged this in a six-year report ending in 1970. In spite of the fact that they reaffirmed the long-range objective of the society to pursue the goal of maintaining or improving genetic potentialities of the human species, they stated that "neither present scientific knowledge, current genetic trends, nor social value justify coercive measures as applied to human reproduction." In fact, the officers wrote, "at this stage the need is for better identification of present and potential directions of changes rather than action to alter these trends in any major way."[22]

Our contemporary population is in a unique situation. The "gene pool" is in fact undergoing a period of stabilization, not change. In an analysis of the demographic trends characterizing the current population in the United States, Dudley Kirk observed that while the tremendous relaxation in the intensity of selection accomplished by modern medical achievements may be inexorably increasing the load of mutations the population carries, the over-all demographic trends are such as to reduce the number of children born with serious congenital abnormalities. He summarized his paper in the following way:

> A relaxation of selection intensity of the degree and durability now existing among Western and American peoples has surely never before been experienced by man. . . . In the short run, demographic trends (in and of themselves) are reducing the incidence of serious congenital anomalies. . . . In the foreseeable future, the possibility of medical and environmental correction of genetic defects will far outrun the effects of the growing genetic load.[23]

Demographic trends such as lowered average age of childbearing, smaller number of children, and the reduction of consanguineous marriages *themselves* effect dramatic changes in the quality of life experienced by the next generation. In the thirteen years between 1947 and 1960 when Japan instituted a revolutionary (if misleadingly termed) "Eugenic Protection Law," there was a 1/3 reduction in the number of children born with Mongolism and a 1/10 reduction in aggregate of all of the other major congenital abnormalities. This startling statistic was accomplished simply as a result of introducing legal abortion and encouraging smaller and earlier families.[24] A similar trend may well be expected in Western countries if we act to encourage the same *non-genetic* changes in our population. The data on the close relationship between higher maternal ages at birth, number of previous offspring, and the high incidence of such devastating congenital defects as anencephaly[25] and Mongolism make the moral imperative of recommending basic changes in childbearing patterns obvious. It is important to note that this kind of recommendation (for example, proscribing childbearing in women over thirty-five) has a universal basis, unlike proscriptions on individual childbearing for genetic reasons.

Societal vs. Individual Costs of Genetic Disease

Statistics such as these do not, however, tell us what specific moral questions are at stake for the future childbearing of individuals who themselves are born with a genetically determined disorder. Society's interest in this question acquires legitimacy only if it is true that society is paying an increasing social (not just monetary) cost for the offspring of the genetically unfit.

The origin of the notion of "societal cost" is rooted in the assumption that the care extended by society to the "unfit," while morally desirable, cannot be accomplished without heavy burden. It is widely accepted, for example, that medical advances have contributed to our genetic load by permitting individuals who are born with genetically determined disorders to survive to childbearing age. Is this in fact the case? The answer appears to be that *some* advances in medicine may have this effect, but that on the whole medical practice is neither generating a race of Orwellian invalids requiring daily injections of insulin, enzymes, and other crucial but absent substances *nor* is it permitting a critical number of the truly "unfit" to procreate.[26] A key but unique case in point would be retinoblastoma, a treatable eye tumor which until recently was fatal. "Treatment" here is understood to entail enucleation of the eye, with an increased residual risk of cancer elsewhere in the body even if initial surgery is successful. It is undeniable that the survival of individuals who can transmit the dominant mutant gene to their children poses grave moral problems to both the parents and society as a whole. Between 1930 and 1960 in the Netherlands, for example, the frequency of this dread cancer *doubled*, probably as a result of the procreation of survivors carrying the gene.[27] Another cogent example would be the legitimate societal interest in counseling or even in regulating childbearing in mothers with phenylketonuria, where there is grave danger of fetal damage and retardation. The moral issue becomes whether or not such statistics establish society's right to intervene in childbearing decisions by parents known to carry genes directly or indirectly causing grave disability in offspring.

With rare exception there is, in my opinion, no compelling case for societal restrictions on childbearing. I am profoundly disturbed by the advocacy of societal intervention in childbearing decisions for genetic reasons, denial of medical care to the congenitally damaged, or sterilization of those identified as likely to pass on the genetic basis for a constitutional disability. Such an advocacy is implicit in the tone of the following excerpt from a letter in *Science*: "Even elementary biology tells us that hereditary disease or susceptibility to disease which leads to death or diminished reproduction rids a population of genes which perpetuate these maladies. Yet modern medical practice is leading to the accumulation of such genes in the most highly advanced society of man."[28]

This statement, like the one of Darwin's one hundred years ago, miscasts the facts of natural selection in human populations. *The con-*

sensus of the best medical and genetic opinion is that whatever genetic deterioration is occurring as a result of decreased natural selection is so slow as to be insignificant when contrasted to "environmental" changes, including those produced by medical innovation.[29] Even where we have identified a disease in which medical advances can be *shown* to have increased the over-all population incidence, as in schizophrenia,[30] few if any competent geneticists would advocate reducing the number of offspring schizophrenic individuals would be permitted to bear. The principal reason is ignorance. We simply do not know what (if any) intellectually desirable attributes are also transmitted with the complex of genes responsible for schizophrenia. Bodmer notes that the conditions which have led to an increase in the frequency of schizophrenia "may also conceivably increase the frequency of some desirable genetic attributes in other individuals."[31]

The variability that we (and geneticists with considerably more perceptivity) "see" in people represents the top of an iceberg of genetic diversity in human populations. Most of the variability which can be found at the genetic level is the result of spontaneous mutations which become fixed in the population. The traditional attitude of geneticists was that these mutations were in the main "undesirable," and the number of mutations and the extent to which a population as a whole was subjected to them constituted society's genetic load. Dobzhansky has been diligent in pointing out that the original definitions of "genetic load" tended to be spurious because they hypothesized a single "best" genotype, specifically one which was "homozygous" (i.e., having the same genes on each chromosome pair) for all of its genes. In Dobzhansky's estimation, this notion was inconsistent with the fact that the nature of human populations is to have a tremendous proportion of their genomes (perhaps as much as 30%) made up of "heterozygous" genes, and thus, to be consistent, geneticists would have to regard genetic uniformity beneficial and genetic heterogeneity inimical to the fitness of the population.[32]

It now appears that the term "genetic load" must be considered as almost synonymous with "genetic variability" and to be similarly bereft of utility. An appreciable portion of the expressed and even greater portion of the concealed variability that we can recognize in man consists of variants that—in most environments—are to some degree unfavorable to the organism.[33] In spite of the tendency to term this unfavorable, deleterious, ostensibly unadaptive part the genetic "load" or "burden" of the population, there is little evidence that it is deleterious to the population as a whole to carry so many variant genes. In fact, the opposite appears true. To be consistent, those who favor this definition must regard genetic uniformity as the *summum bonum*, an attitude incompatible with the adaptive value of genetic diversity in nature. (A sophisticated analysis of the concept of genetic load is available.)[34]

While many would concur that the "load" imposed by novel or recurrent mutations should be minimized, the natural load of variant genes carried by a population is the result of forces exerted by natural selection. The "burden" of variant genes is a "load," according to Dobzhansky, only in the sense in which the expenditures a community makes to bring up and to educate its younger members are a "load" on that community. Genetic diversity is in one sense capital for investment in future adaptations. Since genetic variability represents evolutionary capability, it is a load we should be ready and willing to bear.

It is indeed ironic that just as man is coming to realize the value of the immense genetic diversity of his species,[35] he has embarked in a direction which threatens to restrict or curtail that diversity. For example, it would be unfortunate if the move to reduce the frequencies of specific "deleterious" genes through identification of heterozygotes by carrier detection screening resulted in broad sanctions on the very mating combinations (heterozygous x normal) which tend to perpetuate genetic diversity. Even where the deleteriousness of a *specific* gene is unquestionable, as in the case of the Hemoglobin S gene responsible for sickle-cell anemia, and the "diversity value" of maintaining high frequencies of the gene largely unsubstantiated, I believe that it would *still* be morally unacceptable to restrict childbearing by those heterozygotes married to normals. Part of the conceptual problem underlying the focus on heterozygous individuals as those responsible for ladening us with our "genetic load" is the false assumption that this load is in fact imposed on society only by a select few individuals. Hermann Muller professed this view when he stated:

A conscience that is socially oriented in regard to reproduction will lead many of the persons who are loaded with more than the average share of defects . . . to refrain voluntarily from engaging in reproduction to the average extent, while vice versa it will be considered a social service for those more fortunately endowed to reproduce to more than the average extent.[36]

Such a statement raises but fails to answer the profound moral question of how one identifies the "unfortunately" or "fortunately" genetically endowed. Today we realize that *each* individual bears a small but statistically significant number (variously estimated at 3-8) of deleterious genes. The moral attitude best fitted by our knowledge is that *a genetic burden is not something that a population is laden with, it is what a family is laden with.*

We now know that the very definition of the phrase "genetic load" is fraught with difficulty. As an alternative, Muller would ultimately have preferred to evaluate genetic load in man, as Sewall Wright did, in terms of the balance between the contribution that a carrier of a particular genotype makes to society and his "social cost."[37] Yet even this seemingly enlightened view suffers from the assumption that the worth of a man lies exclusively in his social utility. One quickly gets into the moral dilemma that Robert Gorney proposes when he attempts to assess the relative social worth of mentally defective people on the basis of their mother instincts, or dwarfs on the basis of their "court jestering."[38] Do not individuals have value unto themselves and their own families?

Protecting the Gene Pool or Supporting General Well-Being?

What then are the positions of geneticists themselves on the issue of how genetic knowledge should be used to guide human actions? Virtually all geneticists agree with James Crow that the principal hazard facing the human population stems from the introduction of new mutations through environmental agencies. Thus both James Neel and Joshua Lederberg feel that it is the geneticists' primary obligation to "protect the gene pool against damage." (Presumably, this would mean principally reducing the background levels of radiation and population exposure to mutagens.) However, they differ dramatically in their secondary concerns. Neel emphasizes the importance of stabilizing the gene pool through population control, realizing the genetic potential of the individual, and improving the quality of life through parental choice based on genetic counseling and prenatal diagnosis.[39] In contrast, Lederberg speaks of the crucial need for the detection and "humane containment" of the DNA lesions (sic, mutations) once they are introduced into the gene pool.[40]

There is a profound danger in discussing the need for "containment" or "quarantine," for purportedly genetically "hygienic" reasons, of individuals who by no fault of their own carry genes which place their offspring in jeopardy.[41] The case for society's concern for the genetic welfare of the population and its rights in opposing sanctions on individuals hinges on the demonstration of a clear and present danger of genetic deterioration, which, as I have indicated, is still forthcoming. Yet, a letter I received from a government official rhetorically equated the potential societal threat of genetic disease with that of a highly contagious bacterial one. An individual carrying a deleterious gene was, according to this analysis, analogous to a "Typhoid Mary." Such an attitude is at best naive, and at worst ominously coercive. To equate a genetic disease with one which can be transmitted from person to person is to fail to recognize the salient difference between the two: genetic diseases are transmissible only to offspring of the same family. Contagious diseases not only enjoy a much wider and rapid currency, but also an often fateful degree of anonymity, as in the faceless patrons of Typhoid Mary's restaurant. Only in the case of *genetic* disease do affected siblings and relatives serve as constant reminders of the fate of a subsequent affected child. Those who would argue that legal sanctions are necessary to protect society against genetic disease fail to recognize the basic reality of the deep and enduring bonds that draw a parent to his child. As Montaigne put it, "I have never seen a father who has failed to claim his child, however mangy or hunchbacked he might be. Not that he does not perceive his defect . . . but the fact remains the child is his."[42] A father bearing a heritable disorder himself or having experienced a lifetime of suffering in the genetic disability of his child would be the best judge to make the decision to deny life to his subsequent offspring. I know of no such situation (including retinoblastoma) where the decision to procreate or bear children should be the choice of other than the parents. The moral obligations of parents faced with genetic disease are to conscientiously weigh and act based on the prospects for their *children*, not for society at large. Genetic knowledge does not now justify enjoining any family with the societal obligation to refrain from procreation.

The Peril of a Genetic Imperative

In spite of the weight of evidence which shows that we do not have sufficient information to predict any but the grossest genetic changes following individual or population shifts in childbearing habits, the latent fear remains that to do *nothing* will itself lead to an increase in detrimental genes and thereby compound the genetic problem for future generations.[43] Joshua Lederberg has argued that we are so locked into a genetic double bind that we *should* in fact do nothing. He states:

> Our problem is compounded by every humanitarian effort to compensate for a genetic defect, insofar as this shelters the carrier [of the defective gene] from natural selection. So it must be accepted that medicine, even prenatal care which may permit the fragile fetus to survive, already intrudes on the questions of "Who shall live." . . .
>
> It is so difficult to do only good in such matters that we are best off putting our strongest efforts in the prevention of mutation, so as to minimize the heavy moral and other burdens of decision making once the gene pool has been seeded with them.[44]

Certainly, any decision to act or not to act in the face of the dilemmas posed by human genetics is a moral choice. But one does not escape the moral burden of choosing by rationalizing that intrinsic contradictions in relative goods freeze one into inaction.

As Lederberg rightfully observes, the moral contradictions in choices of this sort are never more clearly visible than in the protection of the "fragile" and by inference damaged fetus. In fact, developments in prenatal and postnatal care now make it possible to ensure the survival of infants burdened with spina bifida and meningomyelocoele, spinal abnormalities which were life-limiting before this decade. To the extent that such abnormalities (like cleft palate or harelip) are heritable, there is an ethical question in encouraging the survival and successful procreation of the affected individuals. What is too often ignored in simplistic analyses of this sort is that the increased survival of the defective and deformed is *not* the result of special and sometimes "precious" care of the weak, but rather is usually accomplished as an indirect result of dramatic improvements in health care to *all* infants. As a recent editorial in the British Medical Journal observed, "Indiscriminate lowering of early mortality may impose terrible burdens on the survivors. But for the overwhelming majority of infants, the normal and healthy, there is hope and increasing evidence that the measures which lower mortality tend to produce a corresponding improvement in the quality of life offered them."[45]

Lederberg's course of nonaction is effectively a course of action, and one which is as morally unacceptable today as bringing newborns to the *Lesch* for sorting and disposal in ancient Sparta. Improvement in prenatal and postnatal care may well encourage the survival of more of those "fragile" and presumably genetically defective fetuses and newborns who would normally succumb, but, as the experience in Britain shows, the cost of that type of action may well be worth paying. Would not mothers in a society which offered the promise of non-discriminative prenatal and postnatal care feel more secure than one (as in ancient Sparta) in which they knew that their children would be subjected to a test of normalcy? If selective care of only the genetically fit leads to a decrease in the survival of the specific few who are congenitally handicapped, it will be at the cost of a general *increase* in the damage wrought by uterine and early environmental deprivation (e.g., cerebral palsy and mental retardation). That would seem a high price for society to pay for its genetic well-being.

Summary

Our knowledge of genes and genetic systems in man shows them to be too complex to readily lend themselves to controlled manipulation. Deep-seated psychologic needs to reduce uncertainty appear to drive our search for genetic control in spite of this complexity. The need for genetic intervention is today justified on the basis of the same unsubstantiated analysis of "genetic deterioration" that characterized the eugenics movement in the late nineteenth century. The notion·of a genetic "burden" imposed on society by individuals carrying deleterious variant genes is a misleading concept: the "burden" of deleterious genes is borne by families, not society. Decisions to have or not have children are best made by parents who have experienced genetic disease in their own families, not by society. Society's obligation is to provide universal maternal and postnatal care, even at the cost of survival of the congenitally handicapped. To do less is both to deprive the healthy of the optimum conditions for their development and to jeopardize the moral tone of society itself.

NOTES

1. Letter from Oswald Avery to Roy Avery, May 17, 1943, in *Readings in Heredity*, ed. John A. Moore (New York: Oxford Univ. Press, 1972) pp. 249-51.

2. Hans Jonas, *The Phenomenon of Life* (New York: Harper & Row, 1966) p. 205.

3. See Marc Lappé, "Biological Warfare," in *Social Responsibility of the Scientist*, ed. Martin Brown (Berkeley: Free Press, 1970).

4. A condition resulting from an enzymatic defect in the ability to metabolize phenylalanine which is usually associated with mental retardation.

5. Joseph Fletcher, "Ethical Aspects of Genetic Controls," *New England Journal of Medicine* 285 (1971) 776-83.

6. See the discussion by Carl Jung in the Introduction to the *I Ching*, tr. Richard Wilhelm (Princeton: Bollingen Series XIX, 1967) p. xix, where he begins: "An incalculable amount of human effort is directed to combating the nuisance and danger represented by chance. . . ."

7. J. F. Crow, "Rates of Genetic Changes under Selection," *Proc. National Academy of Sciences* 59 (1968) 655-61.

8. J. Lederberg, "Experimental Genetics and Human Evolution," *American Naturalist* 100 (1966) 519-26. (Clonal propagation means using the nucleus of a single cell to propagate a whole organism genetically identical with it.)

9. Carl R. Merril, Mark R. Geier, and John Petricciani, "Bacterial Virus Gene Expression in a Human Cell," *Nature* 233 (1971) 398-400.

10. A. G. Schwartz, P. R. Cook, and Henry Harris, "Correction of a Genetic Defect in a Mammalian Cell," *Nature New Biology* 230 (1971) 5-7.

11. Jacques Ellul, *The Technological Society* (New York: Vintage, 1964) p. 11.

12. See J. Lederberg, "Unpredictable Variety Still Rules Human Reproduction," *Washington Post*, Sept. 30, 1967.

13. Eugenics is defined as "an applied science that seeks to maintain or improve the genetic potentialities of the human species" (Gordon Allen, in *International Encyclopedia of the Social Sciences* 5 [1968] 193).

14. Francis Galton, *Hereditary Genius* (London, 1870).

15. Charles Darwin, *The Descent of Man and Selection in Relation to Sex* (1871; New York: Random House Modern Library Edition) pp. 501-2 (italics mine).

16. See in particular Fletcher, *art. cit.*, and Bentley Glass's letter in reply to Leon R. Kass, *Science*, Jan. 8, 1971, p. 23.

17. Theodosius Dobzhansky, "Man and Natural Selection," *American Scientist* 49 (1961) 285-99.

18. H. J. Muller, "The Guidance of Human Evolution," *Perspectives in Biology and Medicine* 1 (1959) 590.

19. W. T. Vukovich, "The Dawning of the Brave New World—Legal, Ethical and Social Issues of Eugenics," *Univ. of Illinois Law Forum* 2 (1971) 189-231; B. Glass, "Human Heredity and Ethical Problems," *Perspectives in Biology and Medicine* 15 (1972) 237-53; R. Gorney, "The New Biology and the Future of Man," *UCLA Law Review* 15 (1968) 273-356.

20. M. Golding, "Our Obligations to Future Generations," *UCLA Law Review* 15 (1968) 443-79.

21. Golding, *ibid.*, p. 463.

22. T. Dobzhansky, D. Kirk, O. D. Duncan, and C. Bajema, *The American Eugenics Society, Inc. Six Year Report, 1965-1970* (published by the Society, New York).

23. Dudley Kirk, "Patterns of Survival and Reproduction in the United States," *Proc. Nat. Acad. Sci.* 59 (1968) 662-70.

24. *Ibid.*

25. Jean Fredrick, "Anencephalus: Variation with Maternal Age, Parity, Social Class and Region in England, Scotland, and Wales," *Ann. Human Genetics* (London) 34 (1970) 31-38.

26. Peter Brian Medawar, "Do Advances in Medicine Lead to Genetic Deterioration?" *Mayo Clinic Proceedings* 40 (1965) 23-33.

27. Anonymous, "The Changing Pattern of Retinoblastoma," *Lancet* 2 (1971) 1016-17.

28. "Biological Unsoundness of Modern Medical Practice," *Science* 165 (1969) 1313.

29. James V. Neel, "Lessons from a 'Primitive' People," *Science* 170 (1970) 815-22. See also John R. G. Turner, "How Does Treating Congenital Disease Affect the Genetic Load?" *Eugenics Quarterly*, 1968, pp. 191-96.

30. Walter F. Bodmer, "Demographic Approaches to the Measurement of Differential Selection in Human Populations," *Proc. Nat. Acad. Sci.* 59 (1968) 690-99.

31. Bodmer, *ibid.*, p. 699.

32. T. Dobzhansky, *Genetics and the Evolutionary Process* (New York: Columbia Univ. Press, 1970) p. 191.

33. Heterozygotes carrying a single dose of a recessive variant gene which is deleterious in the homozygous form are—contrary to popular belief—on the average *less* fit than the person who has both "normal" genes. The sickle-cell heterozygote, for example, is *only* at an advantage in malarial regions, having statistically less fitness than the normal in nonmalarial regions.

34. Bruce Wallace, *Genetic Load: Its Biological and Conceptual Aspects* (Englewood Cliffs, N.J.: Prentice-Hall, 1970).

35. L. C. Dunn, "The Study of Genetics in Man—Retrospect and Prospect," *Birth Defects Original Article Series* (The National Foundation, 1965).

36. H. J. Muller, "The Guidance of Human Evolution," *Perspectives in Biology and Medicine*, 1959, p. 590.

37. Dobzhansky, *Genetics and the Evolutionary Process*, p. 191.

38. Gorney, *art. cit.*, pp. 308-9.

39. Neel, *art. cit.*

40. Joshua Lederberg, "The Amelioration of Genetic Defect—A Case Study in the Application of Biological Technology," *Dimensions* 5 (1971) 13-51.

41. Margery Shaw, "*De jure* and *de facto* Restrictions on Genetic Counseling," *Proceedings of the Airlie House Conference* on "Ethical Issues in the Application of Human Genetic Knowledge," Oct. 10-14, 1971 (Plenum Press, in preparation).

42. Michel de Montaigne, "On the Education of Children," *Selected Essays*, tr. D. M. Frame (New York: Van Nostrand, 1943) chap. 26, p. 5.

43. Bentley Glass, *art. cit.*

44. Lederberg, "The Amelioration of Genetic Defect," p. 15.

45. Anonymous, "Early Deaths," *British Medical Journal*, 1971, pp. 315-16.

24 Recombinant DNA: A Proposal for Regulation

KEY DISMUKES

Of all of the many problems raised by the development of the technology of recombinant DNA and its apparent usefulness as a research tool in genetics, none was more serious than the issue of regulation. Although questions of academic freedom frequently arise, this issue gave rise to the possibility of federal legislation that would directly impact upon the use and development of a research technology. Dismukes reviews several proposals for regulating and overseeing recombinant DNA practices and develops his own model and proposal for how such regulation could balance the interests of the scientist and the public.

Key Dismukes, Ph.D., is Head of Staff of NAS Committee for Vision

The continuing controversy over recombinant DNA research raises many troubling questions. Foremost among them is the responsibility of scientists to society. No one, least of all scientists themselves, doubts that such a responsibility exists, but defining its character and finding ways to implement it is no easy task. The question, of course, is not new, but it has taken on a new dimension. Previous controversies, such as those over nuclear energy or environmental pollution, have concerned technology—how society applies scientific knowledge. With recombinant DNA, public debate has for the first time focused on hazards of the basic research itself.

Ironically, the public controversy was triggered by scientists' own efforts to anticipate and safeguard against potential hazards from this research. In 1974 a group of eminent molecular biologists requested the scientific community to refrain from certain types of

433

experiment until risks could be better evaluated. In the succeeding two years scientists met to assess the hazards and to forge a set of guidelines which prohibit certain high-risk experiments and require stringent safety procedures for others. In June 1976 the National Institutes of Health issued an environmental impact statement, and requested comment on the guidelines.

All these efforts—generally acknowledged to be conscientious—have not allayed public criticism and controversy. Wherever such research is proposed, city councils, state legislatures, and congressional bodies still grapple with the problem. The scientific community has been primarily concerned with its own hot debate over technical questions concerning the adequacy of the guidelines; it has failed to recognize that a different issue lies at the heart of public concern. Who will control this research? Who will determine what is an acceptable level of risk and enforce restrictions? These are not scientific questions, but issues of social policy. The public must have a choice in their resolution, because the public shares the risks.

In the furor of debate, the two issues have become confused. Scientists are legitimately concerned that controversy may distort and exaggerate hazards, but that is not an excuse for neglecting public involvement. It seems clear that this research will go on, and that it must be regulated. Scientists must join public representatives to develop a coherent, rational, and widely acceptable social policy.

I believe that it is possible to devise regulations and means of enforcement that will not be seen as unduly restrictive by scientists and that will protect the public and satisfy their legitimate concerns. To devise such a policy, we must involve all points of view in the decision-making process and accomplish three things: (1) clarify the differences in attitude toward risk-taking and assessing benefits, (2) take into account scientists' discomfort with outside regulation, and (3) examine the experiences of regulatory agencies to learn what might work in this situation and what might not. If these issues are explored, the approach required for regulation becomes clearer, and on this basis I propose a specific model for regulation and control of recombinant DNA research.

WEIGHING RISKS AND BENEFITS

The first question raised in discussions about recombinant DNA concerns the balance between risks and benefits. Scientists do not agree on the answer. Some prominent scientists have decried the continuing public controversy as overblown and a distortion of factual issues. Many regard the probability of inadvertently creating pathogenic bacteria as vanishingly small. Still others—equally competent—

have likened molecular biologists to sorcerers' apprentices who may create an uncontrollable cataclysm from which no magician will be able to rescue us. Small wonder the public is confused about the "facts."

Scientists contribute to the public's confusion by failing to recognize the mixture of fact and value in their own opinions. When an expert calls a risk "overblown" or "terrifying," he is combining his estimate of probability with his values toward risk-taking. Philosophers tell us we can never completely unravel the interwoven threads of fact and value; but in practice we can and must make distinctions.

To break out of this circular argument, we must know more precisely the probabilities of possible mishaps. Since we do not know enough about recombinant processes to make reliable estimates about their danger, we must do safety research which, ironically, involves doing hazardous experiments. A reasonable approach might be to construct a few special laboratories where, under extremely cautious controls, safety questions would be pursued. This suggestion has a subtle difficulty: programmed research tends to lose the power to ferret out the unexpected, and as DeWitt Stetten, the deputy director of NIH pointed out, the biggest dangers of recombinant DNA are those we have not yet thought of. Science is not a linear process; its power to clarify and weed out depends on a pluralistic, cyclic approach in which competing hypotheses are debated, refined, and tested. Thus a tense balance between caution and vigorous inquiry must be struck in bringing the full power of scientific scrutiny to bear on the potential of these techniques.

Yet even when sufficient data are available for scientists to reach some consensus about probabilities of risk and benefits, policy decisions will still have to be made in the face of uncertainty. At the very best, a consensus statement might read something like this: The probability of a certain type of disaster ranges between one in 10 million and one in a billion per laboratory per year, providing certain safety precautions are followed. Should such a disaster occur, the number of resulting cases of cancer would probably lie between 10 and 10,000, depending on the circumstances of release.

Our willingness to take risks is in part predicated on what we stand to gain; great potential benefits might justify large risks. Yet, the unpredictable character of scientific discovery precludes knowing all that we might gain, just as it precludes knowing all that we might risk. The problem is compounded by the fact that social benefits from basic research are generally diffuse, in contrast to the often discrete, traceable character of ill effects. We need to develop principles by which amorphous gains can be evaluated as easily as potential harms. One such principle might be that since progress in medicine and control of the environment hinges on understanding basic biologic function, then potential benefit to society will be proportional to the scientific importance of fundamental discovery.

The calculation of risk and benefit is further complicated by considerable divergence in what people consider beneficial and what their attitudes are toward risk-taking. Some people are gamblers; some are highly conservative. The scientific community considers the pursuit of knowledge worth a considerable amount of risk, but may differ in this from public attitudes. Compare the following statements:

"Since progress in our understanding of eukaryotic genome function is clearly prerequisite to an understanding of the nature of living systems, as well as of cancer, and many other genetic and developmental disorders, to delay these experiments on the basis of highly improbable hazards is unconscionable." (from an unsigned article, "In Defense of Plasmid Engineering," F.A.S. Public Interest Report, April 1976.)

"It is perhaps not irresponsible, but rather an act of enlightened courage, to expose ourselves to an unknown risk of disastrous epidemics in order to give ourselves a chance of lifting some hundreds of millions of our fellow humans out of the degradation of poverty." (Freeman Dyson, letter to *Science 193:* 6 [1976].)

". . . in that respect, scientific activity can be classified with art, music, the study of history or of philosophy—all valuable human enterprises, but not one of them which can be said to impose a duty on human society or on human individuals to pursue them." (Daniel Callahan, "Ethical Responsibility in Science," *Annals of the New York Academy of Sciences 265:* 1 [1976].)

"In the area of public policy, however, it would appear more important that no harm result from government action than that the government action result in benefit . . . government has no obligation to expend its funds for the purpose of bringing benefits, however substantial, to the public." (Harold Green, "Law and Genetic Control," *Annals of the New York Academy of Sciences, 265:* 170 [1976].)

The first two quotations are from scientists; the third and fourth, respectively, from a philosopher and a lawyer involved in bioethical issues. However, the differences in attitude expressed are not due alone to different biases inherent in these professions. As great a range of opinion might be found among congressmen or the public at large. Even if we were to grant an extraordinary degree of objectivity to scientists, their discussion would not explore the full scope of public cost and gain, simply because the scientific community represents only a small part of the spectrum of public perceptions of what is advantageous and what disadvantageous. Full exploration of these advantages and disadvantages requires vigorous public debate. Open dialogue is necessary to provide the public with information to which it has a right. Furthermore, this debate should generate feedback which would sensitize scientists to public concerns and better apprise them of public values about risk-taking. Feedback in turn should influence the evolution of safeguards and stimulate research aimed at questions of safety.

SCIENTISTS' ATTITUDES TOWARD REGULATION

The nastiest word in science is "politicization." Scientists are extremely nervous about outside intervention in the conduct of research. They conjure up specters ranging from uninformed meddling to censorship of politically unacceptable findings. Beyond indignant protestations about inviolability of the "right" of free scientific inquiry,[1] there has been little effort to characterize the problems of politicization.

Science has a close-knit set of rules of evidence and procedures for evaluation. Consensus about basic rules and procedures is sufficiently strong that scientists, even while vociferously disagreeing about interpretation of data, can comfortably accept the validity of each other's findings, and agree which procedures will resolve the question. Methods of resolving dispute in politics, where "fact" has much less compelling priority, dismay scientists. The frequently *ad hominem* nature of criticism is keenly resented. For example, Senator William Proxmire has publicly ridiculed the work of several scientists as frivolous squandering of tax money on absurd projects, without mentioning what sort of information is generated by that research or its relevance to man (a study of aggression and social adaptation in primates was caricatured as a study of how monkeys clench their jaws). That sort of attack has fueled scientists' notions that recombinant DNA is solely a technical question whose nuances would only be obscured by political debate and regulation.

A further and even more fundamental apprehension underlies scientists' protestations on regulation. There are no forbidden questions in science. No area or subject is sacrosanct if it can be approached by scientific procedures. A major factor in advancing scientific understanding and correcting error is the opportunity of critics to challenge prevailing views and, if they can adduce convincing evidence, to modify an existing consensus. This aspect of science is more than a convenient and useful tradition, it is as central to the operation of science as freedom of speech is to the maintenance of democracy. Political control of procedures for controlling recombinant DNA research would tend to isolate these guidelines from this process of challenge and evolution. As the dangers are better elucidated, the restrictions might quickly become archaic and basic research in the area might stagnate.

Obviously there are and always have been limits to what scientists may do. We may not abduct passersby as involuntary subjects of brain experiments. We may not destroy the atmosphere to discover how it was formed. We are not allowed to spend the entire national budget on a single experiment—no matter how elegant. Hans Jonas has said that scientific freedom ends at the boundary between thought and action, because action is always subject to legal and moral restraints. Thus, scientists are free to frame questions using exclusively scientific criteria, but are restricted in carrying out experiments to answer those questions.

Some additional distinctions need to be added to Jonas's principle. Heretofore, society's strictures on scientists' activities have been simply those that apply to all citizens. Experimentation has not been regulated *per se*; rather, scientists are enjoined from performing experiments in such a way that human rights are consequently infringed upon. In other words, the restrictions operate at the level of consequences rather than on the actions themselves. This is an important distinction because it allows scientists great freedom in designing experiments. Further, societal constraints applied in this way are not so different from those internal concerns that the scientist must always integrate into his procedures. He may modify his chemical synthesis to avoid an unstable intermediate product which might demolish his lab; he foregoes a procedure which would cause undue suffering in experimental animals; he looks for a cheaper way to get data that are beyond his budget.

This distinction has another aspect, a psychological one. A scientist, like everyone else, grows up within the context of cultural values, expressed in legal and social restraints. By the time he reaches the laboratory he is accustomed to moving within these constraints and is little aware of the way they delimit and shape his research. In contrast, social restrictions aimed directly at experimental procedures jar his consciousness, both by new impositions focused specifically on his work and because their nature cuts across the grain of the scientific approach.

Even those scientists who argue most vehemently against societally imposed regulation of recombinant DNA research do not deny the principle that protection of the public's health and rights must prevail over science's priorities when there is a conflict. The question is how to protect the public without undermining science.

DEVELOPING PUBLIC POLICY

Lawyers have criticized the procedures by which the NIH guidelines were evolved, on the grounds that neither scientists nor NIH administrators have the authority to establish a major social policy. Where then should such policy be developed and by what mechanisms is it to be enforced? To answer these questions we must consider three principles:

One. Whatever agency is responsible for policy must be capable of identifying and responding to issues that have been brought out in debates in city councils and congressional hearing rooms. There must be a synthesis and summarizing of the on-going dialogue, and an attempt to evaluate and reconcile divergent perspectives.

Two. Two kinds of expertise are required to evaluate technical questions and to assess social impact. The difficulty in separating the two aspects is illustrated by a central question which applies not only to recombinant DNA: how much expense should we take to avoid hazards of extremely low probability, say one in a billion per year, but which if they occurred would be of vast cost, perhaps a hundred thousand fatalities?

Three. Implementing social policy requires authority, which has three components.

(a) *Control.* Whatever agency is responsible, it must be able both to monitor recombinant DNA research and commercial applications and to enforce compliance with regulation.

(b) *Accountability.* Authority for public policy ultimately rests with elected representatives. In practice, much authority is delegated to various agencies. By creating and overseeing regulatory agencies Congress establishes a principle of public accountability.

(c) *Responsiveness.* Any agency deciding policy affecting public welfare must be responsive to public values and attitudes. This is not guaranteed simply by accountability through Congress, as illustrated by the number of government agencies which have been co-opted by the industries they regulate. Several years of public interest activism demonstrate that responsiveness to public concerns is greatly improved when business is conducted in public view and discussion and comment are invited at each stage of policy making. In addition, congressional oversight is usefully supplemented by equipping regulatory agencies with advisory boards which represent various segments of public interest.

MODELS FOR REGULATION

Given the needs we have sketched, what sort of approach should be used to regulate recombinant DNA research?

NIH Model. Most scientists would probably argue that the NIH guidelines were developed through procedures that were both desirable and adequate. In spite of the wide uncertainty about risk, there is a fair consensus among scientists that the guidelines impose a necessary and sufficient caution, while allowing research to continue. This approach follows the traditional model through which basic biomedical research has been publicly supported in this country. Congress has represented the public interest by providing funds and a broad mandate to NIH to pursue whatever basic and applied research seems most likely to promote health and improve medical care. NIH's administration of these funds is primarily guided by the biomedical research community. The utility of this approach is clear from NIH's pre-eminence in advancing biological knowledge and developing new medical techniques.

Proponents of this sort of approach to regulating recombinant procedures argue that it is necessary to bring the full and unfettered power of scientific scrutiny on the difficult technical questions involved. They see the NIH approach as by far the most flexible; and, as our understanding and power expand, this flexibility may be necessary in order to rapidly modify controls and change regulations.

A crucial aspect of any approach to regulation is the willing cooperation of scientists, who are the first line of defense against harm. There must be aggressive vigilance for unrecognized hazard and continuous self-scrutiny for failure of containment. As one of the most respected institutions of science, NIH can be a powerful force for promoting this attitude. In contrast, a heavy-handed external control that produced only letter-correct or begrudging cooperation would create an explosively volatile setting for mischance.

Nonetheless, the difficulties of self-regulation remain. Can the scientific community adequately represent public needs and priorities? Is it capable of framing and answering the questions of public policy? Does NIH have the authority? To expand its consideration NIH held public hearings on the proposed guidelines and set up a public board to advise the director in this area. These measures should create greater awareness of and—I hope—responsiveness to public interest. However, it leaves the director of NIH with conflicting responsibilities: he must both promote an area of research and simultaneously regulate it. The Atomic Energy Commission's performance should make us aware of the difficulties inherent in this kind of situation. Charged for thirty years with both developing nuclear power and safeguarding the public, the AEC failed to adequately explore possible hazards before committing us to a major development program.

If regulation is to be primarily in the hands of the scientific community, it is important that this be made a formal charge laid on a specific institution and administrators who would personally answer for neglect. When responsibility is diffuse and not invested in any particular person it tends to slip through everybody's fingers.

Could self-regulation produce effective enforcement? Self-regulation in other professions, especially law and medicine, is coming under mounting criticism. Scientific research, however, represents a different situation because of its highly communal nature. No individual can carry out a scientific advance alone; there is no impact until he can convince his peers of the correctness of his work. The power of a scientist to influence the process of consensus depends upon the trust of his colleagues, and, in this community, distrust and censure is a devastating weapon. Peer pressure is the most central and effective method of ensuring quality of scientific research. Would scientists apply it with equal vigor to enforcing guidelines? If so, it would provide a powerful force for compliance.

Furthermore, control must be international. Under the most ominous scenarios the entire world could be affected by organisms created in a single lab. Here again the cooperation of scientists is crucial: moral authority of the scientific community cuts across national boundaries to a degree unique in social institutions. Consensus of the scientific community will have more influence on foreign scientists than will any action of our government.

Beyond peer pressure, NIH and other government agencies can further control recombinant research by withholding funds from any scientist who does not give evidence of compliance with guidelines. This would be a fairly effective tool in academic research, which is largely funded through the federal government.

However, the NIH approach fails to meet two critical requirements for regulation of recombinant DNA: enforcement of industrial compliance and monitoring health. Recombinant DNA procedures offer vast commercial potential and are already being applied. Industry, of course, has its own research funds. Furthermore, industrial scientists and technicians operate under pressures different from those of academic peers. Can we assume that there will be no fudging when the product is not just data but a salable chemical product obtained by secret techniques? To monitor and enforce industrial compliance goes far beyond the authority, the expertise, the resources, and—most of all—the interest of NIH, which is strictly a research institute.

It is imperative to monitor the environment and community surrounding recombinant laboratories for signs of escaped pathogens. Without such surveillance a breach of containment might engender epidemic disease before detection. The existing guidelines make no such provisions, and it is doubtful that NIH has facilities for such a program.

From these difficulties it is clear that the NIH model cannot provide the protection the authors of the guidelines attempted to establish. However, NIH and the scientific community obviously must play a central role in establishing safety standards and ensuring compliance.

Statutory Regulation. Senators Kennedy and Javits have proposed that the NIH guidelines be made law. It is not clear from press accounts whether their intent is only to provide a stop-gap measure or to develop a permanent means of ensuring compliance. It may seem surprising that the very scientists who developed the guidelines vigorously object to statutory enactment. To spell out research protocols in statutes would be disastrous, for law is too rigid and too slowly changed. Within a matter of days we might discover some restrictions to be pointless and others to be woefully inadequate. Controls must be flexible and continuously modified as we learn more about the potentials of recombinant DNA.

The present guidelines set forth standard practices for general categories of hazard. The scientist is called upon to assess what category his experiments fall into and to apply his best judgment for means of achieving called-for standards. The uncertainty and potential of recombinant techniques are so great that scientists must not be encouraged to feel that their responsibility is satisfied simply by carefully following detailed procedural regulations; there must be aggressive scrutiny to catch the inevitable cases when the general rule is inadequate for the specific situation.

EPA Model. Regulatory agencies such as the Environmental Protection Agency (EPA) have had considerable experience with problems analogous to those of recombinant DNA research.

Several principles have emerged from the EPA experience and appear to be gaining political and public support.

• Rather than attempting to detail controls for each chemical that might be released into the environment, Congress has authorized EPA to establish administrative regulations. Congress determines EPA's area of responsibility, gives authority for enforcement, sets forth guidelines for fair administrative practice, and ideally outlines basic social policy to guide EPA in deciding particular cases.

• Environmental impact statements are required, forcing corporations and government bureaus to look to social consequences before implementing programs or marketing products.

• There is a shift in the philosophy of burden of proof. More safety research is required before a drug or chemical can be marketed. Recent legislation enables the EPA to suspend industrial production of a chemical if there is good reason to think that it *might* produce harm, whereas previously the agency was required to demonstrate that harm was actively occurring.

Adopting an EPA model for overseeing recombinant DNA practices would offer several advantages. It would provide a clear-cut channel of public accountability, without conflicting responsibilities. At the same time, scientists should appreciate it as a buffer from direct political intervention. By diffusing some of the tendency toward "politicization" it would help guarantee the consistency and continuity of approach that make a healthy atmosphere for good science. The scientific community would retain considerable influence by providing the agency with advisors and study groups. Such an agency would have the expertise to consider both technical aspects and social implications and to integrate with social policy.

On the practical side, a regulatory agency would have the power to enforce and the experience to monitor compliance. At the same time its regulations should be more flexible and adaptive than federal statutes. By providing access to decision-making, this approach should increase the public's confidence that its interests are being considered. By incorporating official policy, this country would be more convincing in persuading other countries to adopt rigorous standards.

In sum, a regulatory agency could deal with all the requirements and difficulties that have been discussed. The regulatory agencies have evolved a great deal of *practical* experience and procedures for dealing with the sorts of problems raised by recombinant DNA research. For example, the EPA has experience in environmental monitoring, monitoring a wide range of industrial processes and procedures, and ultimately of bringing offenders to court. If regulation is assigned to an existing agency, then its mechanisms and expertise can quickly and efficiently be set into motion.

A Proposal

The federal agency whose organization and responsibilities most nearly match the needs we have described is the Public Health Service. I propose that Congress instruct the PHS to consider recombinant DNA research within its purview, to be responsible for establishing and enforcing regulations, to monitor for public health threats, and to establish a research program to ascertain dangers and develop containment facilities. These responsibilities are similar to those already provided for in the congressional mandate to PHS to prevent the spread of communicable diseases.

Strictly speaking, PHS is not a regulatory agency, but performs basically similar functions. It could adopt an EPA approach to regulation of recombinant DNA technology without major reorganization. PHS would receive technical advice from NIH and the Communicable Disease Center (which are presently subordinate arms) and would contract out to these agencies specific responsibilities: NIH to investigate the nature and likelihood of hazards and methods of protection; CDC to monitor health.

The first action of PHS would be to examine the adequacy of the existing guidelines. I suspect that it would find these guidelines to represent our best understanding and our best approach for the time being. I would expect NIH to immediately set up a high-priority research effort, in a single laboratory with the highest containment facilities, to obtain the data we so urgently need to better assess these problems.

Recombinant DNA is not an isolated problem, but the beginning of an era of such problems. There will be other techniques, more hard-to-assess hazards and tantalizing opportunities. Rather than stumbling from one crisis to the next, throwing up stop-gap controls as we go, now is the time to begin to evolve a systematic policy.

The tensions between scientific freedom and public involvement must be balanced. To anticipate and delimit hazards requires the best efforts of a vigorous scientific enterprise. That in turn requires the freedom of scientists to challenge the consensus, with data and argument. We require not just the scientist's data, but also his way of asking questions. It is somehow both ironic and appropriate that the insight of science must be one of our major tools for preventing misapplication of its technologic offspring.

NOTES

1. I am tempted to suggest to scientists who argue against any need for public controls that there should then be no objection to making each principal investigator legally liable for any injury traceable to organisms orginating in his lab. Though impracticable in terms of evidence, the potential dimensions of such liability might introduce a sobering caution.

2. An exception is experimentation on humans, which falls along the boundary of medicine and science.

25 *"Making Babies" Revisited*

LEON R. KASS, M.D.

M79

The birth of Louise Brown whose conception occurred in vitro and was then transferred to her mother's uterus raised a number of questions about genetic engineering. Questions were also raised about the appropriateness of conducting research on the human embryo. Kass reviews a number of the arguments pro and con and then sets forth his own argument against the use of in vitro fertilization and embryo transplantation. Although the Ethics Advisory Board has allowed some research into in vitro fertilization, Kass nonetheless presents an interesting review of all of the arguments concerning this particular technology.

Leon Kass is with the University of Chicago Medical School

And the man knew not Eve his wife; but she conceived without him and bore Cain, and said: I have gotten a man with the help of Dr. Steptoe.

Ectogenesis IV, 1

And Isaac entreated the NIH for his wife, because she was barren; and the NIH let Itself be entreated of him, and Rebekah his wife conceived.

Ectogenesis XXV, 21

Seven years ago in the pages of this journal, in an article entitled "Making Babies—the New Biology and the 'Old' Morality" (Number 26, Winter 1972), I explored some of the moral and political questions raised by projected new powers to intervene in the processes of human reproduction. I concluded that it would be foolish to acquire and use these powers. The questions have since been debated in "bioethical" circles and in college classrooms, and they have received intermittent attention in the popular press and in sensational novels and movies. This past year they have gained the media limelight with the Del Zio suit against Columbia University, and more especially with the birth last

445

summer in Britain of Louise Brown, the first identified human baby born following conception in the laboratory.

Back in 1975, after prolonged deliberations, the National Commission for the Protection of Human Subjects of Biomedical and Behavioral Research issued its report and recommendations for research on the human fetus. And in the *Federal Register* of August 8, 1975, the Secretary of Health, Education, and Welfare published regulations regarding research, development, and related activities involving fetuses, pregnant women, and *in vitro* fertilization. These provided that no Federal monies should be used for *in vitro* fertilization of human eggs until a special Ethics Advisory Board reviews the special ethical issues and offers advice about whether government should support any such proposed research. There has been an effective moratorium on Federal support for human *in vitro* fertilization research since that time. But now the whole matter has once again become the subject of an intensive policy debate, for such a board has been established by HEW to consider whether the United States Government should finance research on human *in vitro* fertilization and embryo transfer.

The question has been placed on the policy table by a research proposal submitted to the National Institute of Child Health and Human Development by Dr. Pierre Soupart of Vanderbilt University. Dr. Soupart requested $465,000, over three years, for a study to define in part the genetic risk involved in obtaining early human embryos by tissue-culture methods. He proposes to fertilize about 450 human ova, obtained from donors undergoing gynecological surgery (i.e., not from women whom the research could be expected to help), with donor sperm, to observe their development for five to six days, and to examine them microscopically for chromosomal and other abnormalities before discarding them. In addition, Dr. Soupart proposes to study whether such laboratory-grown embryos can be frozen and stored without producing abnormalities; it is thought that temporary cold storage of human embryos might improve the success rate in the embryo-transfer procedure used to produce a child. Though Dr. Soupart does not now propose to do embryo transfers to women seeking to become pregnant, his research is intended to serve that goal: He seeks to reassure us that baby-making with the help of *in vitro* fertilization is safe; and he seeks to perfect the techniques introduced by Drs. Edwards and Steptoe in England.

Dr. Soupart's application was approved for funding by the National Institutes of Health review process in October 1977, but because of the administrative regulations it could not be funded without review by an Ethics Advisory Board. The Secretary of HEW, Joseph Califano, has constituted the 13-member Board, and charged it, not only with a decision on the Soupart proposal, but with an inquiry into all the scientific, ethical, and legal issues involved, urging it "to provide recommendations on broad principles to guide the Department in future decision making." The Board, comprising a distinguished group of physicians, academics, and laymen, has invited expert and public testimony on the widest range of questions. By the end of the first phase of its work, it will have held at least 11 meetings and public hearings all over the United States, offering all interested citizens or groups the chance to express their opinions.

I was asked by the Board to discuss the ethical issues raised by the proposed research on human *in vitro* fertilization, laboratory cultures of—and experimentation with—human embryos, and the intrauterine transfer of such embryos for the purpose of assisting human generation. In addition, I was asked to comment on the appropriateness of Federal funding of such research and on the implications of this work for the provision of health care. The present article is based largely on testimony given before the Ethics Advisory Board, at its Boston meeting, October 13–14, 1978.

II

How should one think about the ethical issues, here and in general? There are many possible ways, and it is not altogether clear which way is best. For some people ethical issues are immediately matters of right and wrong, of purity and sin, of good and evil. For others, the critical terms are benefits and harms, risks and promises, gains and costs. Some will focus on so-called rights of individuals or groups, e.g., a right to life or childbirth; still others will emphasize so-called goods for society and its members, such as the advancement of knowledge and the prevention and cure of disease.

My own orientation here is somewhat different. I wish to suggest that before deciding what to do, one should try to understand the implications of doing or not doing. The first task, it seems to me, is not to ask "moral or immoral?" "right or wrong?" but to try to understand fully the meaning and significance of the proposed actions.

This concern with significance leads me to take a broad view of the matter. For we are concerned here not only with the proposed research of Dr. Soupart, and the narrow issues of safety and informed consent it immediately raises, but also with a whole range of implications including many that are tied to definitely foreseeable consequences of this research and its predictable extensions—and touching even our common conception of our own humanity. The very establishment of a special Ethics Advisory Board testifies that we are at least tacitly aware that more is at stake than in ordinary biomedical research, or in experimenting with human subjects at risk of bodily harm. At stake is the *idea* of the *humanness* of our human life and the meaning of our embodiment, our sexual being, and our relation to ancestors and descendants. In reaching the necessarily particular and immediate decision in the case at hand, we must be mindful of the larger picture and must avoid the great danger of trivializing this matter for the sake of rendering it manageable.

III

What is the status of a fertilized human egg (i.e., a human zygote) and the embryo that develops from it? How are we to regard its being? How are we to regard it morally, i.e., how are we to behave toward it? These are, alas, all-too-familiar questions. At least analogous, if not identical, questions are central to the abortion controversy and are also crucial in considering whether and what sort of experimentation is properly conducted on living but aborted fetuses. Would that it were possible to say that the matter is simple and obvious, and that it has been resolved to everyone's satisfaction!

But the controversy about the morality of abortion continues to rage and divide our nation. Moreover, many who favor or do not oppose abortion do so despite the fact that they regard the pre-viable fetus as a living human organism, even if less worthy of protection than a woman's desire not to give it birth. Almost everyone senses the importance of this matter for the decision about laboratory culture of, and experimentation with, human embryos. Thus, we are obliged to take up the question of the status of the embryos, in a search for the outlines of some common ground on which many of us can stand. To the best of my knowledge, the discussion which follows is not informed by any particular sectarian or religious teaching, though it may perhaps reveal that I am a person not devoid of reverence and the capacity for awe and wonder, said by some to be the core of the "religious" sentiment.

I begin by noting that the circumstances of laboratory-grown blastocysts (i.e., 3-to-6-day-old embryos) and embryos are not identical with those of the analogous cases of 1) living fetuses facing abortion and 2) living aborted fetuses used in research. First, the fetuses whose fates are at issue in abortion are unwanted, usually the result of "accidental" conception. Here, the embryos are wanted, and deliberately created, despite certain knowledge that many of them will be destroyed or discarded.[1] Moreover, the fate of these embryos is not in conflict with the wishes, interests, or alleged rights of the pregnant women. Second, though the HEW guidelines governing fetal research permit studies conducted on the not-at-all-viable aborted fetus, such research merely takes advantage of available "products" of abortions not themselves undertaken for the sake of the research. No one has proposed and no one would sanction the deliberate production of live fetuses to be aborted for the sake of research, even very beneficial research.[2] In contrast, we are here considering the deliberate production of embryos for the express purpose of experimentation.

The cases may also differ in other ways. Given the present state of the art, the largest embryo under discussion is the blastocyst, a spherical, relatively undifferentiated mass of cells, barely visible to the naked eye. In appearance it does not look human; indeed, only the most careful scrutiny by the most experienced scientist might distinguish it from similar blastocysts of other mammals. If the human zygote and blastocyst are more like the animal zygote and blastocyst than they are like the 12-week-old human fetus (which already has a humanoid appearance, differentiated organs, and electrical activity of the brain), then there would be a much-diminished ethical dilemma regarding their deliberate creation and experimental use. Needless to say, there are articulate and passionate defenders of all points of view. Let us try, however, to consider the matter afresh.

First of all, the zygote and early embryonic stages are clearly alive. They metabolize, respire, and respond to changes in the environment; they grow and divide. Second, though not yet organized into distinctive parts or organs, the blastocyst is an organic whole, self-developing, genetically unique and distinct from the egg and sperm whose union marked the beginning of its career as a discrete, unfolding being. While the egg and sperm are alive as cells, something new and alive *in a different sense* comes into being with fertilization. The truth of this is unaffected by the fact that fertilization takes time and is not an instantaneous event. For after fertilization is *complete*, there exists a new individual, with its unique genetic identity, fully potent for the self-initiated development into a mature human being, if circumstances are cooperative. Though there is some sense in which the lives of egg and sperm are continuous with the life of the new organism-to-be (or, in human terms, that the parents live on in the child or child-to-be), in the decisive sense there is a discontinuity, a new beginning, with fertilization. *After* fertilization, there is continuity of subsequent development, even if the locus of the embryo alters with implantation (or birth). Any honest biologist must be impressed by these facts, and must be inclined, at least on first glance, to the view that a human life begins at fertilization. Even Dr. Robert Edwards has apparently stumbled over this truth, perhaps inadvertently, in the remark about Louise Brown attributed to him in an article by Peter Gwynne in *Science Digest:* "The last time I saw *her, she* was just eight cells in a test-tube. *She* was beautiful *then,* and she's still beautiful *now!*"[3]

But granting that a human life begins at fertilization, and comes-to-be via a continuous process thereafter, surely—one might say—the blastocyst itself is hardly a human being. I myself would agree that a blastocyst is not, in a *full* sense, a human being—or what the current fashion calls, rather arbitrarily and without clear definition, a person. It does not look like a human being nor can it do very much of what human beings do. Yet, at the same time, I must acknowledge that the human blastocyst is 1) human in origin and 2) *potentially* a mature human being, if all goes well. This too is beyond dispute; indeed it is precisely because of its peculiarly human potentialities that people propose to study *it* rather than the embryos of other mammals. The human blastocyst, even the human blastocyst *in vitro,* is not humanly nothing: it possesses a power to become what everyone will agree is a human being.

Here it may be objected that the blastocyst *in vitro* has today no such power, because there is now no way *in vitro* to bring the blastocyst to that much later fetal stage at which it might survive on its own. There are no published reports of culture of human embryos past the blastocyst stage (though this has been reported for mice). The *in vitro* blastocyst, like the 12-week-old aborted fetus, is *in this sense* not viable (i.e., it is at a stage of maturation before the stage of possible independent existence). But if we distinguish among the *not*-viable embryos, between the *pre*-viable and the *not-at-all* viable—on the basis that the former, though not yet viable is capable of *becoming* or *being made* viable[4]—we note a crucial difference between the blastocyst and the 12-week abortus. Unlike an aborted fetus, the blastocyst is possibly salvageable, and hence *potentially* viable *if it is transferred to a woman for implantation.* It is not strictly true that the *in vitro* blastocyst is *necessarily* not-viable. Until proven otherwise, by embryo transfer and attempted implantation, we are right to consider the human blastocyst *in vitro* as potentially a human being and, in this respect, not fundamentally different from a blastocyst *in utero.* To put the matter more forcefully, the blastocyst *in vitro* is *more* "viable," in the sense of more salvageable, than aborted fetuses at most later stages, up to say 20 weeks.

This is not to say that such a blastocyst is therefore endowed with a so-called right to life, that failure to implant it is negligent homicide, or that experimental touchings of such blastocysts constitute assault and battery. (I myself tend to reject such claims, and indeed think that the ethical questions are not best posed in terms of "rights.") But the blastocyst is not nothing; it is *at least* potential humanity, and as such it elicits, or ought to elicit, our feelings of awe and respect. In the blastocyst, even in the zygote, we face a mysterious and awesome power, a power governed by an immanent plan that may produce an indisputably and fully human being. It deserves our respect not because it has rights or claims or sentience (which it does not have at this stage), but because of what it is, now *and* prospectively.

Let us test this provisional conclusion by considering intuitively our response to two possible fates of such zygotes, blastocysts, and early embryos. First, should such an embryo die, will we be inclined to mourn its passing? When a woman we know miscarries, we are sad—largely for *her* loss and disappointment, but perhaps also at the premature death of a life that might have been. But we do not mourn the departed fetus, nor do we seek ritually to dispose of the remains. In this respect, we do not treat even the fetus as fully one of us.

On the other hand, we would I suppose recoil even from the thought, let alone the practice—I apologize for forcing it upon the reader—of eating such embryos, should someone discover that they would provide a great delicacy, a "human caviar." The human blastocyst would be protected by our taboo against cannibalism, which insists on the humanness of human flesh and which does not permit us to treat even the flesh of the dead as if it were mere meat. *The human embryo is not mere meat; it is not just stuff; it is not a thing.*[5] Because of its origin and because of its capacity, it commands a higher respect.

How much more respect? As much as for a fully developed human being? My own inclination is to say "probably not," but who can be certain? Indeed, there might be prudential and reasonable grounds for an affirmative answer, partly because the presumption of ignorance ought to err in the direction of never underestimating the basis for respect of human life, partly because so many people feel very strongly that even the blastocyst is protectably human. As a first approximation, I would analogize the early embryo *in vitro* to the early embryo *in utero* (because both are potentially viable and human). On this ground alone, *the most sensible policy is to treat the early embryo as a previable fetus, with constraints imposed on early embryo research at least as great as those on fetal research.*

To some this may seem excessively scrupulous. They will argue for the importance of the absence of distinctive humanoid appearance or the absence of sentience. To be sure, we would feel more restraint in invasive procedures conducted on a five-month-old or even 12-week-old living fetus than on a blastocyst. But this added restraint on inflicting suffering on a "look-alike," feeling creature in no way denies the propriety of a prior restraint, grounded in respect for individuated, living, potential humanity. Before I would be persuaded to treat early embryos differently from later ones, I would insist on the establishment of a reasonably clear, naturally grounded boundary that separates "early" and "late," and which provides the basis for respecting "the early" less than "the late." This burden *must* be accepted by proponents of experimentation with human embryos *in vitro* if a decision to permit creating embryos for such experimentation is to be treated as ethically responsible.

IV

Where does the above analysis lead in thinking about treatment of *in vitro* human embryos? I shall indicate, very briefly, the lines toward a possible policy, though that is not my major intent.

The *in vitro* fertilized embryo has four possible fates: 1) *implantation*, in the hope of producing from it a child; 2) *death*, by active "killing" or disaggregation, or by a "natural" demise; 3) use in *manipulative experimentation*—embryological, genetic, etc.; 4) use in attempts at *perpetuation in vitro* beyond the blastocyst stage, ultimately, perhaps, to viability. I will not now consider this fourth and future possibility, though I would suggest that full laboratory growth of an embryo into a viable human being (i.e., ectogenesis), while perfectly compatible with respect owed to its potential humanity as an individual, may be incompatible with the kind of respect owed to its humanity that is grounded in the bonds of lineage and the nature of parenthood.

On the strength of my analysis of the status of the embryo, and the respect due it, no objection would be raised to implantation. *In vitro* fertilization and embryo transfer to treat infertility, as in the case of Mr. and Mrs. Brown, is perfectly compatible with a respect and reverence for human life, including potential human life. Moreover, no disrespect is intended or practiced by the mere fact that several eggs are removed for fertilization, to increase the chance of success. Were it possible to guarantee successful fertilization and normal growth with a single egg, no more would need to be obtained. Assuming nothing further is done with the unimplanted embryos, there is nothing disrespectful going in. The demise of the unimplanted embryos would be analogous to the loss of numerous embryos wasted in the normal *in vivo* attempts to generate a child. It is estimated that over 50 percent of eggs successfully fertilized during unprotected sexual intercourse fail to implant, or do not remain implanted, in the uterine wall, and are shed soon thereafter, before a diagnosis of pregnancy could be made. Any couple attempting to conceive a child tacitly accepts such embryonic wastage as the perfectly acceptable price

to be paid for the birth of a (usually) healthy child. Current procedures to initiate pregnancy with laboratory fertilization thus differ
from the natural "procedure" in that what would normally be spread
over four or five months *in vivo* is compressed into a single effort, using all at once a four or five months' supply of eggs.[6]

Parenthetically, we should note that the natural occurrence of
embryo and fetal loss and wastage does not necessarily or automatically justify all deliberate, humanly caused destruction of fetal life.
For example, the natural loss of embryos in early pregnancy cannot in
itself be a warrant for deliberately aborting them or for invasively experimenting on them *in vitro,* any more than stillbirths could be a justification for newborn infanticide. There are many things that happen
naturally that we ought not to do deliberately. It is curious how the
same people who deny the relevance of nature as a guide for evaluating human interventions into human generation, and who deny that
the term "unnatural" carries any ethical weight, will themselves appeal to "nature's way" when it suits their purposes.[7] Still, in this
present matter, the closeness to natural procreation—the goal is the
same, the embryonic loss is unavoidable and not desired, and the
amount of loss is similar—leads me to believe that we do no more intentional or unjustified harm in the one case than in the other, and
practice no disrespect.

But must we allow *in vitro* unimplanted embryos to die? Why
should they not be either transferred for "adoption" into another infertile woman, or else used for investigative purposes, to seek new
knowledge, say about gene action? The first option raises questions
about the nature of parenthood and lineage to which I will return. But
even on first glance, it would seem likely to raise a large objection
from the original couple, who were seeking a child of their own and
not the dissemination of their "own" biological children for prenatal
adoption.

But what about experimentation on such blastocysts and early
embryos? Is that compatible with the respect they deserve? This is
the hard question. On balance, I would think not. Invasive and manipulative experiments involving such embryos very likely presume
that they are things or mere stuff, and deny the fact of their possible
viability. Certain observational and non-invasive experiments might
be different. But on the whole, I would think that the respect for human embryos for which I have argued—I repeat, not their so-called
right to life—would lead one to oppose most potentially interesting
and useful experimentation. This is a dilemma, but one which cannot
be ducked or defined away. Either we accept certain great restrictions
on the permissible uses of human embryos or we deliberately decide

to override—though I hope not deny—the respect due to the embryos.

.I am aware that I have pointed toward a seemingly paradoxical conclusion about the treatment of the unimplanted embryos: Leave them alone, and do not create embryos for experimentation only. To let them die "naturally" would be the most respectful course, grounded on a reverence, generically, for their potential humanity, and a respect, individually, for their being the seed and offspring of a particular couple who were themselves seeking only to have a child of their own. An analysis which stressed a "right to life," rather than respect, would of course lead to different conclusions. Only an analysis of the status of the embryo which denied both its so-called "rights" or its worthiness of all respect would have no trouble sanctioning its use in investigative research, donation to other couples, commercial transactions, and other activities of these sorts.

V

The attempt to generate a child with the aid of *in vitro* fertilization constitutes an experiment upon the prospective child. It thus raises a most peculiar question for the ethics of human experimentation: Can one ethically choose for a yet hypothetical, unconceived child-to-be the unknown hazards he must face, obviously without his consent, and simultaneously choose to give him life in which to face them? This question has been much debated, as it points to a serious and immediate ethical concern: the hazards of manipulating the embryo as it bears on the health of the child-to-be.

Everyone agrees that human-embryo transfer for the sake of generation should not be performed until prior laboratory research in animals has provided a sound basis for estimating the likely risks to any human beings who will be born as a result of this transfer and gestation. Argument centers on whether a sufficiently sound basis for estimating the likely risks to humans *can be* provided by animal experiments, and, if so, whether adequate experimentation has been done, and what level of risk is acceptable.

There is, it seems to me, good reason for insisting that risk of incidence and likely extent of possible harm be very, very low—lower, say, than in therapeutic experimentation in children or adults. But I do not think that the risk of harm must be positively excluded (and it certainly cannot be). It would suffice if those risks were equivalent to, or less than, the risks to the child from normal procreation. To insist on more rigorous standards, especially when we permit known carriers of genetic disease to reproduce, would seem a denial of equal treatment to infertile couples contemplating *in vitro* assistance. Moreover, it is to give undue weight to the importance of bodily harm over against risks of poor nurture and rearing after birth. Wouldn't the couple's great eagerness for the child count, in the promise of increased parental affection, toward offsetting even a slightly higher but unknown risk of mental retardation?

Finally, to insist on extra-scrupulosity regarding risks in laboratory-assisted reproduction is to attach too much of one's concern to the wrong issue. True, everyone understands about harming children, while very few worry about dehumanization of procreation or problems of lineage. But those are the things that are distinctive about laboratory-assisted reproduction, not the risk of bodily harm to offspring. It should suffice that the risks be comparable to those for ordinary procreation, not greater but no less.

It remains a question whether we now know enough about these risks to go ahead with human-embryo transfer. Here I would defer to the opinions of the cautious experts—for caution is the posture of responsibility toward such prospective children. I would agree with Doctors Luigi Mastroianni, Benjamin Brackett, and Robert Short—all researchers in the field—that the risks for humans have not yet been sufficiently assessed, in large part because the risks in animals have been so poorly assessed (due to the small numbers of such births and to the absence of any *prospective* study to identify and evaluate deviations from the norm).

VI

Many people rejoiced at the birth of Louise Brown. Some were pleased by the technical accomplishment, many were pleased that she was born apparently in good health. But most of us shared the joy of her parents, who after a long, frustrating, and fruitless period, at last had the pleasure and blessing of a child of their own. The desire to have a child of one's own is acknowledged to be a powerful and deep-seated human desire—some have called it "instinctive"—and the satisfaction of this desire, by the relief of infertility, is said to be one major goal of continuing the work with *in vitro* fertilization and embryo transfer. That this is a worthy goal few, if any, would deny.

Yet let us explore what is meant by *"to have a child of one's own."* First, what is meant by *"to have"*? Is the crucial meaning that of gestating and bearing? Or is it "to have" as a possession? Or is it to nourish and to rear, the child being the embodiment of one's activity as teacher and guide? Or is it rather to provide someone who descends and comes after, someone who will replace oneself in the family line or preserve the family tree by new sproutings and branchings?

More significantly, what is meant by *"one's own"*? What sense of one's own is important? A scientist might define "one's own" in terms of carrying one's own genes. Though in some sense correct, this cannot be humanly decisive. For Mr. Brown or for most of us, it would not be a matter of indifference if the sperm used to fertilize the egg were provided by an identical twin brother—whose genes would be, of course, the same as his. Rather, the humanly crucial sense of "one's own," the sense that leads most people to choose their own, rather than to adopt, is captured in such phrases as "my seed," "flesh of my flesh," "sprung from my loins." More accurately, since "one's own" is not the own of one but of *two*, the desire to have a child of "one's own" is *a couple's desire* to embody, out of the conjugal union of their separate bodies, a child who is flesh of their separate flesh made one. This archaic language may sound quaint, but I would argue that this is precisely what is being celebrated by most people who rejoice at the birth of Louise Brown, whether they would articulate it this way or not. Mr. and Mrs. Brown, by the birth of their daughter, fulfill this aspect of their separate sexual natures and of their married life together, they acquire descendants and a new branch of their joined family tree, and the child Louise is given solid and unambiguous roots from which she has sprung and by which she will be nourished.

If this were to be the *only* use made of embryo transfer, and if providing *in this sense* "a child of one's own" were indeed the sole reason for the clinical use of the techniques, there could be no objection. Yet there will almost certainly be other uses, involving third parties, to satisfy the desire "to have" a child of "one's own" in different senses of "to have" and "one's own." I am not merely speculating about future possibilities. With the technology to effect human *in vitro* fertilization and embryo transfer comes the *immediate* possibility of egg donation (egg from donor, sperm from husband), embryo donation (egg and sperm from outside of the marriage), and foster pregnancy (host surrogate for gestation).

Nearly everyone agrees that these circumstances are morally and perhaps psychologically more complicated than the intra-marital case. Here the meaning of "one's own" is no longer so unambiguous; neither is the meaning of motherhood and the status of pregnancy. On the one hand, it is argued that embryo donation, or "prenatal adoption," would be superior to present adoption, because the woman would have the experience of pregnancy and the child would be born of the "adopting" mother, rendering the maternal tie even more close. On the other hand, the mother-child bond rooted in pregnancy and delivery is held to be of little consequence by those who would endorse the use of surrogate gestational "mothers," say for a woman whose infertility is due to uterine disease rather than ovarian disease or oviduct obstruction. Clearly, the "need" and demand for extra-marital embryo transfer are real and probably large, probably even greater than the intra-marital ones. Already, the Chairman of the Ethics Advisory Board has testified in Congress about the need to de-fine the responsibilities of *the donor* and the recipient "parents." Thus the new techniques will not only serve to ensure and preserve lineage, but will also serve to confound and complicate it. The principle truly at work here is not to provide married couples with a child of *their own,* but to provide anyone who wants one with a child, by whatever possible or convenient means.

"So what?" it will be said. First of all, we already practice and enourage adoption. Second, we have permitted artificial insemina-tion—though we have, after some 40 years of this practice, yet to re-solve questions of legitimacy. Third, what with the high rate of divorce and remarriage, identification of "mother," "father," and "child" is already complicated. Fourth, there is a growing rate of ille-gitimacy and husbandless parentages. Fifth, the use of surrogate mothers for foster pregnancy has already occurred, with the aid of ar-tificial insemination.[8] Finally, our age in its enlightenment is no long-er so certain about the virtues of family, lineage, and heterosexuality, or even about the taboos against adultery and incest. Against this background, it will be asked, why all the fuss about some little em-bryos that stray from the nest?

It is not an easy question to answer. Yet, consider. We practice adoption because there are abandoned children who need a good home. We do not, and would not, encourage people deliberately to generate children for others to adopt; partly we wish to avoid baby markets, partly we think it unfair to the child deliberately to deprive him of his natural ties. Recent years have seen a rise in our concern with roots, against the rootless and increasingly homogeneous back-ground of contemporary American life. Adopted children, in particu-lar, are pressing for information regarding their "real parents," and some states now require that such information be made available (on that typically modern ground of "freedom of information," rather than because of the profound importance of lineage for self-identity). The practice of artificial insemination has yet to be evaluated, the se-crecy in which it is practiced being an apparent concession to the dan-gers of publicity.[9] Indeed, most physicians who practice artificial insemination routinely mix in some semen from the husband, to pre-serve some doubt about paternity—again, a concession to the impor-tance of lineage and legitimacy. Finally, what about the changing mores of marriage, divorce, single-parent families, and sexual behavior? Do we applaud these changes? Do we want to contribute further to the confusion of thought, identity, and practice?[10]

Properly understood, the largely universal taboos against incest, and also the prohibition against adultery, suggest that clarity about who your parents are, clarity in the lines of generation, clarity about who is whose, are the indispensable foundations of a sound family life, itself the sound foundation of civilized community. Clarity about your origins is crucial for self-identity, itself important for self-respect. It would be, in my view, deplorable public policy further to erode such fundamental beliefs, values, institutions, and practices. This means, concretely, no encouragement of embryo adoption or especially of surrogate pregnancy. While it would be perhaps foolish to try to proscribe or outlaw such praactices, it would not be wise for the Federal government to foster them. The Ethics Advisory Board should carefully consider whether it should and can attempt to restrict the use of embryo transfer to the married couple from whom the embryo derives.

The case of surrogate wombs bears a further comment. While expressing no objection to the practice of foster pregnancy itself, some people object that it will be done for pay, largely because of their fear that poor women will be exploited by such a practice. But if there were nothing wrong with foster pregnancy, what would be wrong with making a living at it? Clearly, this objection harbors a *tacit* understanding that to bear another's child for pay is in some sense a degradation of oneself—in the same sense that prostitution is a degradation *primarily* because it entails the loveless surrender of one's body to serve another's lust, and *only derivatively* because the woman is paid. It is to deny the meaning and worth of one's body, to treat it as a mere incubator, divested of its human meaning. It is also to deny the meaning of the bond among sexuality, love, and procreation. The buying and selling of human flesh and the dehumanized uses of the human body ought not to be encouraged. To be sure, the practice of womb donation could be engaged in for love not money, as it apparently has been in the case in Michigan. A woman could bear her sister's child out of sisterly love. But to the degree that one escapes in this way from the degradation and difficulties of the *sale* of human flesh and bodily services, and the treating of the body as stuff (the problem of cannibalism), one approaches instead the difficulties of incest and near-incest.

VII

Objections have been raised about the deliberate technological intervention into the so-called natural processes of human reproduction. Some would simply oppose such interventions as "unnatural," and therefore wrong. Others are concerned about the consequences of these interventions, and about their ends and limits. Again, I think it important to explore the meaning and possible significance of such interventions, present and projected, especially as they bear on fundamental beliefs, institutions, and practices. To do so requires that we consider likely future developments in the laboratory study of human reproduction. Indeed, I shall argue that we *must* consider such future developments in reaching a decision in the present case.

What can we expect in the way of new modes of reproduction, as an outgrowth of present studies? To be sure, prediction is difficult. One can never know with certainty what will happen, much less how soon. Yet uncertainty is not the same as simple ignorance. Some things, indeed, seem likely. They seem likely because 1) they are thought necessary or desirable, at least by some researchers and their sponsors, 2) they are probably biologically possible and technically feasible, and 3) they will be difficult to prevent or control (especially if no one anticipates their development or sees a need to worry about them). One of the things the citizenry, myself included, would expect from an Ethics Advisory Board and our policy makers generally is that they face up to reasonable projections of future accomplishments, consider whether they are cause for social concern, and see whether or not the principles *now* enunciated and the practices *now* established are adequate to deal with any such concerns.

I project at least the following:

1. The growth of human embryos in the laboratory will be extended beyond the blastocyst stage. Such growth must be deemed desirable under all the arguments advanced for developmental research *up* to the blastocyst stage; research on gene action, chromosome segregation, cellular and organic differentiation, fetus-environment interaction, implantation, etc., cannot answer all its questions with the blastocyst. Such *in vitro* post-blastocyst differentiation has apparently been achieved in the mouse, in culture; the use of other mammals as temporary hosts for human embryos is also a possibility. How far such embryos will eventually be perpetuated is anybody's guess, but full-term ectogenesis cannot be excluded. Neither can the existence of laboratories filled with many living human embryos, growing at various stages of development.

2. Experiments will be undertaken to alter the cellular and genetic composition of these embryos, at first without subsequent transfer to a woman for gestation, perhaps later as a prelude to reproductive efforts. Again, scientific reasons now justifying Dr. Soupart's research already justify further embryonic manipulations, including

formation of hybrids or chimeras (within species and between species); gene, chromosome, and plasmid insertion, excision, or alteration; nuclear transplantation or cloning, etc. The techniques of DNA recombination, coupled with the new skills of handling embryos, make prospects for some precise genetic manipulation much nearer than anyone would have guessed ten years ago. And embryological and cellular research in mammals is making astounding progress. On the cover of a recent issue of *Science* is a picture of a hexaparental mouse, born after reaggregation of an early embryo with cells disaggregated from three separate embryos. (Note: That sober journal calls this a "handmade mouse"—i.e., literally a *manu-factured* mouse—and goes on to say that it was "manufactured by genetic engineering techniques.")[11]

3. Storage and banking of living human embryos (and ova) will be undertaken, perhaps commercially. After all, commercial sperm banks are already well-established and prospering.

Space does not permit me to do more than identify a few kinds of questions that must be considered in relation to such possible coming control over human heredity and reproduction: questions about the wisdom required to engage in such practices; questions about the goals and standards that will guide our interventions; questions about changes in the concepts of being human, including embodiment, gender, love, lineage, identity, parenthood, and sexuality; questions about the responsibility of power over future generations; questions about awe, respect, humility; questions about the kind of society we will have if we follow along our present course.[12]

Though I cannot discuss these questions now, I can and must face a serious objection to considering them at all. Most people would agree that the projected possibilities raise far more serious questions than do simple fertilization of a few embryos, their growth *in vitro* to the blastocyst stage, and their possible transfer to women for gestation. Why burden the present decision with these possibilities? Future "abuses," it is often said, do not disqualify present uses (though these same people also often say that "future benefits justify present questionable uses"). Moreover, there can be no certainty that "A" will lead to "B." This thin-edge-of-the-wedge argument has been open to criticism.

But such criticism misses the point, for two reasons. *First*, critics often misunderstand the wedge argument. The wedge argument is not primarily an argument of prediction, that A *will* lead to B, say on the strength of the empirical analysis of precedent and an appraisal of the likely direction of present research. It is primarily an argument about the *logic* of justification. Do the principles of justification *now* used to justify the current research proposal already justify *in advance* the further developments? Consider some of these principles:

1. It is desirable to learn as much as possible about the processes of fertilization, growth, implantation, and differentiation of human embryos and about human gene expression and its control.

2. It would be desirable to acquire improved techniques for *enhancing* conception and implantation, for *preventing* conception and implantation, for the treatment of genetic and chromosomal abnormalities, etc.

3. In the end, only research using *human* embryos can answer these questions and provide these techniques.

4. There should be no censorship or limitation of scientific inquiry or research.

This logic knows no boundary at the blastocyst stage, or for that matter, at any later stage. For these principles *not* to justify future extensions of current work, some independent additional principles, limiting such justification to particular stages of development, would have to be found. Here, the task is to find such a biologically defensible distinction that could be respected as reasonable and not arbitrary, a difficult—perhaps impossible—task, given the continuity of development after fertilization. The citizenry, myself included, will want to know *precisely* what grounds our policy makers will give for endorsing Soupart's research, and whether their principles have not already sanctioned future developments. If they do give such wedge-opening justifications, let them do so deliberately, candidly, and intentionally.

A better case to illustrate the wedge logic is the principle offered for the embryo-transfer procedures as treatment for infertility. Will we support the use of *in vitro* fertilization and embryo transfer because it provides a "child of *one's own*," in a strict sense of *one's own*, to a married couple? Or will we support the transfer because it is treatment of involuntary infertility, which deserves treatment in or out of marriage, hence endorsing the use of any available technical means (which would produce a healthy and normal child), including surrogate wombs, or even ectogenesis?

Second, logic aside, the opponents of the wedge argument do not counsel well. It would be simply foolish to ignore what might come next, and to fail to make the *best possible* assessment of the implications of present action (or inaction). Let me put the matter very bluntly: the Ethics Advisory Board, in the decision it must now make, may very well be helping to decide whether human beings will eventually be produced in laboratories. I say this not to shock—and I do not mean to beg the question of whether that would be desirable or not. I say this to make sure that they and we face squarely the full import and magnitude of this decision. Once the genies let the babies into the bottle, it may be impossible to get them out again.

VIII

So much, then, for the meaning of initiating and manipulating human embryos in the laboratory. These considerations still make me doubt the wisdom of proceeding with these practices, both in re-search and in their clinical application, notwithstanding that valuable knowledge might be had by continuing the research and identifiable suffering might be alleviated by using it to circumvent infertility. To doubt the wisdom of going ahead makes one at least a fellow-travel-ler of the opponents of such research, but it does not, either logically or practically, require that one join them in trying to prevent it, say by legal prohibition. Not every folly can or should be legislated against. Attempts at prohibition here would seem to be both ineffec-tive and dangerous—ineffective because impossible to enforce, dan-gerous because the costs of such precedent-setting interference with scientific research might be greater than the harm it prevents. To be sure, we already have legal restrictions on experimentation with hu-man subjects, which restrictions are manifestly not incompatible with the progress of medical science. Neither is it true that science cannot survive if it must take some direction from the law. Nor is it the case that all research, because it is research, is or should be absolutely pro-tected. But it does not seem to me that *in vitro* fertilization and embryo transfer deserve, *at least at present*, to be treated as sufficiently danger-ous for legislative interference.

But if to doubt the wisdom does not oblige one to seek to outlaw the folly, neither does a decision *to permit* require a decision to *encourage or support*. A researcher's freedom to do *in vitro* fertilization, or a wom-an's right to have a child with laboratory assistance, in no way im-plies a public (or even a private) obligation to pay for such research or treatment. A right *against* interference is not an entitlement *for assis-tance*. The question before the Ethics Advisory Board and the Depart-ment of Health, Education, and Welfare is *not* whether to permit such research but whether the Federal government should fund it. This is the policy question that needs to be discussed.

The arguments in favor of Federal support are well known. *First,* the research is seen as continuous with, if not quite an ordinary in-stance of, the biomedical research, which the Federal government supports handsomely; roughly two-thirds of the money spent on bio-medical research in the United States comes from Uncle Sam. Why is this research different from all other research? Its scientific merit has

been attested to by the normal peer-review process at NIH. For some, that is a sufficient reason to support it.

Second, there are specific practical fruits expected from the anticipated successes of this new line of research. Besides relief for many cases of infertility, the research promises new birth-control measures based upon improved understanding of the mechanisms of fertilization and implantation, which in turn could lead to techniques for blocking these processes. Also, studies on early embryonic development hold forth the promise of learning how to prevent some congenital malformations and certain highly malignant tumors (e.g., hydatidiform mole) that derive from aberrant fetal tissue.

Third, as he who pays the piper calls the tune, Federal support would make easy the Federal regulation and supervision of this research. For the government to abstain, so the argument runs, is to leave the control of research and clinical application in the hands of profit-hungry, adventurous, insensitive, reckless, or power-hungry private physicians, scientists, or drug companies; or, on the other hand, at the mercy of the vindictive, mindless, and superstitious civic groups that will interfere with this research through state and local legislation. Only through Federal regulation—which, it is said, can only follow with Federal funding—can we have reasonable, enforceable, and uniform guidelines.

Fourth is the chauvinistic argument that the United States should lead the way in this brave new research, especially as it will apparently be going forward in other nations. Indeed, one witness testifying before the Ethics Advisory Board deplored the fact that the first Louise Brown was British and not American, and complained, in effect, that the existing moratorium on Federal support has already created what one might call an *"in vitro* fertilization gap." The pre-eminence of American science and technology, so the argument implies, is the center of our pre-eminence among the nations, a position which will be jeopardized if we hang back out of fear.

Let me respond to these arguments, in reverse order. Conceding the premise of the importance of American science for American prestige and strength, it is far from clear that failure to support *this* research would jeopardize American science. Certainly the use of embryo transfer to overcome infertility, though a vital matter for the couples involved, is hardly a matter of vital national interest—at least not unless and until the majority of American women are similarly infertile. The demands of international competition, admittedly often a necessary evil, should be invoked only for things that really matter; a missile gap and an embryo-transfer gap are chasms apart. In areas not crucial to our own survival, there will be many things we should allow other nations to develop, if that is their wish, without feeling obliged to join them. Moreover, one should not rush into potential folly to avoid being the last to commit it.

The argument about governmental regulation has much to recommend it. But it fails to consider that there are other safeguards against recklessness, at least in the clinical applications, known to the high-minded as the canons of medical ethics and to the cynical as liability for malpractice. Also, Federal regulations attached to Federal funding will not in any case regulate research done with private monies, say by the drug companies. Moreover, there are enough concerned practitioners of these new arts who would have a compelling interest in regulating their own practice, if only to escape the wrath and interference of hostile citizen groups in response to unsavory goings-on. The available evidence does not convince me that a sensible practice of *in vitro* experimentation requires regulation by the Federal government.

In turning to the argument about anticipated technological powers, we face difficult calculations of unpredictable and more-or-less-likely costs and benefits, and the all-important questions of priorities in the allocation of scarce resources. Here it seems useful to consider separately the techniques for generating children and the anticipated techniques for birth control or for preventing developmental anomalies and malignancies.

First, accepting that providing a child of their own to infertile couples is a worthy goal—and it is both insensitive and illogical to cite the population problem as an argument for ignoring the problem of infertility—one can nevertheless question its rank relative to other goals of medical research. One can even wonder—and I have done so in print—whether it is indeed a *medical* goal, or a worthy goal for *medicine*, that is, whether alleviating infertility, especially in this way, is part of the art of *healing*. [13] Just as abortion for genetic defect is a peculiar innovation in medicine (or in preventive medicine) in which a disease is treated by eliminating the patient (or, if you prefer, a disease is prevented by "preventing" the patient), so laboratory-fertilization is a peculiar treatment for oviduct obstruction, in that it requires the creation of a new life to "heal" an existing one. All this simply emphasizes the uniqueness of the reproductive organs, in that their proper function involves other people, and calls attention to the fact that infertility is not a "disease," like heart disease or stroke, even though obstruction of a normally patent tube or vessel is the proximate cause of each.

However this may be, there is a more important objection to this approach to the problem. It represents yet another instance of our thoughtless preference for expensive, high-technology, therapy-oriented approaches to disease and dysfunctions. What about spending this money on discovering the causes of infertility? What about the prevention of tubal obstruction? We complain about rising medical costs, but we insist on the most spectacular and the most technological—and thereby the most costly—remedies.

The truth is that we do know a little about the causes of tubal obstruction—though much less than we should or could. For instance, it is estimated that at least one-third of such cases are the aftermath of pelvic-inflammatory disease, caused by that uninvited venereal guest, gonococcus. Leaving aside any question about whether it makes sense for a Federally-funded baby to be the wage of aphrodisiac indiscretion,[14] one can only look with wonder at a society that will have "petri-dish babies"[15] before it has found a vaccine against gonorrhea.

True, there are other causes of blocked oviducts, and blocked oviducts are not the only cause of female infertility. True, it is not logically necessary to choose between prevention and cure. But *practically* speaking, with money for research as limited as it is, research funds targeted for the relief of infertility should certainly go first to epidemiological and preventive measures—especially where the costs in the high-technology cure are likely to be great.

What about these costs? I have already explored some of the nonfinancial costs, in discussing the meaning of this research for our images of humanness. Let us, for now, consider only the financial costs. How expensive was Louise Brown? We do not know, partly because Drs. Edwards and Steptoe have yet to publish their results, indicating how many failures preceded their success, how many procedures for egg removal and for fetal monitoring were performed on Mrs. Brown, and so on. One must add in the costs of monitoring the baby's development to check on her "normality" and, should it come, the costs of governmental regulation. A conservative estimate might place the costs of a successful pregnancy of this kind at between five and ten thousand dollars. If we use the conservative figure of 500,000 for estimating the number of infertile women *with blocked oviducts* in the United States whose *only* hope of having children lies in *in vitro* fertilization,[16] we reach a conservative estimated cost of $2.5 to $5 billion. Is it really even fiscally wise for the Federal government to start down this road?

Clearly not, if it is also understood that the costs of providing the service, rendered possible by a successful technology, will also be borne by the taxpayers. Nearly everyone now agrees that the kidney-machine legislation, obliging the Federal government to pay about

$25,000–$30,000 per patient per year for kidney dialysis for anyone in need (cost to the taxpayers in 1978 was nearly $1 billion), is an impossible precedent—notwithstanding that individual lives have been prolonged as a result. But once the technique of *in vitro* fertilization and embryo transfer is developed and available, how should the baby-making be paid for? Should it be covered under medical insurance? If a National Health Insurance program is enacted, will and should these services be included? (Those who argue that they are part of medicine will have a hard time saying no.) Failure to do so will make this procedure available only to the well-to-do, on a fee-for-service basis. Would that be a fair alternative? Perhaps; but it is unlikely to be tolerated. Indeed, the principle of equality—equal access to equal levels of medical care—is the leading principle in the pressure for medical reform. One can be certain that efforts will be forthcoming to make this procedure available equally to all, independent of ability to pay, under Medicaid or National Health Insurance or in some other way. (I have recently learned that a Boston-based group concerned with infertility has obtained private funding to pay for artificial insemination for women on welfare!!)

Much as I sympathize with the plight of infertile couples, I do not believe that they are entitled to the provision of a child at the public expense, especially at this cost, especially by a procedure that also involves so many moral difficulties. Given the many vexing dilemmas that will surely be spawned by laboratory-assisted reproduction, the Federal government should not be misled by compassion to embark on this imprudent course.

In considering the Federal funding of such research for its other anticipated technological benefits, independent of its clinical use in baby-making, we face a more difficult matter. In brief, as is the case with all basic research, one simply cannot predict what kinds of techniques and uses this research will yield. But here, also, I think good sense would at present say that before one undertakes *human in vitro* fertilization to seek new methods of birth control—e.g., by developing antibodies to the human egg that would physically interfere with its fertilization—one should make adequate attempts to do this in animals. One simply can't get large-enough numbers of human eggs to do this pioneering research well—at least not without subjecting countless women to additional risks not for their immediate benefit. Why not test this conceit first in the mouse or rabbit? Only if the results were very promising—and judged also to be relatively safe in practice—should one consider trying such things in humans. Likewise, the developmental research can and should be first carried out in animals, especially in primates. Though *in vitro* fertilization has yet to be achieved in monkeys, embryo transfer of *in vivo* fertilized eggs has been accomplished, thus permitting the relevant research to proceed. Purely *on scientific grounds,* the Federal government ought not *now* to be investing funds in this research for its promised technological benefits—benefits which, in the absence of pilot studies in animals, must be regarded as mere wishful thoughts in the imaginings of scientists.

There remains the first justification, research for the sake of knowledge: knowledge about cell cleavage, cell-cell and cell-environment interactions, and cell differentiation; knowledge of gene action and of gene regulation; knowledge of the effects and mechanisms of action of various chemical and physical agents on growth and development; knowledge of the basic processes of fertilization and implantation. This is all knowledge worth having, and though much can be learned using animal sources—and these sources have barely begun to be sufficiently exploited—the investigation of these matters in man would, sooner or later, require the use of human-embryonic material. Here, again, there are questions of research priority about which there is room for disagreement, among scientists and laymen alike. But these questions of research priority, while not irrelevant to the decision at hand, are not the questions that the Ethics Advisory Board was constituted to answer.

It was constituted to consider whether such research is consistent with the ethical standards of our community. The question turns in large part on the status of the early embryo. If, as I have argued, the early embryo is deserving of respect because of what it is, now and potentially, it is difficult to justify submitting it to invasive experiments, and especially difficult to justify *creating it solely* for the purpose of experimentation. But even if this argument fails to sway the Board, another one should. For their decision, I remind you, is not whether *in vitro* fertilization should be permitted in the United States, but whether *our* tax dollars should encourage and foster it. One cannot, therefore, ignore the deeply held convictions of a sizeable portion of our population—it may even be a majority on this issue—that regards the human embryo as protectable humanity, not to be experimented upon except for its own benefit. Never mind if these beliefs have a religious foundation—as if that should ever be a reason for dismissing them! The presence, sincerity, and depth of these beliefs, and the grave importance of their subject, is what must concern us. The holders of these beliefs have been very much alienated by the numerous court decisions and legislative enactments regarding abortion and research on fetuses. Many who, by and large, share their opinions about the humanity of prenatal life have with heavy heart gone along with the liberalization of abortion, out of deference to the wishes, desires, interests, or putative rights of pregnant women. But will they go along here with what they can only regard as gratuitous and willful assaults on human life, or at least on potential and salvageable human life, and on human dignity? We can ill afford to alienate them further, and it would be unstatesmanlike, to say the least, to do so, especially in a matter so little important to the national health and one so full of potential dangers.

Technological progress can be but one measure of our national health. Far more important is the affection and esteem in which our citizenry holds its laws and institutions. No amount of relieved infertility is worth the further disaffection and civil contention that the lifting of the moratorium on Federal funding is likely to produce. People opposed to abortion and people grudgingly willing to permit women to obtain elective abortion, at their own expense, will not tolerate having their tax money spent on scientific research requiring what they regard as at best cruelty, at worst murder. A prudent Ethics Advisory Board and a prudent and wise Secretary of Health, Education, and Welfare should take this matter most seriously, and refuse to lift the moratorium—at least until they are persuaded that public opinion will overwhelmingly support them. Imprudence in this matter may be the worst sin of all.

AN AFTERWORD

This has been for me a long and difficult exposition. Many of the arguments are hard to make. It is hard to get confident people to face unpleasant prospects. It is hard to get many people to take seriously such "soft" matters as lineage, identity, respect, and self-respect when they are in tension with such "hard" matters as a cure for infertility or new methods of contraception. It is hard to talk about the meaning of sexuality and embodiment in a culture that treats sex increasingly as sport and that has trivialized the significance of gender, marriage, and procreation. It is hard to oppose Federal funding of baby-making in a society which increasingly demands that the Federal government supply all demands, and which—contrary to so much evidence of waste, incompetence, and corruption—continues to believe that only Uncle Sam can do it. And, finally, it is hard to speak about restraint in a culture that seems to venerate very little above man's own attempt to master all. Here, I am afraid, is the biggest question and the one we perhaps can no longer ask or deal with: the question about the reasonableness of the desire to become masters and possessors of nature, human nature included.

Here we approach the deepest meaning of *in vitro* fertilization.

Those who have likened it to artificial insemination are only partly correct. With *in vitro* fertilization, the human embryo emerges for the first time from the natural darkness and privacy of its own mother's womb, where it is hidden away in mystery, into the bright light and utter publicity of the scientist's laboratory, where it will be treated with unswerving rationality, before the clever and shameless eye of the mind and beneath the obedient and equally clever touch of the hand. What does it mean to hold the beginning of human life before your eyes, in your hands—even for 5 days (for the meaning does not depend on duration)? Perhaps the meaning is contained in the following story:

Long ago there was a man of great intellect and great courage. He was a remarkable man, a giant, able to answer questions that no other human being could answer, willing boldly to face any challenge or problem. He was a confident man, a masterful man. He saved his city from disaster and ruled it as a father rules his children, revered by all. But something was wrong in his city. A plague had fallen on generation; infertility afflicted plants, animals, and human beings. The man confidently promised to uncover the cause of the plague and to cure the infertility. Resolutely, dauntlessly, he put his sharp mind to work to solve the problem, to bring the dark things to light. No secrets, no reticences, a full public inquiry. He raged against the representatives of caution, moderation, prudence, and piety, who urged him to curtail his inquiry; he accused them of trying to usurp his rightfully earned power, of trying to replace human and masterful control with submissive reverence. The story ends in tragedy: He solved the problem but, in making visible and public the dark and intimate details of his origins, he ruined his life, and that of his family. In the end, too late, he learns about the price of presumption, of overconfidence, of the overweening desire to master and control one's fate. In symbolic rejection of his desire to look into everything, he punishes his eyes with self-inflicted blindness.

Sophocles seems to suggest that such a man is always in principle—albeit unwittingly—a patricide, a regicide, and a practitioner of incest. We men of modern science may have something to learn from our forebear, Oedipus. It appears that Oedipus, being the kind of man an Oedipus is (the chorus calls him a paradigm of man), had no choice but to learn through suffering. Is it really true that we too have no other choice?

NOTES

1. In the British procedures, several eggs are taken from each woman and fertilized, to increase the chance of success, but only one embryo is transferred for implan-

tation. In Dr. Soupart's proposed experiments, as the embryos will be produced only for the purpose of research and not for transfer, all of them will be discarded or destroyed.

2. A perhaps justifiable exception would be the case of a universal plague on childbirth, say because of some epidemic that fatally attacks all fetuses *in utero* at age 5 months. Faced with the prospect of the end of the race, might we not condone the deliberate institution of pregnancies to provide fetuses for research, in the hope of finding a diagnosis and remedy for this catastrophic blight?

3. Peter Gwynne, "Was the Birth of Louise Brown Only a Happy Accident?" *Science Digest,* October 1978, (emphasis added).

4. For the supporting analysis of the concept of "viability," see my article, "Determining Death and Viability in Fetuses and Abortuses," prepared for the National Commission for the Protection of Human Subjects of Biomedical and Behavioral Research, published in *Appendix: Research on the Fetus,* U.S. Department of Health, Education, and Welfare, HEW Publ. No. (OS) 76-128, 1975.

5. Some people have suggested that the embryo be regarded like a vital organ, salvaged from a newly dead corpse, usable for transplantation or research, and that its donation by egg and sperm donors be governed by the Uniform Anatomical Gift Act, which legitimates pre-mortem consent for organ donation upon death. But though this acknowledges that embryos are not things, it is mistaken in treating embryos as mere organs, thereby overlooking that they are early stages of a *complete, whole* human being. The Uniform Anatomical Gift Act does not apply to, nor should it be stretched to cover, donations of gonads, gametes (male sperm or female eggs), or—especially—zygotes and embryos.

6. There is a good chance that the problem of surplus embryos may be avoidable, for purely technical reasons. Some researchers believe that the uterine receptivity to the transferred embryo might be reduced during the particular menstrual cycle in which the ova are obtained, because of the effects of the hormones given to induce superovulation. They propose that the harvested *eggs* be frozen, and then defrosted one at a time each month for fertilization, culture, and transfer, until pregnancy is achieved. By refusing to fertilize all the eggs at once—i.e., not placing all one's eggs in one uterine cycle—there will not be surplus *embryos,* but at most only surplus eggs. This change in the procedure would make the demise of unimplanted embryos *exactly* analogous to the "natural" embryonic loss in ordinary reproduction.

7. The literature on intervention in reproduction is both confused and confusing on the crucial matter of the meanings of "nature" or "the natural," and their significance for the ethical issues. It may be as much a mistake to claim that "the natural" has *no* moral force as to suggest that the natural way is best, because natural. Though shallow and slippery thought about nature, and its relation to "good," is a likely source of these confusions, the nature of nature may itself be elusive, making it difficult for even careful thought to capture what is natural.

8. An unmarried woman in Dearborn, Michigan, offered to bear a child for her married friend, infertile because of a hysterectomy. She was impregnated by artificial insemination using semen produced by her friend's husband, his wife performing the injection. The threesome lived together all during the pregnancy. The child was delivered at birth by the biological-and-gestational-mother to the wife-and-rearing-mother. The first (pregnancy) mother reports no feelings of attachment to the child she carried and bore. Everyone is reportedly delighted with the event. The trio has publicized its accomplishment and is reported to be considering selling rights to the story for a TV show, a book, and a movie. Their attorney has been swamped with letters requesting similar surrogate "mothers." (*American Medical News,* July 28, 1978, pp. 11–12.)

9. There are today numerous suits pending, throughout the United States, because of artificial insemination with donor semen (AID). Following divorce, the ex-husbands are refusing child support for AID children, claiming, minimally, no paternity, or maximally that the child was the fruit of an adulterous "union." In fact, a few states still treat AID as adultery. The importance of anonymity is revealed in the following bizarre case. A woman wanted to have a child, but abhorred the thought of marriage or of sexual relations with men. She learned a do-it-yourself technique of artificial insemination, and persuaded a male acquaintance to donate his semen. Now some 10 years after this virgin birth, the case has gone to court. The semen donor is suing for visitation privileges, to see his son.

10. To those who point out that the bond between sexuality and procreation has already been effectively and permanently cleaved by "the pill," and that this is therefore an idle worry in the case of *in vitro* fertilization, it must be said that the pill provides only sex without babies. Now the other shoe drops; babies without sex.

11. *Science, 202:5,* October 6, 1978.

12. Some of these questions are addressed, albeit too briefly and polemically, in the latter part of my 1972 "Making Babies" article, to which the reader is referred. It has been pointed out to me by an astute colleague that the tone of the present article is less passionate and more accommodating than the first, which change he regards as an ironic demonstration of the inexorable way in which we get used to, and accept, our technological nightmares. I myself share his concern. I cannot decide whether the decline of my passion is to be welcomed; that is, whether it is due to greater understanding bred of more thought and experience, or to greater callousness and the contempt of familiarity bred from *too much* thought and experience. It does seem to me now that many of the fundamental beliefs and institutions that might be challenged by laboratory growth of human embryos and by laboratory-assisted reproduction are already severely challenged in perhaps more potent and important ways. Here, too, we see the creeping effect of the aggregated powers of modernity and the corrosive power of the familiar. Adaptiveness is our glory and our curse: as Raskolnikov put it, "Man gets used to everything, the beast!"

13. See "Making Babies—the New Biology and the 'Old' Morality," pp. 19–20. See also my "Regarding the End of Medicine and the Pursuit of Health," *The Public Interest,* Number 40, Summer 1975, especially pp. 11–18, and 33–35.

14. Consider the following contributions of Federally-supported programs to rationalizing our sexual and reproductive practices. First, we have Federally-supported programs of sex education in elementary schools, so that the children will know what can happen to them (and what they can make happen). Next, in high school, Uncle Sam provides for teen-age contraception, to prevent the consequences of unavoidable sexual activity. Freed of a major deterrent to unrestricted sexual activity, our teen-agers indulge, but not without consequences: They get gonorrhea, which some of them will have treated, again at the taxpayers' expense through Medicaid. But for some the treatment comes too late to prevent scarring and oviduct obstruction: Federally-supported *in vitro* fertilization research and services come to the rescue, to overcome their infertility. Uncle Sam will, of course, also provide Aid to Dependent Children, if the mother is or goes on welfare.

15. There has been much objection, largely from the scientific community, to the phrase "test-tube baby." More than one commentator has deplored the exploitation of its "flesh-creeping" connotations. They point out that a flat petridish is used, not a test-tube—as if that mattered—and that the embryo spends but a few days in the dish. But they don't ask why the term "test-tube baby" remains the popular designation, and whether it does not embody more of the deeper truth than a more accurate, laboratory appellation. If the decisive difference is between "in the womb" or "in the lab,"

the popular designation conveys it. (See 'Afterword', below.) And it is right on target, and puts us on notice, if the justification for the present laboratory procedures tacitly also *justifies* future extensions, including full ectogenesis—say, if that were the only way a womb-less woman could have a child of her own, without renting a human womb from a surrogate bearer.

16. This figure is calculated from estimates that between 10 and 15 percent of all couples are involuntarily infertile, and that in more than half of these cases the cause is in the female. Blocked oviducts account for perhaps 20 percent of the causes of female infertility. Perhaps 50 percent of these women might be helped to have a child by means of reconstructive surgery on the oviducts; the remainder could conceive *only* with the aid of laboratory fertilization and embryo transfer. These estimates do not include additional candidates with uterine disease (who could "conceive" only by embryo transfer to surrogate-gestators), nor those with ovarian dysfunction who would need egg donation as well, nor that growing population of women who have had tubal ligations and who could later turn to *in vitro* fertilization. It is also worth noting that not all the infertile couples are childless; indeed, a surpassing number are seeking to enlarge an existing family.

Genetic Engineering and Genetic Policy/Further Reading

Berger, Brigitte. "A New Interpretation of the I.Q. Controversy" *The Public Interest.* 50 (Winter, 1978), 29–44.

Caplan, Arthur, Editor, *The Sociobiology Debate.* New York: Harper and Row, 1978.

Gaylin, Willard. "The Frankenstein Factor." *New England Journal of Medicine* 297 (September 22, 1977), 665–67.

Goodfield, June. *Playing God: Genetic Engineering and The Manipulation of Life.* New York: Random House, 1977.

Grobstein, Clifford. "External Human Fertilization." *Scientific American* 240 (June, 1979), 57–67.

Hilton, Bruce, et al., Editors. *Ethical Issues in Human Genetics.* New York: Plenum Publishing Corporation, 1973.

Mertens, Thomas R. *Human Genetics: Readings on The Implications of Genetic Engineering.* New York: John Wiley and Sons, 1975.

Ramsey, Paul. "Fabricated Man: The Ethics of Genetic Control." New Haven: Yale University Press, 1970.

Reilly, Philip. *Genetics, Law and Social Policy.* Cambridge: Harvard University Press, 1977.

Wade, Nicholas. *The Ultimate Experiment.* New York: Walker & Co., 1977.

PART SEVEN

THE ALLOCATION OF SCARCE RESOURCES

Gene Outka
James F. Childress
Willard Gaylin
Aaron Wildavsky

26 *Social Justice and Equal Access to Health Care*[1]

GENE OUTKA, PH.D.

The problem of the allocation of scarce resources is one of the most common but also most difficult problems in medical ethics as well as in other areas of ethical concern. A variety of means of distribution have been proposed. The article by Outka provides a good introduction to this particular problem by considering a variety of formulations of means of distribution based on different concepts of social justice. Although Outka focuses on a particular problem and argues for a particular resolution of that problem, the article formulates a social and institutional way of thinking about this particular issue. As such it provides a framework for helping us evaluate several significant orientations towards a resolution of the issue.

Dr. Gene Outka, Ph.D., teaches in the Department of Religious Studies at Yale University.

I want to consider the following question. Is it possible to understand and to justify morally a societal goal which increasing numbers of people, including Americans, accept as normative? The goal is: the assurance of comprehensive health services for every person irrespective of income or geographic location. Indeed, the goal now has almost the status of a platitude. Currently in the United States politicians in various camps give it at least verbal endorsement (see, e.g., Nixon, 1972:1; Kennedy, 1972:234-252). I do not propose to examine the possible sociological determinants in this emergent consensus. I hope to show that whatever these determinants are, one may offer a plausible case in defense of the goal on reasonable grounds. To demonstrate why appeals to the goal get so successfully under our skins, I shall have recourse to a set of conceptions of social justice. Some of the standard conceptions, found in a number of writings on justice, will do (these writings

477

include Bedau, 1971; Hospers, 1961: 416-468; Lucas, 1972; Perelman, 1963; Rescher, 1966; Ryan, 1916; Vlastos, 1962). By reflecting on them it seems to me a prima facie case can be established, namely, that every person in the entire resident population should have equal access to health care delivery.

The case is prima facie only. I wish to set aside as far as possible a related question which comes readily enough to mind. In the world of "suboptimal alternatives," with the constraints for example which impinge on the government as it makes decisions about resource allocation, what is one to say? What criteria should be employed? Paul Ramsey, in *The Patient as Person* (1970:240), thinks that the large question of how to choose between medical and other societal priorities is "almost, if not altogether, incorrigible to moral reasoning." Whether it is or not is a matter which must be ignored for the present. One may simply observe in passing that choices are unavoidable nonetheless, as Ramsey acknowledges, even where the government allows them to be made by default, so that in some instances they are determined largely by which private pressure groups prove to be dominant. In any event, there is virtue in taking up one complicated question at a time and we need to get the thrust of the case for equal access before us. It is enough to observe now that Americans attach an obviously high priority to organized health care. National health expenditures for the fiscal year 1972 were $83.4 billion (Hicks, 1973:52). Even if such an enormous sum is not entirely adequate, we may still ask: how are we to justify spending whatever we do in accordance as far as possible with the goal of equal access? The answer I propose involves distinguishing various conceptions of social justice and trying to show which of these apply or fail to apply to health care considerations. Only toward the end of the paper will some institutional implications be given more than passing attention, and then in a strictly programmatic way.

Another sort of query should be noted as we begin. What stake does someone in religious ethics have in this discussion? For the reasonable case envisaged is offered after all in the public forum. If the issue is how to justify morally the societal goal which seems so obvious to so many, whether or not they are religious believers, does the religious ethicist then simply participate qua citizen? Here I think we should be wary of simplifying formulae. Why for example should a Jew or a Christian not welcome wide support for a societal goal which he or she can affirm and reaffirm, or reflect only on instances where such support is not forthcoming? If a number of ethical schemes, both religious and humanist, converge in their acceptance of the goal of equal access to health care, so be it. Secularists can join forces with believers, at least at some levels or points, without implying there must be unanimity on every moral issue. Yet it also seems too simple if one claims to wear only the citizen's hat when making the case in question. At least I should admit that a commitment to the basic normative principle which in Christian writings is often called *agape* may influence the account to follow in ways large and small (see Outka, 1972). For example, someone with such a commitment will quite naturally take a special interest in appeals to the generic characteristics all persons share rather than the idiosyncratic attainments which distinguish persons from one another, and in the playing down of desert considerations. As I shall try to show, such appeals are centrally relevant to the case for equal access. And they are nicely in line with the normative pressures agapeic considerations typically exert.

One issue of theoretical importance in religious ethics also emerges in connection with this last point. The approach in this paper may throw a little indirect light on the traditional question, especially prominent in Christian ethics, of how love and justice are related. To distinguish different conceptions of social justice will put us in a better position, I think, to recognize that often it is ambiguous to ask about "*the* relation." There may be different relations to different conceptions. For the conceptions themselves may sometimes produce discordant indications, or turn out to be incommensurable, or reflect, when different ones are seized upon, rival moral points of view. I shall note several of these relations as we proceed.

Which then among the standard conceptions of social justice appear to be particularly relevant or irrelevant? Let us consider the following five:

I.   To each according to his merit or desert.
II.   To each according to his societal contribution.
III.   To each according to his contribution in satisfying whatever is freely desired by others in the open marketplace of supply and demand.
IV.   To each according to his needs.
V.   Similar treatment for similar cases.

In general I shall argue that the first three of these are less relevant because of certain distinctive features which health crises possess. I shall focus on crises here not because I think preventive care is unimportant (the opposite is true), but because the crisis situation shows most clearly the special significance we attach to medical treatment as an institutionalized activity or social practice, and the basic purpose we suppose it to have.

I

To each according to his merit or desert. Meritarian conceptions, above all perhaps, are grading ones: advantages are allocated in accordance with amounts of energy expended or kinds of results achieved. What is judged is particular conduct which distinguishes persons from one another and not only the fact that all the parties are human beings. Sometimes a competitive aspect looms large.

In certain contexts it is illuminating to distinguish between efforts and achievements. In the case of efforts one characteristically focuses on the individual: rewards are based on the pains one takes. Some have supposed, for example, that entry into the kingdom of heaven is linked more directly to energy displayed and fidelity shown than to successful results attained.

To assess achievements is to weigh actual performance and productive contributions. The academic prize is awarded to the student with the highest grade-point average, regardless of the amount of midnight oil he or she burned in preparing for the examinations. Sometimes we may exclaim, "it's just not fair," when person X writes a brilliant paper with little effort while we are forced to devote more time with less impressive results. But then our complaint may be directed against differences in innate ability and talent which no expenditure of effort altogether removes.

After the difference between effort and achievement, and related distinctions, have been acknowledged, what should be stressed I think is the general importance of meritarian or desert criteria in the thinking of most people about justice. These criteria may serve to illuminate a number of disputes about the justice of various practices and institutional arrangements in our society. It may help to explain, for instance, the resentment among the working class against the welfare system. However wrongheaded or self-deceptive the resentment often is, particularly when directed toward those who want to work but for various reasons beyond their control cannot, at its better moments it involves in effect an appeal to desert considerations. "Something for nothing" is repudiated as unjust; benefits should be proportional (or at least related) to costs; those who can make an effort should do so, whatever the degree of their training or significance of their contribution to society; and so on. So, too, persons deserve to have what they have labored for; unless they infringe on the works of others their efforts and achievements are justly theirs.

Occasionally the appeal to desert extends to a wholesale rejection of other considerations as grounds for just claims. The most conspicuous target is need. Consider this statement by Ayn Rand.

A morality that holds *need* as a claim, holds emptiness—

nonexistence—as its standard of value; it rewards an absence, a *Ayn Rand* defect: weakness, inability, incompetence, suffering, disease, disaster, the lack, the fault, the flaw—the *zero*.

Who provides the account to pay these claims? Those who are cursed for being non-zeros, each to the extent of his distance from that ideal. Since all values are the product of virtues, the degree of your virtue is used as the measure of your penalty; the degree of your faults is used as the measure of your gain. Your code declares that the rational man must sacrifice himself to the irrational, the independent man to parasites, the honest man to the dishonest, the man of justice to the unjust, the productive man to thieving loafers, the man of integrity to compromising knaves, the man of self-esteem to sniveling neurotics. Do you wonder at the meanness of soul in those you see around you? The man who achieves these virtues will not accept your moral code; the man who accepts your moral code will not achieve these virtues. (1957:958)

I have noted elsewhere (1972:89-90, 165-167) that *agape*, while it characteristically plays down, need not formally disallow attention to considerations falling under merit or desert; for in the case of merit as well as need it may be possible, the quotation above notwithstanding, to reason solely from egalitarian premises. A major reason such attention is warranted concerns what was called there the differential exercise of an equal liberty. That is, one may fittingly revere another's moral capacities and thus the efforts he makes as well as the ends he seeks. Such reverence may lead one to weigh expenditure of energy and specific achievements. I would simply hold now (1) that the idea of justice is not exhaustively characterized by the notion of desert, even if one agrees that the latter plays an important role; and (2) that the notion of desert is especially ill-suited to play an important role in the determination of policies which should govern a system of health care.

Why is it so ill-suited? Here we encounter some of the distinctive features which it seems to me health crises possess. Let me put it in this way. Health crises seem non-meritarian because they occur so often for reasons beyond our control or power to predict. They frequently fall without discrimination on the (according-to-merit) just and unjust, i.e., the virtuous and the wicked, the industrious and the slothful alike.

While we may believe that virtues and vices cannot depend upon natural contingencies, we are bound to admit, it seems, that many health crises do. It makes sense therefore to say that we are equal in being randomly susceptible to these crises. Even those who ascribe a prominent role to desert acknowledge that justice has also properly to do with pleas of "But I could not help it" (Lucas, 1972:321). One seeks to distinguish such cases from those acknowledged to be praise-worthy or blameworthy. Then it seems unfair as well as unkind to dis-criminate among those who suffer health crises on the basis of their personal deserts. For it would be odd to maintain that a newborn child deserves his hemophilia or the tumor afflicting her spine.

These considerations help to explain why the following rough distinction is often made. Bernard Williams, for example, in his dis-cussion of "equality in unequal circumstances," identifies two dif-ferent sorts of inequality, inequality of merit and inequality of need, and two corresponding goods, those earned by effort and those de-manded by need (1971:126-137). Medical treatment in the event of illness is located under the umbrella of need. He concludes: "Leaving aside preventive medicine, the proper ground of distribution of medi-cal care is ill health: this is a necessary truth" (1971:127). An irratio-nal state of affairs is held to obtain if those whose needs are the same are treated unequally, when needs are the ground of the treatment. One might put the point this way. When people are equal in the rele-vant respects—in this case when their needs are the same and occur in a context of random, undeserved susceptibility—that by itself is a good reason for treating them equally (see also Nagel, 1973:354).

In many societies, however, a second necessary condition for the receipt of medical treatment exists de facto: the possession of money. This is not the place to consider the general question of when inequal-ities in wealth may be regarded as just. It is enough to note that one can plausibly appeal to all of the conceptions of justice we are em-barked in sorting out. A person may be thought to be entitled to a higher income when he works more, contributes more, risks more, and not simply when he needs more. We may think it fair that the indus-trious should have more money than the slothful and the surgeon more than the tobacconist. The difficulty comes in the misfit between the reasons for differential incomes and the reasons for receiving med-ical treatment. The former may include a pluralistic set of claims in which different notions of justice must be meshed. The latter are more monistically focused on needs, and the other notions not accorded a similar relevance. Yet money may nonetheless remain as a causally necessary condition for receiving medical treatment. It may be the power to secure what one needs. The senses in which health crises are distinctive may then be insufficiently determinative for the policies which govern the actual availability of treatment. The nearly automat-ic links between income, prestige, and the receipt of comparatively higher quality medical treatment should then be subjected to critical

scrutiny. For unequal treatment of the rich ill and the poor ill is unjust
if, again, needs rather than differential income constitute the ground
of such treatment.

Suppose one agrees that it is important to recognize the misfit between the reasons for differential incomes and the reasons for receiving medical treatment, and that therefore income as such should not
govern the actual availability of treatment. One may still ask whether
the case so far relies excessively on "pure" instances where desert considerations are admittedly out of place. That there are such pure instances, tumors afflicting the spine, hemophilia, and so on, is not denied. Yet it is an exaggeration if we go on and regard all health crises
as utterly unconnected with desert. Note for example that Williams
leaves aside preventive medicine. And if in a cool hour we examine the
statistics, we find that a vast number of deaths occur each year due to
causes not always beyond our control, e.g., automobile accidents,
drugs, alcohol, tobacco, obesity, and so on. In some final reckoning it
seems that many persons (though crucially, not all) have an effect on,
and arguably a responsibility for, their own medical needs. Consider
the following bidders for emergency care: (1) a person with a heart attack who is seriously overweight; (2) a football hero who has suffered
a concussion; (3) a man with lung cancer who has smoked cigarettes
for forty years; (4) a 60 year old man who has always taken excellent
care of himself and is suddenly stricken with leukemia; (5) a three
year old girl who has swallowed poison left out carelessly by her
parents; (6) a 14 year old boy who has been beaten without provocation by a gang and suffers brain damage and recurrent attacks of uncontrollable terror; (7) a college student who has slashed his wrists
(and not for the first time) from a psychological need for attention; (8)
a woman raised in the ghetto who is found unconscious due to an
overdose of heroin.

These cases help to show why the whole subject of medical treatment is so crucial and so perplexing. They attest to some melancholy
elements in human experience. People suffer in varying ratios the effects of their natural and undeserved vulnerabilities, the irresponsibility and brutality of others, and their own desires and weaknesses.
In some final reckoning then desert considerations seem not irrelevant
to many health crises. The practical applicability of this admission,
however, in the instance of health care delivery, appears limited. We
may agree that it underscores the importance of preventive health care
by stressing the influence we sometimes have over our medical needs.
But if we try to foster such care by increasing the penalties for neglect,
we normally confine ourselves to calculations about incentives. At the
risk of being denounced in some quarters as censorious and puritan-

nical, perhaps we should for example levy far higher taxes on alcohol and tobacco and pump the dollars directly into health care programs rather than (say) into highway building. Yet these steps would by no means lead necessarily to a demand that we correlate in some strict way a demonstrated effort to be temperate with the receipt of privileged medical treatment as a reward. Would it be feasible to allocate the additional tax monies to the man with leukemia before the overweight man suffering a heart attack on the ground of a difference in desert? At the point of emergency care at least, it seems impracticable for the doctor to discriminate between these cases, to make meritarian judgments at the point of catastrophe. And the number of persons who are in need of medical treatment for reasons utterly beyond their control remains a datum with tenacious relevance. There are those who suffer the ravages of a tornado, are handicapped by a genetic defect, beaten without provocation, etc. A commitment to the basic purpose of medical care and to the institutions for achieving it involves the recognition of this persistent state of affairs.

II

To each according to his societal contribution. This conception gives moral primacy to notions such as the public interest, the common good, the welfare of the community, or the greatest good of the greatest number. Here one judges the social consequences of particular conduct. The formula can be construed in at least two ways (Rescher, 1966:79-80). It may refer to the interest of the social group considered collectively, where the group has some independent life all its own. The group's welfare is the decisive criterion for determining what constitutes any member's proper share. Or the common good may refer only to an aggregation of distinct individuals and considered distributively.

Either version accords such a primacy to what is socially advantageous as to be unacceptable not only to defenders of need, but also, it would seem, of desert. For the criteria of effort and achievement are often conceived along rather individualistic lines. The pains an agent takes or the results he brings about deserve recompense, whether or not the public interest is directly served. No automatic harmony then is necessarily assumed between his just share as individually earned and his proper share from the vantage point of the common good. Moreover, the test of social advantage *simpliciter* obviously threatens the agapeic concern with some minimal consideration due each person which is never to be disregarded for the sake of long-range social benefits. No one should be considered as *merely* a means or instrument.

The relevance of the canon of social productiveness to health crises may accordingly also be challenged. Indeed, such crises may cut against it in that they occur more frequently to those whose comparative contribution to the general welfare is less, e.g., the aged, the disabled, children.

Consider for example Paul Ramsey's persuasive critique of social and economic criteria for the allocation of a single scarce medical resource. He begins by recounting the imponderables which faced the widely-discussed "public committee" at the Swedish Hospital in Seattle when it deliberated in the early 1960's. The sparse resource in this case was the kidney machine. The committee was charged with the responsibility of selecting among patients suffering chronic renal failure those who were to receive dialysis. Its criteria were broadly social and economic. Considerations weighed included age, sex, marital status, number of dependents, income, net worth, educational background, occupation, past performance and future potential. The application of such criteria proved to be exceedingly problematic. Should someone with six children always have priority over an artist or composer? Were those who arranged matters so that their families would not burden society to be penalized in effect for being provident? And so on. Two critics of the committee found "a disturbing picture of the bourgeoisie sparing the bourgeoisie" and observed that "the Pacific Northwest is no place for a Henry David Thoreau with bad kidneys" (quoted in Ramsey, 1970:248).

The mistake, Ramsey believes, is to introduce criteria of social worthiness in the first place. In those situations of choice where not all can be saved and yet all need not die, "the equal right of every human being to live, and not relative personal or social worth, should be the ruling principle" (1970:256). The principle leads to a criterion of "random choice among equals" expressed by a lottery scheme or a practice of "first-come, first-served." Several reasons stand behind Ramsey's defense of the criterion of random choice. First, a religious belief in the equality of persons before God leads intelligibly to a refusal to choose between those who are dying in any way other than random patient selection. Otherwise their equal value as human beings is threatened. Second, a moral primacy is ascribed to survival over other (perhaps superior) interests persons may have, in that it is the condition of everything else. ". . . Life is a value incommensurate with all others, and so not negotiable by bartering one man's worth against another's" (1970:256). Third, the entire enterprise of estimating a person's social worth is viewed with final skepticism. ". . . We have no way of knowing how really and truly to estimate a man's societal worth or his worth to others or to himself in unfocused social

situations in the ordinary lives of men in their communities"
(1970:256). This statement, incidentally, appears to allow something
other than randomness in *focused* social situations; when, say, a President or Prime Minister and the owner of the local bar rush for the last
place in the bomb shelter, and the knowledge of the former can save
many lives. In any event, I have been concerned with a restricted
point to which Ramsey's discussion brings illustrative support. The
canon of social productiveness is notoriously difficult to apply as a
workable criterion for distributing medical services to those who need
them.

One may go further. A system of health care delivery which
treats people on the basis of the medical care required may often go
against (at least narrowly conceived) calculations of societal advantage. For example, the health care needs of people tend to rise during
that period of their lives, signaled by retirement, when their incomes
and social productivity are declining. More generally:

> Some 40 to 50 per cent of the American people—the aged,
> children, the dependent poor, and those with some significant
> chronic disability are in categories requiring relatively large
> amounts of medical care but with inadequate resources to pur-
> chase such care. (Somers, 1971a:20)

If one agrees, for whatever reasons, with the agapeic judgment
that each person should be regarded as irreducibly valuable, then one
cannot succumb to a social productiveness criterion of human worth.
Interests are to be equally considered even when people have ceased to
be, or are not yet, or perhaps never will be, public assets.

III

*To each according to his contribution in satisfying whatever is
freely desired by others in the open marketplace of supply and demand.* Here we have a test which, though similar to the preceding one,
concentrates on what is desired de facto by certain segments of the
community rather than the community as a whole, and on the relative
scarcity of the service rendered. It is tantamount to the canon of supply and demand as espoused by various laissez-faire theoreticians (cf.
Rescher, 1966:80-81). Rewards should be given to those who by virtue of special skill, prescience, risk-taking, and the like discern what is
desired and are able to take the requisite steps to bring satisfaction. A
surgeon, it may be argued, contributes more than a nurse because of
the greater training and skill required, burdens borne, and effective
care provided, and should be compensated accordingly. So too per-

haps, a star quarterback on a pro-football team should be remunerated even more highly because of the rare athletic prowess needed, hazards involved, and widespread demand to watch him play.

This formula does not then call for the weighing of the value of various contributions, and tends to conflate needs and wants under a notion of desires. It also assumes that a prominent part is assigned to consumer free-choice. The consumer should be at liberty to express his preferences, and to select from a variety of competing goods and services. Those who resist many changes currently proposed in the organization and financing of health care delivery in the U.S.A.—such as national health insurance—often do so by appealing to some variant of this formula.

Yet it seems health crises are often of overriding importance when they occur. They appear therefore not satisfactorily accommodated to the context of a free marketplace where consumers may freely choose among alternative goods and services.

To clarify what is at stake in the above contention, let us examine an opposing case. Robert M. Sade, M.D., published an article in *The New England Journal of Medicine* entitled "Medical Care as a Right: A Refutation" (1971). He attacks programs of national health insurance in the name of a person's right to select one's own values, determine how they may be realized, and dispose of them if one chooses without coercion from other men. The values in question are construed as economic ones in the context of supply and demand. So we read:

> In a free society, man exercises his right to sustain his own life by producing economic values in the form of goods and services that he is, or should be, free to exchange with other men who are similarly free to trade with him or not. The economic values produced, however, are not given as gifts by nature, but exist only by virtue of the thought and effort of individual men. Goods and services are thus owned as a consequence of the right to sustain life by one's own physical and mental effort. (1971:1289)

Sade compares the situation of the physician to that of the baker. The one who produces a loaf of bread should as owner have the power to dispose of his own product. It is immoral simply to expropriate the bread without the baker's permission. Similarly, "medical care is neither a right nor a privilege: it is a service that is provided by doctors and others to people who wish to purchase it" (1971:1289). Any coercive regulation of professional practices by the society at large is held to be analogous to taking the bread from the baker without his consent. Such regulation violates the freedom of the physician over his own services and will lead inevitably to provider-apathy.

The analogy surely misleads. To assume that doctors autonomously produce goods and services in a fashion closely akin to a baker is grossly oversimplified. The baker may himself rely on the agricultural produce of others, yet there is a crucial difference in the degree of dependence. Modern physicians depend on the achievements of medical technology and the entire scientific base underlying it, all of which is made possible by a host of persons whose salaries are often notably less. Moreover, the amount of taxpayer support for medical research and education is too enormous to make any such unqualified case for provider-autonomy plausible.

However conceptually clouded Sade's article may be, its stress on a free exchange of goods and services reflects one historically influential rationale for much American medical practice. And he applies it not only to physicians but also to patients or "consumers."

> The question is whether the decision of how to allocate the consumer's dollar should belong to the consumer or to the state. It has already been shown that the choice of how a doctor's services should be rendered belongs only to the doctor: in the same way the choice of whether to buy a doctor's service rather than some other commodity or service belongs to the consumer as a logical consequence of the right to his own life. (1971:1291)

This account is misguided, I think, because it ignores the overriding importance which is so often attached to health crises. When lumps appear on someone's neck, it usually makes little sense to talk of choosing whether to buy a doctor's service rather than a color television set. References to just trade-offs suddenly seem out of place. No compensation suffices, since the penalties may differ so much.

There is even a further restriction on consumer choice. One's knowledge in these circumstances is comparatively so limited. The physician makes most of the decisions: about diagnosis, treatment, hospitalization, number of return visits, and so on. In brief:

> The consumer knows very little about the medical services he is buying—probably less than about any other service he purchases. . . . While [he] can still play a role in policing the market, that role is much more limited in the field of health care than in almost any other area of private economic activity. (Schultze, 1972:214-215)

For much of the way, then, an appeal to supply and demand and consumer choice is not quite fitting. It neglects the issue of the value of various contributions. And it fails to allow for the recognition that medical treatments may be overridingly desired. In contexts of catastrophe at any rate, when life itself is threatened, most persons (other than those who are apathetic or seek to escape from the terrifying prospects) cannot take medical care to be merely one option among others.

IV

To each according to his needs. The concept of needs is sometimes taken to apply to an entire range of interests which concern a person's "psycho-physical existence" (Outka, 1972:esp. 264-265). On this wide usage, to attribute a need to someone is to say that the person lacks what is thought to conduce to his or her "welfare"—understood in both a physiological sense (e.g., for food, drink, shelter, and health) and a psychological one (e.g., for continuous human affection and support).

Yet even in the case of such a wide usage, what the person lacks is typically assumed to be basic. Attention is restricted to recurrent considerations rather than to every possible individual whim or frivolous pursuit. So one is not surprised to meet with the contention that a preferable rendering of this formula would be: "to each according to his essential needs" (Perelman, 1963:22). This contention seems to me well taken. It implies, for one thing, that basic needs are distinguishable from felt needs or wants. For the latter may encompass expressions of personal preference unrelated to considerations of survival or subsistence, and sometimes artificially generated by circumstances of rising affluence in the society at large.

Essential needs are also typically assumed to be given rather than acquired. They are not constituted by any action for which the person is responsible by virtue of his or her distinctively greater effort. It is almost as if the designation "innocent" may be linked illuminatingly to need, as retribution, punishment, and so on, are to desert, and in complex ways, to freedom. Thus essential needs are likewise distinguishable from deserts. Where needs are unequal, one thinks of them as fortuitously distributed; as part, perhaps, of a kind of "natural lottery" (see Rawls, 1971:e.g., 104). So very often the advantages of health and the burdens of illness, for example, strike one as arbitrary effects of the lottery. It seems wrong to say that a newborn child deserves as a reward all of his faculties when he has done nothing in particular which distinguishes him from another newborn who comes into the world deprived of one or more of them. Similarly, though crudely, many religious believers do not look on natural events as personal deserts. They are not inclined to pronounce sentences such as, "That evil person with incurable cancer got what he deserved." They are disposed instead to search for some distinction between what they may call the conditions of finitude on the one hand and sin and moral evil on the other. If the distinction is "ultimately" invalid, in this life it seems inscrutably so. Here and now it may be usefully drawn. Inequalities in the need for medical treatment are taken, it appears, to reflect the conditions of finitude more than anything else.

One can even go on to argue that among our basic or essential needs, the case of medical treatment is conspicuous in the following sense. While food and shelter are not matters about which we are at liberty to please ourselves, they are at least predictable. We can plan, for instance, to store up food and fuel for the winter. It may be held that responsibility increases along with the power to predict. If so, then many health crises seem peculiarly random and uncontrollable. Cancer, given the present state of knowledge at any rate, is a contingent disaster, whereas hunger is a steady threat. Who will need serious medical care, and when, is then perhaps a classic example of uncertainty.

Finally, and more theoretically, it is often observed that a need-conception of justice comes closest to charity or *agape* (e.g., Perelman, 1963:23). I think there are indeed crucial overlaps (see Outka, 1972:91-92, 309-312). To cite several of them: the equal consideration *agape* enjoins has to do in the first instance with those generic endowments which people share, the characteristics of a person qua human existent. Needs, as we have seen, likewise concern those things essential to the life and welfare of men considered simply as men (see also Honoré, 1968). They are not based on particular conduct alone, on those idiosyncratic attainments which contribute to someone's being such-and-such a kind of person. Yet a certain sort of inequality is recognized, for needs differ in divergent circumstances and so treatments must if benefits are to be equalized. *Agape* too allows for a distinction between equal consideration and identical treatment. The aim of equalizing benefits is implied by the injunction to consider the interests of each party equally. This may require differential treatments of differing interests.

Overlaps such as these will doubtless strike some as so extensive that it may be asked whether *agape* and a need-conception of justice are virtually equivalent. I think not. One contrast was pointed out before. The differential treatment enjoined by *agape* is more complex and goes deeper. In the case of *agape*, attention may be appropriately given to varying *efforts* as well as to unequal *needs*. More generally one may say that agapeic considerations extend to all of the psychological nuances and contextual details of individual persons and their circumstances. Imaginative concern is enjoined for concrete human beings: for what someone is uniquely, for what he or she—as a matter of personal history and distinctive identity—wants, feels, thinks, celebrates, and endures. The attempt to establish and enhance mutual affection between individual persons is taken likewise to be fitting. Conceptions of social justice, including "to each according to his essential needs," tend to be more restrictive; they call attention to considerations which obtain for a number of persons, to impersonally specified criteria for assessing collective policies and practices. Agape involves more, even if one supposes never less.

Other differences could be noted. What is important now however is the recognition that, in matters of health care in particular, agape and a need-conception of justice are conjoined in a number of relevant respects. At least this is so for those who think that, again, justice has properly to do with pleas of "But I could not help it." It seeks to distinguish such cases from those acknowledged to be praiseworthy or blameworthy. The formula "to each according to his needs" is one cogent way of identifying the moral relevance of these pleas. To ignore them may be thought to be unfair as well as unkind when they arise from the deprivation of some essential need. The move to confine the notion of justice wholly to desert considerations is thereby resisted as well. Hence we may say that sometimes "questions of social justice arise just because people are unequal in ways they can do very little to change and . . . only by attending to these inequalities can one be said to be giving their interests equal consideration" (Benn, 1971:164).

V

Similar treatment for similar cases. This conception is perhaps the most familiar of all. Certainly it is the most formal and inclusive one. It is frequently taken as an elementary appeal to consistency and linked to the universalizability test. One should not make an arbitrary exception on one's own behalf, but rather should apply impartially whatever standards one accepts. The conception can be fruitfully applied to health care questions and I shall assume its relevance. Yet as literally interpreted, it is necessary but not sufficient. For rightly or not, it is often held to be as compatible with no positive treatment whatever as with active promotion of other people's interests, as long as all are equally and impartially included. Its exponents sometimes assume such active promotion without demonstrating clearly how this is built into the conception itself. Moreover, it may obscure a distinction which we have seen agapists and others make: between equal consideration and identical treatment. Needs may differ and so treatments must, if benefits are to be equalized.

I have placed this conception at the end of the list partly because it moves us, despite its formality, toward practice. Let me suggest briefly how it does so. Suppose first of all one agrees with the case so far offered. Suppose, that is, it has been shown convincingly that a need-conception of justice applies with greater relevance than the earlier three when one reflects about the basic purpose of medical care. To treat one class of people differently from another because of income or geographic location should therefore be ruled out, because such reasons are irrelevant. (The irrelevance is conceptual, rather than always, unfortunately, causal.) In short, all persons should have equal access, "as needed, without financial, geographic, or other barriers, to the whole spectrum of health services" (Somers and Somers, 1972a:122).

Suppose however, secondly, that the goal of equal access collides on some occasions with the realities of finite medical resources and needs which prove to be insatiable. That such collisions occur in fact it would be idle to deny. And it is here that the practical bearing of the formula of similar treatment for similar cases should be noticed. Let us recall Williams' conclusion: "the proper ground of distribution of medical care is ill health: this is a necessary truth." While I agree with the essentials of his argument—for all the reasons above—I would prefer, for practical purposes, a slightly more modest formulation. Illness is the proper ground for the *receipt* of medical care. However, the *distribution* of medical care in less-than-optimal circumstances requires us to face the collisions. I would argue that in such circumstances the formula of similar treatment for similar cases may be construed so as to guide actual choices in the way most compatible with the goal of equal access. The formula's allowance of no positive treatment whatever may justify exclusion of entire classes of cases from a priority list. Yet it forbids doing so for irrelevant or arbitrary reasons. So (1) if we accept the case for equal access, but (2) if we simply cannot, physically cannot, treat all who are in need, it seems more just to discriminate by virtue of categories of illness, for example, rather than between the rich ill and poor ill. All persons with a certain rare, noncommunicable disease would not receive priority, let us say, where the costs were inordinate, the prospects for rehabilitation remote, and for the sake of equalized benefits to many more. Or with Ramsey we may urge a policy of random patient selection when one must decide between claimants for a medical treatment unavailable to all. Or we may acknowledge that any notion of "comprehensive benefits" to which persons should have equal access is subject to practical restrictions which will vary from society to society depending on resources at a given time. Even in a country as affluent as the United States there will surely always be items excluded, e.g., perhaps over-the-counter drugs, some teenage orthodontia, cosmetic surgery, and the like (Somers and Somers, 1972b:182). Here too the formula of similar treatment for similar cases may serve to modify the application of a need-conception of justice in order to address the insatiability-problem and limit frivolous use. In all of the foregoing instances of restriction, however, the relevant feature remains the illness, discomfort, etc. itself. The goal of equal access then retains its prima facie authoritativeness. It is imperfectly realized rather than disregarded.

VI

These latter comments lead on to the question of institutional implications. I cannot aim here of course for the specificity rightly sought by policy-makers. My endeavor has been conceptual elucidation. While the ethicist needs to be apprised about the facts, he or she does not, qua ethicist, don the mantle of the policy-expert. In any case, only rarely does anyone do both things equally well. Yet cross-fertilization is extremely desirable. For experts should not be isolated from the wider assumptions their recommendations may reflect. I shall merely list some of the topics which would have to be discussed at length if we were to get clear about the implications. Examples will be limited to the current situation in the United States.

Anyone who accepts the case for equal access will naturally be concerned about de facto disparities in the availability of medical treatment. Let us consider two relevant indictments of current American practice. They appear in the writings not only of those who attack indiscriminately a system seen to be governed only by the appetite for profit and power, but also of those who denounce in less sweeping terms and espouse more cautiously reformist positions. The first shortcoming has to do with the maldistribution of supply. Per capita ratios of physicians to populations served vary, sometimes notoriously, between affluent suburbs and rural and inner city areas. This problem is exacerbated by the distressing data concerning the greater health needs of the poor. Chronic disease, frequency and duration of hospitalization, psychiatric disorders, infant death rates, etc.—these occur in significantly larger proportions to lower income members of American society (Appel, 1970; Hubbard, 1970). A further complication is that "the distribution of health insurance coverage is badly skewed. Practically all the rich have insurance. But among the poor, about two-thirds have none. As a result, among people aged 25 to 64 who die, some 45 to 50 per cent have neither hospital nor surgical coverage" (Somers, 1971a:46). This last point connects with a second shortcoming frequently cited. Even those who are otherwise economically independent may be shattered by the high cost of a "catastrophic illness" (see some eloquent examples in Kennedy, 1972).

Proposals for institutional reforms designed to overcome such disparities are bound to be taken seriously by any defender of equal access. What he or she will be disposed to press for, of course, is the removal of any double standard or "two class" system of care. The viable procedures for bringing this about are not obvious, and comparisons with certain other societies (for relevant alternative models) are drawn now with perhaps less confidence (see Anderson, 1973). One set of commonly discussed proposals includes (1) incentive subsidies to physicians, hospitals, and medical centers to provide services in regions of poverty (to overcome in part the unwillingness—to which no unique culpability need be ascribed—of many providers and their spouses to work and live in grim surroundings); (2) licensure controls to avoid comparatively excessive concentrations of physicians in regions of affluence; (3) a period of time (say, two years) in an underserved area as a requirement for licensing; (4) redistribution facilities which allow for population shifts.

A second set of proposals is linked with health insurance itself. While I cannot venture into the intricacies of medical economics or comment on the various bills for national health insurance presently inundating Congress, it may be instructive to take brief note of one proposal in which, once more, the defender of equal access is bound to take an interest (even if he or she finally rejects it on certain practical grounds). The precise details of the proposal are unimportant for our purposes (for one much-discussed version, see Feldstein, 1971). Consider this crude sketch. Each citizen is (in effect) issued a card by the government. Whenever "legitimate" medical expenses (however determined for a given society) exceed, say, 10 per cent of his or her annual taxable income, the card may be presented so that additional costs incurred will be paid for out of general tax revenues. The reasons urged on behalf of this sort of arrangement include the following. In the case of medical care there is warrant for proportionately equalizing what is spent from anyone's total taxable income. This warrant reflects the conditions, discussed earlier, of the natural lottery. Insofar as the advantages of health and the burdens of illness are random and undeserved, we may find it in our common interest to share risks. A fixed percentage of income attests to the misfit, also mentioned previously, between the reasons for differential total income and the reasons for receiving medical treatment. If money remains a causally necessary condition for receiving medical treatment, then a way must be found to place it in the hands of those who need it. The card is one such means. It is designed effectively to equalize purchasing power. In this way it seems to accord nicely with the goal of equal access. On the other side, the requirement of initial out-of-pocket expenses—sufficiently large in comparison to average family expenditures on health care—is designed to discourage frivolous use and foster awareness that medical care is a benefit not to be simply taken as a matter of course. It also safeguards against an excessively large tax burden while providing universal protection against the often disastrous costs of serious illness. Whether 10 per cent is too great a chunk for the very poor to pay, and whether by itself the proposal will feed price inflation and neglect of preventive medicine are questions which would have to be answered.

Another kind of possible institutional reform will also greatly interest the defender of equal access. This has to do with the "design of health care systems" or "care settings." The prevalent setting in American society has always been "fee-for-service." It is left up to each person to obtain the requisite care and to pay for it as he or she goes along. Because costs for medical treatment have accelerated at such an alarming rate, and because the sheer diffusion of energy and effort so characteristic of American medical practice leaves more and more people dissatisfied, alternatives to fee-for-service have been considered of late with unprecedented seriousness. The alternative care setting most widely discussed is prepaid practice, and specifically the "health maintenance organization" (HMO). Here one finds "an organized system of care which accepts the responsibility to provide or otherwise assure comprehensive care to a defined population for a fixed periodic payment per person or per family . . ." (Somers, 1972b:v). The best-known HMO is the Kaiser-Permanente Medical Care Program (see also Garfield, 1971). Does the HMO serve to realize the goal of equal access more fully? One line of argument in its favor is this. It is plausible to think that equal access will be fostered by the more economical care setting. HMO's are held to be less costly per capita in at least two respects: hospitalization rates are much below the national average; and less often noted, physician manpower is as well. To be sure, one should be sensitive to the corruptions in each type of setting. While fee-for-service has resulted in a suspiciously high number of surgeries (twice as many per capita in the United States as in Great Britain), the HMO physician may more frequently permit the patient's needs to be overridden by the organization's pressure to economize. It may also be more difficult in an HMO setting to provide for close personal relations between a particular physician and a particular patient (something commended, of course, on all sides). After such corruptions are allowed for, the data seem encouraging to such an extent that a defender of equal access will certainly support the repeal of any law which limits the development of prepaid practice, to approve of "front-aid" subsidies for HMO's to increase their number overall and achieve a more equitable distribution throughout the country, and so on. At a minimum, each care setting should be available in every region. If we assume a common freedom to choose between them, each may help to guard against the peculiar temptations to which the other is exposed.

To assess in any serious way proposals for institutional reform such as the above is beyond the scope of this paper. We would eventually be led, for example, into the question of whether it is consistent for the rich to pay more than the poor for the same treatment when, again, needs rather than income constitute the ground of the treatment (Ward, 1973), and from there into the tangled subject of the "ethics of redistribution" in general (see, e.g., Benn and Peters, 1965:155-178; de Jouvenal, 1952). Other complex issues deserve to be considered as well, e.g., the criteria for allocation of limited resources,[2] and how conceptions of justice apply to the providers of health care.[3]

Those committed to self-conscious moral and religious reflection about subjects in medicine have concentrated, perhaps unduly, on issues about care of individual patients (as death approaches, for instance). These issues plainly warrant the most careful consideration. One would like to see in addition, however, more attention paid to social questions in medical ethics. To attend to them is not necessarily to leave behind all of the matters which reach deeply into the human condition. Any detailed case for institutional reforms, for example, will be enriched if the proponent asks soberly whether certain conflicts and certain perplexities allow for more than partial improvements and provisional resolutions. Can public and private interests ever be made fully to coincide by legislative and administrative means? Will the commitment of a physician to an individual patient and the commitment of the legislator to the "common good" ever be harmonized in every case? Our anxiety may be too intractable. Our fear of illness and of dying may be so pronounced and immediate that we will seize the nearly automatic connections between privilege, wealth, and power if we can. We will do everything possible to have our kidney machines even if the charts make it clear that many more would benefit from mandatory immunization at a fraction of the cost. And our capacity for taking in rival points of view may be too limited. Once we have witnessed tangible suffering, we cannot just return with ease to public policies aimed at statistical patients. Those who believe that justice is the pre-eminent virtue of institutions and that a case can be convincingly made on behalf of justice for equal access to health care would do well to ponder such conflicts and perplexities. Our reforms might then seem, to ourselves and to others, less abstract and jargon-filled in formulation and less sanguine and piecemeal in substance. They would reflect a greater awareness of what we have to confront.

NOTES

1. Much of the research for this paper was done during the Fall Term, 1972-73, when I was on leave in Washington, D.C. I am very grateful for the two appointments which made this leave possible: as Service Fellow, Office of Special Projects, Health Services and Mental Health Administration, Department of Health, Education, and Welfare; and as Visiting Scholar, Kennedy Center for Bioethics, Georgetown University.

2. The issue of priorities is at least threefold: (1) between improved medical care and other social needs, e.g., to restrain auto accidents and pollution; (2) between different sorts of medical treatments for different illnesses, e.g., prevention vs. crisis intervention and exotic treatments; (3) between persons all of whom need a single scarce resource and not all can have it, e.g., Ramsey's discussion of how to decide among those who are to receive dialysis. Moreover, (1) can be subdivided between (a) improved medical care and other social needs which affect health directly, e.g., drug addiction, auto accidents, and pollution; (b) improved medical care and other social needs which serve the overall aim of community-survival, e.g., a common defense. In the case of (2), one would like to see far more careful discussion of some general criteria which might be employed, e.g., numbers affected, degree of contagion, prospects for rehabilitation, and so on.

3. What sorts of appeals to justice might be cogently made to warrant, for instance, the differentially high income physicians receive? Here are three possibilities: (1) the greater skill and responsibility involved should be rewarded proportionately, i.e., one should attend to considerations of *desert*; (2) there should be *compensation* for the money invested for education and facilities in order to restore circumstances of approximate equality (this argument, while a common one in medical circles, would need to consider that medical education is received in part at public expense and that the modern physician is the highest paid professional in the country); (3) the difference should benefit the least advantaged more than an alternative arrangement where disparities are less. We prefer a society where the medical profession flourishes and everyone has a longer life expectancy to one where everyone is poverty-stricken with a shorter life expectancy ("splendidly equalized destitution"). Yet how are we to ascertain the minimum degree of differential income required for the least advantaged members of the society to be better off?

Discussions of "justice and the interests of providers" are, I think, badly needed. Physicians in the United States have suffered a decline in prestige for various reasons, e.g., the way many used Medicare to support and increase their own incomes. Yet one should endeavor to assess their interests fairly. A concern for professional autonomy is clearly important, though one may ask whether adequate attention has been paid to the distinction between the imposition of cost-controls from outside and interference with professional medical judgments. One may affirm the former, it seems, and still reject—energetically—the latter.

REFERENCES

Anderson, Odin
    1973   Health Care: Can There Be Equity? The United States, Sweden and England. New York: Wiley.
Appel, James Z.
    1970   "Health care delivery." Pp. 141-166 in Boisfeuillet Jones (ed.), The Health of Americans. Englewood Cliffs, N.J.: Prentice-Hall, Inc.
Bedau, Hugo A.
    1971   "Radical egalitarianism." Pp. 168-180 in Hugo A. Bedau (ed.), Justice and Equality. Englewood Cliffs, N.J.: Prentice-Hall, Inc.
Benn, Stanley I.
    1971   "Egalitarianism and the equal consideration of interests." Pp. 152-167 in Hugo A. Bedau (ed.), Justice and Equality. Englewood Cliffs, N.J.: Prentice-Hall, Inc.
Benn, Stanley I. and Richard S. Peters.
    1965   The Principles of Political Thought. New York: The Free Press.
de Jouvenel, Bertrand
    1952   The Ethics of Redistribution. Cambridge: University Press.
Feldstein, Martin S.
    1971   "A new approach to national health insurance." The Public Interest 23 (Spring):93-105.
Garfield, Sidney R.
    1971   "Prevention of dissipation of health services resources." American Journal of Public Health 61:1499-1506.
Hicks, Nancy
    1973   "Nation's doctors move to police medical care." Pp. 1, 52 in New York Times, Sunday, October 28.
Honoré, A.M.
    1968   "Social justice." Pp. 61-94 in Robert S. Summers (ed.), Essays in Legal Philosophy. Oxford: Basil Blackwell.
Hospers, John
    1961   Human Conduct. New York: Harcourt, Brace and World, Inc.
Hubbard, William N.
    1970   "Health knowledge." Pp. 93-120 in Boisfeuillet Jones (Ed.), The Health of Americans. Englewood Cliffs, N.J.: Prentice-Hall, Inc.
Kennedy, Edward M.
    1972   In Critical Condition: The Crisis in America's Health Care. New York: Simon and Schuster.
Lucas, J. R.
    1972   "Justice." Philosophy 47, No. 181 (July):229-248.
Nagel, Thomas
    1973   "Equal treatment and compensatory discrimination." Philosophy and Public Affairs 2, No. 4 (Summer):348-363.
Nixon, Richard M.
    1972   "President's message on health care system." Document No. 92-261 (March 2). House of Representatives, Washington, D.C.
Outka, Gene
    1972   Agape: An Ethical Analysis. New Haven and London: Yale University Press
Perelman, Ch.
    1963   The Idea of Justice and the Problem of Argument. Trans. John Petrie. London: Routledge and Kegan Paul.
Ramsey, Paul
    1970   The Patient as Person. New Haven and London: Yale University Press.
Rand, Ayn
    1957   Atlas Shrugged. New York: Signet.

Rawls, John
    1971   *A Theory of Justice.* Cambridge, Mass.: Harvard University Press.
Rescher, Nicholas
    1966   *Distributive Justice.* Indianapolis: The Bobbs-Merrill Company, Inc.
Ryan, John A.
    1916   *Distributive Justice.* New York: The Macmillan Company.
Sade, Robert M.
    1971   "Medical care as a right: a refutation." *The New England Journal of Medicine* 285 (December): 1288-1292.
Schultze, Charles L., Edward R. Fried, Alice M. Rivlin and Nancy H. Teeters
    1972   *Setting National Priorities: The 1973 Budget.* Washington, D.C.: The Brookings Institution.
Somers, Anne R.
    1971a   *Health Care in Transition: Directions for the Future.* Chicago: Hospital Research and Educational Trust.
    1971b   (ed.), *The Kaiser-Permanente Medical Care Program.* New York: The Commonwealth Fund.
Somers, Anne R. and Herman M. Somers
    1972a   "The organization and financing of health care: issues and directions for the future." *American Journal of Orthopsychiatry* 42 (January), 119-136.
    1972b   "Major issues in national health insurance." *Milbank Memorial Fund Quarterly* 50, No. 2, Part 1 (April):177-210.
Vlastos, Gregory
    1962   "Justice and equality." Pp. 31-72 in Richard B. Brandt (ed.), *Social Justice.* Englewood Cliffs, N.J.: Prentice-Hall, Inc.
Ward, Andrew
    1973   "The idea of equality reconsidered." *Philosophy* 48 (January):85-90.
Williams, Bernard A.O.
    1971   "The idea of equality." Pp. 116-137 in Hugo A. Bedau (ed.), *Justice and Equality.* Englewood Cliffs, N.J.: Prentice-Hall, Inc.

27 Who Shall Live When Not All Can Live?

1970

JAMES F. CHILDRESS

The problem of the distribution of scarce medical resources is, unfortunately, without a final adequate resolution. Although many approaches have been made to this life-for-life situation, none has fully resolved the entire range of complexities and ethical dilemmas. Childress presents a methodology and ethical argumentation for randomness as the most equitable means of such distribution after medical selection has been made. This argument, together with its critique of social-worth criteria, resolves many problems and avoids many of the seemingly irresponsible approaches to the problem of the distribution of scarce medical resources.

Dr. James F. Childress, Ph.D., is Joseph Kennedy Professor of Ethics at the Kennedy Center for Bioethics at Georgetown University.

Who shall live when not all can live? Although this question has been urgently forced upon us by the dramatic use of artificial internal organs and organ transplantations, it is hardly new. George Bernard Shaw dealt with it in "The Doctor's Dilemma":

SIR PATRICK. Well, Mr. Savior of Lives: which is it to be? that honest decent man Blenkinsop, or that rotten blackguard of an artist, eh?

RIDGEON. It's not an easy case to judge, is it? Blenkinsop's an honest decent man; but is he any use? Dubedat's a rotten blackguard; but he's a genuine source of pretty and pleasant and good things.

SIR PATRICK. What will he be a source of for that poor innocent wife of his, when she finds him out?

RIDGEON. That's true. Her life will be a hell.

501

SIR PATRICK. And tell me this. Suppose you had this choice put before you: either to go through life and find all the pictures bad but all the men and women good, or go through life and find all the pictures good and all the men and women rotten. Which would you choose?[1]

A significant example of the distribution of scarce medical resources is seen in the use of penicillin shortly after its discovery. Military officers had to determine which soldiers would be treated—those with venereal disease or those wounded in combat.[2] In many respects such decisions have become routine in medical circles. Day after day physicians and others make judgments and decisions "about allocations of medical care to various segments of our population, to various types of hospitalized patients, and to specific individuals,"[3] for example, whether mental illness or cancer will receive the higher proportion of available funds. Nevertheless, the dramatic forms of "Scarce Life-Saving Medical Resources" (hereafter abbreviated as SLMR) such as hemodialysis and kidney and heart transplants have compelled us to examine the moral questions that have been concealed in many routine decisions. I do not attempt in this paper to show how a resolution of SLMR cases can help us in the more routine ones which do not involve a conflict of life with life. Rather I develop an argument for a particular method of determining who shall live when not all can live. No conclusions are implied about criteria and procedures for determining who shall receive medical resources that are not directly related to the preservation of life (e.g. corneal transplants) or about standards for allocating money and time for studying and treating certain diseases.

Just as current SLMR decisions are not totally discontinuous with other medical decisions, so we must ask whether some other cases might, at least by analogy, help us develop the needed criteria and procedures. Some have looked at the principles at work in our responses to abortion, euthanasia, and artificial insemination.[4] Usually they have concluded that these cases do not cast light on the selection of patients for artificial and transplanted organs. The reason is evident: in abortion, euthanasia, and artificial insemination, there is no conflict of life with life for limited but indispensable resources (with the possible exception of therapeutic abortion). In current SLMR decisions, such a conflict is inescapable, and it makes them so morally perplexing and fascinating. If analogous cases are to be found, I think that we shall locate them in moral conflict situations.

Analogous Conflict Situations

An especially interesting and pertinent one is U.S. v. Holmes.[5] In

1841 an American ship, the *William Brown*, which was near New-foundland on a trip from Liverpool to Philadelphia, struck an iceberg. The crew and half the passengers were able to escape in the two available vessels. One of these, a longboat, carrying too many passengers and leaking seriously, began to founder in the turbulent sea after about twenty-four hours. In a desperate attempt to keep it from sinking, the crew threw overboard fourteen men. Two sisters of one of the men either jumped overboard to join their brother in death or instructed the crew to throw them over. The criteria for determining who should live were "not to part man and wife, and not to throw over any women." Several hours later the others were rescued. Returning to Philadelphia, most of the crew disappeared, but one, Holmes, who had acted upon orders from the mate, was indicted, tried, and convicted on the charge of "unlawful homicide."

We are interested in this case from a moral rather than a legal standpoint, and there are several possible responses to and judgments about it. Without attempting to be exhaustive I shall sketch a few of these. The judge contended that lots should have been cast, for in such conflict situations, there is no other procedure "so consonant both to humanity and to justice." Counsel for Holmes, on the other hand, maintained that the "sailors adopted the only principle of selection which was possible in an emergency like theirs—a principle more humane than lots."

Another version of selection might extend and systematize the maxims of the sailors in the direction of "utility"; those are saved who will contribute to the greatest good for the greatest number. Yet another possible option is defended by Edmond Cahn in *The Moral Decision*. He argues that in this case we encounter the "morals of the last days." By this phrase he indicates that an apocalyptic crisis renders totally irrelevant the normal differences between individuals. He continues,

> In a strait of this extremity, all men are reduced—or raised, as one may choose to denominate it—to members of the genus, mere congeners and nothing else. Truly and literally, all were "in the same boat," and thus none could be saved separately from the others. I am driven to conclude that otherwise—that is, if none sacrifice themselves of free will to spare the others—they must all wait and die together. For where all have become congeners, pure and simple, no one can save himself by killing another.[6]

Cahn's answer to the question "who shall live when not all can live" is "none" unless the voluntary sacrifice by some persons permits it.

Few would deny the importance of Cahn's approach although

many, including this writer, would suggest that it is relevant mainly as an affirmation of an elevated and, indeed, heroic or saintly morality which one hopes would find expression in the voluntary actions of many persons trapped in "borderline" situations involving a conflict of life with life. It is a maximal demand which some moral principles impose on the individual in the recognition that self-preservation is not a good which is to be defended at all costs. The absence of this saintly or heroic morality should not mean, however, that everyone perishes. Without making survival an absolute value and without justifying all means to achieve it, we can maintain that simply letting everyone die is irresponsible. This charge can be supported from several different standpoints, including society at large as well as the individuals involved. Among a group of self-interested individuals, none of whom volunteers to relinquish his life, there may be better and worse ways of determining who shall survive. One task of social ethics, whether religious or philosophical, is to propose relatively just institutional arrangements within which self-interested and biased men can live. The question then becomes: which set of arrangements—which criteria and procedures of selection—is most satisfactory in view of the human condition (man's limited altruism and inclination to seek his own good) and the conflicting values that are to be realized?

There are several significant differences between the *Holmes* and SLMR cases, a major one being that the former involves *direct* killing of another person, while the latter involve only *permitting* a person to die when it is not possible to save all. Furthermore, in extreme situations such as *Holmes*, the restraints of civilization have been stripped away, and something approximating a state of nature prevails, in which life is "solitary, poor, nasty, brutish and short." The state of nature does not mean that moral standards are irrelevant and that might should prevail, but it does suggest that much of the matrix which normally supports morality has been removed. Also, the necessary but unfortunate decisions about who shall live and die are made by men who are existentially and personally involved in the outcome. Their survival too is at stake. Even though the institutional role of sailors seems to require greater sacrificial actions, there is obviously no assurance that they will adequately assess the number of sailors required to man the vessel or that they will impartially and objectively weigh the common good at stake. As the judge insisted in his defense of casting lots in the *Holmes* case: "In no other than this [casting lots] or some like way are those having equal rights put upon an equal footing, and in no other way is it possible to guard against partiality and oppression, violence, and conflict." This difference should not be exaggerated since self-interest, professional pride, and the like obviously affect the outcome of many medical decisions. Nor do the remaining differences cancel *Holmes'* instructiveness.

Criteria of Selection for SLMR

Which set of arrangements should be adopted for SLMR? Two questions are involved: Which standards and criteria should be used? and, Who should make the decision? The first question is basic, since the debate about implementation, e.g. whether by a lay committee or physician, makes little progress until the criteria are determined.

We need two sets of criteria which will be applied at two different stages in the selection of recipients of SLMR. First, medical criteria should be used to exclude those who are not "medically acceptable." Second, from this group of "medically acceptable" applicants, the final selection can be made. Occasionally in current American medical practice, the first stage is omitted, but such an omission is unwarranted. Ethical and social responsibility would seem to require distributing these SLMR only to those who have some reasonable prospect of responding to the treatment. Furthermore, in transplants such medical tests as tissue and blood typing are necessary, although they are hardly fully developed.

"Medical acceptability" is not as easily determined as many non-physicians assume since there is considerable debate in medical circles about the relevant factors (e.g., age and complicating diseases). Although ethicists can contribute little or nothing to this debate, two proposals may be in order. First, "medical acceptability" should be used only to determine the group from which the final selection will be made, and the attempt to establish fine degrees of prospective response to treatment should be avoided. Medical criteria, then, would exclude some applicants but would not serve as a basis of comparison between those who pass the first stage. For example, if two applicants for dialysis were medically acceptable, the physicians would *not* choose the one with the *better* medical prospects. Final selection would be made on other grounds. Second, psychological and environmental factors should be kept to an absolute minimum and should be considered only when they are without doubt critically related to medical acceptability (e.g., the inability to cope with the requirements of dialysis which might lead to suicide).*

The most significant moral questions emerge when we turn to the final selection. Once the pool of medically acceptable applicants has been defined and still the number is larger than the resources, what other criteria should be used? How should the final selection be made? First, I shall examine some of the difficulties that stem from efforts to make the final selection in terms of social value; these difficulties raise serious doubts about the feasibility and justifiability of the utilitarian approach. Then I shall consider the possible justification for random selection or chance.

Occasionally criteria of social worth focus on past contributions but most often they are primarily future-oriented. The patient's potential and probable contribution to the society is stressed, although this obviously cannot be abstracted from his present web of relationships (e.g., dependents) and occupational activities (e.g., nuclear physicist). Indeed, the magnitude of his contribution to society (as an abstraction) is measured in terms of these social roles, relations, and functions. Enough has already been said to suggest the tremendous range of factors that affect social value or worth.† Here we encounter the first major difficulty of this approach: How do we determine the relevant criteria of social value?

The difficulties of quantifying various social needs are only too obvious. How does one quantify and compare the needs of the spirit (e.g., education, art, religion), political life, economic activity, technological development? Joseph Fletcher suggests that "some day we may learn how to 'quantify' or 'mathematicate' or 'computerize' the value problem in selection, in the same careful and thorough way that diagnosis has been."[7] I am not convinced that we can ever quantify values, or that we should attempt to do so. But even if the various social and human needs, in principle, could be quantified, how do we determine how much weight we will give to each one? Which will have priority in case of conflict? Or even more basically, in the light of which values and principles do we recognize social "needs"?

One possible way of determining the values which should be emphasized in selection has been proposed by Leo Shatin.[8] He insists that our medical decisions about allocating resources are already based on an unconscious scale of values (usually dominated by material worth). Since there is really no way of escaping this, we should be self-conscious and critical about it. How should we proceed? He recommends that we discover the values that most people in our society hold and then use them as criteria for distributing SLMR. These values can be discovered by attitude or opinion surveys. Presumably if fifty-one percent in this testing period put a greater premium on military needs than technological development, military men would have a greater claim on our SLMR than experimental researchers. But valuations of what is significant change, and the student revolutionary who was denied SLMR in 1970 might be celebrated in 1990 as the greatest American hero since George Washington.

Shatin presumably is seeking criteria that could be applied nationally, but at the present, regional and local as well as individual prejudices tincture the criteria of social value that are used in selection. Nowhere is this more evident than in the deliberations and decisions of the anonymous selection committee of the Seattle Artificial Kidney Center where such factors as church membership and Scout leadership have been deemed significant for determining who shall live.[9] As two critics conclude after examining these criteria and procedures, they rule out "creative nonconformists, who rub the bourgeoisie the wrong way but who historically have contributed so much to the making of America. The Pacific Northwest is no place for a Henry David Thoreau with bad kidneys."[10]

Closely connected to this first problem of determining social values is a second one. Not only is it difficult if not impossible to reach agreement on social values, but it is also rarely easy to predict what our needs will be in a few years and what the consequences of present actions will be. Furthermore it is difficult to predict which persons will fulfill their potential function in society. Admissions committees in colleges and universities experience the frustrations of predicting realization of potential. For these reasons, as someone has indicated, God might be a utilitarian, but we cannot be. We simply lack the capacity to predict very accurately the consequences which we then must evaluate. Our incapacity is never more evident than when we think in societal terms.

Other difficulties make us even less confident that such an approach to SLMR is advisable. Many critics raise the spectre of abuse, but this should not be overemphasized. The fundamental difficulty appears on another level: the utilitarian approach would in effect reduce the person to his social role, relations, and functions. Ultimately it dulls and perhaps even eliminates the sense of the person's transcendence, his dignity as a person which cannot be reduced to his past or future contribution to society. It is not at all clear that we are willing to live with these implications of utilitarian selection. Wilhelm Kolff, who invented the artificial kidney, has asked: "Do we really subscribe to the principle that social standing should determine selection? Do we allow patients to be treated with dialysis only when they are married, go to church, have children, have a job, a good income and give to the Community Chest?"[**]

The German theologian Helmut Thielicke contends that any search for "objective criteria" for selection is already a capitulation to the utilitarian point of view which violates man's dignity.[11] The solution is not to let all die, but to recognize that SLMR cases are "borderline situations" which inevitably involve guilt. The agent, however, can have courage and freedom (which, for Thielicke, come from justification by faith) and can

go ahead anyway and seek for criteria for deciding the question
of life or death in the matter of the artificial kidney. Since these
criteria are . . . questionable, necessarily alien to the meaning of
human existence, the decision to which they lead can be little
more than that arrived at by casting lots.[12]

The resulting criteria, he suggests, will probably be very similar to
those already employed in American medical practice.

He is most concerned to preserve a certain *attitude* or *disposition*
in SLMR—the sense of guilt which arises when man's dignity is violated. With this sense of guilt, the agent remains "sound and healthy
where it really counts."[13] Thielicke uses man's dignity only as a judg-
mental, critical, and negative standard. It only tells us how all selection
criteria and procedures (and even the refusal to act) implicate us in the
ambiguity of the human condition and its metaphysical guilt. This
approach is consistent with his view of the task of theological ethics:
"to teach us how to understand and endure—not 'solve'—the borderline situations."[14] But ethics, I would contend, can help us discern
the factors and norms in whose light relative, discriminate judgments
can be made. Even if all actions in SLMR should involve guilt, some
may preserve human dignity to a greater extent than others. Thielicke
recognizes that a decision based on any criteria is "little more than
that arrived at by casting lots." But perhaps selection by chance would
come the closest to embodying the moral and nonmoral values that we
are trying to maintain (including a sense of man's dignity).

The Values of Random Selection

My proposal is that we use some form of randomness or chance
(either natural, such as "first come, first served," or artificial, such as
a lottery) to determine who shall be saved. Many reject randomness as
a surrender to non-rationality when responsible and rational judgments can and must be made. Edmond Cahn criticizes "Holmes'
judge" who recommended the casting of lots because, as Cahn puts it,
"the crisis involves stakes too high for gambling and responsibilities
too deep for destiny."[15] Similarly, other critics see randomness as a
surrender to "non-human" forces which necessarily vitiates human
values. Sometimes these values are identified with the process of decision-making (e.g., it is important to have persons rather than impersonal forces determining who shall live). Sometimes they are identified
with the outcome of the process (e.g., the features such as creativity
and fullness of being which make human life what it is are to be considered and respected in the decision). Regarding the former, it must
be admitted that the use of chance seems cold and impersonal. But

presumably the defenders of utilitarian criteria in SLMR want to make their application as objective and impersonal as possible so that subjective bias does not determine who shall live.

Such criticisms, however, ignore the moral and nonmoral values which might be supported by selection by randomness or chance. A more important criticism is that the procedure that I develop draws the relevant moral context too narrowly. That context, so the argument might run, includes the society and its future and not merely the individual with his illness and claim upon SLMR. But my contention is that the values and principles at work in the narrower context may well take precedence over those operative in the broader context both because of their weight and significance and because of the weaknesses of selection in terms of social worth. As Paul Freund rightly insists, "The more nearly total is the estimate to be made of an individual, and the more nearly the consequence determines life and death, the more unfit the judgment becomes for human reckoning. . . . Randomness as a moral principle deserves serious study."[16] Serious study would, I think, point toward its implementation in certain conflict situations, primarily because it preserves a significant degree of *personal dignity* by providing *equality* of opportunity. Thus it cannot be dismissed as a "non-rational" and "non-human" procedure without an inquiry into the reasons, including human values, which might justify it. Paul Ramsey stresses this point about the *Holmes* case:

> Instead of fixing our attention upon "gambling" as the solution —with all the frivolous and often corrupt associations the word raises in our minds—we should think rather of *equality* of opportunity as the ethical substance of the relations of those individuals to one another that might have been guarded and expressed by casting lots.[17]

The individual's personal and transcendent dignity, which on the utilitarian approach would be submerged in his social role and function, can be protected and witnessed to by a recognition of his equal right to be saved. Such a right is best preserved by procedures which establish equality of opportunity. Thus selection by chance more closely approximates the requirements established by human dignity than does utilitarian calculation. It is not infallibly just, but it is preferable to the alternatives of letting all die or saving only those who have the greatest social responsibilities and potential contribution.

This argument can be extended by examining values other than individual dignity and equality of opportunity. Another basic value in

trust

the medical sphere is <u>the relationship of trust between physician and patient.</u> <u>Which selection criteria are most in accord with this relationship of trust?</u> Which will maintain, extend, and deepen it? My contention is that selection by randomness or chance is preferable from this standpoint too.

Trust, which is inextricably bound to respect for human dignity, is an attitude of expectation about another. It is not simply the expectation that another will perform a particular act, but more specifically that another will act toward him in certain ways—which will respect him as a person. As Charles Fried writes:

> Although trust has to do with reliance on a disposition of another person, it is reliance on a disposition of a special sort: the disposition to act morally, to deal fairly with others, to live up to one's undertakings, and so on. Thus to trust another is first of all to expect him to accept the principle of morality in his dealings with you, to respect your status as a person, your personality.[18]

This trust cannot be preserved in life-and-death situations when a person expects decisions about him to be made in terms of his social worth, for such decisions violate his status as a person. An applicant rejected on grounds of inadequacy in social value or virtue would have reason for feeling that his "trust" had been betrayed. Indeed, the sense that one is being viewed not as an end in himself but as a means in medical progress or the achievement of a greater social good is incompatible with attitudes and relationships of trust. We recognize this in the billboard which was erected after the first heart transplants: "Drive Carefully. Christiaan Barnard Is Watching You." The relationship of trust between the physician and patient is not only an instrumental value in the sense of being an important factor in the patient's treatment. It is also to be endorsed because of its intrinsic worth as a relationship.

Thus the related values of individual dignity and trust are best maintained in selection by chance. But other factors also buttress the argument for this approach. Which criteria and procedures would men agree upon? We have to suppose a hypothetical situation in which several men are going to determine for themselves and their families the criteria and procedures by which they would want to be admitted to and excluded from SLMR if the need arose.†† We need to assume two restrictions and then ask which set of criteria and procedures would be chosen as the most rational and, indeed, the fairest. The restrictions are these: (1) The men are *self-interested*. They are interested in their own welfare (and that of members of their families), and this, of course, includes survival. Basically, they are not motivated by altruism. (2) Furthermore, they are *ignorant* of their own talents, abilities, potential, and probable contribution to the social good. They do not know how they would fare in a competitive situation, e.g., the competition for SLMR in terms of social contribution. Under these conditions which institution would be chosen—letting all die, utilitarian selection, or the use of chance? Which would seem the most rational? the fairest? By which set of criteria would they want to be included in or excluded from the list of those who will be saved? The rational choice in this setting (assuming self-interest and ignorance of one's competitive success) would be random selection or chance since this alone provides equality of opportunity. A possible response is that one would prefer to take a "risk" and therefore choose the utilitarian approach. But I think not, especially since I added that the participants in this hypothetical situation are choosing for their children as well as for themselves; random selection or chance could be more easily justified to the children. It would make more sense for men who are self-interested but uncertain about their relative contribution to society to elect a set of criteria which would build in equality of opportunity. They would consider selection by chance as relatively just and fair.***

An important psychological point supplements earlier arguments for using chance or random selection. The psychological stress and strain among those who are rejected would be greater if the rejection is based on insufficient social worth than if it is based on chance. Obviously stress and strain cannot be eliminated in these borderline situations, but they would almost certainly be increased by the opprobrium of being judged relatively "unfit" by society's agents using society's values. Nicholas Rescher makes this point very effectively:

a recourse to chance would doubtless make matters easier for the rejected patient and those who have a specific interest in him. It would surely be quite hard for them to accept his exclusion by relatively mechanical application of objective criteria in whose implementation subjective judgment is involved. But the circumstances of life have conditioned us to accept the workings of chance and to tolerate the element of luck (good or bad): human life is an inherently contingent process. Nobody, after all, has an absolute right to ELT [Exotic Lifesaving Therapy]—but most of us would feel that we have "every bit as much right" to it as anyone else in significantly similar circumstances.†††

Although it is seldom recognized as such, selection by chance is already in operation in practically every dialysis unit. I am not aware of any unit which removes some of its patients from kidney machines in order to make room for later applicants who are better qualified in terms of social worth. Furthermore, very few people would recommend it. Indeed, few would even consider removing a person from a kidney machine on the grounds that a person better qualified *medically* had just applied. In a discussion of the treatment of chronic renal failure by dialysis at the University of Virginia Hospital Renal Unit from November 15, 1965 to November 15, 1966, Dr. Harry Abram writes: "Thirteen patients sought treatment but were not considered because the program had reached its limit of nine patients."[19] Thus, in practice and theory, natural chance is accepted at least within certain limits.

My proposal is that we extend this principle (first come, first served) to determine who among the medically acceptable patients shall live or that we utilize artificial chance such as a lottery or randomness. "First come, first served" would be more feasible than a lottery since the applicants make their claims over a period of time rather than as a group at one time. This procedure would be in accord with at least one principle in our present practices and with our sense of individual dignity, trust, and fairness. Its significance in relation to these values can be underlined by asking how the decision can be justified to the rejected applicant. Of course, one easy way of avoiding this task is to maintain the traditional cloak of secrecy, which works to a great extent because patients are often not aware that they are being considered for SLMR in addition to the usual treatment. But whether public justification is instituted or not is not the significant question; it is rather what reasons for rejection would be most acceptable to the unsuccessful applicant. My contention is that rejection can be accepted more readily if equality of opportunity, fairness, and trust are preserved, and that they are best preserved by selection by randomness or chance.

This proposal has yet another advantage since it would eliminate the need for a committee to examine applicants in terms of their social value. This onerous responsibility can be avoided.

Finally, there is a possible indirect consequence of widespread use of random selection which is interesting to ponder, although I do *not* adduce it as a good reason for adopting random selection. It can be argued, as Professor Mason Willrich of the University of Virginia Law School has suggested, that SLMR cases would practically disappear if these scarce resources were distributed randomly rather than on social worth grounds. Scarcity would no longer be a problem because the holders of economic and political power would make certain that they would not be excluded by a random selection procedure; hence they would help to redirect public priorities or establish private funding so that life-saving medical treatment would be widely and perhaps universally available.

In the framework that I have delineated, are the decrees of chance to be taken without exception? If we recognize exceptions, would we not open Pandora's box again just after we had succeeded in getting it closed? The direction of my argument has been against any exceptions, and I would defend this as the proper way to go. But let me indicate one possible way of admitting exceptions while at the same time circumscribing them so narrowly that they would be very rare indeed.

An obvious advantage of the utilitarian approach is that occasionally circumstances arise which make it necessary to say that one man is practically indispensable for a society in view of a particular set of problems it faces (e.g., the President when the nation is waging a war for survival). Certainly the argument to this point has stressed that the burden of proof would fall on those who think that the social danger in this instance is so great that they simply cannot abide by the outcome of a lottery or a first come, first served policy. Also, the reason must be negative rather than positive; that is, we depart from chance in this instance not because we want to take advantage of this person's potential contribution to the improvement of our society, but because his immediate loss would possibly (even probably) be disastrous (again, the President in a grave national emergency). Finally, social value (in the negative sense) should be used as a standard of exception in dialysis, for example, only if it would provide a reason strong enough to warrant removing another person from a kidney machine if all machines were taken. Assuming this strong reluctance to remove anyone once the commitment has been made to him, we would be willing to put this patient ahead of another applicant for a vacant machine only if we would be willing (in circumstances in which all machines are being used) to vacate a machine by removing someone from it. These restrictions would make an exception almost impossible.

While I do not recommend this procedure of recognizing exceptions, I think that one can defend it while accepting my general thesis about selection by randomness or chance. If it is used, a lay committee (perhaps advisory, perhaps even stronger) would be called upon to deal with the alleged exceptions since the doctors or others would in effect be appealing the outcome of chance (either natural or artificial). This lay committee would determine whether this patient was so indispensable at this time and place that he had to be saved even by sacrificing the values preserved by random selection. It would make it quite clear that exception is warranted, if at all, only as the "lesser of two evils." Such a defense would be recognized only rarely, if ever, primarily because chance and randomness preserve so many important moral and nonmoral values in SLMR cases.[20]

NOTES

1. George Bernard Shaw, *The Doctor's Dilemma* (New York, 1941), pp. 132-133.

2. Henry K. Beecher, "Scarce Resources and Medical Advancement," *Daedalus* (Spring 1969), pp. 279-280.

3. Leo Shatin, "Medical Care and the Social Worth of a Man," *American Journal of Orthopsychiatry*, 36 (1967), 97.

4. Harry S. Abram and Walter Wadlington, "Selection of Patients for Artificial and Transplanted Organs," *Annals of Internal Medicine*, 69 (September 1968), 615-620.

5. *United States v. Holmes* 26 Fed. Cas. 360 (C.C.E.D. Pa. 1842). All references are to the text of the trial as reprinted in Philip E. Davis, ed., *Moral Duty and Legal Responsibility: A Philosophical-Legal Casebook* (New York, 1966), pp. 102-118.

6. *The Moral Decision* (Bloomington, Ind., 1955), p. 71.

7. Joseph Fletcher, "Donor Nephrectomies and Moral Responsibility," *Journal of the American Medical Women's Association*, 23 (Dec. 1968), p. 1090.

8. Leo Shatin, op. cit., pp. 96-101.

9. For a discussion of the Seattle selection committee, see Shana Alexander, "They Decide Who Lives, Who Dies," *Life*, 53 (Nov. 9, 1962), 102. For an examination of general selection practices in dialysis see "Scarce Medical Resources," *Columbia Law Review* 69:620 (1969) and Harry S. Abram and Walter Wadlington, op. cit.

10. David Sanders and Jesse Dukeminier, Jr., "Medical Advance and Legal Lag: Hemodialysis and Kidney Transplantation," *UCLA Law Review* 15:367 (1968) 378.

11. Helmut Thielicke, "The Doctor as Judge of Who Shall Live and Who Shall Die," *Who Shall Live?* ed. by Kenneth Vaux (Philadelphia, 1970), p. 172.

12. Ibid., pp. 173-174.

13. Ibid., p. 173.

14. Thielicke, *Theological Ethics*, Vol. I, *Foundations* (Philadelphia, 1966), p. 602.

15. Cahn, op. cit., p. 71.

16. Paul Freund, "Introduction," *Daedalus* (Spring 1969), xiii.

17. Paul Ramsey, *Nine Modern Moralists* (Englewood Cliffs, N.J., 1962), p. 245.

18. Charles Fried, "Privacy," In *Law, Reason, and Justice*, ed. by Graham Hughes (New York, 1969), p. 52.

19. Harry S. Abram, M.D., "The Psychiatrist, the Treatment of Chronic Renal Failure,

and the Prolongation of Life: II" *American Journal of Psychiatry* 126:157-167 (1969), 158.

20. I read a draft of this paper in a seminar on "Social Implications of Advances in Biomedical Science and Technology: Artificial and Transplanted Internal Organs," sponsored by the Center for the Study of Science, Technology, and Public Policy of the University of Virginia, Spring 1970. I am indebted to the participants in that seminar, and especially to its leaders, Mason Willrich, Professor of Law, and Dr. Harry Abram, Associate Professor of Psychiatry, for criticisms which helped me to sharpen these ideas. Good discussions of the legal questions raised by selection (e.g., equal protection of the law and due process) which I have not considered can be found in "Scarce Medical Resources," *Columbia Law Review*, 69:620 (1969); "Patient Selection for Artificial and Transplanted Organs," *Harvard Law Review*, 82:1322 (1969); and Sanders and Dukeminier, op. cit.

* For a discussion of the higher suicide rate among dialysis patients than among the general population and an interpretation of some of the factors at work, see H. S. Abram, G. L. Moore, and F. B. Westervelt, "Suicidal Behavior in Chronic Dialysis Patients," *American Journal of Psychiatry* (in press). This study shows that even "if one does not include death through not following the regimen the incidence of suicide is still more than 100 times the normal population."

† I am excluding from consideration the question of the ability to pay because most of the people involved have to secure funds from other sources, public or private, anyway.

** "Letters and Comments," *Annals of Internal Medicine*, 61 (Aug. 1964), 360. Dr. G. E. Schreiner contends that "if you really believe in the right of society to make decisions on medical availability on these criteria you should be logical and say that when a man stops going to church or is divorced or loses his job, he ought to be removed from the programme and somebody else who fulfills these criteria substituted. Obviously no one faces up to this logical consequence" (G.E.W. Wolstenholme and Maeve O'Connor, eds. *Ethics in Medical Progress: With Special Reference to Transplantation*, A Ciba Foundation Symposium [Boston, 1966], p. 127).

†† My argument is greatly dependent on John Rawls's version of justice as fairness, which is a reinterpretation of social contract theory. Rawls, however, would probably not apply his ideas to "borderline situations." See "Distributive Justice: Some Addenda," *Natural Law Forum*, 13 (1968), 53. For Rawls's general theory, see "Justice as Fairness," *Philosophy, Politics and Society* (Second Series), ed. by Peter Laslett and W. G. Runciman (Oxford, 1962), pp. 132-157 and his other essays on aspects of this topic.

*** Occasionally someone contends that random selection may reward vice. Leo Shatin (op. cit., p. 100) insists that random selection "would reward socially disvalued qualities by giving their bearers the same special medical care opportunities as those received by the bearers of socially valued qualities. Personally I do not favor such a method." Obviously society must engender certain qualities in its members, but not all of its institutions must be devoted to that purpose. Furthermore, there are strong reasons, I have contended, for exempting SLMR from that sort of function.

††† Nicholas Rescher, "The Allocation of Exotic Medical Lifesaving Therapy," *Ethics*, 79 (April 1969), 184. He defends random selection's use only after utilitarian and other judgments have been made. If there are no "major disparities" in terms of utility, etc., in the second stage of selection, then final selection could be made randomly. He fails to give attention to the moral values that random selection might preserve.

28 Harvesting the Dead

WILLARD GAYLIN, M.D.

Scarcity of organs and tissue is the most obvious major problem in transplantation. Gaylin suggests a not-too-fanciful solution: declaring people with a flat EEG to be dead and then maintaining them on respirators for the purposes of experimentation, transplantation, training of the physicians, and the production of needed hormones and antibodies. Such a solution has many obvious benefits and few major costs, except as Gaylin suggests, those costs on a level which may eventuate our paying more than we had ever imagined. The technology for such a procedure is with us now; serious ethical analysis of such use of the legally dead yet breathing corpses has not yet been made.

Dr. Willard Gaylin, M.D., is the President of the Institute of Society, Ethics, and the Life Sciences.

Nothing in life is simple anymore, not even the leaving of it. At one time there was no medical need for the physician to consider the concept of death; the fact of death was sufficient. The difference between life and death was an infinite chasm breached in an infinitesimal moment. Life and death were ultimate, self-evident opposites.

Redefining Death

With the advent of new techniques in medicine, those opposites have begun to converge. We are now capable of maintaining visceral functions without any semblance of the higher functions that define a person. We are, therefore, faced with the task of deciding whether that which we have kept alive is still a human being, or, to put it another way, whether that human being that we are maintaining should be considered "alive."

Until now we have avoided the problems of definition and reached the solutions in silence and secret. When the life sustained was unrewarding—by the standards of the physician in charge—it was discontinued. Over the years, physicians have practiced euthanasia on an ad hoc, casual, and perhaps irresponsible basis. They have withheld antibiotics or other simple treatments when it was felt that a life did not warrant sustaining, or pulled the plug on the respirator when they were convinced that what was being sustained no longer warranted the definition of life. Some of these acts are illegal and, if one wished to prosecute, could constitute a form of manslaughter, even though it is unlikely that any jury would convict. We prefer to handle all problems connected with death by denying their existence. But death and its dilemmas persist.

New urgencies for recognition of the problem arise from two conditions: the continuing march of technology, making the sustaining of vital processes possible for longer periods of time; and the increasing use of parts of the newly dead to sustain life for the truly living. The problem is well on its way to being resolved by what must have seemed a relatively simple and ingenious method. As it turned out, the difficult issues of euthanasia could be evaded by redefining death.

In an earlier time, death was defined as the cessation of breathing. Any movie buff recalls at least one scene in which a mirror is held to the mouth of a dying man. The lack of fogging indicated that indeed he was dead. The spirit of man resided in his *spiritus* (breath). With increased knowledge of human physiology and the potential for reviving a nonbreathing man, the circulation, the pulsating heart, became the focus of the definition of life. This is the tradition with which most of us have been raised.

There is of course a relationship between circulation and respiration, and the linkage, not irrelevantly, is the brain. All body parts require the nourishment, including oxygen, carried by the circulating blood. Lack of blood supply leads to the death of an organ; the higher functions of the brain are particularly vulnerable. But if there is no respiration, there is no adequate exchange of oxygen, and this essential ingredient of the blood is no longer available for distribution. If a part of the heart loses its vascular supply, we may lose that part and still survive. If a part of the brain is deprived of oxygen, we may, depending on its location, lose it and survive. But here we pay a special price, for the functions lost are those we identify with the self, the soul, or humanness, i.e., memory, knowledge, feeling, thinking, perceiving, sensing, knowing, learning, and loving.

Most people are prepared to say that when all of the brain is destroyed the "person" no longer exists; with all due respect for the complexities of the mind/brain debate, the "person" (and personhood) is generally associated with the functioning part of the head—the brain. The higher functions of the brain that have been described are placed, for the most part, in the cortex. The brain stem (in many ways more closely allied to the spinal cord) controls primarily visceral functions. When the total brain is damaged, death in all forms will ensue because the lower brain centers that control the circulation and respiration are destroyed. With the development of modern respirators, however, it is possible to artificially maintain respiration and with it, often, the circulation with which it is linked. It is this situation that has allowed for the redefinition of death—a redefinition that is being precipitously embraced by both scientific and theological groups.

The movement toward redefining death received considerable impetus with the publication of a report sponsored by the Ad Hoc Committee of the Harvard Medical School in 1968. The committee offered an alternative definition of death based on the functioning of the brain. Its criteria stated that if an individual is unreceptive and unresponsive, i.e., in a state of irreversible coma; if he has no movements or breathing when the mechanical respirator is turned off; if he demonstrates no reflexes; and if he has a flat electroencephalogram for at least twenty-four hours, indicating no electrical brain activity (assuming that he has not been subjected to hypothermia or central nervous system depressants), he may then be declared dead.

What was originally offered as an optional definition of death is, however, progressively becoming *the* definition of death. In most states there is no specific legislation defining death;[1] the ultimate responsibility here is assumed to reside in the general medical community. Recently, however, there has been a series of legal cases which seem to be establishing brain death as a judicial standard. In California in May of this year an ingenious lawyer, John Cruikshank, offered as a defense of his client, Andrew D. Lyons, who had shot a man in the head, the argument that the cause of death was not the bullet but the removal of his heart by a transplant surgeon, Dr. Norman Shumway. Cruikshank's argument notwithstanding, the jury found his client guilty of voluntary manslaughter. In the course of that trial, Dr. Shumway said: "The brain in the 1970s and in the light of modern day medical technology is the sine qua non—the criterion for death. I'm saying anyone whose brain is dead is dead. It is the one determinant that would be universally applicable, because the brain is the one organ that can't be transplanted."

This new definition, independent of the desire for transplant, now permits the physician to "pull the plug" without even committing an act of passive euthanasia. The patient will first be defined as dead; pulling the plug will merely be the harmless act of halting useless treatment on a cadaver. But while the new definition of death avoids one complex problem, euthanasia, it may create others equally difficult which have never been fully defined or visualized. For if it grants the right to pull the plug, it also implicitly grants the privilege *not* to pull the plug, and the potential and meaning of this has not at all been adequately examined.

These cadavers would have the legal status of the dead with none of the qualities one now associates with death. They would be warm, respiring, pulsating, evacuating, and excreting bodies requiring nursing, dietary, and general grooming attention—*and could probably be maintained so for a period of years.* If we chose to, we could, with the technology already at hand, legally avail ourselves of these new cadavers to serve science and mankind in dramatically useful ways. The autopsy, that most respectable of medical traditions, that last gift of the dying person to the living future, could be extended in principle beyond our current recognition. To save lives and relieve suffering—traditional motives for violating tradition—we could develop hospitals (an inappropriate word because it suggests the presence of living human beings), banks, or farms of cadavers which require feeding and maintenance, in order to be harvested. To the uninitiated the "new cadavers" in their rows of respirators would seem indistinguishable from comatose patients now residing in wards of chronic neurological hospitals.

Precedents

The idea of wholesale and systematic salvage of useful body parts may seem startling, but it is not without precedent. It is simply magnified by the technology of modern medicine. Within the confines of one individual, we have always felt free to transfer body parts to places where they are needed more urgently, felt free to reorder the priorities of the naturally endowed structure. We will borrow skin from the less visible parts of the body to salvage a face. If a muscle is paralyzed, we will often substitute a muscle that subserves a less crucial function. This was common surgery at the time that paralytic polio was more prevalent.

It soon becomes apparent, however, that there is a limitation to this procedure. The person in want does not always have a second-best substitute. He may then be forced to borrow from a person with a surplus. The prototype, of course, is blood donation. Blood may be seen as a regeneratable organ, and we have a long-standing tradition of blood donation. What may be more important, and perhaps dangerous, we have established the precedent in blood of commercialization—not only are we free to borrow, we are forced to buy and, indeed, in our country at least, permitted to sell. Similarly, we allow the buying or selling of sperm for artificial insemination. It is most likely that in the near future we will allow the buying and selling of ripened ova so that a sterile woman may conceive her baby if she has a functioning uterus. Of course, once *in vitro* fertilization becomes a reality (an imminent possibility), we may even permit the rental of womb space for gestation for a woman who does manufacture her own ova but has no uterus.

Getting closer to our current problem, there is the relatively long-standing tradition of banking body parts (arteries, eyes, skin) for short periods of time for future transplants. Controversy has arisen with recent progress in the transplanting of major organs. Kidney transplants from a near relative or distant donor are becoming more common. As heart transplants become more successful, the issue will certainly be heightened, for while the heart may have been reduced by the new definition of death to merely another organ, it will always have a core position in the popular thinking about life and death. It has the capacity to generate the passion that transforms medical decisions into political issues.

The ability to use organs from cadavers has been severely limited in the past by the reluctance of heirs to donate the body of an individual for distribution. One might well have willed one's body for scientific purposes, but such legacies had no legal standing. Until recently, the individual lost control over his body once he died. This has been changed by the Uniform Anatomical Gift Act. This model piece of legislation, adopted by all fifty states in an incredibly short period of time, grants anyone over eighteen (twenty-one in some states) the right to donate en masse all "necessary organs and tissues" simply by filling out and mailing a small card.

Beyond the postmortem, there has been a longer-range use of human bodies that is accepted procedure—the exploitation of cadavers as teaching material in medical schools. This is a long step removed from the rationale of the transplant—a dramatic gift of life from the dying to the near-dead; while it is true that medical education will inevitably save lives, the clear and immediate purpose of the donation is to facilitate training.

It is not unnatural for a person facing death to want his usefulness to extend beyond his mortality; the same biases and values that influence our life persist in our leaving of it. It has been reported that the Harvard Medical School has no difficulty in receiving as many donations of cadavers as they need, while Tufts and Boston Universities are usually in short supply. In Boston, evidently, the cachet of getting into Harvard extends even to the dissecting table.

The way is now clear for an ever-increasing pool of usable body parts, but the current practice minimizes efficiency and maximizes waste. Only a short period exists between the time of death of the patient and the time of death of his major parts.

Uses of the Neomort

In the ensuing discussion, the word *cadaver* will retain its usual meaning, as opposed to the new cadaver, which will be referred to as a *neomort*. The "ward" or "hospital" in which it is maintained will be called a *bioemporium* (purists may prefer *bioemporion*).

Whatever is possible with the old embalmed cadaver is extended to an incredible degree with the neomort. What follows, therefore, is not a definitive list but merely the briefest of suggestions as to the spectrum of possibilities.

TRAINING: Uneasy medical students could practice routine physical examinations—auscultation, percussion of the chest, examination of the retina, rectal and vaginal examinations, et cetera—indeed, everything except neurological examinations, since the neomort by definition has no functioning central nervous system.

Both the student and his patient could be spared the pain, fumbling, and embarrassment of the "first time."

Interns also could practice standard and more difficult diagnostic procedures, from spinal taps to pneumoencephalography and the making of arteriograms, and residents could practice almost all of their surgical skills—in other words, most of the procedures that are now normally taught with the indigent in wards of major city hospitals could be taught with neomorts. Further, students could practice more exotic procedures often not available in a typical residency—eye operations, skin grafts, plastic facial surgery, amputation of useless limbs, coronary surgery, etc.; they could also practice the actual removal of organs, whether they be kidneys, testicles, or what have you, for delivery to the transplant teams.

TESTING: The neomort could be used for much of the testing of drugs and surgical procedures that we now normally perform on prisoners, mentally retarded children, and volunteers. The efficacy of a drug as well as its toxicity could be determined beyond limits we might not have dared approach when we were concerned about permanent damage to the testing vehicle, a living person. For example, operations for increased vascularization of the heart could be tested to determine whether they truly do reduce the incidence of future heart attacks before we perform them on patients. Experimental procedures that proved useless or harmful could be avoided; those that succeed could be available years before they might otherwise have been. Similarly, we could avoid the massive delays that keep some drugs from the marketplace while the dying clamor for them.

Neomorts would give us access to other forms of testing that are inconceivable with the living human being. We might test diagnostic instruments such as sophisticated electrocardiography by selectively damaging various parts of the heart to see how or whether the instrument could detect the damage.

EXPERIMENTATION: Every new medical procedure demands a leap of faith. It is often referred to as an "act of courage," which seems to me an inappropriate terminology now that organized medicine rarely uses itself as the experimental body. Whenever a surgeon attempts a procedure for the first time, he is at best generalizing from experimentation with lower animals. Now we can protect the patient from too large a leap by using the neomort as an experimental bridge.

Obvious forms of experimentation would be cures for illnesses which would first be induced in the neomort. We could test antidotes by injecting poison, induce cancer or virus infections to validate and compare developing therapies.

Because they have an active hematopoietic system, neomorts would be particularly valuable for studying diseases of the blood. Many of the examples that I draw from that field were offered to me by Dr. John F. Bertles, a hematologist at St. Luke's Hospital Center in New York. One which interests him is the utilization of marrow transplants. Few human-to-human marrow transplants have been successful, since the kind of immunosuppression techniques that require research could most safely be performed on neomorts. Even such research as the recent experimentation at Willowbrook—where mentally retarded children were infected with hepatitis virus (which was not yet culturable outside of the human body) in an attempt to find a cure for this pernicious disease—could be done without risking the health of the subjects.

BANKING: While certain essential blood antigens are readily storable (e.g., red cells can now be preserved in a frozen state), others are not, and there is increasing need for potential means of storage. Research on storage of platelets to be used in transfusion requires human recipients, and the data are only slowly and tediously gathered at great expense. Use of neomorts would permit intensive testing of platelet survival and probably would lead to a rapid development of a better storage technique. The same would be true for white cells.

As has been suggested, there is great wastage in the present system of using kidney donors from cadavers. Major organs are difficult to store. A population of neomorts maintained with body parts computerized and catalogued for compatibility would yield a much more efficient system. Just as we now have blood banks, we could have banks for all the major organs that may someday be transplantable—lungs, kidney, heart, ovaries. Beyond the obvious storage uses of the neomort, there are others not previously thought of because there was no adequate storage facility. Dr. Marc Lappé of the Hastings Center has suggested that a neomort whose own immunity system had first been severely repressed might be an ideal "culture" for growing and storing our lymphoid components. When we are threatened by malignancy or viral disease, we can go to the "bank" and withdraw our stored white cells to help defend us.

HARVESTING: Obviously, a sizable population of neomorts will provide a steady supply of blood, since they can be drained periodically. When we consider the cost-benefit analysis of this system, we would have to evaluate it in the same way as the lumber industry evaluates sawdust—a product which in itself is not commercially feasible but which supplies a profitable dividend as a waste from a more useful harvest.

The blood would be a simultaneous source of platelets, leukocytes, and red cells. By attaching a neomort to an IBM cell separator, we could isolate cell types at relatively low cost. The neomort could also be tested for the presence of hepatitis in a way that would be impossible with commercial donors. Hepatitis as a transfusion scourge would be virtually eliminated.

Beyond the blood are rarer harvests. Neomorts offer a great potential source of bone marrow for transplant procedures, and I am assured that a bioemporium of modest size could be assembled to fit most transplantation antigen requirements. And skin would, of course, be harvested—similarly bone, corneas, cartilage, and so on.

MANUFACTURING: In addition to supplying components of the human body, some of which will be continually regenerated, the neomort can also serve as a manufacturing unit. Hormones are one obvious product, but there are others. By the injection of toxins, we have a source of antitoxin that does not have the complication of coming from another animal form. Antibodies for most of the major diseases can be manufactured merely by injecting the neomort with the viral or bacterial offenders.

Perhaps the most encouraging extension of the manufacturing process emerges from the new cancer research, in which immunology is coming to the fore. With certain blood cancers, great hope attaches to the use of antibodies. To take just one example, it is conceivable that leukemia could be generated in individual neomorts—not just to provide for in vivo (so to speak) testing of antileukemic modes of therapy but also to generate antibody immunity responses which could then be used in the living.

Cost-Benefit Analysis

If seen only as the harvesting of products, the entire feasibility of such research would depend on intelligent cost-benefit analysis. Although certain products would not warrant the expense of maintaining a community of neomorts, the enormous expense of other products, such as red cells with unusual antigens, would certainly warrant it. Then, of course, the equation is shifted. As soon as one economically sound reason is found for the maintenance of the community, all of the other ingredients become gratuitous by-products, a familiar problem in manufacturing. There is no current research to indicate the maintenance cost of a bioemporium or even the potential duration of an average neomort. Since we do not at this point encourage sustaining life in the brain-dead, we do not know the limits to which it could be extended. This is the kind of technology, however, in which we have previously been quite successful.

Meantime, a further refinement of death might be proposed. At present we use total brain function to define brain death. The source of electroencephalogram activity is not known and cannot be used to distinguish between the activity of higher and lower brain centers. If, however, we are prepared to separate the concept of "aliveness" from "personhood" in the adult, as we have in the fetus, a good argument can be made that death should be defined not as cessation of total brain function but merely as cessation of cortical function. New tests may soon determine when cortical function is dead. With this proposed extension, one could then maintain neomorts without even the complication and expense of respirators. The entire population of decorticates residing in chronic hospitals and now classified among the incurably ill could be redefined as dead.

But even if we maintain the more rigid limitations of total brain death it would seem that a reasonable population could be maintained if the purposes warranted it. It is difficult to assess how many new neomorts would be available each year to satisfy the demand. There are roughly 2 million deaths a year in the United States. The most likely sources of intact bodies with destroyed brains would be accidents (about 113,000 per year), suicides (around 24,000 per year), homicides (18,000), and cerebrovascular accidents (some 210,000 per year). Obviously, in each of these categories a great many of the individuals would be useless—their bodies either shattered or scattered beyond value or repair.

And yet, after all the benefits are outlined, with the lifesaving potential clear, the humanitarian purposes obvious, the technology ready, the motives pure, and the material costs justified—how are we to reconcile our emotions? Where in this debit-credit ledger of limbs and livers and kidneys and costs are we to weigh and enter the repugnance generated by the entire philanthropic endeavor?

Cost-benefit analysis is always least satisfactory when the costs must be measured in one realm and the benefits in another. The analysis is particularly skewed when the benefits are specific, material, apparent, and immediate, and the price to be paid is general, spiritual, abstract, and of the future. It is that which induces people to abandon freedom for security, pride for comfort, dignity for dollars.

William May, in a perceptive article,[2] defended the careful distinctions that have traditionally been drawn between the newly dead and the long dead. "While the body retains its recognizable form, even in death, it commands a certain respect. No longer a human presence, it still reminds us of that presence which once was utterly inseparable from it." But those distinctions become obscured when, years later, a neomort will retain the appearance of the newly dead, indeed, more the appearance of that which was formerly described as living.

Philosophers tend to be particularly sensitive to the abstract needs of civilized man; it is they who have often been the guardians of values whose abandonment produces pains that are real, if not always quantifiable. Hans Jonas, in his *Philosophical Essays*, anticipated some of the possibilities outlined here, and defended what he felt to be the sanctity of the human body and the unknowability of the borderline between life and death when he insisted that "Nothing less than the maximum definition of death will do—brain death plus heart death plus any other indication that may be pertinent—before final violence is allowed to be done." And even then Jonas was only contemplating *temporary* maintenance of life for the collection of organs.

The argument can be made on both sides. The unquestionable benefits to be gained are the promise of cures for leukemia and other diseases, the reduction of suffering, and the maintenance of life. The proponents of this view will be mobilized with a force that may seem irresistible.

They will interpret our revulsion at the thought of a bioemporium as a bias of our education and experience, just as earlier societies were probably revolted by the startling notion of abdominal surgery, which we now take for granted. The proponents will argue that the revulsion, not the technology, is inappropriate.

Still there will be those, like May, who will defend that revulsion as a quintessentially human factor whose removal would diminish us all, and extract a price we cannot anticipate in ways yet unknown and times not yet determined. May feels that there is "a tinge of the inhuman in the humanitarianism of those who believe that the perception of social need easily overrides all other considerations and reduces the acts of implementation to the everyday, routine, and casual."

This is the kind of weighing of values for which the computer offers little help. Is the revulsion to the new technology simply the fear and horror of the ignorant in the face of the new, or is it one of those components of humanness that barely sustain us at the limited level of civility and decency that now exists, and whose removal is one more step in erasing the distinction between man and the lesser creatures—beyond that, the distinction between man and matter?

Sustaining life is an urgent argument for any measure, but not if that measure destroys those very qualities that make life worth sustaining.

[1] Kansas and Maryland have recently legislated approval for a brain definition of death.

[2] "Attitudes Toward the Newly Dead," *The Hastings Center Studies*, volume 1, number 1, 1973.

29 Doing Better and Feeling Worse: The Political Pathology of Health Policy

AARON WILDAVSKY

This entertaining but extremely provocative article provides an analysis of a number of slogans
that have characterized most discussions on health
policy as well as pro and con arguments for a national health insurance system. Wildavsky also
analyzes a number of financial and programmatic
issues related to the development of health policy
and evaluates their outcomes.

*Aaron Wildavsky is Dean of the Graduate School of
Public Policy at the University of California, Berkeley.*

According to the great equation, Medical Care
equals Health. But the Great Equation is
wrong. More available medical care does not equal better health. The
best estimates are that the medical system (doctors, drugs, hospitals)
offers about 10 per cent of the usual indices for measuring health
whether you live at all (infant mortality), how well you live (days lost
due to sickness), how long you live (adult mortality). The remaining
90 per cent are determined by factors over which doctors have little or
no control, from individual life style (smoking, exercise, worry), to
social conditions (income, eating habits, physiological inheritance), to
the physical environment (air and water quality). Most of the bad
things that happen to people are at present beyond the reach of medicine.

Everyone knows that doctors do help. They can mend broken
bones, stop infections with drugs, operate successfully on swollen ap-
pendices. Inoculations, internal infections, and external repairs are
other good reasons for keeping doctors, drugs, and hospitals around.
More of the same, however, is counterproductive. Nobody needs un-

necessary operations; and excessive use of drugs can create dependence or allergic reactions or merely enrich the nation's urine.

More money alone, then, cannot cure old complaints. In the absence of medical knowledge gained through new research, or of administrative knowledge to convert common practice into best practice, current medicine has gone as far as it can. It will not burn brighter if more money is poured on it. No one is saying that medicine is good for nothing, only that it is not good for everything. Thus the marginal value of one—or one billion—dollars spent on medical care will be close to zero in improving health. And, for purposes of public policy, it is not the bulk of present medical expenditures, which do have value, but the proposed future spending, which is of dubious value, that should be our main concern.

When people are polled, they are liable, depending on what they are asked, to say that they are getting good care but that there is a crisis in the medical-care system. Three-quarters to four-fifths of the population, depending on the survey, are satisfied with their doctors and the care they give; but one-third to two-thirds think the system that produces these results is in bad shape. Opinions about the family doctor, of course, are formed from personal experience. "The system," on the other hand, is an abstract entity—and here people may well imitate the attitudes of those interested and vocal elites who insist the system is in crisis. People do, however, have specific complaints related to their class position. The rich don't like waiting, the poor don't like high prices, and those in the middle don't like both. Everyone would like easier access to a private physician in time of need. As we shall see, the widespread belief that doctors are good but the system is bad has a plausible explanation. That's the trouble: everyone behaves reasonably; it is only the systemic effects of all this reasonable behavior that are unreasonable.

If most people are healthier today than people like themselves have ever been, and if access to medical care now is more evenly distributed among rich and poor, why is there said to be a crisis in medical care that requires massive change? If the bulk of the population is satisfied with the care it is getting, why is there so much pressure in government for change? Why, in brief, are we doing better but feeling worse? Let us try to create a theory of the political pathology of health policy.

PARADOXES, PRINCIPLES, AXIOMS, IDENTITIES, AND LAWS

The fallacy of the Great Equation is based on the Paradox of Time: past successes lead to future failures. As life expectancy increases and as formerly disabling diseases are conquered, medicine is faced with an older population whose disabilities are more difficult to defeat. The cost of cure is higher, both because the easier ills have already been dealt with and because the patients to be treated are older. Each increment of knowledge is harder won; each improvement in health is more expensive. Thus time converts one decade's achievements into the next decade's dilemmas. Yesterday's victims of tuberculosis are today's geriatric cases. The Paradox of Time is that success lies in the past and (possibly) the future, but never the present.

The Great Equation is rescued by the *Principle of Goal Displacement*, which states that any objective that cannot be attained will be replaced by one that can be approximated. Every program needs an opportunity to be successful; if it cannot succeed in terms of its ostensible goals, its sponsors may shift to goals whose achievement they can control. The process subtly becomes the purpose. And that is exactly what has happened as "health" has become equivalent to "equal access to" medicine.

When government goes into public housing, it actually provides apartments; when it goes into health, all it can provide is medicine. But medicine is far from health. So what the government can do then is try to equalize access to medicine, whether or not that access is related to improved health. If the question is, "Does health increase with government expenditure on medicine?," the answer is likely to be "No." Just alter the question—"Has access to medicine been improved by government programs?"—and the answer is most certainly, with a little qualification, "Yes."

By "access," of course, we mean quantity, not quality, of care. " Access " Access, moreover, can be measured, and progress toward an equal number of visits to doctors can be reported. But better access is not the same as better health. Something has to be done about the distressing stickiness of health rates, which fail to keep up with access. After all, if medical care does not equal health, access to medical care is irrelevant to health—unless, of course, health is not the real goal but merely a cover for something more fundamental, which might be called "mental health" (reverently), or "shamanism" (irreverently), or "caring" (most accurately).

Any doctor will tell you, say sophisticates, that most patients are not sick, at least physically, and that the best medicine for them is reassurance. Tranquilizers, painkillers, and aspirin would seem to be the functional equivalents, for these are the drugs most often prescribed. Wait a minute, says the medical sociologist (the student not merely of medicine's manifest, but also of its latent, functions), pain is just as real when it's mental as when it's physical. If people want to know somebody loves them, if today they prefer doctors of medicine to doctors of theology, they are entitled to get what they want.

Once "caring" has been substituted for (or made equivalent to) "doctoring," access immediately becomes a better measure of attainment. The number of times a person sees a doctor is probably a better measure of the number of times he has been reassured than of his well-being or a decline in his disease. So what looks like a single goal substitution (access to medicine in place of better health) is actually a double displacement: caring instead of health, and access instead of caring.

This double displacement is fraught with consequences. Determining how much medical care is sufficient is difficult enough; determining how much "caring" is, is virtually impossible. The treatment of physical ills is partially subjective; the treatment of mental ills is almost entirely subjective. If a person is in pain, he alone can judge how much it hurts. How much caring he needs depends upon how much he wants. In the old days he took his tension chiefly to the private sector, and there he got as much attention as he could pay for. But now with government subsidy of medicine looming so large, the question of how much caring he should get inevitably becomes public.

By what standard should this public question be decided? One objective criterion—equality of access—inevitably stands out among the rest. For if we don't quite know what caring is or how much of it there should be, we can always say that at least it should be equally distributed. Medicaid has just about equalized the number of doctor visits per year between the poor and the rich. In fact, the upper class is showing a decrease in visits, and the life expectancy of richer males is going down somewhat. Presumably, no one is suggesting remedial action in favor of rich men. Equality, not health, is the issue.

EQUALITY

One can always assert that even if the results of medical treatment are illusory, the poor are entitled to their share. This looks like a powerful argument, but it neglects the *Axiom of Inequality*. That axiom states that every move to increase equality in one dimension necessarily decreases it in another. Consider space. The United States has unequal rates of development. Different geographic areas vary considerably in such matters as income, custom, and expectation. Establishing a uniform national policy disregards these differences; allowing local variation means that some areas are more unequal than others. Think of time. People not only have unequal incomes, they also differ in the amount of time they are prepared to devote to medical care. In equalizing the effects of money on medical care—by removing money as a consideration—care is likely to be allocated by the distribution of available time. To the extent that the pursuit of money takes time, people with a monetary advantage will have a temporal disadvantage. You can't have it both ways, as the Axiom of Allocation makes abundantly clear.

"No system of care in the world," says David Mechanic, summing up the *Axiom of Allocation*, "is willing to provide as much care as people will use, and all such systems develop mechanisims that ration . . . services." Just as there is no free lunch, so there is *no free medicine*. Rationing can be done by time (waiting lists, lines), by distance (people farther from facilities use them less than those who are closer), by complexity (forms, repeated visits, communications difficulties), by space (limiting the number of hospital beds and available doctors), or by any or all of these methods in combination. But why do people want more medical service than any system is willing to provide? The answer has to do with uncertainty.

If medicine is only partially and imperfectly related to health, it follows that doctor and patient both will often be uncertain as to what is wrong or what to do about it. Otherwise—if medicine were perfectly related to health—either there would be no health problem or it would be a very different one. Health rates would be on one side and health resources on the other; costs and benefits could be neatly compared. But they can't, because we often don't know how to produce the desired benefits. Uncertainty exists because medicine is a quasi-science—more science than, say, political science; less science than physics. How the participants in the medical system resolve their uncertainties matters a great deal.

The *Medical Uncertainty Principle* states that there is always one more thing that might be done—another consultation, a new drug, a different treatment. Uncertainty is resolved by doing more: the patient asks for more, the doctor orders more. The patient's simple rule for resolving uncertainty is to seek care up to the level of his insurance. If everyone uses all the care he can, total costs will rise; but the individual has so little control over the total that he does not appreciate the connection between his individual choice and the collective result. A corresponding phenomenon occurs among doctors. They can resolve uncertainty by prescribing up to the level of the patient's insurance, a rule reinforced by the high cost of malpractice. Patients bringing suit do not consider the relationship between their own success and higher medical costs for everyone. The patient is anxious, the doctor insecure; this combination is unbeatable until the irresistible force meets the immovable object—the Medical Identity.

The *Medical Identity* states that use is limited by availability. Only so much can be gotten out of so much. Thus, if medical uncertainty suggests that existing services will be used, Identity reminds us to add the words "up to the available supply." That supply is primarily doctors, who advise on the kind of care to provide and the number of hospital beds to maintain. But patients, considering only their own desires in time of need, want to maximize supply, a phenomenon that follows inexorably from the *Principle of Perspective*.

That principle states that social conditions and individual feelings are not the same thing. A happy social statistic may obscure a sad personal situation. A statistical equilibrium may hide a family crisis. Morbidity and mortality, in tabulating aggregate rates of disease and death, describe you and me but do not touch us. We do not think of ourselves as "rates." Our chances may be better or worse than the aggregate. To say that doctors are not wholly (or even largely) successful in alleviating certain symptoms is not to say that they don't help some people and that one of those people won't be me. Taking the chance that it will be me often seems to make sense, even if there is reason to believe that most people can't be helped and that some may actually be harmed. Most people, told that the same funds spent on other purposes may increase social benefits, will put their personal needs first. This is why expenditures on medical care are always larger than any estimate of the social benefit received. Now we can understand, by combining into one law the previous principles and Medical Identity, why costs rise so far and so fast.

The *Law of Medical Money* states that medical costs rise to equal the sum of all private insurance and government subsidy. This occurs because no one knows how much medical care ought to cost. The patient is not sure he is getting all he should, and the doctor does not want to be faulted for doing less than he might. Consider the triangular relationship between doctor, patient, and hospital. With private insurance, the doctor can use the hospital resources that are covered by the insurance while holding down his patient's own expenditures. With public subsidies, the doctor may charge his highest usual fee, abandon charitable work, and ignore the financial benefits of eliminating defaults on payments. His income rises. His patient doesn't have to pay, and his hospital expands. The patient, if he is covered by a government program or private insurance (as about 90 per cent are) finds that his out-of-pocket expenses have remained the same. His insurance costs more, but either it comes out of his paycheck, looking like a fixed expense, or it is taken off his income tax as a deduction. Hospitals work on a cost-plus basis. They offer the latest and the best, thus pleasing both doctor and patient. They pay their help better; or, rather, they get others to pay their help. It's on the house—or at least on the insurance.

Perhaps our triangle ought to be a square: maybe we should include insurance companies. Why are they left out of almost all discussions of this sort? Why don't they play a cost-cutting role in medical care as they do in other industries? After all, the less the outlay, the more income for the company. Here the simplest explanation seems the best: insurance companies make no difference because they are no different from the rest of the healty-care industry. The largest, Blue Cross *and* Blue Shield, are run by the hospital establishment on behalf of doctors. After all, hospitals do not so much have patients as they have doctors who have patients. Doctors run hospitals, not the other way around. Insurance companies not willing to play this game have left the field.

What process ultimately limits medical costs? If the Law of Medical Money predicts that costs will increase to the level of available funds, then that level must be limited to keep costs down. Insurance may stop increasing when out-of-pocket payments exceed the growth in the standard of living; at that point individuals may not be willing to buy more. Subsidy may hold steady when government wants to spend more on other things or when it wants to keep its total tax take down. Costs will be limited when either individuals or governments reduce the amount they put into medicine.

No doubt the Law of Medical Money is crude, even rude. No doubt it ignores individual instances of self-sacrifice. But it has the virtue of being a powerful and parsimonious predictor. Costs have risen (and are continuing to rise) to the level of insurance and subsidy.

WHY THERE IS A CRISIS

If more than three-quarters of the population are satisfied with their medical care, why is there a crisis? Surveys on this subject are inadequate, but invariably they reveal two things: (one) the vast majority are satisfied, but (two) they wish medical care didn't cost so much and they would like to be assured of contact with their doctor. So far as the people are concerned, then, the basic problems are cost and access. Why, to begin at the end, aren't doctors where patients want them to be?

To talk about physicians being maldistributed is to turn the truth upside down: it is the potential patients who are maldistributed. For doctors to be in the wrong place, they would have to be where people aren't, and yet they are accused of sticking to the main population centers. If distant places with little crowding and less pollution, far away from the curses of civilization, attracted the same people who advocate their virtues, doctors would live there, too. Obviously, they prefer the amenities of metropolitan areas. Are they wrong to live where they want to live? Or are the rural and remote wrong to demand that others come where they are?

Doctors can be offered a government subsidy—more money, better facilities—on the grounds that it is a national policy for medical care to be available wherever citizens choose to live. Virtually all students in medical schools are heavily subsidized, so it would not be entirely unjust to demand that they serve several years in places not of their own choosing. The reason such policies do not work well—from Russia to the "Ruritanias" of this world—is that people who are forced to live in places they don't like make endless efforts to escape.

Because the distribution of physicians is determined by rational choice—doctors locate where their psychic as well as economic income is highest—there is no need for special laws to explain what happens. But the political pathology of health policy—the more the government spends on medicine, the less credit it gets—does require explanation.

The syndrome of "the more, the less" has to be looked at as it developed over time. First we passed Medicare for the elderly and Medicaid for the poor. The idea was to get more people into the mainstream of good medical care. Following the Law of Medical Money, however, the immediate effect was to increase costs, not merely for the poor and elderly but for all the groups in between. You can't simply add the costs of the new coverage to the costs of the old; you have to multiply them both by higher figures up to the limits of the joint coverage. This is where the *Axiom of Inequality* takes over. The wealthier aged, who can afford to pay, receive not merely the same benefits as the aged poor, but even more, because they are better able to negotiate the system. Class tells. Inequalities are immediately created within the same category. Worse still is the "notch effect" under Medicaid, through which those just above the eligibles in income may be worse off than those below. Whatever the cutoff point, there must always be a "near poor" who are made more unequal. And so is everybody else who pays twice, first in taxes to support care for others and again in increased costs for themselves. Moreover, with increased utilization of medicine, the system becomes crowded; medical care is not only more costly but harder to get. So there we have the Paradox of Time—as things get better, they get worse.

The politics of medical care becomes a minus-sum game in which every institutional player leaves the table poorer than when he sat down. In the beginning, the number of new patients grows arithmetically while costs rise geometrically. The immediate crisis is cost. Medicaid throws state and federal budgets out of whack. The talk is all about chiselers, profiteers, and reductions. Forms and obstacles multiply. The Medical Identity is put in place. Uncle Sam becomes Uncle Scrooge. One would hardly gather that billions more are actually being spent on medicine for the poor. But the federal government is not the only participant who is doing better and feeling worse.

Unequal levels of development within states pit one location against another. A level of benefits adequate for New York City would result in coverage of half or more of the population in upstate areas as well as nearly all of Alaska's Eskimos and Arizona's Indians. The rich pay more; the poor get hassled. Patients are urged to take more of their medicine only to discover they are targets of restrictive practices. They are expected to pay deductibles before seeing a doctor and to contribute a co-payment (part of the cost) afterward. Black doctors are criticized if their practice consists predominantly of white patients, but they are held up to scorn if they increase their income by treating large numbers of the poor and aged in the ghettos. Doctors are urged to provide more patients with better medicine, and then they are criticized for making more money. The *Principle of Perspective* leads each patient to want the best for himself disregarding the social cost; and, at the same time, doctors are criticized for giving high-cost care to people who want it. The same holds true for hospitals: keeping wages down is exploitation of workers; raising them is taking advantage of insurance. Vast financial incentives are offered to encourage the establishment of nursing homes to serve the aged, and the operators are then condemned for taking advantage of the opportunity.

Does anyone win? Just try to abolish Medicare and Medicaid. Crimes against the poor and aged would be the least of the accusations. Few argue that the country would be better off without these programs than with them. Yet, as the programs operate, the smoke they generate is so dense that their supporters are hard to find.

By now it should be clear how growing proportions of people in need of medicine can be getting it in the midst of what is universally decried as a crisis in health care. Governments face phenomenal increases in cost. Administrators alternately fear charges of incompetence for failing to restrain real financial abuse and charges of niggardliness toward the needy. Patients are worried about higher costs, especially as serious or prolonged illnesses threaten them with financial catastrophe. That proportionally few people suffer this way does not decrease the concern, because it *can* happen to anyone. Doctors fear federal control, because efforts to lower costs lead to more stringent regulations. The proliferation of forms makes them feel like bureaucrats; the profusion of review committees threatens to keep them permanently on trial. New complaints increase faster than old ones can be remedied. Specialists in public health sing their ancient songs—you are what you eat, as old as you feel, as good as the air you breathe—with more conviction and less effect. True but trite: what can be done isn't worth doing; what is worth doing can't be done. The watchwords are malaise, stasis, crisis.

If money is a barrier to medicine, the system is discriminatory. If money is no barrier, the system gets overcrowded. If everyone is insured, costs rise to the level of the insurance. If many remain underinsured, their income drops to the level of whatever medical disaster befalls them. Inability to break out of this bind has made the politics of health policy pathological.

POLITICAL PATHOLOGY

Health policy began with a laudable effort to help people by resolving the polarized conflict between supporters of universal, national health insurance ("socialized" medicine) and the proponents of private medicine. Neither side believed a word uttered by the other. The issue was sidestepped by successfully implementing medical care for the aged under Social Security. Agreement that the aged needed help was easier to achieve than consensus on any overall medical system. The obvious defect was that the poor, who needed financial help the most, were left out unless they were also old and covered by Social Security. The next move, therefore, was Medicaid for the poor, at least for those reached by state programs.

Even if one still believed that medicine equaled health, it became impossible to ignore the evidence that availability of medical services was not the same as their delivery and use. Seeing a doctor was not the same as actually doing what he prescribed. It is hard to alleviate stress in the doctor's office when the patient goes back to the same stress at home and on the street.

"Health delivery" became the catchword. At times it almost seemed as if the welcome wagon was supposed to roll up to the door and deliver health, wrapped in a neat package. One approach brought services to the poor through neighborhood health centers. The idea was that local control would increase sensitivity to the patients' needs. But experience showed that this "sensitivity" had its price. Local "needs" encompassed a wider range of services, including employment. The costs per patient-visit for seeing a doctor or social worker were three to four times those for seeing a private practitioner. Achieving local control meant control by inside laymen rather than outside professionals, a condition doctors were loath to accept. Innovation both in medical practice and in power relationships proved a greater burden than distant federal sponsors could bear, so they tried to co-opt the medical powers by getting them to sponsor health centers. The price was paid in higher costs and lower local control. Amid universal complaints, programs were maintained where feasible, phased out where necessary, and forgotten where possible.

By now the elite participants have exceeded their thresholds of pain: government can't make good on its promises to deliver services;

administrators are blamed for everything from malpractice by doctors to overcharges by hospitals; doctors find their professional prerogatives invaded by local activists from below and by state and federal bureaucrats from above. From the left come charges that the system is biased against the poor because local residents are unable to obtain, or maintain, control of medical facilities, and because the rates by which health is measured are worse for them than for the better off. Loss of health is tied to lack of power. From the right come charges that the system penalizes the professional and the productive: excessive governmental intervention leads to lower medical standards and higher costs of bureaucracy, so that costs go up while health does not.

As neighborhood health centers (NHCs) phased out, the new favorites, the health-maintenance organizations (HMOs), phased in. If the idea behind the NHCs was to bring services to the people, the idea behind the HMOs is to bring the people to the services. If a rationale for NHCs was to exert lay control over doctors, the rationale for HMOs is to exert medical control over costs. The concept is ancient. Doctors gather together in a group facility. Individuals or groups, such as unions and universities, join the HMO at a fixed rate for specified services. Through efficiencies in the division of labor and through features such as bonuses to doctors for less utilization, downward control is exerted on costs.

Since the basic method of cutting costs is to reduce the supply of hospital beds and physician services (the Medical Identity), HMOs work by making people wait. Since physicians are on salary, they must be given a quota of patients or a cost objective against which to judge their efforts. Both incentives may have adverse effects on patients. HMO patients complain about the difficulty of building up a personal relationship with a doctor who can be seen quickly when the need arises. Establishing such a relationship requires communication skills most likely to be found among the middle class. The patient's ability to shop around for different opinions is minimized, unless he is willing to pay extra by going outside the system. Doctors are motivated to engage in preventive practices, though evidence on the efficacy of these practices is hard to come by. They are also motivated to engage in bureaucratic routines to minimize the patients' demands on their time; and they may divert patients to various specialties or ask them to return, so as to fit them into each physician's assigned quota. In a word, HMOs are a mixed bag, with no one quite sure yet what the trade-off is between efficiency and effectiveness. Turning the Great Equation into an Identity—where Health = Health Maintenance Organization—does, however, solve a lot of problems by definition.

HMOs may be hailed by some as an answer to the problem of medical information. How is the patient-consumer to know whether he is getting proper care at reasonable cost? If it were possible to rate HMOs, and if they were in competition, people might find it easier to choose among them than among myriads of private doctors. Instead of being required to know whether all those tests and special consultations were necessary, or how much an operation should cost, the patients (or better still, their sponsoring organizations) might compare records of each HMO's ability to judge. Our measures of medical quality and cost, however, are still primitive. Treatment standards are notoriously subjective. Health rates are so tenuously connected to medicine that they are bound to be similar among similar populations so long as everyone has even limited access to care.

If health is only minimally related to care, less expertise may be about as good as more professional training. If by "care" many or most people mean simply a sympathetic listener as much as, or more than, they mean a highly trained, cold diagnostician, cheaper help may be as good as, or even better than, expensive assistance. Enter the nurse-practitioner or the medical corpsman or the old Russian *feldsher*—medical assistants trained to deal with emergencies, make simple diagnoses, and refer more complicated problems to medical doctors. They cost less, and they actually make home visits. The main disadvantage is their apparent challenge to the prestige of doctors, but it could work the other way around: doctors might be elevated because they deal with more complicated matters. But the success of the medical assistant might nonetheless raise questions about the mystique of medical doctors. In response the doctors might deny that anyone else can really know what is going on and what needs to be done, and they might then use assistants as additions to (but not substitutes for) their services. That would mean another input into the medical system and therefore an additional cost. The politics of medicine is just as much about the power of doctors as it is about the authority of politicians.

Now we see again, but from a different angle, why the medical system seems in crisis although most people are satisfied with the care they are receiving. At any one time, most people are reasonably healthy. When they do need help, they can get it. The quality of care is generally impressive; or whatever ails them goes away of its own accord. But these comments apply only to the mass of patients. The elite participants—doctors, administrators, politicians—are all frustrated. Anything they turn to rebounds against them. Damned if they do and cursed if they don't, it is not surprising that they feel that any future position is bound to be less uncomfortable than the one they hold today. Things can always get worse, of course, but it is not easy for them to see that.

GOVERNMENTAL LEGITIMACY: CURING THE SICKNESS OF HEALTH

Why should government pay billions for health and get back not even token tribute? If government is going to be accused of abusing the poor, neglecting the middle classes, and milking the rich; if it is to be condemned for bureaucratizing the patient and coercing the doctor, it can manage all that without spending billions. Slanders and calumnies are easier to bear when they are cost-free. Spending more for worse treatment is as bad a policy for government as it would be for any of us. The only defendant without counsel is the government. What should it do?

The Axiom of Inequality cannot be changed; it is built into the nature of things. What government can do is to choose the kinds of inequalities with which it is prepared to live. Increasing the waiting time of the rich, for instance—that is, having them wait as long as everybody else—may not seem outrageous. Decreasing subsidies in New York City and increasing them in Jacksonville may seem a reasonable price to pay for national uniformity. From the standpoint of government, however, the political problem is not to achieve equal treatment but to get support, at least from those it intends to benefit. Government needs gratitude, not ingrates.

The Principle of Goal Displacement, through the double-displacement effect, succeeds only in substituting access to care for health; it by no means guarantees that people will value the access they get. Equal access to care will not necessarily be equated with the best care available or with all that patients believe they require. Government's task is to resolve the Paradox of Time so that, as things get better, people will see themselves as better off.

Proposals for governmental support of medical care have ranged from modest subsidies to private insurance (the AMA's Medicredit) to public control of the medical industry on the British model. The latter has never had much support in this country, because of widespread opposition to socializing doctors by turning them into de facto government employees. The former has lost whatever support it once had as respect for the AMA has declined, its internal unity has diminished, and its congressional supporters have nearly vanished. Private insurance seems as much the problem as the solution.

The two most prominent proposals would resolve the political problems of medical care in contrasting ways, but substantively they are similar. Both the Comprehensive Health Insurance Plan (CHIP), introduced in the last days of the Nixon administration, and the Kennedy-Mills proposal would involve billions of dollars in additional expenditures. Estimates put each of them at $42 billion to start, less substantial existing expenditures—but then no estimates in this field have ever come remotely close to reality. Both proposals would provide health insurance for virtually everyone and would cover almost everything (including catastrophic and long-term illness) except for prolonged mental illness and nursing-home care. Both include a string of deductibles and co-insurance mechanisms, with CHIP so complex as almost to defy description. Both seek to hold down costs by giving individuals a financial incentive to limit use. Neither provides incentives for the medical community to contain costs, other than the importunings of insurance companies and state governments (CHIP) or the federal government (Kennedy-Mills), which have not been noticeably effective in the past.

CHIP would be financed largely through employer-employee contributions, with employees making a per capita payment; Kennedy-Mills substitutes a more (though by no means entirely) progressive proportionate tax. CHIP mandates insurance and gives a choice of private plans supervised through state agencies. Kennedy-Mills works largely through a special fund collected and administered by the federal government. The basic difference between them is that more of the cost of Kennedy-Mills shows up in the federal budget, while most of the cost of CHIP, as its acronym suggests, is diffused through the private sector.

The most likely consequence of both proposals would be a vast inflation of costs without a corresponding increase in services. Since medical manpower and facilities could not increase proportionately with demand, prices would rise. It would be Medicaid all over again, only worse because so many new things would be attempted and so many old things expanded. Almost before the ink dried on the legislation, efforts would be under way to delay this provision, lessen the cost of that one, introduce rationing in nonmonetary ways, find more forms for doctors and patients to fill out, and on and on. Cries of systemic crisis would be replaced by prophecies of systemic failure. But enough. My purpose is not to predict the medical consequences of these proposals but to analyze their political rationale.

Based on the political premise that some form of national health insurance was inevitable, CHIP sought to limit the government's liability. By joining the opposition, the Nixon administration hoped to control the apparatus so as to lessen its impact on the federal budget and bureaucracy. If people were determined to have something that wasn't going to help them, the government could at least see to it that the totals did not swamp its budget or overload its administration.

The costs of failure would be spread around among the states, the various insurance companies, and innumerable individual and group medical practices. Just as revenue sharing was designed to channel demands to state and local governments, instead of the national government (here's a little money and lot of trouble, and don't bother me!) so CHIP was devised to diffuse responsibility.

What the Republican administration did not foresee was that the rapid breakdown of the existing medical system would inevitably lead to demands for a federal takeover. When a company goes bankrupt, it is usually returned, not to its owners, but to its creditors. This insight belongs to the sponsors of the Kennedy-Mills bill. They seized on the Nixon plan to advance one that would load additional clients, services, and billions onto the shoulders of government. Wouldn't this proposal be too expensive and cumbersome? The worse the better, politically! For then the stage would be set for a national health service.

Under the Kennedy-Griffith (now Kennedy-Corman) proposal, which was the senator's original preference, every person in the United States would, without personal payment, be covered for a wide variety of services, thus replacing all public programs and private insurance with an all-inclusive federal system. Every public and personal medical expense would be transferred to the federal government, paid for half by additional payroll taxes and higher taxes on unearned income and half from general revenues. Obviously, as the sole direct payer, the federal government would have control over costs, but, by the same token, it would have to make all the decisions on how much of what service would be provided to which people in what way for how long.

The difference between Kennedy-Mills and the Kennedy-Corman Health Society Act (HSA) is that the latter would work directly on the Law of Medical Money by limiting the financial resources flowing into the medical system. Whatever the federal government allocated would be all that could be spent, except for the sums spent by those people choosing to pay extra to go outside the system. To put HSA in proper perspective, it is useful to contrast it with another proposal, one that would also limit supply but from a different direction. Senators Long and Ribicoff proposed to deal with the costs of catastrophic illness by setting individual-expenditure limits beyond which costs would be paid by the government. But Long-Ribicoff did not relate individual payments to income. For our comparison, therefore, it is more helpful to concentrate on Martin Feldstein's proposal for an income-graded program in which each person pays medical costs up to a specified proportion of his income, after which the government picks up the remaining (defined as catastrophic) expenses. Medicare and Medicaid are replaced, as all benefits are related to income. The poor pay less, the rich pay more, but everyone is protected against the costs of catastrophe. Although the catastrophic portion would rise in cost, especially for long-term disability, it would represent a relatively small proportion of medical expenditures. Total costs would be determined by overall financial inputs, which would be limited by the willingness of people to pay instead of inflated by using up their insurance or subsidy.

At first glance it might appear strange for national health insurance (whether through private intermediaries or direct government operation) to be conceived of as a method for limiting costs; but experience in practice, as well as deduction from theory, bears out that conception. The usual complaint in Britain, for example, is that the National Health Service is being starved for funds: hospital construction has been virtually nil; the number of doctors per capita has hardly increased; long queues persist for hospitalization in all but emergency cases. Why? Because health care accounts for a sizable proportion of both government expenditure and gross national product and must compete with family allowances, housing, transportation, and all the rest. While there are pressures to increase medical expenditures, they are counterbalanced by demands from other sectors. In times of extreme financial stringency, all too frequent as government expenditure approaches half of the GNP, it is not likely that priority will go to medicine.

So much for current trends. In the future, the nation will probably move toward (and vacillate between) three generic types of health care policies: (1) a mixed public and private system like the one we have now, only bigger; (2) total coverage through a national health service; and (3) income-graded catastrophic health insurance. It will be convenient to refer to these approaches as "mixed," "total," and "income."

The total and income approaches have weaknesses. The income-catastrophic approach might encourage a "sky's the limit" attitude toward large expenditures; the other side of the coin is that resources would flow to those chronically and/or extremely ill people who most need help. The total approach would strain the national budget, putting medical needs at the mercy of other concerns, such as tax increases; on the other hand, making medicine more political might have the advantage of providing more informed judgment on its relative priority. The two approaches, however, are more interesting for their different strengths than for their weaknesses.

The income approach would magnify individual choice until the level of catastrophic cost is reached. Holding ability to pay relatively constant, each person would be able to decide how much (in terms of what money can buy) he is willing to give up to purchase medical services. There would be no need to regulate the medical industry as to cost and service: supply and demand would determine the price. Paperwork would be minimized. So would bureaucracy. Under- or over-utilization could be dealt with by raising or lowering the percentage limits at each level of income, rather than by dealing with tens of thousands of doctors, hospitals, pharmacies, and the like. The total approach, by contrast, could promise a kind of collective rationality in the sense that the government would make a more direct determination of how much the nation wanted to spend on health versus other desired expenditures.

How might we choose between an essentially administrative and a primarily market-oriented mechanism? Each is as political as the other, but they come to their politics in different ways. An income approach would be simpler to administer and easier to abandon. If it didn't work, more ambitious programs could readily be subsidized. A total approach could promise more, because no one under existing programs would be worse off (except taxpayers), and everyone with insufficient coverage would come under its comprehensive umbrella. The backers of totality fear that the income approach would preempt the health field for years to come. The proponents of income grading fear that, once a comprehensive program is begun, there will be no getting out of it—too many people would lose benefits they already have, and the medical system would have unalterably changed its character. The choice (not only now but in the future) really has to be made on fundamental grounds of a modified-market versus an almost entirely administrative approach. Which proposal would be not only proper for the people but good for the government?

Market versus Administrative Mechanisms

At the outset, I should state my conviction that doing either one consistently would be better than mixing them up. Both methods would give government a better chance to know what it is doing and to get credit for what it does. Expenditures on the medical system, whether too high or too low for some tastes, would be subject to overall control instead of sudden and unpredictable increases. Patients would have a system they could understand and would therefore be able to hold government accountable for how it was working. Under one system they would know that care was comprehensive, crediting government with the program and criticizing it for quality and cost. Under the other, they would know they were being encouraged to exercise discretion, but within boundaries guaranteeing them protection against catastrophe. Under the present system, they can't figure out what's going on (who can?); or why their coverage is inadequate; or why, if there is no effective government control, there are so many governmental forms. Mixed approaches will only exacerbate these unfortunate tendencies, multiplying ambiguities about deductibles and co-payment amid startling increases in cost. If we want our future to be better than our past, then let us look more closely at the bureaucratic and market models for medical care.

What do we know about medical care in a bureaucratic setting? Distressingly little. But there may be just enough collected from studies of HMOs and of systems in other countries, especially Britain, to provide a few clues. Doctors in HMOs work fewer hours than do doctors in private practice. This is not surprising. One of the attractions of HMOs for doctors is the limit on the hours they can be put on call. Market physicians respond to increases in patient load by increasing the hours they see patients; physicians working in a bureaucratic context respond by spending less time with patients. Two consequences of a public system are immediately apparent: more doctors will be needed, and less time will be spent listening and examining. Patients' demands for more time with the doctor will be met by repeated visits rather than longer ones. But will doctors be distributed more equally over the nation? The evidence suggests not. Britain has failed to achieve this goal in the quarter-century since the National Health Service began. The reason is that not only economic but also political allocations are subject to biases, one of which, incidentally, is called majority rule. The same forces that gather doctors in certain areas are reflected in the political power necessary to supply funds to keep them there.

Surely the ratio of specialists to general practitioners could be better controlled by central direction than by centrifugal market forces. Agreed. But a price is paid that should be recognized. The much higher proportion of general practitioners in Britain is achieved through a class bias that values "consultants" (their "specialists") more highly than ordinary doctors. (Consultants are called "Mister," as if to emphasize their individual excellence, while general practitioners are given the collective title of "Doctor.") The much higher proportion of specialists in America may stem in part from a desire to maintain equality among doctors—a nice illustration of the Axiom of Inequality. One result of the British custom is to lower the quality of general practice; another is to deny general practitioners access to hospitals. They lose control of their patients at the portal, leaving them without the comfort they may need in a stressful time and subjecting them to a bewildering maze of specialists and subspecialists, separated by custom and procedure, none of whom may be in charge of the whole person.

Would a bureaucratic system based on fixed charges and predetermined salaries place more emphasis on cheaper prevention than on more expensive maintenance, or on outpatient rather than hospital service? Possibly. (No one knows for sure whether preventive medicine actually works.) Doctors, in any event, do not cease to be doctors once they start operating in a bureaucratic setting. Cure, to doctors, is intrinsically more interesting than prevention; it is also something they know they can attempt, whereas they cannot enforce measures such as "no smoking." If it were true, moreover, that providing ample opportunities to see doctors outside the hospital would reduce the need to use hospitals, then providing outpatient services should hold down costs. The little evidence available, however, suggests otherwise. A natural experiment for this purpose takes place when patients have generous coverage for both in- and out-patient medical services. Visits to the doctor go up, but so does utilization of hospitals. More frequent visits generate awareness of more things wrong, for which more hospitalization is indicated. The way to limit hospital costs, if that is the objective, is to limit access to hospitals by reducing the number of available beds.

The great advantage of a comprehensive health service is that it keeps expenditures in line with other objectives. The Principle of Perspective works both ways: if an individual is not an aggregate, neither is an aggregate an individual. Left to our own devices, at near zero cost, you and I use as much as we and ours need. At the governmental level, however, it is not a question of personal needs and desires but of collective choice among different levels of taxation and expenditure. Hence, it should not be surprising that our collective choice would be less than the sum total of our individual preferences.

The usual complaint about the market method is money. Poor people are kept out of the medical system by not having enough. No one disputes this. And whatever evidence exists also suggests that the use of deductibles and co-payment exerts a disproportionate effect in deterring the poor from acquiring medical care. Therefore, to preserve as much of the market as possible, the response is to provide the poor with additional funds they can use for any purpose they desire, including (but not limited to) medical care. This immediately raises the issue of services in kind versus payment in cash. Enabling the poor to receive medical services without financial cost to themselves means they cannot choose alternative expenditures. A negative way of looking at this is to say that it reveals distrust of the poor: presumably, the poor are not able to make rational decisions for themselves, so the government must decide for them. A positive approach is to say that health is so important that society has an interest in assuring that the poor receive access to care. I almost said, "whether they want it or not," but, the argument continues, the choice of seeking or not seeking health care is neither easy nor simple: the poor—because they are poor, because money means more to them, because they have so many other vital needs—are under great temptation to sacrifice future health to present concerns. The alleged short-sighted psychology of the poor requires that they be protected against themselves.

The problem is not with the intellectually insubstantial (though politically potent) arguments that medical care is a right and that money should have nothing to do with medicine. The Axiom of Allocation assures us that medical care must be allocated in some way, and that, if it is not done at the bottom through individual income, it will be done at the top through national income. If medicine is a right, so is education, housing, food, employment (without which other rights can no longer be enjoyed), and so on, until we are led to the same old problems of resource allocation. The real question is whether care will be allocated by governmental mechanisms, in which one-man-one-vote is the ideal, or by the distribution of income, in which one-dollar-one-preference is the ideal, modified to assist the poor.

The problem for market men is not to demonstrate resource scarcity but to show that one of the essential conditions of buying and selling really is operating. I refer to consumer information about the cost and quality of care. The same problems crop up in many other areas involving technical advice: without knowing as much as the lawyer, builder, garage mechanic, or television repair man, how can the consumer determine whether the advice is good and the work performed properly and at reasonable cost?

The image in the literature is amateur patient versus professional doctor: the patient is not sure what is wrong, who the best doctor is, and how much the treatment should cost. Worse still, doctors deliberately withhold information by making it unethical to advertise prices or criticize peers. Should the doctor be less than competent or more than usually inclined to run up a bill, there is little the patient can do.

There are elements of reality in this picture, as all of us will recognize, but it is exaggerated. People can and do ask others about their experiences with various doctors; mothers endlessly compare pediatricians, for example. The abuses with which we are concerned are more likely to occur when patients lack a stable relationship with at least one doctor, and when there is no community whose opinions the doctor values and the patient learns to consult.

Nevertheless, it is obvious that patient-consumers do lack full information about the medical services they are buying. So, in fact, do doctors lack full knowledge of the services they are selling. How, then, might the imperfect medical market be improved? Would some alternative provision of medical services ensure better information?

Since all costs would be paid by taxpayers, government would *National Health* have an incentive to keep the expenditures on a national health service in proportion to the expenditures for other vital activities. The very feature that has so far made a national health service politically unpalatable—it would take over about $50 billion of now private expenditures, thus requiring a massive tax increase—would immediately make the government financially responsible. Under a total governmental program, central authorities would have to determine how much should be spent and how these funds should be allocated to regional authorities. Basing the formula on numbers would put remote places at a disadvantage; basing it on area would put populous places at a disadvantage. How would regional authorities decide how much money to put toward hospital beds versus outpatient clinics, versus drugs, versus long-term care? There are few objective criteria. Would teams of medical specialists make the decisions? Professional boundaries would cause problems. Would administrators? Lack of medical expertise would cause problems. Administrative committees would have to decide who receives how much treatment, given the limited resources available from the central authority. Would their collective judgment be better or worse than that of individuals negotiating with doctors and hospitals? No one knows. But something can be said about the trade-off between quality and cost.

Suppose the question is: Under which type of system are costs *Costs* likely to be highest per capita? The answer is: first, mixed public and private; second, mostly private; third, mostly public. Costs are greater under a mixed system because potential quality is valued over real cost: it pays each individual to use up his insurance and subsidy, because the quality-cost ratio is set high. Under the mostly private system, the individual has an incentive to keep his costs down. Under the largely public one, the government has an incentive to keep its costs within bounds. Because each individual regards his personal worth more than his social value, however, a series of individual payments will add up to something more than the payments determined by the very same people's collective judgment. At the margins, then, the economic market, preferring quality over cost, would produce somewhat larger expenditures than would the political arena.

Who would value a public medical system? Those who want government to exert maximum control over at least cost. The term "cost" here may be used in two ways—financial and political. Government does more, is able to allocate more resources, and has more of a chance of getting support for what it does. People who are more concerned with equality than with quality of care—though, of course, they want both—also should prefer public financing. It assures reasonably equal access, and it also places medical care in the context of other public needs. Doctors who value independence and patients who value responsiveness would be less in favor of a public system.

Who would prefer a private system, providing the effects of income were mitigated? People who want less governmental direction and more personal control over costs. These include doctors who want less governmental control, patients who want more choice, and politicians who want more leeway in resource allocation and less blame for bureaucratizing medicine.

I would prefer the income approach, because it is readily reversible; it means less bureaucracy and more choice. The total approach, however, could be infused with choice: Under the rubric of a single national health service, there could be three to six competitive and alternative programs, each organized on a different basis. There could be HMOs, foundation plans (under which individual doctors contract with a central service), and other variants. Patients could use any of these programs, all of which would be competing for their favor. The total sum to be spent each year would be fixed at the federal level, and each service would be paid its proportionate share according to the number and type of patients it had enrolled. Thus, we could mitigate the worst features of a bureaucratic system while maintaining its strengths.

THOUGHT AND ACTION

Let us summarize. Basically there are two sites for relating cost to quality—that is, for disciplining needs, which may be infinite, by controlling resources, which are limited. One is at the level of the individual; the other, at the level of the collectivity. By comparing his individual desires with his personal resources, through the private market, the individual internalizes an informal cost-effectiveness analysis. Since incomes differ, the break-even point differs among individuals. And if incomes were made more equal, individuals would still differ in the degree to which they choose medical care over other goods and services. These other valued objects would compete with medicine, leading some individuals to choose lower levels of medicine and thus reducing the inputs into (and cost of) the system. This creative tension can also be had at the collective level. There it is a tension between some public services, such as medicine, and others, such as welfare, and a tension between the resources left in private hands and those devoted to the public sector. The fatal defect of the mixed system, a defect that undermines the worth of its otherwise valuable pluralism, is that it does not impose sufficient discipline either at the individual or at the collective level. The individual need not face his full costs, and the government need not carry the full burden.

My purpose in writing this essay has not been to assess current political feasibility but to determine longer-lasting political virtue. The proposals I believe to be the worst for sustaining the legitimacy of government are at present the most popular. Proposals that deserve the most serious attention are ignored. The falsely assumed excessive cost of total care and the falsely believed inequality of the income approach have removed them from serious consideration. Perhaps this is the way it has to be. But I believe there is still time to change our ways of thinking about medical care. Medicine is by no means the only field where how we think affects what we believe, where what we believe is the key to how we feel, and where how we feel determines how we act.

If politicians did not believe that better health would emerge from greater effort, could they justify pouring billions more into the medical system? It could be argued that belief in medicine—doctor as witchdoctor—is so deeply ingrained that no evidence to the contrary would be accepted. Maybe. But this argument does not reach the question of what politicians would do if they believed otherwise.

Suppose the people were told that additional increments devoted to medicine would not improve their collective health but would give them more opportunity to express their individual feelings to doctors. How much more would they pay for this "caring"? Would it be as much as $10 billion? Would it be that high if the program contained no guarantee—and none do—that doctors would care more or be more available?

In any event, after the mixed approach fails, as it surely will, this country will be faced with the same alternatives—putting together the pieces administratively through a national health service, or dismantling what exists in favor of a modified market mechanism. But this is all too neat.

It could be, of course, that the future will find the worst is really the best. The three systems I have separated for analytical convenience—private, public, and mixed—may in practice refuse to reveal their pristine purity. What life has joined together no abstraction may be able to put asunder. A national health service, for instance, might quickly lose its putatively public character as numerous individuals opt for private care. In Scandinavian countries, even those in the professional strata who are convinced supporters of public medicine often prefer to use private doctors. They pay out to jump queues, so as to be treated when they wish, and to have private hospital rooms to carry on business or just to receive extra attention. By paying twice, once through taxes and once through fees for services, they raise the total cost of medicine to society. Would not a public system that was 20 or 30 per cent private be, in reality, mixed?

Consider an income-graded catastrophic system. It would, to begin with, have to pay all costs for those below the poverty line. As time passed, political pressure might increase the proportion of the population subsidized to 25 or 30 per cent. As costs increased, administrative action might be undertaken to limit coverage of expensive long-term illness. How different, then, would this presumably private system be from the mixed system it was designed to replace?

The present as future may be replaced by the future as future only to be superseded by the future as past. First the mixed system (the present as future) will be intensified by pouring billions into it (à la Kennedy-Mills). When that fails, an income-graded catastrophic plan or a national health service (the future as future) will be tried. Efforts to make the former system wholly private will be unfeasible, because public sentiment is against rationing medical care solely by money. Efforts to make the latter system wholly public will fail, because forbidding private fees for service will appear to citizens as an intolerable restraint on their liberty. Then we can expect the future as past. By the next century, we may have learned that a mixed system is bad in every respect except one—it mirrors our ambivalence. Whether we will grow up by learning to live with faults we do not wish to do without is a subject for a seer, not a social scientist.

Health policy is pathological because we are neurotic and insist on making our government psychotic. Our neurosis consists in knowing what is required for good health (Mother was right: Eat a good breakfast! Sleep eight hours a day! Don't drink! Don't smoke! Keep clean! *And* don't worry!) but not being willing to do it. Government's ambivalence consists in paying coming and going: once for telling people how to be healthy and once for paying their bills when they disregard this advice. Psychosis appears when government persists in repeating this self-defeating play. Maybe twenty-first-century man will come to cherish his absurdities.*

* I wish to thank Eli Ginzberg, Osler Peterson, Jack Fein, Lee Friedman, William Niskanen, Marc Pauley, Otto Davis, and Merlin DuVal for their helpful comments on various drafts of this paper. Responsibility for the final version, however, is mine.

BIBLIOGRAPHY

Eugene Feingold, *Medicare: Policy and Politics, A Case Study and Policy Analysis* (San Francisco, 1966).

Martin S. Feldstein, *The Rising Cost of Hospital Care* (a publication of the National Center for Health Services Research and Development, Information Resources Press, Washington, D.C., 1971).

Elliot Friedson, *Doctoring Together* (New York, 1976).

Victor R. Fuchs, ed., *Essays in the Economics of Health and Medical Care* (National Bureau of Economic Research, New York, 1972).

Eli Ginzberg, "Preventive Health: No Easy Answers," *The Sight-Saving Review*, Winter, 1973–74, pp. 187–93.

Edward Hughes, et al., "Utilization of Surgical Manpower in a Prepaid Group Practice," *The New England Journal of Medicine*, October 10, 1974, pp. 759–63.

Herbert E. Klarman, "Application of Cost-Benefit Analysis to the Health Services and the Special Case of Technologic Innovation," *International Journal of Health Services*, 4:2 (1974), pp. 325–52.

Theodore R. Marmor, "Can the U.S. Learn from Canada?," in S. Andreopoulos, ed., *National Health Insurance: Can We Learn from Canada?* (New York, 1975).

Thomas McKeown, *Medicine in Modern Society* (London, 1965).

David Mechanic, *The Growth of Bureaucratic Medicine: An Inquiry into the Dynamics of Patient Behavior and the Organization of Medical Care* (New York, 1976).

Osler Peterson, M.D., "Is Medical Care Worth the Price?," *Bulletin of the American Academy of Arts and Sciences*, 29:1 (October, 1975), pp. 17–23.

Robert Stevens and Rosemary Stevens, *Welfare Medicine in America: A Case Study of Medicaid* (New York, 1974).

Alan Williams, "Measuring the Effectiveness of Health Care Systems," *British Journal of Preventive and Social Medicine*, 28:3 (August, 1974), pp. 196–202.

Warren Winkelstein, Jr., "Epidemiological Considerations Underlying the Allocation of Health and Disease Care Resources," *International Journal of Epidemiology*, 1:1 (1972), pp. 69–74.

The Allocation
of Scarce Resources/Further Reading

Fox, Renee and Swazey, Judith P. *The Courage to Fail: A Social View of Organ Transplant and Dialysis.* Chicago: University of Chicago Press, 1974.

Fost, Norman. "Children as Renal Donors." *New England Journal of Medicine* 296 (February 17, 1977), 363–67.

Jonson, A. R. "The Totally Implantable Artificial Heart." *Hastings Center Report* 3 (November, 1973), 1–4.

Katz, J. and Capron, Alexander Morgan. *Catastrophic Diseases: Who Decides What? A Psychological and Legal Analysis of the Problems Posed by Hemodialysis and Organ Transplantation.* New York: Russell Sage Foundation, 1975.

Rettig, Richard A. "The Policy Debate on Patient Care Financing for Victims of End-Stage Renal Disease." *Law and Contemporary Problems* 40 (Autumn, 1976), 196–230.

Robertson, John A. "Organ Donations by Incompetents and the Substituted Judgement Doctrine." *Columbia Law Review* 76 (January, 1976), 48–78.

"The Sale of Human Body Parts" *Michigan Law Review* 72 (May, 1974), 1182–1264.

Shapiro, Michael H. "Who Merits Merit? Problems in Distributive Justice and Utility Posed by the New Biology." *Southern California Law Review* 49 (November, 1974), 318–70.

Simmons, Roberta G. et al., *Gift of Life: The Social and Psychological Impact of Organ Transplantation.* New York: John Wiley and Sons, 1977.

Titmuss, Richard. *The Gift Relationship: From Human Blood to Social Policy.* New York: Pantheon Books, 1971. (Penguin Paperback, 1973.)

PART EIGHT

BEHAVIOR MODIFICATION

Seymour L. Halleck
Oscar M. Ruebhausen and
Orville G. Brim Jr.
Gerald Klerman
Barton L. Ingraham and
Gerald W. Smith

30 Legal and Ethical Aspects of Behavior Control

SEYMOUR L. HALLECK, M.D.

Many of our images and knowledge of behavior modification have come through fictional and dramatic sources; yet some contemporary therapies and programs of treatment make us realize that such fictional presentations are with us today. Halleck responds to some of the problems posed by this new technology by focusing on the issues of informed consent and the problem of giving a patient necessary treatment in the absence of consent. Because of the wide ranging implications of behavior modification and its far reaching effects, Halleck feels some regulations are necessary to meet the needs of both the physician and the patient. Such concerns are intensified when one thinks of the potential socio-political uses of behavior modification.

Dr. Seymour L. Halleck, M.D., is Professor of Psychiatry at the School of Medicine of the University of North Carolina.

For more than a decade, the practice of involuntary and indeterminate commitment of the mentally ill has been rigorously criticized by those who fear that psychiatrists are too arbitrary in depriving people of liberty. More recently the use of treatments such as lobotomy, behavior therapy, and drug therapy has been questioned on the grounds that such treatment deprives the patient of the right to choose his own course of action (1-3). The new critiques go beyond questioning the imposition of treatment upon involuntary patients. Some treatments offered to voluntary patients are also being attacked as repressive and dehumanizing. A few critics even fear that the psychiatric profession has involved itself in a gigantic conspiracy to control the behavior of citizens who deviate from social norms (4).

In the new climate of concern over the powers of psychiatrists to shape behavior, there is a real likelihood that treatment decisions psychiatrists have come to view as routine medical decisions will be rigorously challenged by the courts. In one jurisdiction, for example, a judge recently upheld the argument of a plaintiff who asked the court to restrain state hospital doctors from giving him phenothiazines. Decisions such as this, which I believe will become common, raise critical questions as to the future practice of psychiatry.

If psychiatrists wish to continue to practice in an effective manner, I am convinced they must develop a system of internal control and monitoring of psychiatric practices that will protect the rights of patients and still make it possible to treat those who will benefit from treatment.

The Potentialities of Behavior Control

The new fear of psychiatric treatment is best understood in terms of the concept of behavior control. Perry London has defined behavior control quite simply as getting people to do someone else's bidding (5). I would like to expand slightly on this definition. Most psychiatric treatments are designed to change the patient's behavior. In its broadest sense behavior control can be viewed as a special form of behavioral change. It is treatment imposed on or offered to the patient that to a large extent is designed to satisfy the wishes of others. Such treatment may lead to the patient's behaving in a manner which satisfies his community or his society. Of course, behavioral change that satisfies the wishes of others may also satisfy many of the wishes of the patient. For reasons that will become clear later, it is sometimes necessary to include even this category of behavior change under the heading of behavior control.

The question of behavior control has been made more critical by the growing effectiveness of psychiatric treatment. The newer drugs and new behavior therapy techniques make it possible to change behavior in a relatively efficient and rapid manner. Long-term psychotherapeutic techniques can, of course, also modify behavior. However, traditional psychotherapy works slowly. It gives the patient time to contemplate the meaning of behavioral change and to resist such change. It also offers the patient the opportunity to learn to behave in ways which do not meet the needs of others. Some traditional psychotherapists even welcome changes that leave the patient more aggressive, more rebellious, and perhaps more abrasive to those around him. (This may, of course, just be a variant of behavior control in which the patient does the bidding of his therapist rather than the bidding of members of his community.)

Biological and behavior therapies, on the other hand, seem to be peculiarly adaptable to serving the needs of society. Lobotomy, electric shock, and tranquilization are likely to increase conformity and to decrease assertiveness. Behavior therapies are used somewhat more flexibly. Some forms of behavior therapy may help the patient develop greater assertiveness. In practice, however, there is little evidence that they have been used to promote assertiveness. Biological and behavior therapies also work quickly. They can be used without giving the patient a chance to contemplate the meaning of his behavior. Once the patient agrees to or is coerced into treatment, he is unlikely to consider the interpersonal or social causes of his behavior or the social consequences of his treatment.

While recognizing that almost any psychiatric treatment can be a form of behavior control, I will focus this discussion on some of the more ethically controversial therapies, including biological therapies (i.e., lobotomy, electric shock therapy, and drug therapy); behavior therapies (including aversive therapy, desensitization, and operant conditioning); and the use of physical or chemical restraint.

There are three classes of situations in which the implications of behavior control will be considered.

1. Situations in which the patient does not verbally consent to receive treatment. Sometimes the patient verbally (or physically) resists treatment. Sometimes he merely acquiesces without giving verbal consent. Usually these situations arise when the patient is civilly or criminally committed. However, treatments can also be imposed upon voluntary patients without their knowing about it.

2. Situations in which the patient consents to treatment under some duress. These situations are most blatant when the patient is involuntarily committed, civilly or criminally, and is informed that certain actions that he might consider punitive will take place unless he consents to treatment. They also arise in a more subtle form when the patient's family or community pressures him into accepting treatment either through threats of sanction or threats of withdrawal of love or status.

3. Situations in which the patient consents to treatment or may even request treatment. At present these situations are the least controversial, but in the long run they may raise the most complex ethical questions for our profession and our society.

The Nonconsenting Patient

Nonconsenting treatment is most likely to be imposed upon those who are civilly committed as mentally ill or who are criminally committed and are later certified psychotic. In such situations it is usually assumed that the patient desperately needs to be treated but is too confused to understand what is best for him.

Currently electric shock, lobotomy, and drug therapy are used with some frequency in treating nonconsenting patients. Behavior therapy is used somewhat less frequently. For example, aversive therapy (with some troubling exceptions) is not regularly used with nonconsenting patients. The use of operant methods with the nonconsenting patient, however, is more common. Some patients are required to live in units where a token system that rewards socially approved behavior is enforced without the patient's ever having agreed to participate in such a program. Behavior-shaping programs are also used in more traditional hospital wards without consultation with the patient and therefore without his consent. In such instances the staff merely decides what types of behavior it wishes to reinforce and creates an environment that provides such reinforcement. Since there is no informed consent here, the treatment can be viewed as coercive.

I do not feel there is ever any ethical justification for deceiving the patient. All patients, even the most disturbed, should be informed of what will be done to them, why it will be done, and what effects the treatment is likely to have. If after having such treatment, the patient still does not consent to treatment, coercive treatment is justified only if the following sets of conditions are met:

First, the patient must be judged to be dangerous to himself or others. In the case of civilly committed patients, this judgment has often been made in the process of commitment.

Second, those who are providing treatment must believe there is a reasonable probability that treatment will be of benefit to the patient as well as to those around him.

Third, the patient must be judged to be incompetent to evaluate the necessity for treatment.

It should be obvious that each of these criteria calls for highly value-laden and sometimes arbitrary judgments. Without embarking on a prolonged discussion of dangerousness, it may be sufficient to note that psychiatrists have modest skills in predicting dangerousness to self and quite limited skills in predicting dangerousness to others (6); in both instances we tend to overdiagnose dangerousness.

With regard to the issue of treatment, too often the only criterion used in evaluating the helpfulness of a treatment is whether it makes the patient more placid. There is an implied assumption that if the patient ceases to be abrasive, he will benefit from the favorable reactions of others and will feel better. The words "help" or "treatment" in this context should mean something more than docility. As a result of treatment the patient should experience a greater sense of psychological well-being, and his feeling better should not be totally dependent on the reactions of those who initially disapproved of his behavior. The doctor who treats should have a reasonable belief that the treatment imposed will produce changes that it can be assumed the patient might have sought if he had been more rational.

Decisions as to the patient's competency to evaluate the usefulness of treatment are also difficult to make. There are many patients who are too confused to understand that certain potentially frightening treatments might be helpful to them. But there is at the same time ample evidence that psychiatrists tend to overdiagnose incompetency (7). Many patients who resist psychiatric treatment, including some severely handicapped persons, may be making quite rational decisions. The patient's competency to understand the usefulness of treatment should be evaluated in rather straightforward terms. I would suggest that to be competent in this situation, the patient need only appreciate the possible results of being treated and the possible results of not being treated.

There is little problem justifying coercive treatment if all three criteria of dangerousness, treatability, and incompetence are obviously present. In some rare instances, however, imposition of treatment upon nonconsenting patients might be desirable if only two of the three criteria were met. This is a highly controversial issue. Should a very dangerous incompetent person be involuntarily treated even when considerable doubt exists as to the efficacy of treatment? Should a highly treatable incompetent person be involuntarily treated if he is not dangerous? Or should a highly dangerous, easily treatable person who is competent to resist treatment ever be treated against his will? Each of these questions could be debated endlessly. There can be no standard formula for answering them; the most that can be said here is that the psychiatric profession and society need to thoroughly consider these questions whenever an involuntary treatment decision is being made.

The decision to impose treatment upon a nonconsenting patient requires extraordinarily complex medical and ethical judgments. I do not believe that such decisions should be made by a single doctor except in emergencies. For that matter, such decisions should not be made by groups of doctors working in the same institutional setting *without* the benefit of some outside monitoring and feedback.

I propose that any treatment recommendation for nonconsenting patients involving brain surgery, electric shock therapy, prolonged use of tranquilizers, or behavior therapy be reviewed and approved by a monitoring agency. A board consisting of one of the therapists who recommended treatment, a psychiatrist who is affiliated with the institution (who ideally should represent a different school of thought than the doctor who made the recommendation), and an attorney should review and pass on the desirability of each therapeutic recommendation. Ideally, the consulting psychiatrist and the attorney should be replaced by new people at regular intervals so that there is less danger of stagnant attitudes and collusion.

Under this plan I would find nothing objectionable in the non-consenting patient's requesting that his privately hired psychiatrist and/or attorney also participate in these proceedings. If the case for treatment is good, even the patient's own agents should concur in the recommendation. If the case for treatment is weak, an adversary procedure might help in making the right decisions.

The kind of review board I am recommending would certainly add to the administrative burdens of psychiatric treatments and could conceivably be expensive. However, its advantages outweigh these deficits. First of all, such an approach would give patients a greater sense of safety and security. It would serve as a message to society that even the most disturbed individuals are not to be subjected to behavior control without careful consideration. The existence of such a board would encourage doctors to be more precise and thoughtful in making recommendations for treatment. Finally, this approach would continue to leave medical decisions primarily in the hands of doctors and in the long run would diminish the possibility that treatment decisions might be made by the courts.

It must be emphasized that emergency situations would have to be excluded from committee review. When the patient is so disorganized that his life or the physical well-being of those around him is threatened, there is not time to review treatment decisions. The committee would become involved only if the physician wished to continue involuntary treatment after the emergency situation had passed. An arbitrary period of time for invoking committee review of what were initially emergency treatment decisions might be two weeks. In 14 days most "life or death" issues will have been resolved and treatment decisions can be reviewed with due regard to the patient's rights.

Consent Under Duress

The ethical problems involved in treating patients who consent to treatment under pressure are almost as excruciating as those involving patients who do not consent. One immediate problem is assessing the nature of the pressure imposed upon the patient. Pressure to consent may be quite subtle. Often the patient's family or community may threaten loss of love or sanctions if the patient does not enter a hospital and cooperate in receiving certain treatments.

The authority of the doctor is also a powerful influence. Most people will take the advice of doctors even if they fear that the consequences of such passivity might not be in their own interests. Unfortunately, some doctors are unnecessarily authoritarian. Sometimes they enunciate or imply, without good reason, that terrible consequences will befall the patient if he does not consent to a certain treatment. Sometimes the patient is not fully informed as to alternative treatments that might be available.

The ethical issues involved in consent under subtle pressure are in many ways similar to those involved when the patient consents willingly. These will be discussed later. It should be stated here, however, that when the patient is reluctant to undergo treatment, the doctor has an ethical obligation not to frighten or threaten him. He should also provide the patient with all possible information as to the effects of the treatment he is recommending and as to the possibilities of alternative treatments.

Consent obtained under severe and direct pressure is a different matter. Sometimes patients in mental hospitals are told (or correctly perceive) that they must undergo certain treatments if they are ever going to be released. In some correctional programs, particularly indeterminate programs for sexual and "psychopathic" offenders, the patient knows that release is primarily dependent upon his cooperation in treatment. If he accepts treatment the meaning of such "consent" is difficult to ascertain.

The problem is poignantly illustrated in a recent film based on the novel *A Clockwork Orange*. A violent and sexually assaultive young man serving a long prison sentence is informed that he can be released in only a few weeks if he agrees to undergo a new treatment. He consents and is given a sophisticated form of aversive therapy that "cures" him of his sexuality and aggressiveness. He leaves the prison a free man but finds himself unable to enjoy life or even to function effectively when deprived of his old behavior patterns. Eventually he is driven to attempt suicide.

The film raises many ethical and political questions. Assuming that we can change people so drastically—and it is likely that we will soon be able to do so—do we have a right to alter human beings in a manner that so seriously impairs their capacity to choose? What if the patient's assaultiveness was politically motivated? What would be the political consequences of "curing" the aggressiveness of a Malcolm X or an Eldridge Cleaver? These are fascinating questions, but for our purposes here the most important questions are: "To what extent can we view the patient's consent as freely chosen if the only alternative to consent is harsh punishment?" "Does the consent of a man who has no way of knowing what effect the treatment will have upon him really mean very much?"

As long as there are potentially dangerous individuals confined to prisons or hospitals and as long as new treatments are available that might change these individuals, the pressures of incarceration will motivate many of them to accept treatment, whether or not this is the conscious intent of social agencies. Since many patients can and probably will benefit from such treatment, consent under pressure is not altogether undesirable. It would help, however, if in the process of motivating and treating such patients, certain rules were carefully followed.

First, the patient should be given a clear explanation of the possible consequences of treatment.

Second, the patient should be told what other treatments are available and he should be given an opportunity to choose among them, rather than having only one treatment option open to him.

Third, no special punishment should be imposed on the patient if he refuses treatment. His conditions of confinement should never be made worse in order to persuade him to accept treatment, nor should he be denied release from an institution if he improves without undergoing treatment.

Fourth, treatments to which the patient consents under direct threat of sanction, like treatments imposed upon a nonconsenting patient, should be reviewed and approved by a committee that should include an attorney and at least one doctor not directly involved with the treatment.

The Voluntary Patient

The most fascinating and in some ways the most insidious aspect of behavior control involves truly voluntary patients who are neither treated without consent nor coerced into giving consent. People are usually unwilling to tolerate the psychological pain associated with anxiety or depression. If treatments that alleviate these unpleasant emotions are available, such treatments will be sought eagerly. The problem is that, while alleviation of suffering through treatment may serve the patient's short-term needs, the behavioral changes produced by treatment may not serve the patient's long-term needs and may eventually be of more value to those around him than to the patient himself. By gratifying a short-term need for comfort, the patient may find himself in a situation where his long-term needs for power, autonomy, and status are compromised.

This concept can be dramatized by considering the use of heroin in the ghetto. Heroin usage is a highly effective, albeit short-lived and dangerous, treatment for human despair. Many oppressed blacks use heroin to make life tolerable, to add meaning to life, or to blot out the psychological suffering resulting from poverty and discrimination. The drug is not forced upon them; rather, they seek it eagerly. The immediate effect of heroin is to make the user feel better. But in the long run the willing use of this drug strengthens the position of those who oppress blacks. The contentment and euphoria produced by heroin diminishes the militancy of the user and makes him less likely to do anything to change his situation.

The psychiatrist is usually called upon to treat symptoms that have less powerful political implications than the militancy of oppressed blacks. Nevertheless, treatment of common psychiatric symptoms has important implications for the patient's relationship to his environment and to the ethics of behavior control. Symptoms can be viewed as signals or as efforts on the part of an individual to communicate personal distress to his environment. A person is viewed as having a symptom if he behaves in a manner that reveals he is anxious, depressed, confused, or angry. The behavior tends to be viewed as abnormal because the basis for the behavior is not apparent either to the patient or to those around him. Yet such behavior often arises from a need to influence and attempt to change what the patient perceives as an oppressive environment.

Symptoms always elicit some response from the patient's environment. Often they have a powerful influence. The husband of the frigid wife must deal with his partner's unresponsiveness. The community must respond to the aggressive child's delinquency. There is no way of knowing the extent to which any symptomatic behavior is an effort to influence an oppressive environment and to what extent it is an autonomous happening that has little social meaning. It is likely, however, that even the sickest person who suffers profoundly is in part seeking to change his environment. To the extent that we treat and extinguish behaviors that are designed to influence the environment which is bothering the patient, we tend to preserve the stability of social systems and risk becoming agents of the status quo.

In the case of voluntary as well as involuntary treatments, it is the biological and behavioral therapies that have the most significant ethical implications. In traditional psychotherapy (individual, family, or group) considerable effort is made to help the patient understand the meaning of his symptom. If he is aware of the manner in which his environment may be oppressing him, the patient at least has the option to do something about changing his environment. Biological therapies and behavior therapies, however, do not expand awareness. As a matter of fact, their principal merit is that they can be used efficiently and impersonally without the necessity to deal with the troubling implications of what the patient's behavior might mean. They can be superior instruments of behavior control without the slightest pretense of coercion.

This issue has profound implications for psychiatric practice. The overwhelming majority of patients accept biological and behavioral treatments without having sufficient information as to how these treatments may affect their future capacities. I believe the psychiatrist has an ethical responsibility to help the patient find this information.

Except in emergencies, if the therapist is to use drugs or behavior therapy he should accompany such treatment with an effort to help the patient explore the meaning of his symptom. The patient should be encouraged to seek an awareness of the extent to which his symptom is an effort to influence the environment and also to become aware of how the alleviation of the symptom might change his relationship to his environment. If a reasonable effort (perhaps only an hour or two of investigation) has been made to help the patient explore these variables, and if he still wants the biological or behavioral treatment, he should receive it. All of this is of course time-consuming and might impress some "hard-thinking" clinicians as too compulsively ethical. I would argue, however, that efforts to increase the patient's awareness of a social situation are not merely an ethical necessity but are also an essential part of good psychiatric treatment. The patient's capacity to understand and then either to try to accept or to change his environment may in the long run be a greater force in promoting his psychological well-being than the sometimes temporary comfort he might obtain from symptomatic treatment.

This does not mean that the physician should withhold treatment on the basis of the unproven assumption that to withhold it might be best for the patient in the long run. Rather, physicians should be committed to helping the patient make a rational choice as to the desirability of the treatment. I am convinced that the usefulness and reasonableness of the patient's choice will be positively correlated with the amount of accurate information he has about himself and about the stressful factors in his environment.

Conclusions

While there are many people in our country who are wary of the psychiatrist's power to control behavior, there are also many who would encourage psychiatrists to use their power to shape citizens in such a way that they are more conforming. I predict that psychiatrists will soon experience an increase in pressure from both groups. If we are to respond to these pressures rationally and humanistically, we must familiarize ourselves with the legal and ethical implications of behavior control. And we must develop a system of internal regulations of our activities that will satisfy the needs of our patients, ourselves, and the general public.

REFERENCES

1. Szasz T: Psychiatric Justice. New York, Macmillan, 1965

2. Halleck S: The Politics of Therapy. New York, Science House, 1971

3. Kittrie N: The Right To Be Different. Baltimore, Johns Hopkins Press, 1971

4. Ennis B: Prisoners of Psychiatry: Mental Patients, Psychiatry, and the Law. New York, Harcourt, Brace, Jovanovich, 1972

5. London P : Behavior Control. New York, Harper & Row, 1970

6. Wenk E, Robinson O, Smith G: Can violence be predicted? Crime and Delinquency 18:393-402, 1972

7. Szasz T: Law, Liberty, and Psychiatry. New York, Macmillan, 1963

31 Privacy and Behavioral Research

OSCAR M. RUEBHAUSEN
ORVILLE G. BRIM, JR.

One prerequisite condition of freedom is privacy. Recent disclosures have made clear how little privacy we may have left and with what little respect it is treated by some agencies. A similar problem also exists within the sciences, especially those dealing with behavioral research. Often persons are made subjects of research without being so informed or data obtained for one purpose may be used for quite another. This article, after evaluating key issues in this area, proposes a code of ethics for behavioral research. While all may not agree with the elements of this code, it does provide a point of departure for a discussion of a serious problem.

Mr. Oscar M. Ruebhausen is a member of the New York Bar and the Chairperson of the Special Committee on Science and Law of the Association of the Bar of the City of New York.

Mr. Orville G. Brim, Jr. is the President of the Russell Sage Foundation.

A successful society is marked by an ability to maintain a productive equilibrium between numerous competing forces. The goal of our own federal political system is to assure for the individual an ample range of freedom, and an ample opportunity for diversity. By tradition and conviction our form of democracy jealously seeks to protect the individual from accumulations of power. This protection finds its expression, for example, in the separation of powers in government, the divorce of church and state, the civilian control over the military, and in the working of both the labor and antitrust laws against the concentration of economic power.

The familiar and constructive tension which exists between science, with its need to be free and open, and society, with its need for restrictions on individual freedom, is thus only one of many examples of conflicting forces that must be held in balance to assure individual dignity, creativity and well-being in our society. This tension between society and science extends to all the disciplines in the social, physical and life sciences. It affects the practitioner as well as the research investigator.

Examples of this tension are many, and one of the most familiar is the conflict of secrecy for purposes of national security with the free dissemination of knowledge. This conflict is especially complex since dissemination of knowledge is essential to the very developments in science, in industry, and in government upon which the security of the nation ultimately rests. Additionally, there is the equally familiar conflict between proprietary interests and the disclosure of scientific knowledge. The private property interest at odds with disclosure may be personal or institutional, commercial or nonprofit, but the conflict is essentially the same. In each of these two illustrative areas of conflict, tension still exists, but accommodations, imperfect as they may be, have been worked out to balance the competing needs and to serve the public interest.

There is, however, another area of tension involving the freedom of science which is not nearly so well recognized. This is the conflict of science and scientific research with the right, not of private property, but of private personality.[1] And it is to this particular conflict in values that this article is addressed.

I. THE MORAL CLAIM TO PRIVATE PERSONALITY

Although scholars may trace its origins into antiquity, the recognition of a moral claim to private personality is relatively modern. For most of our recorded history, privacy was not physically possible in either the home, or the place of work or of public accommodation. Furthermore, privacy of belief or opinion clearly was not respected until the last few centuries. The record of Robert Bolt, in his moving drama, *A Man for All Seasons*, had the doomed Sir Thomas More say to his inquisitors: "What you have hunted me for is not my actions, but the thoughts of my heart. It is a long road you have opened. For first men will disclaim their hearts and presently they will have no hearts. God help the people whose statesmen walk your road."[2]

Three of the great forces that have nourished the modern claim to privacy are science, the secularization of government, and political democracy. It was, for example, science that brought about the industrial revolution and made privacy physically possible. Consider, as a small sample, what steam heat and plumbing have done to the design of our homes and to the manner of our living in them. Further, the separation of church and state encouraged pluralism as well as diversity in religious belief. And it was political democracy that in the last analysis truly elevated the concept of the essential worth and dignity of the individual to the place it now holds in the western world.

It is therefore only in the last few centuries that the primacy of the individual has emerged, has been articulated by philosophers, reflected in political institutions, and implemented in law. Although the moral claim to a private personality has developed along with the claim to individual freedom and dignity, such development has proceeded at a slower rate, perhaps because the western preoccupation with private property as the tangible expression of the dignity of the individual has tended, for more than a century, to obscure the claim to private personality on which the claim to private property was based. Not only did the interest in private property obscure the human claim to privacy but, over the years, it tended to define the claim itself.

Thus, in the absence of trespass, bodily injury, theft, or tangible damage measurable in money, as in the case of defamation of reputation, our law has often failed to perceive injury to the private personality. This has led to such legal anomalies as now exist with electronic eavesdropping devices. Thus, if an eavesdropping device is placed next to a wall by a police officer, or brought into one's room concealed on the person of an invitee, then, under present federal law, there has been no affront to an individual's constitutional rights. Yet, should the device be a spike microphone and penetrate an apartment wall by only a few inches, then a trespass has been committed and the fourth amendment violated.[3]

Just fifty years ago Dean Roscoe Pound published a paper in the *Harvard Law Review* on "Interests of Personality."[4] There he identified the claim to private personality as "the demand which the individual may make that his private personal affairs shall not be laid bare to the world."[5] But though he thought the interest was clear, the law, he found, had been slow to recognize such an interest and raise it to the dignity of a legal right.[6]

Even had society's developing awareness of the claim to privacy not been blunted by the then dominant commercial concern for tangible property as evidence of personal worth, the establishment of a right of private personality was destined to be slow. For this there are a number of reasons. The right of privacy is largely a subjective, incorporeal right, difficult to identify and incapable of measurement. Other more definable values—such as freedom of speech—loomed larger a century and less ago. Until recently, furthermore, science had not provided the devices which, circumventing the old concepts of property, make surveillance possible without an actual trespass. In addition, the modest range of governmental activities of a half century and more ago made the threat to the individual from government seem negligible. The formidable attributes of concentrated economic power were, also, only beginning to be appreciated. Indeed, the aggressive spirit of individual self reliance which prevailed in America would have made society's concern for the private personality seem incongruous.

It is reasonable, moreover, that the claim to privacy should evolve slowly, for privacy is in conflict with other valued social interests, such as informed and effective government, law enforcement, and free dissemination of the news. Whenever competing rights and values confront each other, it is always a slow and arduous process to evaluate the claim and counterclaim in real life situations. This process, however, is a classic function of the law. In time, therefore, the boundaries between the permissible and unreasonable interferences with privacy will be delineated just as hosts of similar conflicts have been resolved in the past.

Although the claim to private personality has yet to reach its destined stature in our law,[7] it has become a moral imperative of our times. Reflecting the ethical values of our civilization, it flows, as do most of our values, from our concept of the essential dignity and worth of the individual. In discussing this concept in 1958, Pope Pius XII made the following perceptive observations:

'There is a large portion of his inner world which the person discloses to a few confidential friends and shields against the intrusion of others. Certain [other] matters are kept secret at any price and in regard to anyone. Finally, there are other matters which the person is unable to consider.'[8]

Pope Pius then concluded:

"And just as it is illicit to appropriate another's goods or to make an attempt on his bodily integrity, without his consent, so it is not permissible to enter into his inner domain against his will, whatever is the technique or method used."[9]

While Pope Pius' ethics and logic seem persuasive, it is nonetheless a fact that the protections afforded private personality are not yet comparable to those granted private property.

The rules for the protection of private property—whether in ideas, creative works, goods or real estate—have over many decades received extensive legislative and judicial attention. These rules are imbedded in the common law and they have often been elaborately developed, as in our systems of copyright and patent law. Moreover, the manner of the taking of private property for a paramount public purpose has been a matter of intense and continuing national concern. Early evidence of the reverence with which private property has been viewed is found in the constitutional provisions against "unreasonable searches and seizures,"[10] against the quartering of soldiers "in any house without the consent of the Owner,"[11] against the deprivation of property without due process of law, and against the taking of "private property . . . for public use, without just compensation."[12] These constitutional protections have been judicially elaborated over decades of concentrated attention to the proper equilibrium between an identified public need and the claim to private property.

There has been no comparable abundance of legislative or judicial attention to the balance between the public need and the claim to private personality. The application of the first, fourth and fifth amendments of the federal constitution to the claim to private personality is in a very early stage of evolution.[13] More than thirty states have now recognized some form of a common law right of privacy: four have created at least a limited right by statute.[14] Yet, another four states have rejected the existence of a right of privacy at common law,[15] although the rejection may be more verbal than substantive.[16] Thus, in terms of a sophisticated system of protections for the claim to private personality—protections discriminatingly balanced to permit reasonable interference with privacy in appropriate circumstances—it is clear that our law has not yet matured.

II. The Nature of Privacy

What then is this emerging claim to private personality?

Private personality is as complex and many-faceted as human beings themselves, but two principal aspects of the claim to privacy are clear. The one most frequently expressed is the "right to be let alone." This facet of the claim to privacy, first formulated by scholars[17] and repeated by judges,[18] was given widest currency by Justice Brandeis in his magnificent dissent in the Olmstead case.[19] But there is another, and obverse facet of the claim to privacy which has yet to receive equal attention: it is the right to share and to communicate.[20]

Each and every one of us is well aware of this complicated, ambivalent personal need to communicate and, the correlative need, even while communicating, to hold back some area, at least for the moment, for ourselves. Our personal experience is supported by the behavioral scientists. They have documented our need both to share and to withhold.[21]

We need to share in order to feel a useful part of the world in which we live; we need to share in order to test what we truly believe, to obtain the feedback from others which will shape our thoughts, support our egos, and reduce our anxiety. Communication is a form of nourishment, essential to growth and, indeed, to survival. In fact, we are told that if an individual is deprived of all sensory intake and thus isolated from all meaningful association with his environment, he promptly becomes thoroughly disoriented as a person.

Yet, as human beings we also need to withhold—and this for a variety of reasons. There are some things we cannot face and therefore suppress. There are other facts or fears that, although not suppressed, we neither prefer to know nor wish to discuss. Then, too, there are ideas or beliefs or behavior that we are not sure we understand or, even if we do, fear that the world may not. So to protect ourselves, or our processes of creativity, or our minority views, or our self-respect, all of us seek to withhold at least certain things from certain people at certain times.

Psychologically, then, privacy is a two-way street consisting not only of what we need to exclude from or admit into our own thoughts or behavior, but also of what we need to communicate to, or keep from, others. Both of these conflicting needs, in mutually supportive interaction are essential to the well-being of individuals and institutions, and any definition of privacy, or of private personality, must relfect this plastic duality: sharing and concealment.

It follows that the right of privacy does not deal with some fixed area of personal life that has been immutably ordained by either law, or divinity, or science, or culture, to be off-limits and private.[22] The essence of privacy is no more, and certainly no less, than the freedom of the individual to pick and choose for himself the time and circumstances under which, and most importantly, the extent to which, his attitudes, beliefs, behavior and opinions are to be shared with or withheld from others. The right to privacy is, therefore, a positive claim to a status of personal dignity—a claim for freedom, if you will, but freedom of a very special kind.

The way in which the choice between disclosure and non-disclosure is exercised, and the extent to which it is exercised, will vary with each individual, and with each institution. Indeed, the choice will vary in the same individual from day to day, and even on the same day, in differing circumstances. Thus, flexibility and variety are faithful companions of the concept of privacy.

III. The Scientific Challenge

The claim to privacy will always be embattled—its collision with the community's need to know is classic and continuous. Man has always lived in a community, and the community has always required some forfeiture of freedom, including that of privacy. It is, indeed, a fact of life that there has never been a condition of complete privacy for the individual insofar as he is a normal man living with other men. At one time or another, privacy has yielded—as it must—to the positive group needs for security, for order, for sustenance, for survival. The degree of privacy granted throughout history to an individual by one or another community has varied markedly with the nature of the political system, the economic level, the population density, and the characteristics of the environment.

It should also be recognized that not every threat to private personality is a matter of sufficient concern to warrant social protection. Similarly, not every technical trespass is serious enough to warrant social redress. The test is always this: is the threat or the invasion unreasonable, or intolerable?

Today, there are those who point an accusing finger at science and argue that science now poses an unprecedented and grievous threat to the privacy of personality.[23] The argument, while clearly exaggerated, is not implausible. Modern acoustics, optics, medicine and electronics have exploded most of our normal assumptions as to the circumstances under which our speech, beliefs and behavior are safe from disclosure, and these developments seem to have outflanked the concepts of property and physical intrusion, and presumed consent—concepts which have been relied on by the law to maintain the balance between the private personality and the public need. The miniaturized microphone and tape recorder, the one-way mirror, the sophisticated personality test, the computer with its enormous capacity for the storage and retrieval of information about individuals and groups, the behavior-controlling drugs, the miniature camera, the polygraph, the directional microphone (the "big ear"), hypnosis, infra-red photography—all of these, and more, exist today.

All of these significant advances are capable of use in ways that can frustrate an individual's freedom to choose not only what shall be disclosed or withheld about himself, but also his choice as to when, to whom and the extent to which such disclosure shall be made. Notwithstanding the large contribution made by each of these scientific developments to the well-being of man, each is, quite clearly, capable of abuse in its application. And such abuse can occur in industry,[24] in commerce,[25] in the law and by law enforcement agencies,[26] in medicine,[27] in government,[28] and in a myriad of other fields.[29]

So may abuse be found in the area with which we are primarily concerned—scientific research. The one-way mirror is a common fixture in facilities designed for bio-medical and behavioral research. Personality and ability tests are as familiar to researchers in these fields as a stethoscope is to the family doctor. The computer and electronic data storage and retrieval have become crucial to the intelligent and efficient use of research data. Socio-active and psycho-active drugs are ever more tempting research tools, as are the concealed camera and the hidden microphone. When these and other scientific and technological advances are used by scientists, they are used by highly trained, well-motivated, professional people for a social purpose on which the community places a high value. But this fact by itself, obviously, does not warrant the invasion of private personality any more than it would warrant the taking of private property or the administration of live cancer cells to a non-consenting patient.[30]

The recent advances in science have made it clear that society must now work out some reasonable rules for the protection of private personality. It is, perhaps, becoming imperative now to define how the interests of the community—whether in scientific research or law enforcement or economic growth—can be accommodated with the need for privacy. The necessity for such an accommodation poses no idle problem. The consequences of the failure to resolve it are predictable: they begin with the recoil and revulsion of the community;[31] they conclude with arbitrary legislation.

There is no doubt as to the community reaction to the administration, even in the name of research, of live cancer cells to unwitting patients. Nor should we expect that the community will be any more tolerant of behavioral research that subjects non-consenting persons to the risk of injurious, though non-fatal, after-effects. Indeed, community sensitivity as to what is reasonable, or tolerable, is not limited to situations where physical or psychic injury may be involved.

While neither the most representative nor serious intrusion, a well known example of privacy invasion in the field of behavioral research is the so-called "jury bugging" experiment conducted by the University of Chicago. Financed by the Ford Foundation, this was a scientific inquiry conceived and carried out with the best of professional motivation and skill. Although the consent, in advance, of the court and of opposing counsel was obtained, the surreptitious probing of the individual and institutional[32] privacy of the members of the jury shocked the community when the experiment became public knowledge in October, 1955. Federal and state statutes were promptly passed, in 1956 and 1957, to ban all attempts to record or observe the proceedings of a jury.[33] The New York statute, for example, reads as follows:

"A person: . . . who, not a member of a jury, records or listens to by means of instrument the deliberations of such jury or who aids, authorizes, employs, procures, or permits another to do so; is guilty of eavesdropping."[34]

And in New York eavesdropping is a felony punishable by imprisonment![35]

Another example where neither physical injury nor emotional trauma is necessarily involved is found in personality testing.[36] It requires no Cassandra to predict lawsuits by parents, and a spate of restrictive legislation,[37] if those who administer these tests in schools —even for the most legitimate of scientific purposes—do not show a sensitive appreciation for both individual and group claims to a private personality.

The lesson is plain. Unless the advances of science are used with discrimination by scientists engaged in behavioral research—as well as by other professions, by industry and by government—the constructive and productive uses of these advances may be drastically and unnecessarily restricted by a fearful community.[38]

IV. The Need for Equilibrium

Obviously, as Samuel Messick wrote recently:

"Absolute rules forbidding the use of [personality tests] . . . because they delve into contents beyond the bounds of decent inquiry would be an intolerable limitation both to scientific freedom and to professional freedom".[39]

It should be equally obvious—yet it may not be[40]—that absolute rules permitting professional license, in the name of scientific research, to probe beyond the bounds of decent inquiry are equally intolerable to a free society and to free men. Absolute rules do not offer useful solutions to conflicts in values. What is needed is wisdom and restraint, compromise and tolerance, and as wholesome a respect for the dignity of the individual as the respect accorded the dignity of science.

If discrimination and discernment are in fact brought to bear, then we can be confident that the advances in science and technology pose no intolerable threat to privacy. Indeed, they promise to contribute more to an understanding of the claim to private personality, to the recognition of its proper limits, and to the protection of its creative integrity than anything in our recorded experience. Worthy of note is Dr. Robert Morison's reminder that: ". . . the sciences are providing more accurate ways of describing moral problems, and are actually calling attention to types of moral problems which heretofore have not been recognized."[41]

It is not enough to be optimistic about the consequences of the tensions between science and privacy. It is incumbent upon lawyer and scientist to accommodate the goals of science with the claim to privacy, and to help articulate the rules and concepts that will maintain both the productivity of science and the integrity of personality.

In his well-known essay *On Liberty*, John Stuart Mill, while concluding that "over himself, over his own body and mind, the individual is sovereign," continued:

"There is a limit to the legitimate interference of collective opinion with individual independence: and to find that limit, and maintain it against encroachment, is as indispensable to a good condition of human affairs, as protection against political despotism.

"But though this proposition is not likely to be contested in general terms, the practical question, where to place the limit—how to make the fitting adjustment between individual independence and social control—is a subject on which nearly everything remains to be done. . . . Some rules of conduct, therefore must be imposed, by law in the first place, and by opinion on many things which are not fit subjects for the operation of law. What these rules should be, is the principal question in human affairs; but if we except a few of the most obvious cases, it is one of those in which least progress has been made in resolving."[42]

Although more than a century has passed since this pessimistic estimate was made, its essential validity remains.

Our purpose is to identify some of the rules of conduct which, by providing balance and sensitive awareness, can in this century accommodate, and perhaps even resolve, the confrontation of the values of privacy with other values. While the focus here is on behavioral research, it should be emphasized again, that this clash with the values of privacy is not unique to behavioral research.[43] The rules of conduct which can accommodate behavioral research to the claims of private personality may, it is hoped, provide useful parallels in other areas.

V. BEHAVIORAL RESEARCH AND INDIVIDUAL PRIVACY

The traditional methods of behavioral research may, on occasion, involve a violation of the individual claim to private personality.[44] These traditional research methods can be grouped into three broad types: first, self-descriptions elicited by interviews, questionnaires, and personality tests; secondly, direct observations and recording of individual behavior; and thirdly, descriptions of a person by another serving as an informant, or the use of secondary data such as school, hospital, court or office records.

These three major research methods do not necessarily lead to a violation of the claim to privacy. All may be, and most often are, used under conditions of anonymity or individual consent and with strict control over confidentiality. Nevertheless, each method, if improperly employed, can make serious inroads on personal privacy. Thus, some personality tests induce the subject unwittingly to reveal more about himself than he wishes to; carefully designed questionnaires and interview procedures can be used to trap the individual into making public those facts and feelings about himself or others that he would not wish to disclose. Direct observational methods similarly can involve privacy invasion; as, for example, in the use of one-way glass for the observation of children without their knowledge, or in the use of an unidentified participant observer such as a social scientist pretending to be either a patient in a mental hospital or a member of a minority group, or a drug addict among troubled juveniles. Descriptions of one individual by another, either oral or in the form of written records, can also be used in ways that invade the individual's privacy. Illustrative is information elicited from children about their parents' life together, or the description of husbands by wives, or the use of institutional records, originally compiled for one purpose, for quite another. An example of the latter is found when school data are made available to outsiders for research not related to the administration of the educational program. It is the same when welfare data are made available for purposes not connected with the welfare objectives for which they were obtained.

Each of these three basic research methods may engage one or both of the two central—and ethical—issues which are at the core of the relationship between research and personal privacy. These are first, the degree of individual consent that exists and, second, the degree of confidentiality that is maintained. The former concerns the conditions under which information is obtained from a person, the latter, the conditions under which the information is used.

Let us consider some of the ways in which these two issues are raised by behavioral research.

In the use of self-description, a privacy issue arises if the individual respondent does not participate willingly, or if he participates without knowledge of the information being elicited from him, or without an understanding of the purposes for which such information will be used. The nature of the private information being yielded can be obscured from the respondent either by direct artifice, by reliance on the respondent's ignorance or his lack of sophistication, or by some form of coercion, employed to enlist his cooperation. Similarly, with direct observations, a privacy issue arises if the examinee does not know he is being observed, or if he is put off by misleading instructions as to the nature or purpose of the observation or the identity of the observer, or if he is an unwitting participant in a deceptively constructed test situation. An examinee, for example, might be the only person not to know that a group of which he is a part is behaving in a planned abnormal manner so as to test his desire to conform. Where informants, or secondary data, are employed, privacy questions can arise in several ways. An inducement to a breach of faith or confidence may be involved; naivete may be purposefully and systematically exploited. Alternatively, the information may have been supplied only because its nature, or the subsequent use to be made of it, was not known to the respondent.

In each of these three research techniques, an additional point of some complexity can be involved: was the privacy-related data obtained originally for a different purpose? For example, we may consent to yielding vital data for the purpose of being admitted to practice law, or society may properly insist on some loss of individual privacy in order to combat disease or other hazards to life or tranquility.[45] In any such case, however, the individual should not then be deemed to have consented, without qualification, to the subsequent use of such data by a credit agency, or by a member of the school board, or even a scientist engaged in bona fide research.[46]

Lawyers are persuaded that they must not talk about their clients' affairs. While this is now a matter of professional ethics, this restraint is rooted in a recognition that any other state of affairs would corrode the trust which is of the very essence of the professional relationship. The effectiveness of the doctor, plainly, is similarly vulnerable if patients ever believed they could not rely on their physicians to respect imparted confidences. In quite another area: what would happen to the process of education if student attitudes, as revealed in the Socratic interchanges of the classroom, were recorded and reported by the teacher and then used for scientific research or for other purposes— such as responding to inquiries by potential employers?

The point, then, is that consent and confidentiality have a pragmatic as well as a moral importance to the pursuit of any profession. The quality and effectiveness of behavioral research will depend, accordingly, on the confidence the public has in the behavioral scientists and in the way they pursue their science.[47]

VI. THE CONCEPT OF CONSENT

The essence of the claim to privacy is the choice of the individual as to what he shall disclose or withhold, and when he shall do so. Accordingly, the essential privacy-respecting ethic for behavioral research must revolve around the concept of consent.[48] Taken literally, the concept of consent would require that behavioral research refuse to engage in the probing of personality, attitudes, opinions, beliefs, or behavior without the fully informed consent, freely given, of the individual person being examined. There are, however, several reasons why the concept of consent cannot be so literally invoked in the name of privacy.

In the first place, a rigid and literal insistence on formal consent, in a research context, can readily become unrealistic. In some instances, insistence on consent would shake the validity of the research itself. The very selectivity involved in consent would ensure that the research was based on a biased sample and therefore could not be generalized to a wider population. And where subtle attitudes are being measured, knowledge of, and consent to, what is being sought is almost certain to distort the results. In other instances, the requirement of consent might frustrate the project at the outset.[49] Finally, in many instances a full appreciation of the nature of the research, the purposes to be achieved and the risks involved would be impossible to convey fully, either because of their essential complexity, or because they involve unknown factors, or because they are beyond the capacity of the subject to understand.

Any application of the concept of consent as a privacy-protecting test for scientific research is further complicated by the difficult factual problem of assessing, in each particular case, what constitutes consent. When is it informed; when is it freely given; who is entitled to give it? In research situations consent may be given by tacit acquiescence, by explicit oral avowal, by written statement, or it may be implied from the totality of the circumstances. While each of these methods of consent can raise troublesome issues, implied consent is by far the most difficult.

Obviously, in many situations, consent can be fairly implied. Certainly, public figures, particularly those who appeal to the public for elective office, have impliedly consented to the yielding up of some areas of private personality. The comings and goings of a Mayor or Governor, or Hollywood starlet, and a public evaluation and discussion of their strengths and weaknesses in their public roles, are proper subjects of news report, analysis, and research. Similarly when a client seeks occupational counseling from a psychologist, or a parent seeks educational guidance for his child, or when a patient seeks psychotherapy he has consented to some probing, and revelation, of his private personality.[50] While the combination of circumstances that will warrant the implication of informed consent are myriad, restraint must be exercised not to imply such consent in the absence of reasonably compelling facts. Otherwise, the whole requirement of consent can too readily be rationalized away through implication.

Moreover, consent to the revelation of private personality for one purpose, or under one set of circumstances, is not license to publish or use the information so obtained for different purposes or under different conditions. This is especially so when the operative consent is implied or when it would be reasonable to assume that the initial consent would not have been given for the new purpose or the different situation. Further, varying degrees of consent must be recognized. Consent, however given, may be restricted in numerous ways—as to the methods to be used, the risks to be taken, the degree of information the subject wishes to give or receive, the type of data to be obtained, or the uses to which it may be put.

Another complicating factor in the concept of consent is the determination of whether consent has been freely given or coerced. Torture is an old and well-tried technique for extracting private information—and torture need not be physical. Mental anguish can be just as searing and difficult to endure. The prospect of release from suffering, therefore, is a powerful lever for access to the private area. Its uses for the manipulation of behavior or the probing for knowledge are not unknown to sheriffs and prosecutors, to personnel directors, school teachers, and parents—indeed, to virtually anyone who has experienced authority. Conversely, its uses are very well known by the jobless, the hungry, the homeless, the ambitious and the young. The obvious cases of physical, mental, economic, or social duress are readily identifiable; but when does a subtle inducement such as the regard of your boss or even of your peers, or some inducement, not quite so subtle, such as an extra point added to your college grade in return for participation in psychological experiments—when do these become tantamount to duress? What about the vast prestige of scientific research itself as a means of persuasion upon the unsophisticated? And when does the relative disproportion between the knowledge, sophistication and talents of the investigator and his subject make the consent of the respondent questionable, however freely and explicitly given? It is all too apparent that the distinction between consent and concealed coercion may often be difficult to establish. This is however, the type of distinction with which our social institutions, in particular our law and our courts, have a demonstrated competence to deal.

As compared with the complexities of coercion, the problem of identifying the person whose consent must be obtained can, in most cases, be more readily resolved. Normally, when a competent adult is the examinee, or the subject of research, he is the person whose consent must be obtained. If he is not an adult, or if he is not legally competent, then the consent must be obtained from the person legally responsible, namely, a guardian or parent. In the case of children, however, while the legal principles may be clear, a lingering ethical question remains. Should not a child, even before the age of full legal responsibility, be accorded the dignity of a private personality? Considerations of healthy personal growth, buttressed with reasons of ethics, seem to command that this be done. If so, then, in the case of adolescents (and probably even earlier), some form of prior consent to privacy probing should be obtained from *both* the parent and the respondent child.[51]

A special word should be said about anonymity in behavioral research. Frequently it is possible to obtain data of value for behavioral research where the subjects need never be identified by name. National opinion surveys are one example; the use of students in a college classroom may be another. Where anonymity in fact exists, the invasion of privacy involved in behavioral research might well be regarded as *de minimis*. Nevertheless, it must be stressed that anonymity is not a complete substitute for consent. On occasion an individual may feel that his privacy is being invaded when asked to reveal his thoughts or feelings, or to describe his actions, even though he remains quite anonymous to the researcher. It is a fact that many people even under conditions of anonymity resist such revelation to others. So it would seem that, wherever possible, both consent and anonymity should be sought in behavioral research.

The condition of anonymity sometimes is used as a justification for the invasion of privacy in psychological experiments where the subject is deceived as to the meaning of the experiment, or where false information is given to the person so as experimentally to arouse or decrease self-esteem, motivation, or other similar feelings. That the subject remains anonymous, however, can not justify the failure to obtain his consent prior to any such purposeful manipulation of his personality.[52]

Behavioral scientists need no reminder that the concept of consent is not now universally operative as a condition of the research projects on which they are engaged. The use of human guinea pigs is not confined to prisons. Examples of "forced" submission to privacy probes can be found in our hospitals, our schools, our colleges, our social welfare programs, our research institutes, and our institutions for the disturbed, handicapped, or retarded. Such a disregard for the dignity of personality—occasional though it may be—must be guarded against and eliminated by the social scientists themselves.[53] If they fail or refuse to exercise self-control, then the community will inevitably feel compelled to act for itself and legislate for the protection of personal privacy.

While the knowledgeable, freely-given consent of a participant should be a basic ground rule for all behavioral research, there is, of course, a need for exceptions. There must be, indeed, a fundamental exception to cover the many instances where society will accept the invasion of privacy as permissible and reasonable. Thus, when the general welfare requires it and due process is observed, our society permits the taking of private property without consent. There is no reason to doubt that, under similar circumstances, society will permit at least a limited invasion, or taking, of private personality. Circumstances under which the community tolerates the probing into private areas without the consent, and if necessary, without the knowledge of the examinee do, in fact, exist. A number of examples can be easily found in law enforcement, in selection for military service, in social and welfare work, in the protection of the public health, in the national census, and in the selection of employees for the Central Intelligence Agency or as airline pilots.

A public trial may also invade the privacy of the individuals involved in the litigation. Yet since our society is persuaded that a public hearing is essential to a fair trial and to social order, it finds entirely reasonable that the individual claim to privacy must yield in this instance. Even here, however, the equilibrium between the competing values is sensitively preserved and there are occasions when the court is cleared, or the testimony sealed.[54]

Even where the public interest may warrant the taking of private property or of private personality, no absolute license is justified. The taking should be reasonable, it should be conducted with due process, and it should be limited to no more than what is necessary for the fulfillment of the public purpose which, in fact, warranted the invasion.

If we apply these principles to behavioral research, it is clear that, in determining whether the interference with the right of private personality is reasonable, one must appraise many diverse factors. They include such matters as whether the research is necessary, or simply desirable; whether the identification of the individual is in fact required for the successful conduct of the research; whether the invasion of privacy is being limited to the narrowest extent possible; whether artifice and the risk of physical or psychological injury are being avoided; whether the research is being conducted by trained professionals under controlled conditions; whether the paramount public interest favors the research at the risk of a reduction in individual privacy; and whether the paramount nature of the public interest has been explicity recognized, or otherwise accepted, by the community in its laws, by its codes, through its political action, or in such other laborious ways as social consensus is reached and expressed in a free society.

The analogy between behavioral research in the public interest and investigative visits by welfare agents administering public assistance is pertinent. So are the words of the Deputy Commissioner of the New York City Department of Welfare:

"The fact that public assistance is a statutory right means, therefore, that it is subject to conditions imposed by the Legislature. . . . It means that the Legislature may require that the applicant waive his right to privacy to permit a thorough investigation of his eligibility for public assistance. It means that the applicant must open his home to admit representatives of the Welfare Department to enter and to inquire and to observe. It does not mean, of course, that this permissible and necessary invasion of privacy may go so far as to violate the constitutional right against unreasonable search and seizure. It does not mean that the investigator may enter forcibly and without the consent of the applicant nor does it mean that the investigator may come in the dead of night, but it does mean that the applicant must submit to an investigation and, therefore, to an invasion of privacy which falls short of being unreasonable and that if he refuses to submit and refuses to permit such infringement upon his right of privacy, then he may not exercise his right to receive public assistance. The question, therefore, is wholly one of reasonableness and in this respect there may well be a difference of opinion among people of good will. . . .[55]

A clear and paramount public interest in a particular behavioral research inquiry, in spite of a high cost in human privacy, can no doubt frequently be established. However, the recent emergence of behavioral science knowledge as a potential contribution to human welfare has yet to be matched with an explicitly recognized set of laws or codes or otherwise publicly expressed agreements on the value of different kinds of research. Thus, there are and will be many occasions in which conflict between the individual's claim to privacy and the larger community interest in research for the general good must be resolved—and the method of resolution must be an expression of community consensus.

This concept of consensus is not employed in any formal mechanistic way. In a sense, what is meant is that the issue of paramountcy as between private personality and a particular program of scientific research should not be left solely to the decision of the research investigator. There should be some strong element of community approbation; the delicate balancing of the colliding values involved should reflect more than a single point of view.

Community consensus can obviously be expressed in laws, judicial decisions, or political constitutions. But it demands no such formal manifestation, and can also be expressed in far more subtle but equally pervasive ways. For example, consensus can be expressed in the values of our peers as they are articulated to us. Consensus can be formed through the stated views of our opinion leaders whether they be leaders in government or industry, in labor, the professions or the clergy. Consensus can also be reflected in the provisions of collective bargaining contracts between labor and management. In the executive orders or instructions issued by Presidents, cabinet officers, personnel directors, and administrators of all kinds.

Yet, most appropriate for scientific research—as it is for all the professions—is the expression of a consensus on values in a published and operative code of ethics. Such a code yields a triple return—it articulates the values involved, uplifts thereby the awareness and standards not only of the profession but the entire community, and can provide a means for disciplining transgressions within the profession.

Thus, in launching any behavioral research project, the investigator should first determine whether voluntary, informed consent, as well as anonymity, can be accommodated with the integrity of the research. If not, the investigator should then ascertain whether the community consensus approves the conduct of the research, under the proposed conditions, without the actual consent and anonymity of the subjects. As a minimum, this means the knowledgeable concurrence of those responsible for both the research project (for example, the financing institution) and for the well being of the subject (as, for example, the administration of the college he attends). The history of public health and medicine in this country, and earlier in Europe, gives many illustrations of the establishment of just such a community consensus on the invasion of privacy for the general welfare.[56]

One may anticipate that, as behavioral science develops and its contributions to society increase, the democratic process may afford to it more occasions of publicly approved invasions of personal privacy.

VII. THE CONCEPT OF CONFIDENTIALITY

Whether private data are collected with consent, or without consent, but with society's permission because of the perceived public interest involved, the minimal requirements of privacy seem to call for the retention of the private data in a manner that assures its maximum confidentiality consistent with the integrity of the research. Thus, the second privacy issue presented by behavioral research, as it is with all inroads on the private personality, is the issue of confidentiality.

One of the most important ways in which the concept of confidentiality in behavioral research can be served is to seek to design the research so that the responses of the persons providing the data can be anonymous; the design should avoid identifying any individual respondent with a particular response. While this should be possible in all opinion surveys, in many instances the nature of the research will require an ability to identify each respondent with the data elicited from him. This would of course be true in longitudinal studies—as of child growth and development—where respondents must be examined or interviewed a number of times, or in studies of several diverse sets of records which must be matched up to a particular individual.

If full anonymity is not possible in the research design,[57] then there are several other safeguards which should be stressed to provide some degree of anonymity or confidentiality. The first, needing no more than a passing mention, is the integrity of the behavioral research scientist, which, along with his interest in science, must be assumed as a basic prerequisite. The integrity of the professional scientist will assure both his informants and society at large that he will be responsible and will maintain the confidence of any information given to him by identifiable informants. That there are occasional breaches of professional confidence at this level underscores the significance of putting stress on the responsibility of the investigator both during his professional training and throughout his research career.

Another important safeguard for confidentiality can be provided through control techniques. For example, the identity of the respondent may be coded and separated from his response except for the code number. The code, in turn, may be made accessible only to a few of the most responsible officials, or perhaps, only on two signatures or by the use of double keys, even as elementary a safeguard as a locked file can make for substantial improvement. Penalties within the profession may also be devised for any breach of the confidentiality which should be of the very essence of professionalism.

Another readily available step is the destruction of research data. At the very least, that part of the data which would identify any individual with any portion of it should be destroyed, and destroyed at the earliest moment it is possible to do so. Today, it is quite rare for an institution or an individual scientist to take what is now viewed as a radical step and destroy data which potentially have value over a longer time span. Indeed, behavioral scientists have strong incentives to retain all original research data.[58] Such data can provide information of a longitudinal nature about the development of personality or organizations over time, the early childhood antecedents of career success, the degree of change in interest and attitude from one age to another, the effects of marriage upon personality characteristics and other fascinating problems. There are now great repositories of such data in the United States collected about individuals in schools, both secondary and college, and other institutional settings, which have been maintained because of this natural resistance of the research scientist to discard anything of such potential value. Nevertheless, the maintenance and use of this information for purposes other than that originally agreed to, and the threat to confidentiality inherent in its continued maintenance, strongly suggest that the proper course of the person or institution possessing such data is either to obtain the consent of the individual involved to its continued preservation, or to destroy the data, painful as the latter prospect may be.

It should be emphasized that neither the integrity of the scientist nor the technical safeguards of locks and codes can protect research data against a valid subpoena; such data are at present quite clearly subject to subpoena. In the last analysis, therefore, unless our laws are changed to accord a privileged status to privately given research information, confidentiality can be assured only by destruction of the data. The change in the law required to accord a privileged status to research data can be accomplished by statute. Thus, by statute in eighteen states,[59] a privilege has already been afforded to information received by a psychologist from his client. That statutory privilege does not, however, seem to extend to psychological research.[60]

While statutes may be desirable, they may not always be necessary. A privilege status has been afforded by the common law to communications between husband and wife,[61] and attorney and client;[62] privilege also inheres a constitutional doctrine—as in the privilege against self-incrimination. Thus, it is conceivable that privilege could be extended by the courts to other situations—perhaps in a persuasive case, where a research scientist was willing to resist a subpoena and risk imprisonment, in order to protect the private research data in his possession. While there is a role for the martyr both in science and in law, privilege should not be viewed as a status symbol for the scientist.[63] It should, rather, be a protective shield for his informant. As the law now stands, however, it is apparent that the research scientist who probes in the realm of the private personality, without consent, bears a special and heavy responsibility to the subjects of his research. It is a responsibility for confidentiality which, at present, in the face of a subpoena he may find himself powerless to discharge.

Of crucial importance also to the protection of confidentiality is a sensitivity on the part of the scientist to the limited purpose for which the research data were originally obtained. It is generally accepted that research data should not be published by the investigator with identities of the individual subjects attached to the data, and there is no reason why this same ethical sense of the confidentiality, or the privacy, of the data cannot be extended to other forms of publication. Thus, it should be part of the responsibility of the research scientist not to make this research data, in which individuals are identifiable, available to others, whether such others be personnel directors, private detectives, police officers, journalists, government agents, or even other scientists.

Assuredly, one can visualize situations in which the release of research data for a use not initially contemplated would, because of the great public interest involved, be socially tolerable. But, just as certainly, it is possible to visualize situations in which it clearly would not. In the latter category, for example, obviously falls the sale of personal information to commercial organizations for subscription or mailing lists.

In determining the proper limits to be placed on the availability of research data, a workable proposition may well be to confine such data to the particular research purpose for which permission was initially obtained, or to a reasonably equivalent purpose. At the least, such a proposition might be accepted as an operative rule in the absence of persuasive considerations to the contrary. Of course, it must be recognized that as an individual may consent to an initial privacy invasion, so may he waive a limitation of that consent to the original research purpose. Care must, however, be taken in such instances not to imply a waiver in situations where it may not have been intended.

As in other affairs, there is, unquestionably a happy mean between excessive privacy and indecent exposure in behavioral research. One way to begin to establish such a mean is for the behavioral scientists themselves to demonstrate, by codes of ethics and research standards, their own acute sensitivity and concern for the problem. Psychologists have made a start on an enforceable code of ethical standards directed primarily to the client relationship.[64] Other disciplines can learn from their example and all can extend such codes more broadly to behavioral research.

VIII. An Ethical Code

From the foregoing there emerges an outline of the contest between the values of privacy and those of behavioral research. The community is sensitive to both values. Our society will support, and indeed, will insist on, a decent accommodation between them. An accommodation which takes into account the ethical and legal obligations of the investigating scientist can be achieved without diminishing the effectiveness of the scientific inquiry. Scientists who are responsive to the claim of privacy will find themselves pressed to develop better and more rational research techniques. Their innate inventiveness can be expected to yield new and better research methods.

Not only will the behavioral scientists be inventive in accommodating the competing values of privacy and research, but in doing so they will be more sensitive to the complexities and nuances involved than either courts or legislatures. To be sure, however, judges and legislators do have a supportive role and can be expected to fill it either by correcting abuses or protecting the responsible investigator who operates in accordance with the ethical consensus of the community.

The supportive measures available to the law, several of which have already been mentioned, are numerous and varied. One is the extension of a privileged status to the confidential communication of private information to a behavioral scientist. Another is the provision of civil or criminal remedies for the breach of the right of privacy.[65] A third is to assess and define the context in which, or the conditions under which, the cost in privacy is either marginal or *de minimis*, or permissible, because outweighed by the positive gains perceived for society in particular research. A fourth measure is to preclude public officials or employees from disclosing confidential information acquired in the course of employment.[66] A fifth approach is to develop "disciplinary proceedings" to enforce the claim to privacy against public officials in some form of mandamus or contempt,[67] and against private professional persons through disbarment or loss of license. Still another possible supportive legal measure is to require registration for the possession of all privacy-invading devices.[68] The alternatives are clearly varied. It should be noted, however, that the existing legislative attempts to prohibit eavesdropping by use of devices have been uniformly defective. The current statutes are either inadequate in scope or indiscriminate in application, or both.

A precondition for the development of a proper balance between the values of privacy and those of behavioral research is the growth, among behavioral scientists themselves, of a heightened sense of their own confidential professional relationship with their informants. One of the best ways of articulating and developing this heightened sense of the confidential professional relationship is through the development and observance of codes of ethics in which the claim to privacy is recognized.

Codes of ethics for the several disciplines of scholarship and research are sound and sensible, and such codes should be general rather than specific, simple rather than complex. A workable code of ethics should be subject to expansion interpretation, and application in specific cases according to the distinctive character of the research situation.

In accord with this view, seven principles are suggested for inclusion in a general code of ethics for behavioral research:

One: There should be a recognition, and an affirmation, of the claim to private personality.

Two: There should be a positive commitment to respect private personality in the conduct of research.

Three: To the fullest extent possible, without prejudicing the validity of the research, the informed, and voluntary, consent of the respondents should be obtained.

Four: If consent is impossible without invalidating the research, then before the research is undertaken, the responsible officials of the institutions financing, administering and sponsoring the research should be satisfied that the social good in the proposed research outweighs the social value of the claim to privacy under the specific conditions of the proposed invasion. These officials in turn are responsible, and must be responsive, to the views of the larger community in which science and research must work.

Five: The identification of the individual respondent should be divorced as fully and as effectively as possible from the data furnished. Anonymity of the respondent to a behavioral research study, so far as possible, should be sought actively in the design and execution of the study as a fundamental characteristic of good research.

Six: The research data should be safeguarded in every feasible and reasonable way, and the identification of individual respondents with any portion of the data should be destroyed as soon as possible, consistent with the research objectives.

Seven: The research data obtained for one purpose should not thereafter be used for another without the consent of the individual involved or a clear and responsible assessment that the public interest in the newly proposed use of the data transcends any inherent privacy transgression.

Neither these seven suggested principles, nor any other set, will resolve, nor should be expected to resolve, the productive tension between the needs and advancement of science and the vibrant diversity of human personality. If it is correct, however, that there has been a growing imbalance in the relation of science and research to the values of privacy, then either the dignity, diversity and strength of the individual in our free democratic society will be diminished, or society will correct the balance. If the balance is to be corrected—as it will and must be—the lead should be taken by the scientific community through its own codes, its own attitudes, and its own behavior.

NOTES

1. See generally Shils, *Social Inquiry and the Autonomy of the Individual* in THE HUMAN MEANING OF THE SOCIAL SCIENCES 114 (Lerner ed. 1959).

2. BOLT, A MAN FOR ALL SEASONS, ACT II, at 157 (Random House 1962).

3. Lack of trespass was cited by the Supreme Court in refusing to invalidate the use of a detectaphone on the outer wall of a hotel room, Goldman v. United States, 316 U.S. 129 (1942); see United States v. Pardo-Bolland, 348 F.2d 316 (2d Cir. 1965), *peti-*

tion for cert. filed, 34 U.S.L. Week 3081 (U.S. Sept. 2, 1965) (No. 521); in allowing the use of a concealed transmitter by a government undercover agent in a suspect's laundry, On Lee v. United States, 343 U.S. 747 (1952); and in upholding the use of a concealed recorder by a tax agent in a suspect's place of business, Lopez v. United States, 373 U.S. 427 (1963). In Silverman v. United States, 365 U.S. 505 (1961), the decision excluding evidence was based on the actual penetration of an apartment wall by a spike micro-phone which, by making contact with a heating conduit, enabled the police to overhear every word spoken within the house.

4. Pound, *Interests of Personality*, 28 Harv. L. Rev. 343 (1915).

5. *Id*. at 362.

6. To the extent that the claim to privacy has not yet been recognized or protected by law it cannot, at least in a technical sense, be called a "right."

7. By contrast with American legal development, it has been said that ". . . the trend in the foreign legislation is towards an outspoken protection of the rights of personality. We find the expression of this common concern in the Civil Code of Liechtenstein (1926), in the Italian (1942) and Greek (1946) codes, in the reformed Japanese code (1948) and the recent Egyptian and Philippine codes, and in a project of law in the German Federal Republic. Janssens, *European Law Includes Rights of Personality*, Va. L. Weekly, April 29, 1965, p. 1. See also Krause, *The Right to Privacy in Germany— Pointers for American Legislation*, 1965 Duke L.J. 481.

8. Address to the Congress of the International Association of Applied Psychology, April 10, 1958.

9. *Ibid*.

10. U.S. Const. amend. IV.

11. U.S. Const. amend. III.

12. U.S. Const. amend. V.

13. The law on this issue appears, however, to be in an active phase of transition. See *e.g.*, Judge Sobel's opinion in People v. Grossman, 45 Misc. 2d 557, 257 N.Y.S.2d 266 (1965) and Justice Brennan's dissent in Lopez v. United States, 373 U.S. 427, 446 (1963). See also the new constitutional right of privacy announced by Justice Douglas in Griswold v. Connecticut, 381 U.S. 479 (1965), and Massiah v. United States, 377 U.S. 201 (1964) (sixth amendment held to have been violated when an eavesdropping device was used to elicit information from a defendant in the absence of counsel).

14. See *e.g.* the listing in Prosser, *Privacy*, 48 Calif. L. Rev. 383, 386-89. (1960). For a better analysis, see Bloustein, *Privacy as an Aspect of Human Dignity: An Answer to Dean Prosser*, 39 N.Y.U.L. Rev. 962 (1964). See also Hamberger v. Eastman, 206 A.2d 239 (N.H. 1964); Truxes v. Kenco Enterprises, Inc., 119 N.W.2d 914 (S.D. 1963).

15. See Prosser, *supra* note 14.

16. In New York, for example, where the common law right to privacy is thought not to exist, the same result may be reached by more tortuous routes—*e.g.*, actions for libel, slander, trespass, or unfair labor practice, or the common-law remedy to safeguard mental tranquility from the intentional infliction of distress. See Battalla v. State, 10 N.Y.2d 237, 176 N.E.2d 729, 219 N.Y.S.2d 34 (1961); Scheman v. Schlein, 35 Misc. 2d

581, 231 N.Y.S.2d 548 (Sup. Ct. N.Y. Co. 1962). See also RESTATEMENT (SECOND), TORTS § 46 (1965), and especially the caveat and comment thereon. Consider also the possibility of basing civil remedies on criminal statutes such as N.Y. PEN. LAW § 738 (eavesdropping) or § 834 (holding a person up to ridicule). See RESTATEMENT (SECOND), TORTS § 286; see also Reitmaster v. Reitmaster, 162 F.2d 691 (2d Cir. 1947).

17. See COOLEY, TORTS 29 (2d ed. 1888).

18. See, e.g., Roberson v. Rochester Folding Box Co., 171 N.Y. 538, 544, 64 N.E. 442, 443 (1902).

19. Olmstead v. United States, 277 U.S. 438, 478 (1927). See also Warren & Brandeis, The Right to Privacy, 4 HARV. L. REV. 193 (1890).

20. See Shils, supra note 1, at 156.

21. On the importance of individual (and collective) secrecy in social relationships, see THE SOCIOLOGY OF GEORG SIMMEL 307-44 (Wolff ed. 1950).

23. Yet, it is to be expected that particular cultures will, from time to time, reach a consensus on definable areas that are deemed to be private. Such a consensus is likely, however, to be both temporary and limited.

23. See, e.g., PACKARD, THE NAKED SOCIETY 5 (1964).

24. (a) For example in personnel selection or retention, compare Town & Country Food Co., 39 Lab. Arb. 332 (1962), with McCain v. Sheridan, 160 Cal. App. 2d 174, 324 P.2d 923 (1958) (refusal of employees to take "lie detector" tests). Several state statutes prohibit employers from making certain uses of lie detector tests. See, e.g., ALASKA STAT. § 23:10.037 (Supp. 1965); CAL. LABOR CODE § 432.2; MASS. ANN. LAWS ch. 149, § 19B (Supp. 1963); ORE. REV. STAT. § 659.225 (1963); R.I. GEN. LAWS ANN. § 28-6.1-1 (Supp. 1964). In New York, bills to preclude the use of lie detectors as a condition of initial or continued employment are introduced in the Legislature with regularity. In the 1965 session, seven such bills were introduced, see 1965 N.Y. LEG. RECORD & INDEX 1337, and two, after reaching the Governor, were vetoed for "technical defects." See N.Y. Assembly Bill Print No. 4439, passed June 7, 1965, vetoed June 28, 1965 (1965 N.Y. LEG. RECORD & INDEX 865; N.Y. Sen. Bill Print No. 279, passed April 27, 1965, vetoed May 24, 1965 (1965 N.Y. LEG. RECORD & INDEX 29). See also 111 CONG. REC. 15378 (daily ed. July 8, 1965) (a resolution of the Communications Workers of America on invasions of privacy).

(b) For examples, in labor relations, compare Chesapeake & Potomac Tel. Co., 98 N.L.R.B. 1122 (1952) (monitoring an employee's home telephone), with Eico Inc., 44 Lab. Arb. 563 (1965) (television surveillance of production floor) and Thomas v. General Elec. Co., 207 F. Supp. 792 (W.D. Ky. 1962) (in-plant movies for time, motion and safety studies). See also N.Y. LAB. LAW § 704.

25. See McDaniel v. Atlanta Coca-Cola Bottling Co., 60 Ga. App. 92, 2 S.E.2d 810 (1939) (use of eavesdropping device to obtain evidence for defense of civil action); Schmukler v. Ohio-Bell Tel. Co., 66 Ohio L. Abs. 213, 116 N.E.2d 819 (Ohio C.P. 1953) (use of telephone monitoring to ascertain breach of contract). For the statutes of those states making at least some form of eavesdropping a crime, see note 65 infra. For a discussion of some of the ethical issues in personality testing in business, see CRONBACH, ESSENTIALS OF PSYCHOLOGICAL TESTING 459-62 (2d ed. 1960).

26. (a) For examples in the practice of law, see Matter of Wittner, 264 App. Div. 576, 35 N.Y.S.2d 773 (1st Dep't 1942), aff'd per curiam, 291 N.Y. 574, 50 N.E.2d 660 (1943) (lawyer suspended from practice for surreptitious use of recording device). The

Committee on Professional Ethics of the Association of the Bar of the City of New York has concluded that the use of recording devices by lawyers, without the consent of the person whose conversation is being recorded, violates the Canon of Ethics. See, e.g., Opinions Nos. 832, 836, 13 N.Y.C.B.A. RECORD 36, 568 (1958); No. 813, 11 N.Y.C.B.A. RECORD 207 (1956).

(b) In law enforcement: see DASH, THE EAVESDROPPERS: (1959); Symposium, 44 MINN. L. REV. 811 (1960). See also N.Y. Times, July 14, 1965, p. 1, col. 3 (use of two-way mirrors and other eavesdropping devices by Internal Revenue Service).

27. (a) In medical research: see Lewis, Restrictions on the Use of Drugs, Animals and Persons in Research (paper delivered at the Rockefeller Institute Conference on Law and the Social Role of Science, April 8, 1965).

(b) In medical practice: see Rheingold, Products Liability—The Ethical Drug Manufacturer's Liability, 18 RUTGERS L. REV. 947, 957, 1009 (1964).

28. See STAFF OF HOUSE COMM. ON GOV'T OPERATIONS USE OF POLYGRAPHS BY THE FEDERAL GOVERNMENT (Preliminary Study 1964). 88th CONG., 2D SESS. (Comm. Print 1964); House Comm. on Post Office and Civil Service, Use of Electronic Data Processing Equipment in the Federal Government, H.R. REP. No. 858, 88th Cong., 1st Sess. (1963); Hearings Before the House Comm. on Post Office and Civil Service, Confidentiality of Census Reports, 87th Cong., 2d Sess. (1962); cf. United States v. Rickenbacker, 309 F.2d 462 (2d Cir. 1962), cert. denied, 371 U.S. 962 (1963).

29. (a) In newsgathering: see the charge of Alex Rose that a New York Herald Tribune reporter had rented an adjoining hotel room to eavesdrop on a political meeting. N.Y. Times, June 20, 1965, § 1, p. 46, col. 1.

(b) In public safety: consider the number of apartments, office buildings, hospitals, laboratories, jails, and other public buildings that have electronic systems to cover entrances, elevators, reception rooms, conference rooms, corridors and tellers' windows with television cameras or sound monitoring and recording systems; also the FAA rule on the installation of voice recorders in the cockpits of large airplanes as proposed, 28 Fed. Reg. 13786 (1963). For the regulation as enacted, see 29 Fed. Reg. 19209 (1964).

(c) In education: see authorities cited in notes 31, 37 infra, for some aspects of the use of personality tests in schools; consider also the two-way communication system that enables a school principal to speak directly to a class or, at his choice, to monitor, unobserved and unannounced, the classroom proceedings.

(d) In social welfare: see Reich, Individual Rights and Social Welfare: The Emerging Legal Issues, 74 YALE L.J. 1245, 1254 (1965); Sokol, Due Process in the Protection of Adults and Children (paper presented Sept. 11, 1964, at the Northeast Regional Conference of the American Public Welfare Association).

(e) In entertainment: consider the television programs which have used hidden cameras to photograph unsuspecting subjects; see N.Y. PEN. LAW § 834 dealing with exhibitions, and particularly the prohibition of "any act . . . whereby any . . . citizen . . . is held up to contempt or ridicule.

30. See Matter of Hyman v. Jewish Chronic Disease Hosp., 15 N.Y.2d 317, 206 N.E.2d 338, 258 N.Y.S.2d 397 (1965). See also, Carley, Research and Ethics, Wall Street Journal, June 10, 1965, p. 1, col. 1; N.Y. Times March 20, 1965, p. 56, col. 1.

31. See Eron & Walder, Test Burning II, 16 AMERICAN PSYCHOLOGIST 237-44 (1961); Nettler, Test Burning in Texas, 14 AMERICAN PSYCHOLOGIST 682-83 (1959).

32. Although this article is concerned with individual privacy, the claim to institutional and collective (or group) privacy should be noted. Institutional privacy is more than the sum of the claims to privacy of the members of a particular institution. For example, even had each of the members of the jury in the University of Chicago experiment consented to the recording of the jury room proceedings, the tone of the public

response indicates that such recording would still have been viewed as tampering with a sacred institution and, therefore, offensive. See Shils, *supra* note 1, at 132-39. The individual claim to privacy is plainly paralleled by the institutional claim, and both are rooted in the need of an organism to learn and grow by quiet trial and error (sometimes called practice) without loss of dignity or public accountability, or risk of punishment. Both involve the concepts of consent and confidentiality discussed later in this article. But the conditions under which the claim may be asserted—by private institutions as well as public—and the determination of who may consent (if the judge cannot consent for the jury, can the President consent to the disclosure of his cabinet discussions?) raise the privacy issues in a different context worthy of separate analysis. The public accountability of institutions (both government and private) must be weighed and balanced with the institutional need for privacy to maintain their effectiveness and integrity. This is well appreciated by all who are responsible for the destiny of an institution and who have dealt, for example, with journalists' inquiries, congressional investigations, government questionnaires, judicial subpoenas, FBI interviews or stockholders' demands. A recent illustration of a lack of sensitivity to this claim of institutions for privacy is afforded by a bill introduced in the New York State Senate on March 9, 1965 (Senate Print 2832, Intro. 2691) which would have declared "all books . . . bills, vouchers, checks, contracts or other *papers connected with or used or filed* in the office of every authority or commission . . . or with any officer acting for or on its behalf . . . public records . . . *open to public inspection* at all times. . . ." (Emphasis added.)

33. 18 U.S.C. § 1508 (1964); see, *e.g.*, Mass. Ann. Laws ch. 272, § 99A (Supp. 1964).

34. N.Y. Pen. Law § 738. The new penal law, effective Sept. 1, 1967, replaced Section 738 with a general provision prohibiting "wiretapping or mechanical overhearing of a conversation." N.Y. Sess. Laws 1965, ch. 1030, § 250.05. The memory of the Chicago experiment lingers on. See the anti-eavesdropping bill introduced in the Minnesota Legislature on March 4, 1965, S.F. No. 915, § 2(d) (Phillips Legislative Service).

35. N.Y. Pen. Law § 740. The new penal law makes no substantial change in this provision. N.Y. Sess. Laws 1965, ch. 1030, § 250.05.

36. Lee J. Cronbach, one of the nation's outstanding authorities on psychological testing, in his book, *Essentials of Psychological Testing* (2d ed. 1960) observes:
"Any test is an invasion of privacy for the subject who does not wish to reveal himself to the psychologist. While this problem may be encountered in testing knowledge and intelligence of persons who have left school, the personality test is much more often regarded as a violation of the subject's rights. Every man has two personalities: the role he plays in his social interactions and his 'true self.' In a culture where open expression of emotion is discouraged and a taboo is placed on aggressive feelings, for example, there is certain to be some discrepancy between these two personalities. The personality test obtains its most significant information by probing deeply into feelings and attitudes which the individual normally conceals. One test purports to assess whether an adolescent boy resents authority. Another tries to determine whether a mother really loves her child. A third has a score indicating the strength of sexual needs. These, and virtually all measures of personality, seek information on areas which the subject has every reason to regard as private, in normal social intercourse. He is willing to admit the psychologist into these private areas only if he sees the relevance of the questions to the attainment of his goals in working with the psychologist. The psychologist is not 'invading privacy' where he is freely admitted and where he has a genuine need for the information obtained."
Id. at 459-60.

37. See S. Rep. No. 553, 88th Cong., 1st Sess. 41 (1963) for the legislative proposal (H.R. 4955) of Representative Ashbrook of Ohio. In New York, Assemblyman

Russo introduced a bill in 1964 (A.I. 1701) to preclude the testing of a school child without the consent of a parent or guardian.

38. In addition to the restrictions that may be imposed on the uses of science and technology, there should also be considered the prospect of legal liability for any injury that may be suffered from their use. See Rheingold, *supra* note 27; Comment, *Legal Implications of Psychological Research with Human Subjects,* 1960 DUKE L.J. 265. See also note 65 *infra* for statutes which make eavesdropping—including eavesdropping by behavioral scientists in the course of research—a crime.

39. Messick, *Personality Measurement and the Ethics of Assessment,* 20 AMERICAN PSYCHOLOGIST 136, 140 (1965).

40. See a not unrelated discussion in WEST, THE NEW MEANING OF TREASON 158-61 (1965).

41. Morison, *Foundations and Universities,* 93 DAEDALUS 1109, 1137 (1964).

42. MILL, ON LIBERTY 7-8 (Bobbs-Merrill 1956).

43. See notes 24-29 *supra* and accompanying text.

44. They may also involve the invasion of group or institutional privacy. One example is provided by research on minority groups or associations. See note 32 *supra.*

45. The Public Health Law of New York, for example, requires physicians, and others, to report communicable diseases to the local health officer (§ 2101), permits health officers to seek court orders to compel persons to be examined for venereal diseases (§ 2301), and requires vaccination of school children for smallpox (§ 2130).

46. The New York statute, for example, contains provisions designed to preserve the confidentiality of the private information obtained about the venereal diseases with which a person may be infected. See N.Y. PUB. HEALTH LAW § 2306.

47. See Gross, *Social Science Techniques: a Problem of Power and Responsibility,* 83 THE SCIENTIFIC MONTHLY 242 (1956); Mead, *The Human Study of Human Beings,* 133 SCIENCE 163 (1961).

48. The tribunal in the Nuremberg trials considered at some length the circumstances under which medical research conducted with human beings would conform to the ethics of the medical profession. It evolved ten basic principles that "all agree . . . must be observed in order to satisfy moral, ethical and legal concepts." The first of these ten Nuremberg commandments was that: "The voluntary consent of the human subject is absolutely essential." II TRIALS OF WAR CRIMINALS BEFORE THE NUREMBERG MILITARY TRIBUNALS UNDER CONTROL COUNCIL LAW NO. 10, THE MEDICAL CASE (United States v. Brandt) 181 (U.S. Gov't Printing Office 1949). See generally Lewis, *supra* note 27.

49. How many people, for example, could be expected to participate willingly in a test to devise a standard of homosexual tendencies? Or to measure intra-family hostility?

50. See CRONBACH, *op. cit. supra* note 25, at 459-62.

51. For an interesting commentary on some of the subtle ethical problems involved, see Mace, *Privacy in Danger,* 171 THE TWENTIETH CENTURY 173, 176-77 (1962). Compare State v. Kinderman, 136 N.W.2d 577 (Minn. 1965), where the court held that

an adult home owner could effectively consent to a search of his *adult* child's room not-withstanding the absence of *both* a court warrant and the consent of the adult child. This is another instance of a judicial preoccupation with the concepts of property when the claim to privacy is involved. See cases cited note 3 *supra* and accompanying text.

52. It is apparent that this view is not yet fully shared by the behavioral scientists. For example, Dr. Lee J. Cronbach, who has given thoughtful consideration to the problems of ethics in psychological testing, and who sensitively perceives the ethical issues involved in the use of psychological tests in other contexts, with respect to scientific research, has stated:

"No ethical objection can be raised to the use of subtle techniques and even of misleading instructions when the information so obtained will be used entirely for research purposes, the subject's identity being concealed in any report."

Cronbach, *op cit. supra* note 25, at 461. Even for research purposes, however, Cronbach raises a caution where the investigator occupies a position of authority over the person being tested. *Id.* at 462.

53. An excellent example of a responsible attitude toward behavioral research in schools is to be found in Kohn & Beker, *Special Methodological Considerations in Conducting Field Research in a School Setting*, 1 PSYCHOLOGY IN THE SCHOOLS 31 (1964). See also Castaneda & Fahel, *The Relationship between the Psychological Investigator and the Public Schools*, 16 AMERICAN PSYCHOLOGIST 201-03 (1961). While neither of these articles deals with the claim to privacy as such, Messrs. Kohn and Beker show a lively appreciation of it, and recognize the importance of consent, anonymity and confidentiality in, and for, behavioral research.

54. Examples of the range of protections available in the judicial process are:

(a) Court orders to protect confidential information obtained for evidentiary purposes from being improperly used for other purposes. See Covey Oil Co. v. Continental Oil Co., 340 F.2d 993 (10th Cir. 1965), *cert. denied*, 380 U.S. 964 (1965); United States v. Lever Brothers Co., 193 F. Supp. 254 (S.D.N.Y. 1961), *appeal dismissed*, 371 U.S. 207 (1962), *cert. denied*, 371 U.S. 932 (1962). See also N.Y. CPLR § 3103 (preventing the abuse of pre-trial disclosure proceedings).

(b) Statutory provisions relating to the disposition of the evidence submitted to the Tax Court, see INT. REV. CODE OF 1954, § 746; or the reception of certain evidence by the Civil Rights Commission. See Civil Rights Act of 1957, 102(g), as amended, 78 Stat. 249 (1964), 42 U.S.C. § 1975a(e) (1964).

(c) Statutory provisions for the sealing of records in judicial proceedings and limiting access thereto. See N.Y. DOM. REL. LAW §§ 114 (adoption), 235 (matrimonial actions); N.Y. FAMILY CT. ACT § 166 (privacy of records); N.Y. SOC. WELFARE LAW § § 372(4) (records as to children), 132, 136 (welfare records).

(d) Statutory provisions for the exclusion of the public from court proceedings. See N.Y. JUDICIARY LAW § 4; N.Y. FAMILY CT. ACT § 531 (paternity proceedings).

(e) Statutory provisions restricting the availability of information obtained by the Department of Justice under a Civil Investigative Demand, see Antitrust Civil Process Act § 4(c), 76 Stat. 550 (1962). 15 U.S.C. § 1313(c) (1964), or obtained by the Department of Commerce. See 13 U.S.C. § 9 (1964).

(f) Statutory prohibitions against televising or broadcasting of judicial proceedings, such as N.Y. CIV. RIGHTS LAW § 52.

55. See Sokol, *supra* note 29; see also Coser, *The Sociology of Poverty*, 13 SOCIAL PROBLEMS (Oct. 1965).

56. See note 45 *supra.*

57. It should be borne in mind that there are various degrees of anonymity in the gathering of research data, and it may be useful to distinguish between them in balan-

cing the values of particular research with the costs in privacy that may be involved. Dr. Isidor Chein, Professor of Psychology at New York University's Graduate School of Arts and Science, in a letter to the authors making this point, identified, among the possible levels of anonymity, the following six:

(a) the particular subject is never identifiable, not even by the investigator or his agents; (b) the particular subject is temporarily identifiable, but his identity is never ascertained up to and including the point at which the data that he has provided are consolidated in some meaningful and interpretable form; (c) the particular subject is temporarily identifiable and his identity is known up to, but not including, the point at which the data that he has provided are consolidated in some meaningful and interpretable form; (d) the particular subject is temporarily identifiable and can be associated with data that are in themselves meaningful and interpretable, but his identity is not ascertained; (e) the identity of the particular subject is known in conjunction with meaningful and interpretable data, but his identifiability and identity are submerged in the treatment of the data from many subjects and his own data are never scrutinized from the point of view of interpreting or drawing any inferences about him or his behavior; and (f) the identity of the particular subject is known in conjunction with meaningful and interpretable data and these data are scrutinized from the point of view of interpreting some aspect of the individual or his behavior, but his identity is thereafter submerged in the collection of similar processes of interpretation for many subjects.

58. See, e.g., Johnson, *Retain the Original Data!,* 19 AMERICAN PSYCHOLOGIST 350-51 (1964). See also de Mille, *Central Data Storage,* 19 AMERICAN PSYCHOLOGIST 772-73 (1964). The prospect of the use of computers for central recording, storage and retrieval of research data in the behavioral sciences adds a troublesome new dimension to the protection of privacy. Computerized central storage of information would remove what surely has been one of the strongest allies of the claim to privacy—the inefficiency of man and the fallibility of his memory.

59. The eighteen states are: Alabama, ALA. CODE tit. 46, § 297(36) (Supp. 1963); Arkansas, ARK. STAT. ANN. § 72-1516 (1957); California, CAL. BUS. & PROF. CODE § 2904; Colorado, COLO. REV. STAT. ANN. § 154-1-7(8) (1963); Delaware, DEL. CODE ANN. tit. 24, §3534 (Supp. 1964); Georgia, GA. CODE ANN. § 84-3118 (1955); Idaho, IDAHO CODE ANN. § 54-2314 (Supp. 1965); Illinois, ILL. ANN. STAT. ch. 91½, § 406 (Smith-Hurd Supp. 1964); Kentucky, KY. REV. STAT ANN. § 319.111 (Supp. 1965); Michigan, MICH. COMP. LAWS § 338.1018 (Supp. 1961); Nevada, NEV. REV. STAT. § 48.085 (1963); NEW HAMPSHIRE, N.H. REV. STAT. ANN. § 330-A:19 (Supp. 1963); New Mexico, N.M. STAT. ANN. § 67-30-17 (Supp. 1965); New York, N.Y. EDUC. LAW § 7611; Oregon, ORE. REV. STAT. § 44.040 (1963); Tennessee, TENN. CODE ANN. §63-1117 (1955); Utah, UTAH CODE ANN. § 58-25-9 (1963); Washington, WASH. REV. CODE § 18.83.110 (1957).

60. A Montana statute does, however, seem to extend a limited privilege to certain types of behavioral research if conducted by a person teaching psychology in a school. The Montana statute reads as follows:

"Any person engaged in teaching psychology in any school, or who acting as such is engaged in the study and observation of child mentality, shall not without the consent of the parent or guardian of such child being so taught or observed testify in any civil action as to any information so obtained."

MONT. REV. CODES ANN. § 93-701-4(6) (1964).

61. See generally 8 WIGMORE, EVIDENCE §§ 2332-41 (McNaughten rev. 1961).

62. See, e.g., Hurlburt v. Hurlburt, 128 N.Y. 420, 424, 28 N.E. 651, 652 (1891) (dictum). See also Louisell, *Confidentiality, Conformity and Confusion: Privileges in Federal Court Today,* 31 TUL. L. REV. 101 (1956). See generally 8 WIGMORE, *op. cit. supra* note 61, §§ 2290-2329. It is unlikely that testimonial privilege will be judicially extended to situations that do not fully satisfy Dean Wigmore's four conditions for the existence of a privilege: (1) the privileged communication must originate in a confidence

that it will not be disclosed, (2) the element of confidentiality must be essential to the relationship of the parties to the communication, (3) the relationship is one which is to be assiduously fostered, and (4) the injury that would inure to the relationship by disclosure of the communication must be greater than the benefit to be gained from its contribution to the disposition of the litigation. *Id.* § 2285.

63. This, nevertheless, seems to be the situation in those eighteen states which accord the privilege only to licensed or registered psychologists. See Geiser & Rheingold, *Psychology and the Legal Process: Testimonial Privileged Communications,* 19 AMERICAN PSYCHOLOGIST 831 (1964).

64. See *Ethical Standards of Psychologists,* 18 AMERICAN PSYCHOLOGIST 56 (1963).

65. Remedies for the breach of this right are already availably in many states:
(a) See the list of states which recognize a common-law right of privacy in Prosser, *supra* note 14, at 386-89.
(b) Oregon and Maryland have statutes which make eavesdropping, without the consent of all persons being overheard, a crime. Neither accords any exemption for behavioral research. Thus, in Oregon, it is unlawful to obtain any part of a conversation by an eavesdropping device "if all participants in the conversation are not specifically informed that their conversation is being obtained." ORE. REV. STAT. ¶ 165.540(1) (c) (1963). Violation of this Oregon statute is punishable by fine or imprisonment and renders the violator liable for damages in a civil suit. ORE. REV. STAT. ¶¶ 30.780, 165.540(6) (1963). In Maryland it is unlawful to use any device "to overhear or record any part of the conversation or words spoken to or by any person in private conversation without the knowledge or consent, expressed or implied, of that other person." MD. ANN. CODE art. 27, ¶ 125A(a) (Supp. 1964).
(c) See the statutes in five other states which make eavesdropping unlawful without the consent of a party to the conversations—again without an exemption for scientific research: CAL PEN. CODE § 653j; ILL. ANN. STAT. ch. 38, §§ 14-2, 14-4 (Smith-Hurd (1964); MASS. GEN. LAWS ANN. ch. 272, § 99 (Supp. 1964); NEV. REV. STAT. § 200.650 (1957); N.Y. PEN LAW § 738
(d) See also the comparable but more limited statutes in six other states: ARK. STAT. ANN. § 41-1426 (1964) (loitering for purposes of invading privacy); GA. CODE ANN. § 26-2001 (1953) (peeping or similar acts tending to invade privacy); N.D. CENT. CODE § 12-42-05 (Supp. 1965) (using any mechanical or electronic device to overhear or record and to repeat with intent to vex or injure); OKLA. STAT. tit. 21, § 1202 (1941) (loitering with intent to overhear and repeat to vex or injure); S.C. CODE ANN. § 16-554 (1962) (peeping or similar acts tending to invade privacy); S.D. CODE, § 13.1425 (1939) (loitering with intent to overhear and repeat to vex or injure).
(e) See RESTATEMENT (SECOND), TORTS § 286 (1965), which reflects the judicial acceptance of such statutory standards as a basis for civil liability.

66. See, *e.g.,* Antitrust Civil Process Act § 4(c), 76 Stat. 550 (1962), 15 U.S.C. § 1313(c) (1964); N.Y. EDUC. LAW § 1007; N.Y. LAB. LAW § 537; N.Y. PEN. LAW § 762; N.Y. PUB. OFFICERS LAW § 74(b).

67. The Swedish Ombudsman suggests another interesting possibility. See *A State Statute to Create the Office of Ombudsman,* 2 HARV. J. LEGIS. 213 (1965).

68. Maryland, by House Bill 1197, approved by the Governor on April 8, 1965, added a new § 125D to Article 27 of its Annotated Code and thereby became the first state to require "every person possessing any eavesdropping and/or wiretapping device" to register such device with the State Police. Unless registered it is unlawful to manufacture or possess any such device. It will be interesting to see how vigorously and effectively this new statute is enforced. Will it be applied, for example, as it would seem was intended, to the manufacturers of tape recorders or dictaphones? Or to the lawyers or scientists who use them?

32 Behavior Control and the Limits of Reform

GERALD L. KLERMAN

The treatment and rehabilitation of the mentally ill and prisoners raise a number of professional, social, and ethical problems. Klerman provides a review of three strategies for dealing with such individuals: the total institution as a therapeutic community; decarceration and deinstitutionalization; the use of behavior control technologies. He then provides a thematic discussion of a variety of social and ethical dilemmas that arise when such strategies are used either singly or in concert.

Gerald Klerman, M.D., is Deputy Chief of Psychiatry at Massachusetts General Hospital

In what has subsequently come to be recognized as the classic description of "total institutions," Erving Goffman in 1955 identified a number of similarities among prisons, mental hospitals, monasteries, military garrisons, tuberculosis sanitariums, and other institutions.[1] In total institutions, two distinct groups, usually labeled as staff and inmates, live together in a setting physically isolated and socially demarcated from the larger society. The inmates or patients have all their needs supplied around the clock by the staff, who may or may not live in the institution or on its grounds; but the staff's lives, almost as much as their clients', are determined by the quality of chronicity and de-individualization that characterizes the total institution. Although the two groups live in close physical proximity, they are separated by rigid social boundaries that approach those of a caste system.

Goffman's concept of a total institution was offered as a sociological analysis of a unique type of social organization, but there is little doubt that he was, in addition, proposing a profound moral critique of these institutions. He described a number of processes be-

601

tween staff and inmates (privilege systems, social barriers, "mortification" of the person) by which the institution subverted its own long-term goals. He concluded with a pessimistic verdict that, whatever may be the publicly mandated purposes of these institutions and the private intentions of their staffs, professional and nonprofessional, the results in terms of the lives of the patients and inmates were loss of individuality, dehumanization, and depersonalization.

Current social theorists and policy analysts would recast Goffman's insights by invoking the concept of "unintended consequences." Just as the pharmacologist has learned to regard drugs as having side effects and complications in addition to their therapeutic benefits, social planners and administrators have come to realize that reforms and innovations carry with them unintended consequences of a social nature, effects often contrary to the stated purpose of the institutional goals. In the case of mental hospitals, the original intent was to provide for asylum—a protective setting for the mentally ill that would offer treatment and humane incarceration. In the case of modern prisons, incarceration, while providing physical restriction and punishment, was also justified as a preliminary step toward rehabilitation. But researches by Goffman, David Rothman, Gerald Grob, Barbara Rosenkrantz, and other social historians have demonstrated that whatever may have been the intent of the humanitarian reformers who founded these institutions, the eventual consequence until very recent years has been to create human warehouses that produce humiliation, degradation, and rebellion rather than treatment and rehabilitation.[2]

In the 1940s and 1950s, crusading journalists, humanitarian professionals and social critics focused attention on scandalous conditions in the mental hospitals, the "snake pits" of that era. Today, similar groups are converging on the prisons and mental hospitals, bringing public awareness and social change. Professionals and policy makers in both mental health and criminology are now questioning many basic assumptions held indisputable for over a century. Throughout the society, law and order, deinstitutionalization, and decarceration have become matters of public concern and national policy.

There is an interesting reciprocal relationship between mental hospitals and prisons as regards innovation and reform. In the late 1950s and early 1960s, the prisons benefited from the new therapeutic ethic as it developed in the mental hospitals. Today, mental patients benefit from legal decisions establishing prisoners' rights. Debate over the proper response of society to deviant behaviors has moved to the courtroom, where the new "public interest" lawyers are bringing actions to bear on the use of behavior modification techniques, psychosurgery, involuntary treatment, and other behavior control methodologies.

New technologies, derived from the biological and behavioral sciences, are being studied for their applicability in modifying the deviant behavior of patients and prisoners. Ironically, the immediate successes of these techniques have become the source of controversy. The controversies are best understood in the larger context of social dilemmas common to both prisons and mental hospitals. Three dilemmas they share are the following:

1. To what extent does the internal social structure of total institutions inevitably doom efforts at treatment and rehabilitation? Goffman and other sociologists describe the forces which militate against the success of rehabilitation and therapy efforts. Can these forces be countered by such reforms as the therapeutic community structures devised in the mental hospitals of the 1950s and now being carried over to the prisons of the 1970s?

2. How can these institutions simultaneously meet the demands of the society and the needs of the individuals they serve? Society wants the incarceration of individuals whom it regards as not only deviant but potentially dangerous either in themselves or to others. The same society has promised to promote changes beneficial to the individual by rehabilitation or treatment. Are these two goals compatible?

3. What sort of control should be exercised over the development and application of behavior control technology? Prisons and mental hospitals can be expected to make increasing use of new physiological and psychological techniques in pursuit of their stated goals of immediate behavior control and long-term behavior change. Incarceration, of course, brings immediate behavior control by restricting the movement of the deviant individual. The promise of recent technology is that newly learned personality, skills, and attitudes will endure; that when the individual returns to the larger society, his or her behavior will in fact be changed, not simply controlled. But this rehabilitative hope is troubled by reservations about the manipulative capacities of the new techniques. The current impetus for regulation has come less from within the professions than from groups concerned to bring principles of civil rights and personal liberties to bear on the inner functioning of total institutions.

Thus, over the past two decades, the recognition that both kinds of total institutions were failing the goals society set for them—indeed, that their intrinsic structures tended to undermine their avowed purposes—spurred three kinds of change: first, internal reorganization along lines of greater personalization and the "therapeutic community"; second, advocacy and some measures toward deinstitutionalization; and third, the introduction of new techniques for behavior control, capable both of enhancing individual change and of permitting release from closed institutions. Reforms of the first two sorts are made more possible by the last, but paradoxically, behavior control technology raises its own new problems. Without social regulation, the threats it poses may be unacceptable to society.

I. The Liberal Reform: The Total Institution Becomes a Therapeutic Community

Goffman's work synthesized a large body of research on the social and psychological characteristics of mental hospitals that emerged after World War II. A number of studies had described the custodial, authoritarian, and bureaucratic nature of the social structures of the mental hospitals which, it appeared, operated to undermine—if not completely negate—the publicly mandated therapeutic goal and the conscious intent of their leadership and staff.[3] Stimulated by these investigations and by public concern, major efforts at therapeutic reform were initiated in the mental hospital field during the 1950s and 1960s. Within mental hospitals in the United States and Great Britain, progressive superintendents and their staffs opened doors and eliminated restraints; they took steps to decrease the social distance between patients and staff, to facilitate communication among staff groups, and to upgrade employee morale and training. These reforms soon led to the development of alternatives to hospitalization such as halfway houses and crisis intervention centers, the forerunners of today's community mental health and community corrections innovations.

The reforms were crystallized in the ideal of the "therapeutic community" enunciated by Maxwell Jones.[4] Jones united the spirit of egalitarian democracy with the techniques of group dynamics to level unequal status, to facilitate communication, and to share decision making throughout the "total institution." His term, the "therapeutic community," became the rallying slogan of the mental hospital reform movements of the 1950s and 1960s. Initially developed in Great Britain for the treatment of individuals with personality disorders, the "therapeutic community" was rapidly extended to in-patient settings for acute psychotics and later to the treatment of adolescent drug addicts in residential settings. Therapeutic community approaches are now being proposed and evaluated for juvenile delinquents, criminal offenders, parolees, and other deviant groups.

Implicit in the idea of therapeutic community is a belief in the human potential for change. This potential is to be sought not only among the professional and non-professional staff, but among the patients, inmates, and clients themselves. Stimulated by these ideas, many ex-drug addicts and ex-alcoholics have become effective counselors for peers with problems where professionals have often been unsuccessful. Similar efforts are now under way in the corrections field.

The "therapeutic community" movement within mental health institutions has parallels in the dramatic innovations in out-patient psychotherapy during recent decades. There are many similarities between the group techniques employed by the therapeutic community and the expressive techniques of encounter groups, Gestalt therapy, primal scream, and other new forms of psychotherapy. Applied to institutional settings, these techniques have had early success in psychiatric units with large numbers of adolescents or young adults. Soon they were also applied in self-help residential treatment settings for drug addicts, such as Synanon, Daytop, and Phoenix House. Now they are being tried in peer programs for offenders and delinquents.

Disturbance and Change

Turbulence, and in some cases even violence, accompanied these reform movements. In some prisons, riots and collective disturbances have reached public awareness through dramatic incidents like Attica. Less well-known to the public, however, are the numerous disruptions and "collective disturbances" which occurred in many mental hospitals during the 1950s. Some of these events were described in the professional literature, but most never reached the attention of the public. The inmates' reaction to these collective disturbances illustrates an important difference between the prisons and the mental hospitals. The oppressive, punitive features of the prison, in combination with the characteristics of the prison population, have generated the particular inmate subculture with its own values, its high degree of organization, and its effective group resistance to established authority. Within mental hospitals, however, comparable cohesion and action have been fairly uncommon; they appear more frequently among adolescents and young adults with personality disorders, than among the psychotic or neurotic residents. Patients suffering from depression, organic brain disease, and schizophrenia are apparently less capable of forming social groups for collective action. For the mentally ill, apathy, dependency, depersonalization, and social isolation continue. Some observers have felt that inmate agitation and staff conflicts in such settings produced the unintended consequence of disturbance and withdrawal in these patients, but it is probably unjust to impute long-term effects of that kind to the rebel-lions when such states are already heavily determined by individual pathology and the nature of the institution.

II. THE RADICAL CRITIQUE: DECARCERATION AND DEINSTITUTIONALIZATION

Attempts at internal reform, even when successful as with the therapeutic community movement within the mental hospital, have only partially silenced the critics and skeptics. In the current demands for redress of social grievances, professionals like Thomas Szasz and R. D. Laing, jurists like David Bazelon, and public interest lawyers like Charles Halpern have joined with spokespersons for the New Left, black militants, and even some right-wing conservatives in criti-cizing prisons and mental hospitals as being both ineffective and un-just.[5] Their remedy is radical: deinstitutionalization and the dismantling of these institutions. Decarceration is the battle cry of the new abolitionists.

These critics argue that the ineffectiveness of the total institu-tions results only in part from their internal social structure and its effect on otherwise noble attempts at rehabilitation and treatment. According to these critics, such efforts as building a therapeutic com-munity within the institution are only partial solutions. The feature of total institutions that most inescapably generates their failure, they assert, is their relationship to the larger society: that is, their status as "double agents." Modern total institutions claim to be committed both to social control and to personal change; yet whenever conflicts arise between the two goals, the social-control mandate takes priority over rehabilitation. Why? Because, they claim, ultimately the prison or public mental hospital is under the legal aegis, administrative con-trol, and fiscal dominion of the larger society—the legislatures, com-missions, and agencies whose highest priority is controlling deviance rather than meeting the needs of the individual client.

The reformers propose the abolition of total institutions and their replacement by alternative forms of voluntary treatment for the mentally ill and community corrections for offenders.[6]

Into the Community

The community mental health movement gained public recogni-tion in the 1960s with President Kennedy's message to Congress. The movement for community corrections facilities is only now gaining public attention. The movements share many features. Each arises from an awareness of the evils of institutionalization and the complex forces, intended or otherwise, operating within large institutions. Day centers, half-way houses, community residences, and rehabilitation programs first used in the mental health field are now being created in the corrections field, and both fields can anticipate new problems as they succeed in changing public policy towards support of community programs.

Within the mental health field, these reforms have had significant success. In contrast to the prisons whose populations are growing and where internal discontent is great, the mental health field has seen a reduction in resident population and a radical restructuring of its delivery system. The relative place of the public mental hospital in the overall delivery system has changed. The majority of mentally ill patients are now being treated as out-patients. In 1950, eighty percent of all mental health episodes were treated in public hospitals; today, the public hospital provides the minority of treatment. For in-patient treatment, psychiatric units in general hospitals and private hospitals, rather than public mental health institutions, are providing the settings for the majority of admissions.[7]

The mental health system, then, has experienced a major shift away from almost total reliance upon public institutions employing involuntary incarceration and treatment, towards a voluntaristic and pluralistic system. The system is voluntaristic in that increasing numbers of patients have the choice of going for treatment; it becomes pluralistic as patients have more options as to the type of treatment that they will receive.

No longer is the mental health system a monopoly owned and operated by state governments as is still the case with correctional institutions. In the health field, the availability of increased numbers of individual practitioners, non-state hospitals, and voluntary agencies has greatly diversified the treatment alternatives. The availability of a voucher system, in the form of health insurance, gives the individual patient and his or her family increasing degrees of freedom. Should we design a similar voucher system for criminals or delinquents whereby they could choose the type of setting to which they would go? Since the eighteenth century, we have assumed that the state will provide the resources for correction. This was the case for the mentally ill, but is no longer.

Efforts are under way to create after-care facilities run by community groups for youthful offenders and adult parolees. Treatment and rehabilitation programs for alcoholics and drug addicts are often operated by non-profit corporations supported by government through grants or contracts. Soon some state will extend contracts or grants to residential treatment for adult offenders.

Prospects for Acceptability

However, we can anticipate resistance to the establishment of community corrections facilities, resistance similar to that encountered by the community mental health movement. The mental health field is now experiencing a backlash as legislatures, local communities, and neighborhoods fight against the placement of halfway houses, community residences, and day programs. Since the apprehension aroused by mental illness is often less great than that aroused by delinquency and crime, the formation of halfway houses and similar centers for parolees, criminal offenders, or juvenile delinquents is likely to meet greater resistance than those for mental patients, the retarded, or even drug addicts.

Another problem for deinstitutionalization has been the emergence of a new profile of chronicity in the community. The hope, verging perhaps on magical thinking, that simply returning ambulatory mental patients to the community would give them access to vitalizing patterns of work, recreation, and neighborhood has for the most part not been borne out. What we see instead are pockets of slightly mitigated pathology, great numbers of people living ineffectual and isolated lives in neighborhoods that neither integrate nor eject them, outside any systems of care beyond the periodic issuance of their medication and occasional follow-up visits. It is not clear what the equivalent forms of chronicity might be for criminal offenders released into community care. Failure to elaborate social programs to receive the deinstitutionalized could produce more diffuse but as difficult problems as the institutions themselves.

These reflections on backlash and chronicity, as problems for community mental health and community corrections, are intended to place current controversies in some perspective. We do not know how to gauge the capacity or willingness of the larger society or of individual communities and neighborhoods to tolerate, accept, and integrate deviant behavior. In practice, the limits of community acceptance are determined by trial and error, with periods of reform followed by reaction and retrenchment, usually a return to some form of institutionalization.

However, an important ingredient of the current scene is the emergence of new technologies. In the mental health field it is generally, if not universally, agreed that the present decreases in patient populations in public mental hospitals and the recent growth of community programs are dependent upon the efficacy of the psychotropic drugs introduced in the 1950s.[8] New technologies, whether biological or psychological, offer promise of changing the delicate balance between reform and reaction that has characterized the institutionalized "control" of two major classes of deviant behavior, mental illness, and criminal offenses. Yet as these new technologies develop, they have generated new controversies.

III. Behavior Control Technology

The introduction of new technology has generated a quantum change in the current situation. Since World War II, behavior control technologies have become more sophisticated, more effective, and more efficient. Even during the nineteenth century, mental hospitals had always used sedative and narcotic drugs. Electroshock therapy and psychosurgery were widely used in the 1940s and 1950s but with limited effect. Then in the 1950s, the new psychoactive drugs were developed; they were many in number, more diverse in their effects, and far more efficacious. Whereas psychosurgery and shock treatments are in many ways anathema to large numbers of professionals as well as to the public, pills are "easier to swallow." Consequently, the potential ethical issues raised by their efficacy have been slower to emerge.

New behavior modification and other psychological techniques raise many of the same issues. The use of token economies, graded privileges, aversive stimuli, and other methods of behavior modification have raised basic ethical and legal concerns about the rights of inmates and the powers of staff and patients.

Is Behavior Control Cruel and Unjust Punishment?

With both psychosurgery and behavior modification, issues of cruel and unjust punishment come to the forefront. To what extent do the techniques employed, either behavioral or biological, represent a violation of the constitutional protection against cruel and unjust punishment? In the case of therapeutic institutions like Patuxent in Maryland or the Operation Start program at the Federal prison in Springfield, Missouri, the lowest level of reward involves conditions close to classic solitary confinement. This confinement has its counterpart in mental hospitals in the form of seclusion and restraints. In both instances, the physical movement of the client is restricted and access to the amenities of civilized existence, that is, toilet facilities, food, washing, privacy, adequate temperature, clothing, is limited. The courts are now challenging these procedures and the internal practices of total institutions are coming under greater scrutiny.

Rights or Privileges

Related to the issue of whether certain practices constitute cruel and unjust punishment is a major legal and ethical issue. What are the rights and privileges of individuals within such an institution? As Benjamin Bursten has pointed out in a recent discussion of decision making in psychiatric therapeutic communities, there is often an unrecognized difference of opinion between the staff and patients as to what constitute rights and what constitute privileges.[9] The staff tend to define the normal amenities and freedoms of life "outside" as privileges rather than as rights. Once they are defined as privileges, access to personal clothing, telephone, TV, mail, and visitors must be earned by "appropriate" behavior. Declaring these activities privileges makes them powerful levers for behavior control. If free access to these activities is contingent upon behaviors specified by the institution, then the behaviors of the residents are readily molded and controlled. Within the paradigm of behavior modification, what are regarded as the freedoms of everyday life outside become the positive reinforcers for life inside the total institutions.

If, however, such activities are regarded as rights, then they are subject to the inmate's wishes, independent of whether or not his behavior conforms to the rules of the institution. Where to draw the line between rights and privileges is a constant source of conflict within the staff, between staff and patients, and between the institutions and outside agencies. In a number of states, patients' bills of rights have been promulgated by law or administrative decree, and similar efforts are now underway in prison reform.

Let us assume that the courts will soon decide that all inmates in total institutions—patients or prisoners—and participants in community programs—probationers, parolees, or discharged mental patients—have specified rights. Let us further assume that these rights will be defined by the courts, by legislation, and by administrative regulations, and that a minimum number of rights will be established to which clients will be entitled without the requirement to maintain behavioral conformity. These rights include certain freedoms—access to mail, visitors, clothing, sanitary facilities, food of good quality, and so on. This development will eliminate many of the ethical and moral concerns about behavior control.

Programs such as those at Patuxent, Maryland and Operation Start may have made blunders in using the access to activities and personal prerogatives as their minimal level of reward. The courts will probably place a "floor" under reward contingencies. This restriction will not, however, limit the potential power of the new technology to alter behavior. We will have to come to grips with other serious issues of the resistance to acceptance of these new technologies. These include the adequacy of the social system for their development and delivery, and the sanctity of the soul.

The Body and the Sanctity of the Soul

The issue of sanctity of the soul may at first reading appear irrelevant in a discussion of scientific behavior control; yet discussions of drugs, psychosurgery, or conditioning almost always elicit a revulsion against proposals for biological or pharmacological treatments for behavioral deviance. In liberal circles the implicit assumption seems to be that even incarceration for life is more "acceptable" than the implantation of brain electrodes or the use of psychotropic drugs. Why this reservation? I suspect it is because we have some feeling that deep within the psyche there is some essence of humanity or individuality, traditionally called the soul, which in its biological integrity can somehow resist the indignity, pain, or restricted movement of the incarceration; but that it is more vulnerable to chemical or electrophysiological tampering, or to the covert manipulations of the Skinnerian reinforcement contingencies. Even with torture, brutality, verbal and physical indignities, incarceration is believed to be only partially successful at injuring this core. Yet, despite a moving literature of heroic individuals resisting the effects of imprisonment, we do not know how representative the sampling is; we cannot compare it with the unwritten memoirs of the many thousands of prisoners or inmates who have succumbed to the behavior control techniques of past and present incarcerations.

Some such assumption has been with Western society for many centuries. Perhaps its modern implications need to be reviewed. Until there is articulate discussion of this powerful sense of resistance to science-based manipulations, there will be a strong bias against the new biological and behavioral technologies.

Needed: An Alternative to the Medical Model

Behavior control technology is a frightening term because there is little precedent for its guidance and regulation. It has been easier to label certain deviant behaviors as "sick" and to invoke the techniques, protections, and institutions of the medical model. The health system does contain some established procedures for the development, delivery, and monitoring of effective medical technologies. Over the centuries, medicine has developed professional codes and practices for the selection, training, and review of its practitioners, and for some regulation of new technology through control of prescriptions and the activities of agencies like the FDA.

No comparable system has yet emerged for regulating behavioral modification or other technologies for "non-sick" forms of deviance, such as delinquency. Within the correction system, we lack accepted professional organization and mechanisms for researching, developing, and testing technologies, be they biological, social, behavioral or psychological. Nor does there exist an organized group of practitioners, within or without the institutions, schooled in the development and applications of socially sanctioned behavioral technologies and mandated to perform these functions. It has been convenient to extend the medical model and thus to assume that the safeguards of medicine apply. To a great extent, the medical model has been successfully used with mental illness, and it is now being extended to alcoholism and some forms of drug addiction, but there are signs that its limits are being reached.

On occasion, I have imagined myself as a warden of a correctional institution and wondered under what conditions I could feel comfortable in seeking to utilize these new techniques for behavior control. After all, the stated mandate of a prison is to control the behavior of individuals whom the rest of society has regarded as deviant—at least through temporary incarceration, and usually in the hope that the facility will mobilize other rehabilitative forces to produce enduring change. Because this stated intent has generally not been achieved with existing methods, one would expect support for an intensified search for new technologies. Yet, when these technologies have been applied, they have not been universally hailed, but rather have become subjects of controversy. The fear is not that the technologies will be ineffective, but that they will be too effective; that they will restrict the individuality and political freedom not only of the inmates in publicly created institutions, but also of the citizens outside who helped to create them.

IV. Conclusion: Beyond the Total Institution

It is probable that we have come to the end of an era in our society's attempt to deal with deviance. For over one hundred years, we have been guided in our control efforts by the utopia of the "good" total institution. We have separated the bad from the mad. We have held the "bad" responsible for their actions and, therefore, subject to judgment by the criminal justice system. The mad we have said need treatment; we have expected them to be unresponsive to the usual mechanisms of social control, particularly punishment. For both categories of deviance, the bad and the mad, we have attempted to create institutions that would simultaneously provide incarceration and treatment. We have wanted to satisfy the needs of both constituencies—those of the society at large that has demanded retribution and restriction of the physical movement of deviants, and those of the deviant for rehabilitation.

At issue in the theory of total institutions is a concept of individuality. Goffman seems to be saying that if we live in a social setting where all our social roles—working, learning, playing, sleeping, eating—are conducted with the same persons and in the same physical locale, we will easily lose our individuality and become less than truly human. Although unstated, one corollary of these observations on the total institutions is the notion that opportunities for individual freedom are greatest in a social system where we can choose which of our multiple roles are played with which individuals and in what locales. Thus, the large modern urban center with its opportunities for multiple personal groupings and freedom of movement is potentially far more individualizing than the small rural village or the commune. However, in the modern city, we also meet counterforces which suggest new limits to individuality; ironically, the price that individuals often pay for this potential freedom of choice is actual social isolation, personal loneliness, and the *anomie* that characterize so much of modern urban life, especially for the poor, the marginal, and the mentally ill. Thus the urban community may generate exactly the same psychological consequences that are likely to occur in a total institution. One of the unintended consequences of decarceration and deinstitutionalization may be new forms of *anomie* and isolation.

The hope is that the new technologies will allow for the restraint of deviant behavior and the return of individuals to the community. Without new technologies, the total institutions probably cannot be dismantled. Ironically, at a time when the technical potential for behavior control treatment is at its greatest, civil libertarians and other critics are more and more concerned about its ethical aspects. Until these issues of rights, due process, and appropriate mandates are settled, the full potential of behavior control technology for enhancing individual freedom and social cohesion cannot begin to be realized.

The ultimate fear is that unless they are regulated, these technologies together with the *anomie* and isolation of urban life could convert the community into the ultimate total institution, a totalitarian society. Hence the visions—or nightmares—of *1984* and *A Clockwork Orange*. The ultimate dilemma is that without these new technologies, long-term changes in total institutions are unlikely; and even community alternatives are probably dependent on new technologies. Thus the issues of behavior control are not only ethical in the usual sense of governing personal relationships—here, between professional practitioner and client; they are also social, political, and ethical, in the largest sense of those terms.

REFERENCES

1. Erving Goffman, *Asylums* (New York: Doubleday, 1961), pp. 43–85.

2. David J. Rothman, *The Discovery of the Asylum: Social Order and Disorder in the New Republic* (Boston: Little, Brown and Co., 1971); Gerald N. Grob, *Mental Institutions in Amer-*

ica: Social Policy to 1875 (New York: The Free Press, 1973); and Barbara Gutmann Rosenkrantz, *Public Health and the State: Changing Views in Massachusetts* (Cambridge: Harvard University Press, 1972).

3. Alfred H. Stanton and Morris S. Schwartz, *The Mental Hospital* (New York: Basic Books, 1954); William A. Caudill, *A Psychiatric Hospital as a Small Society* (Cambridge: Harvard University Press, 1958); Milton Greenblatt, Daniel J. Levinson, and Gerald L. Klerman, eds., *Mental Patients in Transition* (Springfield, Ill.; Charles C. Thomas, Inc., 1961); and Milton Greenblatt, Daniel J. Levinson, and Richard Williams, eds., *The Patient and the Mental Hospital* (Glencoe, Ill.: The Free Press, 1957).

4. Maxwell Jones, *The Therapeutic Community* (New York: Basic Books, 1953).

5. Thomas S. Szasz, *The Myth of Mental Illness: Foundations of a Theory of Personal Conduct* (New York: Harper and Row, 1974); Robert Boyers and Robert Orrill, *R. D. Laing and Anti-Psychiatry* (New York: Harper and Row, 1971). For a description of this critique, see Gerald L. Klerman, "Public Trust and Professional Confidence," *Smith College Studies in Social Work* 17 (2:1972), 115–24.

6. David J. Rothman, "Decarcerating Prisoners and Patients," *Civil Liberties Review* 1 (1973), 8–30.

7. Leon Eisenberg, "Psychiatric Intervention," *Scientific American* 229 (1973), 116–27.

8. Gerald L. Klerman, "Psychotropic Drugs as Therapeutic Agents," *Hastings Center Studies* 2 (1974), 81–93.

9. Benjamin Bursten, "Decision Making in the Hospital Community." *Archives of General Psychiatry* 29 (1973), 732–35.

33 The Use of Electronics in the Observation and Control of Human Behavior and Its Possible Use in Rehabilitation and Parole

BARTON L. INGRAHAM
GERALD W. SMITH

This article discusses a possible use of a telemetric device united with a computer system to maintain observation and control of human behavior. Several reasons for using this type of advanced electronic monitoring system, and its applications for criminology, are discussed. The authors specifically focus on a variety of ethical issues that come into sharp focus as a result of this type of technological development and application. The questions raised are important and significant and they make us evaluate carefully how much control we actually wish to have over other human beings.

Barton L. Ingraham is a professor at the Institute of Criminal Justice and Criminology, at the University of Maryland.

Gerald W. Smith is presently a professor in the Department of Sociology at the University of Utah.

In the very near future, a computer technology will make possible alternatives to imprisonment. The development of systems for telemetering information from sensors implanted in or on the body will soon make possible the observation and control of human behavior without actual physical contact. Through such telemetric devices, it will be possible to maintain twenty-four-hour-a-day surveillance over the subject and to intervene electronically or physically to influence and control selected behavior. It will thus be possible to exercise control over human behavior and from a distance without physical contact. The possible implications for criminology and corrections of such telemetric systems is tremendously significant.

The purpose of this paper is: (1) to describe developments during the last decade in the field of telemetry and electrophysiology as they relate to the control of human behavior; (2) to dispel, if possible, some of the exaggerated notions prevalent amongst legal and philosophical Cassandras as to the extent of the power and range of these techniques in controlling human behavior and thought; (3) to discuss some applications of these techniques to problem areas in penology and to show how they can make a useful contribution, with a net gain, to the values of individual freedom and privacy; and (4) to examine critically "ethical reservations" which might impede both valuable research in these areas and the application of their results to solving the problem of crime control.

I. Electronic Techniques for Observing and Controlling Behavior in Humans

A telemetric system consists of small electronic devices attached to a subject that transmit via radio waves information regarding the location and physiological state of the wearer. A telemetry system provides a method whereby phenomena may be measured or controlled at a distance from where they occur—i.e., remotely (Grisamore, 1965). The great benefit derived from the use of such systems in studying animals (including man) lies in the ability to get data from a heretofore inaccessible environment, thus avoiding the experimental artifacts which arise in a laboratory setting (Slater, 1965; Schwitzgebel, 1967b). It also provides long-range, day-to-day, continuous observation and control of the monitored subject, since the data can be fed into a computer which can act as both an observer and a controller (Konecci, 1965a).

Telemetry has been put to many and diverse uses. In aerospace biology, both man and animal have been telemetered for respiration, body temperature, blood pressure, heart rate (ECG's), brain waves (EEG's) and other physiological data (Konecci, 1965b; Slater, 1965; Barr, 1960). Telemetric devices have been placed on and in birds, animals and fish of all kinds to learn about such things as migration patterns, hibernation and spawning locations, respiration rates, brain wave activity, body temperatures, etc. (Slater, 1965; Lord, 1962; Sperry, 1961; Mackay, 1961; Young, 1964; Epstein, 1968). Telemetry has also been used in medicine to obtain the EEG patterns of epileptics during seizures, and to monitor heart rhythms and respiration rates in humans, for purposes of diagnosis and rescue in times of emergency (Slater, 1965; Caceres, 1965). The technology has proceeded so far that one expert in the field remarked (Mackay, 1965):

"It appears that almost any signal for which there is a sensor can be transmitted from almost any species. Problems of size, life, and accuracy have been overcome in most cases. Thus, the future possibilities are limited only by the imagination."

Telemetric systems can be classified into two types of devices—"external devices" and "internal devices."

External devices.—For the past several years, Schwitzgebel (1967a, b: Note: *Harvard Law Review*, 1966) at Harvard has been experimenting with a small, portable transmitter, called a Behavior Transmitter-Reinforcer (BT-R), which is small enough to be carried on a belt and which permits tracking of the wearer's location, transmitting information about his activities and communicating with him (by tone signals). The tracking device consists of two containers, each about the size of a thick paperback book, one of which contains batteries and the other, a transmitter that automatically emits radio signals, coded differently for each transmitter so that many of them may be used on one frequency band. With a transmitting range of approximately a quarter of a mile under adverse city conditions and a receiving range of two miles, the BT-R signals are picked up by receivers at a laboratory base station and fed into a modified missile-tracking device which graphs the wearer's location and displays it on a screen. The device can also be connected with a sensor resembling a wristwatch which transmits the wearer's pulse rate. In addition, the wearer can send signals to the receiving station by pressing a button, and the receiver can send a return signal to the wearer.

At present, the primary purpose of the device is to facilitate medical and therapeutic aid to patients, i.e., to effectuate the quick location and rescue of persons subject to emergency medical conditions that preclude their calling for help, such as cases of acute cardiac infarction, epilepsy or diabetes (Schwitzgebel, 1967a). Also, so far, the use of the device has been limited to volunteers, and they are free to remove the device whenever they wish (Schwitzgebel, 1967b). Schwitzgebel has expressed an interest in applying his device to monitoring and rehabilitating chronic recidivists on parole.

At the University of California, Los Angeles, Ralph Schwitzgebel's brother, Robert Schwitzgebel, has perfected a somewhat similar device in which a miniature two-way radio unit, encased in a wide leather belt containing its own antenna and rechargeable batteries, is worn by volunteer experimental subjects (R. Schwitzgebel, 1969). Non-voice communication is maintained between a central communications station and the wearer by means of a radio signal which, when sent, activates a small coil in the wearer's receiver unit that makes itself felt as a tap in the abdominal region, accompanied by a barely audible tone and a small light. Information is conveyed to the subject by a coded sequence of taps. In turn, the wearer can send simple coded signal messages back to the central station, indicating his receipt of the signal, his general state of well being, or the lack of it, and many other matters as well. So far, this device and its use depend entirely upon a relationship of cooperation and trust between experimenter and subject.

Another use of radiotelemetry on humans which has reached a high level of sophistication is the long-distance monitoring of ECG (electro-cardiogram) waves by Caceres (1965) and his associates (Cooper, 1965; Hagan, 1965). They have developed a telemetry system by which an ambulatory heart patient can be monitored continuously by a central computer in another city. The patient has the usual electrocardiograph leads taped to his chest, which are connected to a small battery powered FM radio transmitter on the patient's belt. The ECG waves are transmitted, as modulated radio frequencies, to a transceiver in the vicinity which relays them via an ordinary telephone (encased in an automated dialing device called a Dataphone). The encoded signals of the ECG can then be transmitted to any place in the world which can be reached by telephone. On the receiving end, there is an automatic answering device that accepts the call and turns on the appropriate receiving equipment. In the usual case this will be an analog-to-digital converter, which quantizes the electrical waves and changes them to a series of numbers, representing amplitudes at certain precise times. The computer then analyzes the numerical amplitude values and, when an abnormal pattern appears, it not only warns the patient's physician (with a belt or light) but will produce, on request, some or all of the previous readings it has stored. The computer can monitor hundreds of patients simultaneously by sharing computer time among hundreds of input signals, and produce an "analysis" of ECG activity for each in as little as 2.5 minutes—the time required for the signal to get into the computer's analytical circuits. Although this "analysis" does not yet amount to a diagnosis of heart disease or the onset of an attack, there is no reason why computers could not be taught to read ECG patterns as well as any heart specialist, and with their ability to make stochastic analyses, in time they should become better at it than most doctors.

The third area where external telemetry has been used to advantage is also in the medical field. For several years, Vreeland and Yeager (1965) have been using a subminiature radiotelemeter for taking EEG's of epileptic children. The device is glued to the child's scalp with a special preparation and electrodes extend from it to various places on the child's scalp. A receiver is positioned in an adjoining room of the hospital and sound motion pictures record the child's behavior, his voice and his EEG on the same film. Some of the benefits derived from the use of this equipment are: (1) that it permits readings to be taken of an epileptic seizure as it occurs; and (2) it allows studies to be made of EEG patterns of disturbed children without encumbering them in trailing wires. At present, however, the device is "external" in the sense that the electrodes do not penetrate into the brain, and only surface cortical brain wave patterns are picked up by the transmitter. It is believed, however, that many epileptic seizures originate in areas deep in the subcortical regions of the brain (Walker, 1961), and to obtain EEG readings for these areas, it would be necessary to implant the electrodes in these areas stereotaxically. The significance of such a modification would be that if the transmitter were transformed into a transceiver (a minor modification), it would then be possible to stimulate the same subcortical areas telemetrically. This would, then, convert the telemetry system into an "internal" device, such as the ones we are now about to describe.

Internal devices.—One of the leaders in the field of internal radiotelemetry devices is Mackay (1961). He has developed devices which he calls "endoradiosondes." These are tiny transmitters that can be swallowed or implanted internally in man or animal. They have been designed in order to measure and transmit such physiological variables as gastrointestinal pressure, blood pressure, body temperature, bioelectrical potentials (voltage accompanying the functioning of the brain, the heart and other muscles), oxygen levels, acidity and radiation intensity (Mackay, 1965). In fact, in many cases for the purposes of biomedical and physiological research, internal telemetry is the only way of obtaining the desired data. In the case where the body functions do not emit electrical energy (as the brain, heart and other neuromuscular structures do), these devices have been ingeniously modified in order to measure changes in pressure, acidity, etc., and to transmit electrical signals reflecting these changes to receivers outside the body. In this case the transmitters are called "transducers." Both "active" and "passive" transmitters have been developed, "active" transmitters containing a battery powering an oscillator, and "passive" transmitters not containing an internal power source, but having instead tuned circuits modulated from an outside power source. Although "passive" systems enjoy the advantage of not being concerned with power failure or battery replacement, they do not put out as good a signal as an "active" system. Both transmitter systems, at present, have ranges of a few feet to a dozen—just enough to bring out the signal from inside the body (Mackay, 1965). Thus, it is generally necessary for the subject to carry a small booster transmitter in order to receive the weak signal from inside the body and increase its strength for rebroadcasting to a remote laboratory or data collection point. However, with the development of integrated circuits, both transmitters and boosters can be miniaturized to a fantastic degree.

Electrical Stimulation of the Brain.—The technique employed in electrophysiology in studying the brain of animals and man by stimulating its different areas electrically is nothing new. This technique was being used by two European physiologists, Fritsch and Hitzig, on dogs in the latter half of the 19th Century (Sheer, 1961; Krech, 1966). In fact, much of the early work in experimental psychology was devoted to physiological studies of the human nervous system. During the last twenty years, however—perhaps as a result of equipment which allows the implantation of electrodes deep in the subcortical regions of the brain and the brain stem by stereotaxic instruments—the science of electrophysiology has received new impetus, and our understanding of neural activity within the brain and its behavioral and experiential correlates has been greatly expanded.

The electrical stimulation of various areas of the brain has produced a wide range of phenomena in animals and humans. An examination of published research in electrical stimulation of the brain suggests two crude methods of controlling human behavior: (1) by "blocking" of the response, through the production of fear, anxiety, disorientation, loss of memory and purpose, and even, if need be, by loss of consciousness; and (2) through conditioning behavior by the manipulation of rewarding and aversive stimuli (Jones, 1965). In this regard, the experiments of James Olds (1962; 1967) on animals and Robert G. Heath (1960) and his associates at Tulane on humans are particularly interesting. Both have shown the existence in animals and humans of brain areas of or near the hypothalamus which have what may be very loosely described as "rewarding" and "aversive" effects. The interesting thing about their experiments is that both animals and man will self-stimulate themselves at a tremendous rate in order to receive stimulation "rewards" regardless of, and sometimes in spite of, the existence of drives such as hunger and thirst. Moreover, their experiments have put a serious dent in the "drive-reduction" theory of operant conditioning under which a response eliciting a reward ceases or declines when a point of satiation is reached, since in their experiments no satiation point seems ever to be reached (the subject losing consciousness from physical exhaustion unless the stimulus is terminated beforehand by the experimenter). Thus their experiments indicate that there may be "pleasure centers" in the brain which are capable of producing hedonistic responses which are independent of drive reduction. In humans, however, the results of hypothalamus stimulation have not always been as clear as those with animals, and some experimenters have produced confusing and inconsistent results (King, 1961; Sem-Jacobsen, 1960).

Current research in the field of electrophysiology seems to hold out the possibility of exerting a limited amount of external control over the emotions, consciousness, memory and behavior of man by electrical stimulation of the brain. Krech (1966) quotes a leading electrophysiologist, Delgado of the Yale School of Medicine, as stating that current researches "support the distasteful conclusion that motion, emotion and behavior can be directed by electrical forces and that humans can be controlled like robots by push buttons." Although the authors have the greatest respect for Delgado's expertise in this field, they believe he overstates the case in this instance. None of the research indicates that man's every action can be directed by a puppeteer at an electrical keyboard; none indicates that thoughts can be placed into the heads of men electrically; none indicates that a man can be directed like a mechanical robot. *At most,* they indicate that some of man's activities can possibly be deterred by such methods, that certain emotional states might be induced (with very uncertain consequences in different individuals), and that man might be conditioned along certain approved paths by "rewards" and "punishments" carefully administered at appropriate times. Techniques of direct brain stimulation developed in electrophysiology thus hold out the possibility of influencing and controlling selected human behavior within limited parameters.

The use, then, of telemetric systems as a method of monitoring man, of obtaining physiological data from his body and nervous system, and of stimulating his brain electrically from a distance, seems in the light of present research entirely feasible and possible as a method of control. There is, however, a gap in our knowledge which must be filled before telemetry and electrical stimulation of the brain could be applied to any control system. This gap is in the area of interpretation of incoming data. Before crime can be prevented, the monitor must know what the subject is doing or is about to do. It would not be practical to attach microphones to the monitored subjects, nor to have them in visual communication by television, and it would probably be illegal (Note: *Harvard Law Review,* 1966). Moreover, since the incoming data will eventually be fed into a computer,[1] it will be necessary to confine the information transmitted to the computer to such non-verbal, non-visual data as location, EEG patterns, ECG patterns and other physiological data. At the present time, EEG's tell us very little about what a person is doing or even about his emotional state (Konecci, 1965a). ECG's tell us little more than heart rhythms. Certain other physiological data, however, such as respiration, muscle tension, the presence of adrenalin in the blood stream, combined with knowledge of the subject's location, may be particularly revealing—e.g., a parolee with a past record of burglaries is tracked to a downtown shopping district (in fact, is exactly placed in a store known to be locked up for the night) and the physiological data reveals an increased respiration rate, a tension in the musculature and an increased flow of adrenalin. It would be a safe guess, certainly, that he was up to no good. The computer in this case, *weighing the probabilities*, would come to a decision and alert the police or parole officer so that they could hasten to the scene; or, if the subject were equipped with an implanted radio-telemeter, it could transmit an electrical signal which could block further action by the subject by causing him to forget or abandon his project. However, before computers can be designed to perform such functions, a greater knowledge derived from experience in the use of these devices on human subjects, as to the correlates between the data received from them and their actual behavior, must be acquired.

II. CONDITIONS UNDER WHICH TELEMETRY TECHNIQUES MIGHT INITIALLY BE APPLIED IN CORRECTIONAL PROGRAMMING

The development of sophisticated techniques of electronic surveillance and control could radically alter the conventional wisdom regarding the merits of imprisonment. It has been the opinion of many thoughtful penologists for some time that prison life is not particularly conducive to rehabilitation (Sutherland, 1966; Sykes, 1966; Vold, 1954; Morris, 1963). Some correctional authorities, such as the Youth and Adult Corrections Agency of the State of California, have been exploring the possibilities of alternatives to incarceration, believing that the offender can best be taught "to deal lawfully with the given elements of the society while he functions, at least partially, in that society and not when he is withdrawn from it (Geis, 1964). Parole is one way of accomplishing that objective, but parole is denied to many inmates of the prison system, not always for reasons to do with their ability to be reformed or the risk of allowing them release on parole. The development of telemetric control systems could help increase the number of offenders who could safely and effectively be supervised within the community.

Schwitzgebel suggests (1967b) that it would be safe to allow the release of many poor-risk or nonparolable convicts into the community provided that their activities were continuously monitored by some sort of telemetric device. He states:

"A parolee thus released would probably be less likely than usual to commit offenses if a record of his location were kept at the base station. If a two-way tone communication were included in this system, a therapeutic relationship might be established in which the parolee could be rewarded, warned, or otherwise signalled in accordance with the plan for therapy."

He also states:

"Security equipment has been designed, but not constructed that could insure the wearing of the transmitting equipment or indicate attempts to compromise or disable the system."

He further states that it has been the consistent opinion of inmates and parolees interviewed about the matter that they would rather put up with the constraints, inconveniences and annoyances of an electronic monitoring system, while enjoying the freedom outside an institution, than to suffer the much greater loss of privacy, restrictions on freedom, annoyance and inconveniences of prison life.

The envisioned system of telemetric control while offering many possible advantages to offenders over present penal measures also has several possible benefits for society. Society, through such systems, exercises control over behavior it defines as deviant, thus insuring its own protection. The offender, by returning to the community, can help support his dependents and share in the overall tax burden. The offender is also in a better position to make meaningful restitution. Because the control system works on conditioning principles, the offender is habituated into non-deviant behavior patterns—thus perhaps decreasing the probability of recidivism and, once the initial cost of development is absorbed, a telemetric control system might provide substantial economic advantage compared to rather costly correctional programs. All in all, the development of such a system could prove tremendously beneficial for society.

The adequate development of telemetric control systems is in part dependent upon their possible application. In order to ensure the beneficial use of such a system, certain minimal conditions ought to be imposed in order to forestall possible ethical and legal objections:

1. The consent of the inmate should be obtained, after a full explanation is given to him of the nature of the equipment, the limitations involved in its usage, the risks and constraints that will be placed upon his freedom, and the option he has of returning to prison if its use becomes too burdensome.

2. The equipment should not be used for purposes of gathering

evidence for the prosecution of crimes, but rather should be employed as a crime prevention device. A law should be passed giving the users of this equipment an absolute privilege of keeping confidential all information obtained therefrom regardless of to whom it pertains, and all data should be declared as inadmissible in court. The parole authorities, if they be the users of this equipment, should have the discretionary power to revoke parole whenever they see fit without the burden of furnishing an explanation, thus relieving them of the necessity of using data obtained in this fashion as justification for their actions. The data should be destroyed after a certain period of time, and, if the system is hooked up with a computer, the computer should be programmed to erase its tapes after a similar period of time.

By employing the above safeguards, the use of a telemetric system should be entirely satisfactory to the community and to the convicts who choose to take advantage of it. Nevertheless there are a number of ethical objections which are bound to arise when such a system is initially employed that deserve special discussion.

III. ETHICAL OBJECTIONS

The two principal objections raised against the use of modern technology for surveillance and control of persons deemed to be deviant in their behavior in such a degree as to warrant close supervision revolve around two issues: privacy and freedom (Note: *Harvard Law Review*, 1966; King, 1964; Miller, 1964; Fried, 1968; Ruebhausen, 1965).

Privacy.—It has often been said that privacy, in essence, consists of the "right to be let alone" (Warren, 1890; Ernst, 1962). This is a difficult right to apply to criminals because it is precisely their inability to leave their fellow members of society alone that justifies not leaving them alone. This statement, however, might be interpreted to mean that there is a certain limited area where each man should be free from the scrutiny of his neighbors or his government and from interference in his affairs. While most people would accept this as a general proposition, in point of fact it is not recognized in prison administration, where surveillance and control are well-nigh absolute and total (Sykes, 1966; Clemmer, 1958). Therefore, it is difficult to see how the convict would lose in the enjoyment of whatever rights or privacy he has by electronic surveillance in the open community. If the watcher was a computer, this would be truer still, as most people do not object to being "watched" by electric eyes that open doors for them. It is the scrutiny of humans by humans that causes embarrassment—the knowledge that one is being judged by a fellow human.

Another definition of privacy is given by Ruebhausen (1965).

"The essence of privacy is no more, and certainly no less, than the freedom of the individual to pick and choose for himself the time and the circumstances under which, and most importantly, the extent to which, his attitudes, beliefs, behavior and opinions are to be shared with or withheld from others."

To this statement the preliminary question might be raised as to the extent to which we honor this value when we are dealing with convicts undergoing rehabilitation, mental patients undergoing psychiatric treatment, or even minors in our schools. Certainly it is not a statement that can be generally applied, especially in those cases where every society deems itself to have the right to shape and change the attitudes, beliefs, behavior and opinions of others when they are seriously out of step with the rest of society. But a more fundamental objection can be raised, in that the statement has little or no relevance to what we propose. Not only does the envisioned equipment lack the power to affect or modify directly the "attitudes," "beliefs" and "opinions" of the subject, but it definitely does not force him to share those mental processes with others. The subject is only limited in selected areas of his behavior—i.e., those areas in which society has a genuine interest in control. The subject is consequently "free" to hold any set of attitudes he desires. Of course, on the basis of behavioral psychology, one would expect attitudes, beliefs and opinions to change to conform with the subject's present behavior (Smith, 1968).

Still a third definition of privacy has been proposed by Fried (1968) in a recent article in the Yale Law Journal, an article which specifically discusses Schwitzgebel's device. He advances the argument that privacy is a necessary context for the existence of love, friendship and trust between people, and that the parolee under telemetric supervision who never feels himself loved or trusted will never be rehabilitated. While this argument might have some validity where the device is used as a therapeutic tool—a point that Schwitzgebel (1967b) recognizes since he would use it partly for that purpose—it is not particularly relevant where no personal relationship is established between the monitors and the subject and where the emphasis is placed upon the device's ability to control and deter behavior, rather than to "rehabilitate." Rehabilitation, hopefully, will follow once law-abiding behavior becomes habitual.

As far as privacy is concerned, most of the arguments are squarely met by the conditions and safeguards previously proposed. However, when one begins to implant endoradiosondes subcutaneously or to control actions through electrical stimulation of the brain, one runs into a particularly troublesome objection, which is often included within the scope of "privacy," although perhaps it should be separately named as the "human dignity" or "sacred vessel of the spirit" argument. This is the argument that was raised when compulsory vaccination was proposed, and which is still being raised as to such things as birth control, heart transplants, and proposals for the improvement of man through eugenics. The argument seems to stem from an ancient, well-entrenched belief that man, in whatever condition he finds himself, even in a state of decrepitude, is as Nature or God intended him to be and inviolable. Even when a man consents to have his physical organism changed, some people feel uneasy at the prospect, and raise objections.

Perhaps the only way to answer such an argument is to rudely disabuse people of the notion that there is any dignity involved in being a sick person, or a mentally disturbed person, or a criminal person whose acts constantly bring him into the degrading circumstances, which the very persons praising human dignity so willingly inflict upon him. Perhaps the only way to explode the notion of man as a perfect, or perfectible, being, made in God's image (the Bible), a little lower than the angels (Disraeli), or as naturally good but corrupted by civilization (Rousseau), is to review the unedifying career of man down through the ages and to point to some rather interesting facets of his biological make-up, animal-like behavior, and evolutionary career which have been observed by leading biologists and zoologists (Lorenz, 1966; Morris, 1967; Rostand, 1959). Unfortunately, there is not time here to perform such a task or to rip away the veil of human vanity that so enshrouds these arguments.

Freedom.—The first thing that should be said with regard to the issue of human freedom is that there is none to be found in most of our prisons. As Sykes (1966) remarks:

". . . the maximum security prison represents a social system in which an attempt is made to create and maintain total or almost total social control."

This point is so well recognized that it need not be belabored, but it does serve to highlight the irrelevancy of the freedom objection as far as the prison inmate is concerned. Any system which allows him the freedom of the open community, which maintains an unobtrusive surveillance and which intervenes only rarely to block or frustrate his activities can surely appear to him only as a vast improvement in his situation.

Most discussions of freedom discuss it as if man were the inhabitant of a natural world, rather than a social world. They fail to take into account the high degree of subtle regulation which social life necessarily entails. As Hebb (1961) put very well:

"What I am saying implies that civilization depends on an all-pervasive thought control established in infancy, which both maintains and is maintained by the social environment, consisting of the behavior of the members of society. . . . What we are really talking about in this symposium is mind in an accustomed social environment, and more particularly a social environment that we consider to be the normal one. It is easy to forget this, and the means by which it is achieved. The thought control that we object to, the 'tyranny over the mind of man' to which Jefferson swore 'eternal hostility,' is only the one that is imposed by some autocratic agency, and does not include the rigorous and doctrinaire control that society itself exercises, by common consent, in moral and political values. I do not suggest that this is undesirable. Quite the contrary, I argue that a sound society must have such a control, but let us at least see what we are doing. We do not bring up our children with open minds and then, when they can reason, let them reason and make up their minds as they will concerning the acceptability of incest, the value of courtesy in social relations, or the desirability of democratic government. Instead we tell them what's what, and to the extent that we are successful as parents and teachers, we see that they take it and make it part of their mental processes, with no further need of policing.

"The problem of thought control, or control of the mind, then, is not how to avoid it, considering it only as a malign influence exerted over the innocent by foreigners, Communists, and other evil fellows. We all exert it; only, on the whole, we are more efficient at it. From this point of view the course of a developing civilization is, on the one hand, an increasing uniformity of aims and values, and thus also of social behavior, or on the other, an increasing emotional tolerance of the stranger, the one who differs from me in looks, beliefs, or action— a tolerance, however, that still has narrow limits."

Discussions of freedom that one customarily finds in law journals also fail to take into account the distinction between objective and subjective freedom. Objective freedom for each man is a product of power, wealth or authority, since it is only through the achievement of one or more of these that one can control so as not to be controlled— i.e., it is only through these that one can, on one hand, guard against the abuses, infringements, and overreaching of one's fellow man which limit one, and, on the other hand, commit those very offenses against one's neighbor and, by doing so, obtain all one's heart desires. This is not to neglect the role of the law in preventing a war of all against all, in providing the freedom that goes with peace, and with ensuring that all share to a certain extent in the protections and benefits of a well-ordered society. But laws are themselves limitations imposed upon objective freedom. Radical objective freedom is inconsistent with social life, since in order for some to have it, others must be denied it. Such a radical freedom may also be intolerable psychologically; one may actually feel "constrained" by an excess of options (Fromm, 1963).

Subjective freedom, on the other hand, is a sense of not being pressed by the demands of authority and nagged by unfulfilled desires. It is totally dependent on *awareness*. Such a concept of freedom is easily realizable within the context of an ordered society, whereas radical objective freedom is not. Since society cannot allow men too much objective freedom, the least it can do (and the wise thing to do) is to so order its affairs that men are not aware or concerned about any lack of it. The technique of telemetric control of human beings offers the possibility of regulating behavior with precision on a subconscious level, and avoiding the cruelty of depriving man of his subjective sense of freedom.

IV. Conclusion

Two noted psychologists, C. R. Rogers and B. F. Skinner, carried on a debate in the pages of *Science* magazine (1956) over the issue of the moral responsibility of behavioral scientists in view of the ever-widening techniques of behavior control. Skinner said:

"The dangers inherent in the control of human behavior are very real. The possibility of misuse of scientific knowledge must always be faced. We cannot escape by denying the power of a science of behavior or arresting its development. It is no help to cling to familiar philosophies of human behavior simply because they are more reassuring. As I have pointed out elsewhere, the new techniques emerging from a science of behavior must be subject to the explicit counter control which has already been applied to earlier and cruder forms."

Skinner's point was that the scientific age had arrived; there was no hope of halting its advance; and that scientists could better spend their time in explaining the nature of their discoveries so that proper controls might be applied (not to stop the advance, but to direct it into the proper channels), rather than in establishing their own set of goals and their own *ne plus ultra* to "proper research." This is a ʾ alid point. Victor Hugo once said: "Nothing is as powerful as an idea whose time has arrived." The same holds true for a technology whose time is upon us. Those countries whose social life advances to keep pace with their advancing technology will survive in the world of tomorrow; those that look backward and cling to long-outmoded values will fall into the same state of degradation that China suffered in the 19th and early 20th Centuries because she cherished too much the past. These are not inappropriate remarks to make here, because the nations that can so control behavior as to control the crime problem will enjoy an immense advantage over those that do not. Whether we like it or not, changes in technology require changes in political and social life and in values most adaptable to those changes. It would be ironic indeed if science, which was granted, and is granted, the freedom to invent weapons of total destruction, were not granted a similar freedom to invent methods of controlling the humans who wield them.

Rogers agreed with Skinner that human control of humans as practiced everywhere in social and political life, but framed the issues differently. He said (1956):

". . . They can be stated very briefly: Who will be controlled? Who will exercise control? What type of control will be exercised? Most important of all, toward what end or what purposes, in pursuit of what values, will control be exercised?"

These are very basic questions. They need to be answered, and they should be answered.

Jean Rostand (1959), a contemporary French biologist of note, asks: can man be modified? He points to the fact that, since the emergence of *homo sapiens* over 100,000 years ago, man has not evolved physically in the slightest degree. He has the same brain now that he had then, except that now it is filled up with the accumulated knowledge of 5,000 years of civilization—knowledge that has not seemed to be adequate to the task of erasing certain primitive humanoid traits, such as intraspecific aggression, which is a disgusting trait not even common to most animals. Seeing that man now possesses the capabilities of effecting certain changes in his biological structure, he asks whether it isn't a reasonable proposal for man to hasten evolution along by modifying himself into something better than what he has been for the last 100,000 years. We believe that this is a reasonable proposal, and ask: What better place to start than with those individuals most in need of a change for the better?

NOTE

1. Obviously, no system monitoring thousands of parolees would be practical if there had to be a human monitor for every monitored subject on a 24-hour-a-day, seven-day-a-week basis. Therefore, computers would be absolutely necessary.

REFERENCES

1. Barr, N. L., 1960. "Telemetering Physiological Responses During Experimental Flight." American Journal of Cardiology 6:54.

2. Caceres, C. A. and James K. Cooper, 1965. "Radiotelemetry: A Clinical Perspective." Biomedical Telemetry. Edited by C. A. Caceres. New York: Academic Press.

3. Clemmer, Donald, 1958. The Prison Community. New York: Holt, Rinehart and Winston.

4. Cooper, James K. and C. A. Caceres, 1965. "Telemetry by Telephone." Biomedical Telemetry. Edited by C. A. Caceres. New York: Academic Press.

5. Ernst, Morris L. and Alan U. Schwartz, 1962. Privacy: The Right to Be Let Alone. New York: Macmillan.

6. Epstein, R.J., J. R. Haumann and R. B. Kenner, 1968. "An Implantable Telemetry Unit for Accurate Body Temperature Measurements." Journal of Applied Physiology 24(3): 439.

7. Fried, Charles, 1968. "Privacy." Yale Law Journal 77:475.

8. Fromm, Erich, 1963. Escape From Freedom. New York: Holt, Rinehart and Winston.

9. Geis, Gilbert, 1964. "The Community-Centered Correctional Residence." Correction in the Community: Alternatives to Incarceration. Sacramento, California: Youth and Adult Corrections Agency, State of California.

10. Grisamore, N. T., James K. Cooper and C. A. Caceres, 1965. "Evaluating Telemetry." Biomedical Telemetry. Edited by C. A. Caceres. New York: Academic Press.

11. Hagan, William K., 1965. "Telephone Applications." Biomedical Telemetry. Edited by C. A. Caceres. New York: Academic Press.

12. Heath, R. G. and W. A. Mickle, 1960. "Evaluation of Seven Years' Experience with Depth Electrode Studies in Human Patients." Electrical Studies of the Unanesthetized Brain. Edited by E. R. Ramey and D. S. O'Doherty. New York: Paul B. Hoeber, Inc.

13. Hebb, D. O., 1961. "The Role of Experience." Man and Civilization: Control of the Mind; A Symposium. Edited by Seymour M. Farber and R. H. L. Wilson. New York: McGraw-Hill.

14. Jones, H. G., Michael Gelder and H. M. Holden, 1965. "Behavior and Aversion Therapy in the Treatment of Delinquency." British Journal of Criminology 5(4): 355-387.

15. King, D. B., 1964. "Electronic Surveillance and Constitutional Rights: Some Current Developments and Observations." George Washington Law Review 33:240.

16. King, H. E., 1961. "Psychological Effects of Excitation in the Limbic System." Electrical Stimulation of the Brain. Edited by Daniel E. Sheer. Austin: University of Texas Press.

17. Konecci, E. B. and A. James Shiner, 1965a. "The Developing Challenge of Biosensor and Bioinstrumentation Research." Biomedical Telemetry. Edited by C. A. Caceres. New York: Academic Press. 1965b. "Uses of Telemetry in Space." Biomedical Telemetry. Edited by C. A. Caceres. New York: Academic Press.

18. Krech, David, 1966. "Controlling the Mind-Controllers." Think 32 (July-August); 2.

19. Lord, R. D., F. C. Bellrose and W. W. Cochran, 1962. "Radiotelemetry of the Respiration of a Flying Duck." Science 137:39.

20. Lorenz, Konrad, 1966. On Aggression. New York: Harcourt, Brace and World, Inc.

21. Mackay, R. S., 1961. "Radiotelemetering From Within the Body." Science 134:1196. 1965. "Telemetry From Within the Body of Animals and Man: Endoradiosondes." Biomedical Telemetry. Edited by C. A. Caceres. New York: Academic Press.

22. Miller, A. S., 1964. "Technology, Social Change and the Constitution." George Washington Law Review 33:17.

23. Morris, Desmond, 1967. The Naked Ape. New York: McGraw-Hill.

24. Morris, Terrence and Pauline Morris, 1963. Pentonville: A Sociological Study of an English Prison. London: Routledge and Kegan Paul.

25. Note, 1966. "Anthropotelemetry: Dr. Schwitzgebel's Machine." Harvard Law Review 80:403.

26. Olds, James, 1962. "Hypothalamic Substrates of Reward." Physiological Reviews 42:554, 1967. "Emotional Centers in the Brain." Science Journal 3(5):87.

27. Reubhausen, O. M. and O. G. Brim, 1965. "Privacy and Behavior Research." Columbia Law Review 65:1184.

28. Rogers, C. R. and B. F. Skinner, 1956. "Some Issues Concerning the Control of Human Behavior." Science 124:1057.

29. Rostand, Jean, 1959. Can Man Be Modified? London: Secker and Warburg.

30. Schwitzgebel, Ralph, Robert Schwitzgebel, W. N. Pahnke and W. S. Hurd, 1964. "A Program of Research in Behavioral Electronics." Behavioral Science 9:233.

31. Schwitzgebel, Ralph, 1967a. "Electronic Innovation in the Behavioral Sciences: A Call to Responsibility." American Psychologist 22(5):364. 1967b. "Issues in the Use of an Electronic Rehabilitation System with Chronic Recidivists." (unpublished paper).

32. Schwitzgebel, Robert L., 1969. "A Belt From Big Brother." Psychology Today 2(11): 45-47, 65.

33. Sem-Jacobsen, C. W. and Arne Torkildsen, 1960. "Depth Recording and Electrical Stimulation in the Human Brain." Electrical Studies of the Unanesthetized Brain. Edited by E. R. Ramey and D. S. O'Doherty. New York: Paul B. Hoeber, Inc.

34. Sheer, Daniel, 1961. "Brain and Behavior: The Background of Interdisciplinary Research." Electrical Stimulation of the Brain. Edited by Daniel Sheer. Austin: University of Texas Press.

35. Slater, Lloyd E., 1965. "A Broad-Brush Survey of Biomedical Telemetric Progress." Biomedical Telemetry. Edited by C. A. Caceres. New York: Academic Press.

36. Smith, Gerald W., 1968. "Electronic Rehabilitation and Control: An Alternative to Prison." Paper read at the American Correctional Association Meeting, San Francisco.

632   BARTON L. INGRAHAM, GERALD W. SMITH

37. Sperry, C. J., C. P. Gadsden, C. Rodriguez and L. N. N. Bach, 1961. "Miniature Subcutaneous Frequency-Modulated Transmitter for Brain Potentials" Science 134:1423.

38. Sutherland, Edwin and Donald R. Cressey, 1966. Principles of Criminology. Seventh Edition. Philadelphia: Lippincott.

39. Sykes, Gresham, 1966. The Society of Captives. New York: Atheneum.

40. Vold, George B., 1954. "Does Prison Reform?" Annals of the American Academy of Political and Social Science 293:42-50.

41. Vreeland, Robert and C. L. Yeager, 1965. "Application of Subminiature Radio Telemetry Equipment to EEG Analysis from Active Subjects." Paper delivered at Sixth International Congress of Electroencephalography and Clinical Neurophysiology, Vienna.

42. Walker, A. E. and Curtis Marshall, 1961. "Stimulation and Depth Recording in Man." Electrical Stimulation of the Brain. Edited by Daniel Sheer. Austin: University of Texas Press.

43. Warren, Samuel D. and Louis D. Brandeis, 1890. "The Right to Privacy." Harvard Law Review 4:193.

44. Young, I. J. and W. S. Naylor, 1964. "Implanted Two-Way Telemetry in Laboratory Animals." American Journal of Medical Electronics 3:28.

Behavior Modification / Further Reading

Bloch, Sidney, and Reddaway, Peter. *Psychiatric Terror: How Psychiatry Is Used To Suppress Dissent.* New York: Basic Books, 1977.

Cook, J. W., et al. "Consent for Adversive Treatment: A Model Farm." *Mental Retardation* 16 (February, 1978), 47–51.

Dworkin, Gerald, "Autonomy and Behavior Control." *Hastings Center Report* 6 (February, 1976), 23–28.

Gaylin, Willard M.: Meister, Joel S.: and Neville, Robert C., Editors. *Operating On The Mind.* New York: Basic Books, 1975.

Halleck, Seymour L. *The Politics of Therapy.* New York: Jason Aronson, 1971.

London, Perry. *Behavior Control.* Second Edition. New York: The New American Library, 1977.

Miller, Harry L. "The Single 'Right to Treatment': Can the Courts Rehabilitate and Cure?" *The Public Interest,* Winter, 1977, pp. 96–118.

"New Technologies and Strategies for Social Control: Ethical and Practical Limits." *American Behavioral Scientist* 18 (May–June 1977).

Smith, W. Lynn, and King, Arthur, Editors. *Issues in Brain/Behavior Control.* New York: Spectrum Publications, 1976.

Spece, Roy G., Jr. "Conditioning and Other Technologies Used to 'Treat?' 'Rehabilitate?' 'Demolish?' Prisoners and Mental Patients." *Southern California Law Review* 45 (Spring, 1972), 616–84.

9 Is useful for explaining "Brain cleant"
10 + 11 Make a nice pair.
12 is unnecessary